Crop Diseases

Identification, Treatment and Management

An Illustrated Handbook

Crop Diseases

Identification, Treatment and Management

An Illustrated Handbook

L. DARWIN CHRISTDHAS HENRY
Senior Lecturer in Plant Pathology
Annamalai University
Chidambaram, Tamil Nadu

&

H. LEWIN DEVASAHAYAM
Associate Professor of Plant Pathology (*Retd.*)
Tamil Nadu Agricultural University
Coimbatore, Tamil Nadu

New India Publishing Agency
Pitam Pura, New Delhi-110 088

Published by
Sumit Pal Jain *for*
New India Publishing Agency
101, Vikas Surya Plaza, CU Block, L.S.C. Mkt.,
Pitam Pura, New Delhi- 110 088, (India)
Phone: 011-27341717, Fax: 011-27341616
Mobile : 09717133558
E-mail: newindiapublishingagency@gmail.com
Web: www.bookfactoryindia.com

ISBN : 978-93-80235-46-2

Composed and Designed by NIPA

CONTENTS

FOREWORD

Plant diseases are among the important factors that are responsible for causing yield loss in crop production. The loss due to diseases alone is estimated to be around 26 per cent. Diseases may attack at any stage of the standing crop, from seedlings till maturity of the crop. They may affect different parts of the plants, such as foliage, stem, root, flowers or seed and cause various types of symptoms, while the diseases such as wilt affect the entire plant. All these ultimately result in the reduction of yield and poor quality of the produce. Further, many pathogens continue to attack the stored grains and stored produce, and cause spoilage. To save the crops from diseases caused by pathogens and thereby to increase crop production, it is imminent that diseases have to be controlled by any means.

To adopt various strategies for the control of pathogens, one should have some basic knowledge about the symptoms produced by the pathogens, their life cycle, mode of survival and spread, and the stage at which the host is most vulnerable to attack by the pathogens. Most of the cultivated varieties of different crops are susceptible to one disease or another, while some others are susceptible to many diseases. Even resistant cultivars of some of the crop species may become susceptible to some specific diseases in course of time as a result of development of new physiologic races of the pathogen by hybridization or natural mutation or when the environmental conditions are highly favorable for the pathogen and not quite favorable for the host.

In this book, the authors have given a detailed account of the major diseases of important field crops and horticultural crops, and their management. The text is substantiated with many hand-drawn illustrations, which are of excellent quality and in fact it is the highlight of the book.

A chapter on important edible mushrooms commonly grown in India, methods of cultivation of different mushrooms, diseases and pests attacking mushroom beds and mushrooms is also included in the book. This may be quite useful to emerging entrepreneurs.

The book, which has been compiled as per the undergraduate syllabus of Agricultural Institutions, will also be of use to Postgraduate students and to those working in the Department of Agriculture. I congratulate the authors, Dr.L.Darwin Christdhas Henry and Thiru.H.Lewin Devasahayam for their effort in bringing out this very useful book.

Chidambaram
July, 2010

(J. VASANTHAKUMAR)
Dean
Faculty of Agriculture
Annamalai University

PREFACE

Food, clothing and shelter are the basic needs for the survival and existence of all humans, and plants are the chief source for providing all these things. Not only man, all other living beings also require food for their survival, which is derived from plants. Crop husbandry deals with increasing the production of crop produce to meet these requirements.

The world population is increasing at an alarming rate, especially in the developing and under-developed countries. Though there is slight reduction in the birth rate due to adoption of various family planning measures, the longevity of human life has gone up due to advancement in various fields of medical sciences, which has also contributed to the increase in population. The population in India as on 2008 has crossed 100 crores. This has resulted in increase in the demand for food and other commodities.

Cultivable land is dwindling gradually. Vast areas of arable land, which were under cultivation, have been converted into housing plots and industrial establishments. Further, many sources of irrigation, such as tanks and lakes have been converted into bus stands and play grounds. Even forest areas have been encroached to a considerable extent for purposes other than agriculture. With limitations in the availability of more land for crop husbandry, several methods, such as introduction of high yielding, high fertilizer responsive varieties, use of better and quality seeds, soil testing and application of balanced dose of fertilizers and micro-nutrients, adopting improved cultural operations and better irrigation practices, cultivation of disease and pest resistant varieties etc. are being advocated to boost production to meet the ever increasing demand for food and other materials derived from plants. Sophisticated life style and high living standards have also indirectly increased the demand for various products derived mostly from plants.

Diseases take a heavy toll in the production of crop produce. Plant diseases are caused by various pathogenic microorganisms, such as fungi, bacteria, viruses, mycoplasma, protozoa and nematodes; plant parasites, such as algae, lichens and phanerogamic parasites; abiotic agents, such as deficiency due to lack of macro and micronutrients, toxicity due to excess of macro and micronutrients, light and heat stress, as well as presence of some toxic chemicals in the air, soil or water.

No plant species on this planet is free from attack by any disease. Some plant species are vulnerable to attack by several diseases. Diseases may attack specific plant parts, such as roots, stems, foliage, fruits, seeds etc. or the entire plant and all these diseases ultimately result in the reduction of yield or quality of the produce or death of some parts or the entire plant. Some pathogens cause extensive damage and death of seedlings, both in the nurseries and main fields. Even stored grains or other produce are subjected to attack by certain microbes leading to spoilage. Some microorganisms produce certain toxins, such as mycotoxins in the grains or food products, which may be lethal to humans and other live stocks.

To obtain maximum yield from plants according to their genetic potential, they *have* to be protected from attack by pathogenic organisms. Similarly stored grains and other food products have to be protected from ravages by microorganisms. Plant protection is a special-branch of agricultural science, which aims at protecting the plants from attack by pathogenic organisms through various methods, such as exclusion, evasion, eradication, protection and immunization. An integrated approach including all the above methods is most effective in controlling the diseases. However, under certain circumstances, especially when the disease intensity crosses the economic threshold level, direct control measures by the use of plant protection chemicals have to be resorted to, so as to control the disease and avoid economic loss.

For the survival, perpetuation, spread and initiation of diseases by pathogens, nature has provided a number of means. So, for a concerted effort for the control of diseases, a basic knowledge about the disease symptoms, the causal organisms responsible for causing the diseases, their life cycle, reproductive capacity, mode of survival and spread of the pathogens etc. go a long way in adopting appropriate control measures to combat the diseases.

In plant protection, the various measures to be taken to combat the different diseases and the plant protection chemicals to be applied, whenever necessary, differ depending upon the causal organisms and other characteristics of the pathogens. Some of the recently introduced systemic fungicides and antibiotics, as well as some other chemicals are pathogen-specific and are capable of controlling only specific diseases, while some of the broad-spectrum fungicides can control many diseases. So, selection of appropriate plant protection chemicals, the correct dosage of the chemical formulation to be applied, the time and method of application of the chemicals are very important.

Several virus and mycoplasma diseases are transmitted by various insect and non-insect vectors in a persistent, semi-persistent or non-persistent manner and for the control of these vectors, specific pesticides having quick knock-down effect have to be applied at the proper time. So, a knowledge about the vectors capable of transmitting such diseases and the pesticides to be used for controlling them are also of vital importance.

In the recent past, cultivation of edible mushrooms is gaining momentum. Mushroom cultivation requires less of land and agricultural wastes can be used as substrates for preparing mushroom beds. Further, mushrooms serve as an excellent protein substitute. Mushroom beds and mushrooms are subjected to attack by several fungal, bacterial and virus pathogens, resulting in poor production and spoilage of mushrooms. Cultivation methods of some of the popular mushroom varieties and measures of controlling pathogens attacking mushroom beds and mushrooms have also been included.

In this book, which has been compiled as per the syllabus of B.Sc.(Ag.) degree course of Agricultural Universities, the authors have endeavored to give a detailed account of all the major diseases that occur on important field crops and horticultural crops, including control measures to combat the diseases. Besides that, a list of diseases considered to be of minor importance, but which may assume serious proportions, when environmental conditions become favorable for the causative pathogens have also been included. A large number of illustrations, all hand-drawn by the authors have been incorporated to substantiate and highlight the text. The authors hope that this book will be of immense help and guidance, not only to the under graduate students, but also to students in pursuit of higher studies, people working in the Department of Agriculture, people involved in scientific agriculture and the general Public.

The authors are extremely thankful to Dr.J. Vasanthakumar, Dean, Faculty of Agriculture, Annamalai University, Chidambaram, Tamil Nadu for being kind enough to give the foreword to the book.

Acknowledgements are also due to friends, colleagues and well wishers, who have inspired, helped and encouraged the authors in taking up this venture and complete the work successfully.

The authors wish to express their sincere thanks and appreciation to M/s New India Publishing Agency, 101, Vikas Surya Plaza, Pitam Pura, New Delhi-88 for the excellent manner in which the book has been brought out.

L.Darwin Christdhas Henry
H. Lewin Devasahayam

LIST OF FIGURES

LIST OF TABLES

COLOUR PLATES

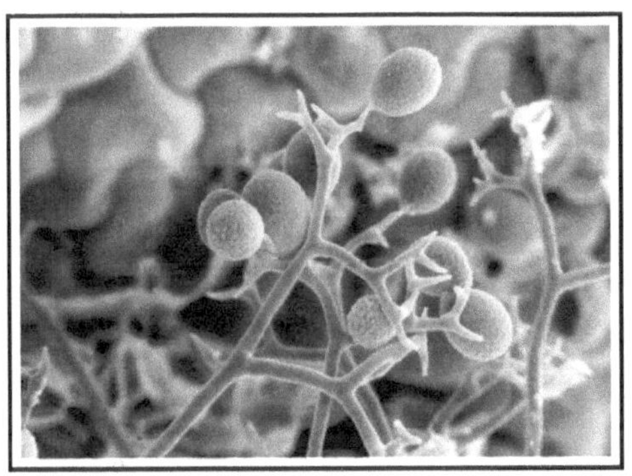

Fungi

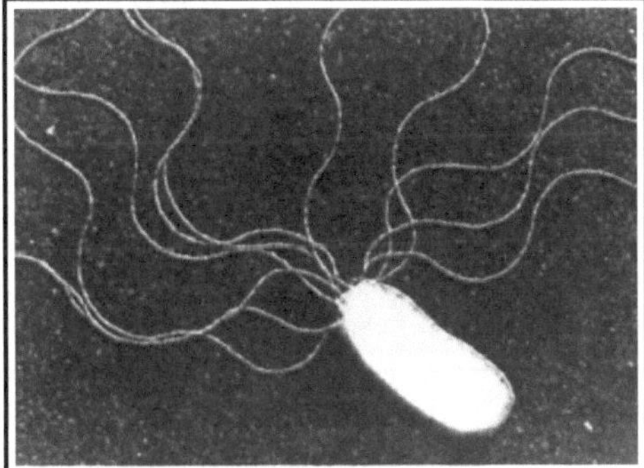

Bacteria

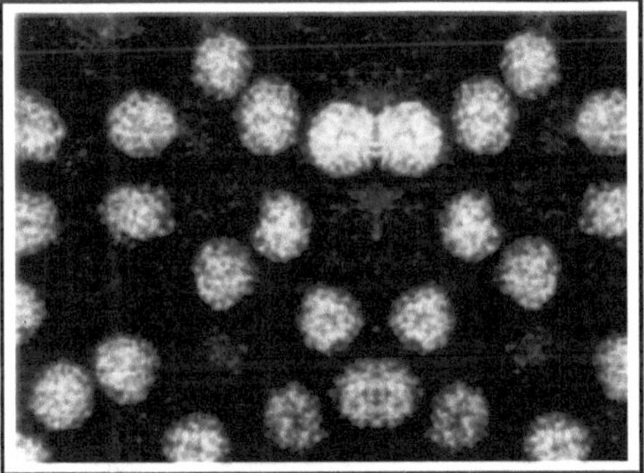

Spherical virus

Rod shaped virus

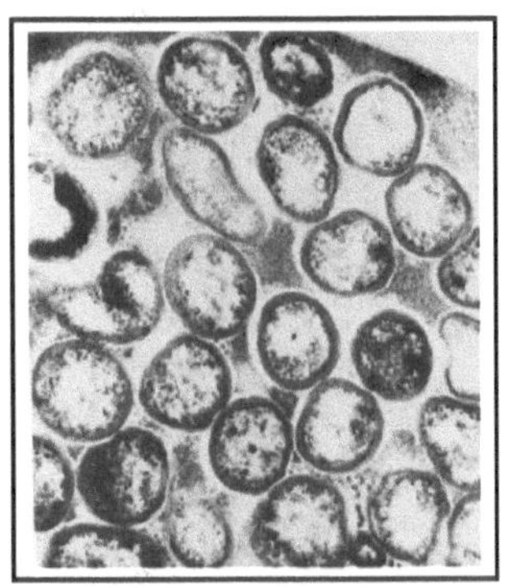

Rickettsia like Bacteria

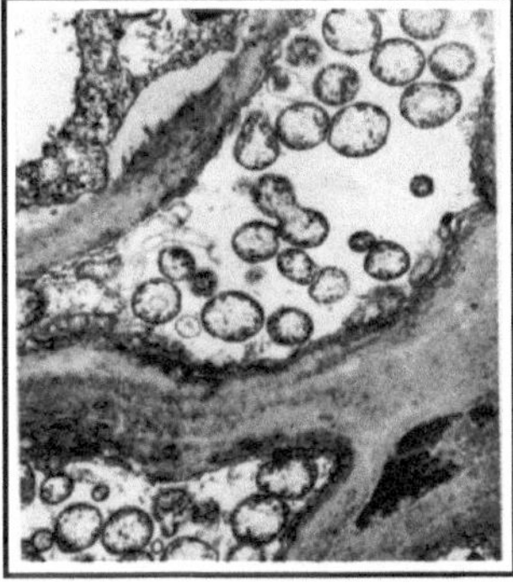

Mycoplasma like organism

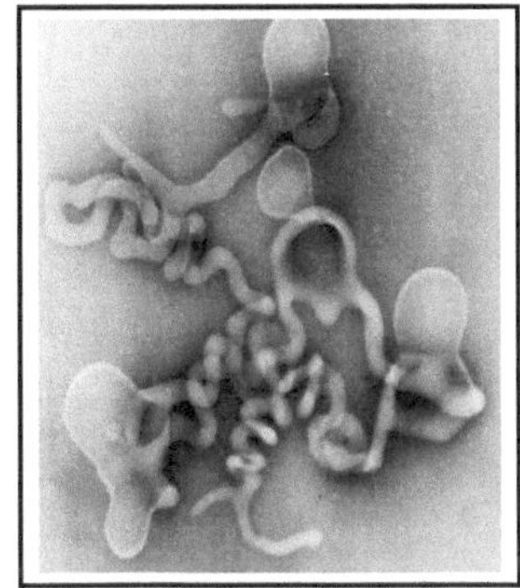

Spiroplasma

PLANT PATHOGENS

Rice brown leaf spot

Rice blast

Grey leaf spot of coconut

Alternaria leaf spot/
Early blight

Early 'tikka' leaf spot

Late 'tikka' leaf spot

Cercospora leaf spot

Sapota leaf spot

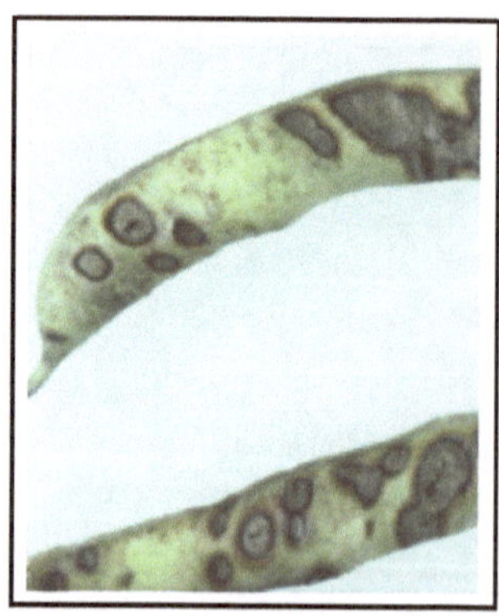

Bean anthracnose

LEAF SPOTS

Black/ Stem rust- Wheat

Orange/ leaf rust- Wheat

Yellow/ striped rust- Wheat

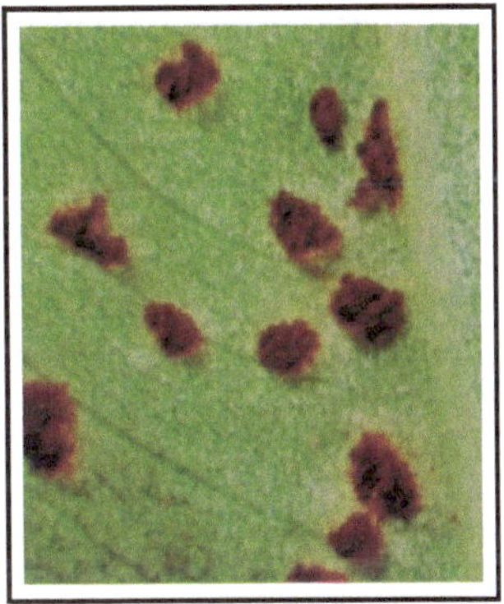

Groundnut rust

Soybean rust

Coffee rust

Algal red rust

White rust of amaranthus

RUST DISEASES

Downy mildew of cucurbits

Downy mildew of cucurbits

Downy mildew of onion

Downy mildew of sorghum

Downy mildew of pearl millet

Downy mildew of maize

Downy mildew of grapes

Downy mildew of grapes

DOWNY MILDEW

Powdery mildew of cucurbits

Powdery mildew of grapes

Powdery mildew of grapes

POWDERY MILDEW

Cumbu smut

Sorghum smut

Corn smut

Whip smut of sugarcane

Loose smut of wheat

Wheat bunt

Wheat bunt

Honeydew stage of cumbu

Ergot of cumbu

SMUT DISEASES

Bacterial leaf blight of rice

Ciitrus canker

Ciitrus canker

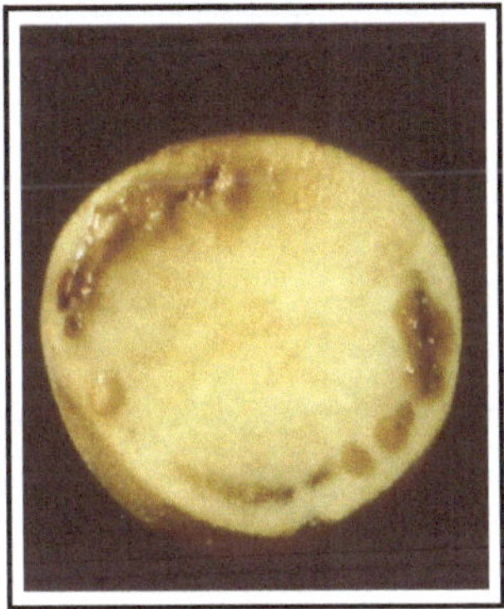
Ring rot of potato

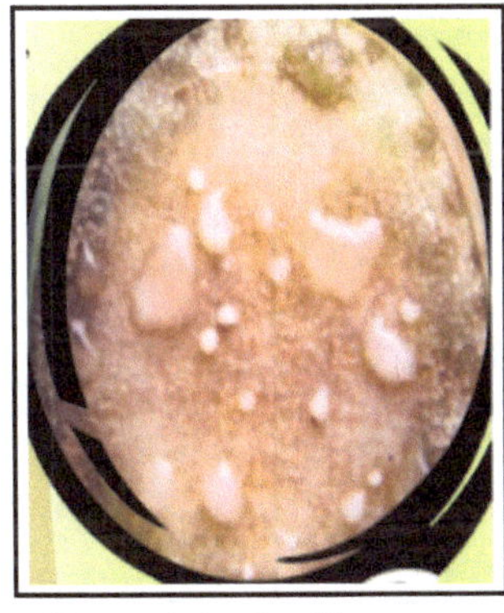
Bacterial wilt of banana

Bacterial rot of cabbage

Crown gall of rose

Bacterial speck of tomato

Bacterial wilt of ginger

BACTERIAL DISEASES

Bean golden yellow mosaic

Tomato spotted wilt

Casava mosaic

Bean leaf crinkle

Rice tungro

Banana bunchy top

VIRAL DISEASES

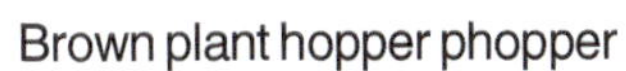
Brown plant hopper phopper

Thrips

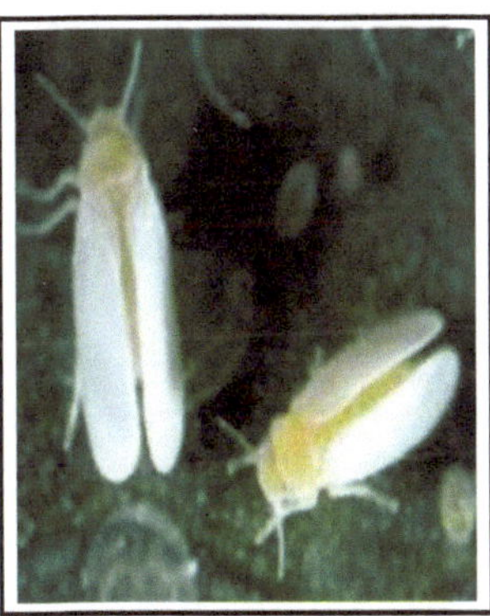
White fly

Leaf hopper

Aphid

Mite

Tree hopper

Beetle

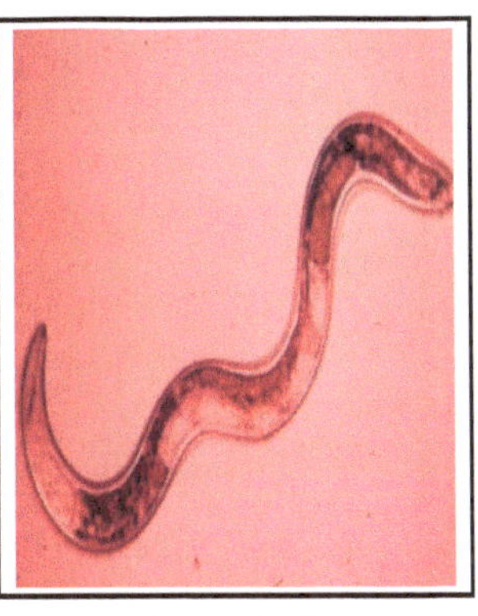
Nematode

VECTORS OF VIRAL DISEASES

Cuscuta

Loranthus

Orabanche

Striga

PHANEROGAMIC (FLOWERING) PLANT PARASITES

Rice - Boron deficiency

Sorghum- iron deficiency

Soy bean – Mn deficiency

Sugarcane – iron deficiency

Cotton – Mg deficiency

Apple – Pottasium deficiency

Oilpalm – Boron deficiency

Nitrogen deficiency

Citrus – Zinc deficiency

NUTRITIONAL DEFICIENCY (DISORDERS)

Oyster mushrooms (*Pleurotus* sp.)

Button mushrooms (*Agaricus* sp.)

Milky mushrooms (*Calocybe indica*)

Paddy straw mushrooms (*Volvariella* sp.)

Morels (*Morchella esculenta*)

Reishi (*Ganoderma lucidum*)

EDIBLE MUSHROOMS

Chapter 1

Diseases of Field Crops

CEREALS

Rice *(Oryza sativa)*

1. Rice blast

Pyricularia oryzae

'Rice blast' occurs in more than 80 countries across the world. The disease is more prevalent in Thailand, Japan, America, Guyana, Philippines and India. It is particularly serious in places, where rice is irrigated or receives heavy rainfall and where high levels of nitrogenous fertilizers are applied. Rice blast was first reported in China in 1637, in Japan in 1704 and in India in 1913. In India, it is found in all rice growing states, particularly in Telungu Desam, Karnataka, Maharashtra, Uttar Pradesh. Assam, Tripura, West Bengal, Bihar, Orissa, Himachal Pradesh, Punjab, Madhya Pradesh, Gujarat, Kerala, Tamil Nadu and Pondicherry and causes considerable damage to rice crop. In Tamil Nadu, the disease occurs in all rice growing districts and attains epidemic proportions in some seasons. Several rice blast epidemics have occurred in different parts of the world, causing yield losses ranging from 50 to 90 %. In Tamil Nadu, a serious epidemic of the disease occurred in 1919 in the Thanjavur Delta.

Symptoms. The disease, though chiefly a foliage disease, attacks other parts of rice plants, such as leaf sheath, rachis, joints of the culms and even the glumes. On the leaves, the initial symptoms appear as small bluish or grayish specks, which remain circular in older leaves. On young leaves, they enlarge and become eye-shaped, diamond-shaped or spindle-shaped spots, extending over several centimeters in length and about 1.0 cm. in width. The central portion of the lesion appears pale green or grayish green and water-soaked in the beginning. In older spots, the center becomes white to gray with a dark-brown margin surrounded by a yellow hallow. The lesions may coalesce, cover large area of the leaves and eventually the entire leaves are killed. The shape, size and color of the spots vary depending upon the variety, growth stage of the plants and environmental conditions. Similar spots may be formed on the leaf sheath also. During severe incidence, large areas of the leaves dry and die and the entire crop presents a burnt appearance. Hence the disease is known as **'blast disease'**. Leaf blast is more severe during the seedling and maximum tillering stages. In the nursery, the seedlings show numerous spots on the leaves, which wither and die. In case of severe infection, large number of seedlings may die. Blast also affects the leaf collar, which results in the death of the entire leaf.

The disease attacks the stem-nodes at the heading stage leading to production of **'white panicles'** or the stem may break at the infected node. Rarely the internodes are affected. Further, at heading the fungus also attacks the panicle neck, which is girdled and becomes shining black and shrivelled. This is the most destructive phase of the disease and is known as **'neck blast'**, **'neck rot'** or **'panicle blast'**. If infection of the panicle neck occurs early, the grains become chaffy and the panicle remains erect. If panicle neck infection occurs late, the grains become partially filled and because of the weight of the grains, the panicle breaks at the neck-node and droops. Sometimes, only parts of the panicle and joints of the culms and some glumes are infected and develop brown to black spots **(Fig.1)**.

Under conditions highly favorable for the pathogen, even the varieties resistant to the fungus under normal conditions are also attacked and small, round or short elliptical spots develop. However, the spots normally do not enlarge as typical blast spots.

The causal organism. The mycelium of the fungus is hyaline or grayish to sub-hyaline, branched, septate, mostly uninucleate, inter- and intracellular. The mycelium is mostly localized and does not spread far from the place of infection. Before sporulation, the mycelium aggregates below the epidermis of the host. From this mycelium, conidiophores are produced in clusters, which come out of the leaf either through the stomata or by breaking the epidermis. The conidiophores, which emerge in clusters of a few to many are erect, unbranched or rarely branched, septate, hyaline to sub-hyaline and produce conidia in succession, one at a time at the tip of the conidiophores. When the conidium matures, it is pushed aside and the conidiophore continues to grow further and produces another conidium at the tip. The conidiophore forms a slight bend at the point of attachment of each of the conidium and appears geniculate. Fully matured conidium falls off, leaving a scar on the conidiophore at the point of attachment of the conidium. From each conidiophore, up to 20 conidia or more are produced at an interval of about 30 minutes. From each leaf spot 4,000 - 6,000 conidia are produced each night for about two weeks. The spores are scattered by rain or dewsplashes and are carried up to a distance of about 20 meters. They fall on the leaves of adjacent plants or nearby fields and produce fresh infection.

The conidia are pear-shaped, mostly 2 septate, rarely single septate, hyaline to pale olive and measure 19.0 - 23.0 x 7.0 - 9.0µ in size. The basal cell has a projection at the bottom known as **'hilum'** and the top cell is slightly elongated and narrows down towards the tip. The conidia germinate from the apical or basal cell or from both by issuing a germ tube. From the tip of the germ tube an appressorium is formed, which attaches itself to the host. From the appressorium, infection hyphae are produced, which enter the host tissue, either through the stomata or by direct penetration of the host epidermis. It takes 7 - 8 hours for the spores to germinate and enter the host. Blast lesions appear in about 4 days after the entry of the pathogen into the host and in about 6 - 7 days spores are produced from these lesions **(Fig.1).**

The perfect stage or teleomorph stage of this fungus has been found to be *Magnaporthe grisea,* but it has not been found to occur in nature.

The fungus produces some toxins, such as **'piricularin'**, **'á-picolinic acid'** etc., which affect the rice crop adversely. The fungus also produces **'proteolytic enzymes'**, which cause breakdown of the host cell walls.

Mode of survival, spread and epidemiology. The pathogen overseasons as mycelium and conidia on diseased rice straw, haystacks and seeds. The mycelium can survive in the infected straw for one or two years under dry conditions and in infected seeds during storage. In the tropics, conidia are present in the air throughout the year. These conidia may be produced from rice crop grown throughout the year or from collateral hosts of this fungus.

The pathogen is known to infect a number of weed hosts, such as *Brachiaria mutica, Echinochloa crusgalli, Digitaria marginata, Panicum repens, Arundo donax* etc. and these hosts may act as primary source of inoculum. The conidia produced on the primary lesions on the nursery seedlings may also be spread by wind and lead to secondary spread of the disease.

The weather factors, such as temperature, humidity and moisture play a vital role in the occurrence, development and spread of the disease. Production of conidia, germination of conidia, entry into the host, all these happen during the night time, when there are droplets of water on the plant surface. Rainfall and dewdrops provide the water required for these functions. A night temperature of 20° - 26°C, alternating with a day temperature of 30°C, high relative humidity of 88 - 90 %, gentle breeze, day light for about 14 hours and darkness for 10 hours, besides water droplets on the plant surface favor rapid spread of the disease. When the relative humidity is below 88 %, the fungus does not produce spores and when there is bright sunshine, the spores do not germinate. Cloudy, overcast weather encourages blast occurrençe. Maximum spore load is found in the atmosphere in the mornings between 4 - 6 a.m.

Even in susceptible varieties of rice, cultivars having upright or vertical leaves trap less number of spores and are affected to a lesser extent by the fungus than in the case of varieties having slanting or prostrate leaves. Application of high levels of nitrogenous fertilizer favor the occurrence of blast. The pathogen exists as numerous physiological races, each carrying different genes for virulence. More than 13 genes for resistance to blast have been identified in different rice cultivars, but each such gene is readily overcome within a few years by the appearance of new pathogenic races. More than 54 pathogenic races have been identified in India.

Disease management

Agronomic practices (i) Diseased crop debris should be destroyed by burning (ii) Collateral hosts in and around rice fields should be removed and destroyed (iii) Seeds for sowing should be collected from disease-free crops (iv) Application of excessive dose of nitrogenous fertilizer should be avoided. Recommended dose of nitrogenous fertilizer should be applied in split doses, 50 % as basal, 25 % at active tillering phase and the balance 25 % at the booting stage (v) While transplanting, closer planting should be avoided (vi) The crops should be planted early, as early planted crops escape disease occurrence.

Chemical control (i) Spraying the crop with edifenphos - 200 ml. or kitazin - 200 ml. or carbendazim - 200 gm. or thiophanate methyl - 200

Fig. 1: Rice blast-*Pyricularia oryzae*

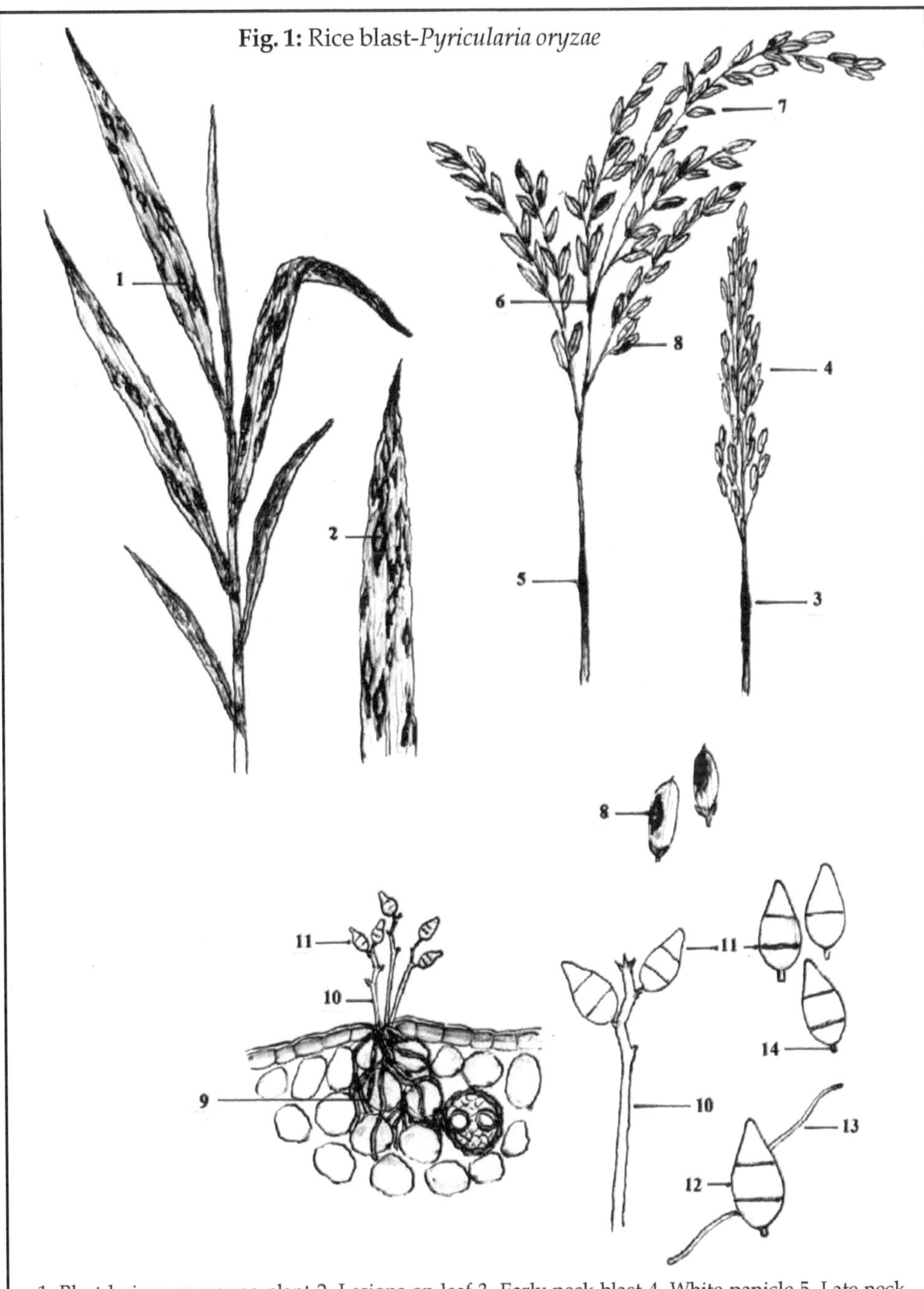

1. Blast lesions on young plant 2. Lesions on leaf 3. Early neck blast 4. White panicle 5. Late neck blast 6. Infection at joints of culms 7. Partially filled grains 8. Spots on grains 9. Mycelium 10. Conidiophore 11. Conidia 12. Germinating conidium 13. Germ tube 14. Hilum

gm. in 200 lit. of water per acre gives adequate control of the disease. In case of severe incidence of the disease, a second spraying may be given after an interval of 10 days. More recently, spraying with the systemic fungicide pyroquilon or tricyclazole at 200 gm. in 200 lit. of water per acre has been found to afford very good control of the disease. These fungicides inhibit penetration by the fungus by interfering with the melanin production in the appressorium of the fungus.

If the disease appears in the nursery, spraying with any of the above fungicidal fluid at 20 lit. per 8 cent nursery gives effective control of the disease.

Biological control. Two to three sprayings with the bacterial biocontrol agent, *Pseudomonas fluorescens* at 200 gm. in 200 lit. of water per acre controls the disease effectively.

Seed treatment. Blast fungus is externally seed-borne and seed treatment with suitable fungicides helps to eradicate the fungal inoculum from the seed. (i) For dry seed treatment, 1.0 kg. of seeds are treated with captan - 4 gm. or thiram - 4 gm. or carboxin - 2 gm., 24 hours prior to sowing. (ii) For wet seed treatment, carbendazim - 2 gm. or pyroquilon - 2 gm. or tricyclazole - 2 gm. is mixed with 1.0 lit. of water and 1.0 kg. of seeds are kept immersed in the fungicidal fluid for 24 hours. Then the seeds are allowed to sprout and sown in the nursery. This treatment protects the seedlings from blast occurrence for about 20 days (iii) Seed treatment with the biocontrol agent *Pseudomonas fluorescens* at 10 gm./ kg. of seeds prevents occurrence of the disease.

Resistant varieties. The most practical and cheapest method of control of blast disease is to grow resistant varieties. Breeding for resistance to blast is being carried out in India for the past 80 years with International collaboration and periodically, varieties resistant to blast are being released. However, because of the presence of many pathogenic races and continuous evolution of new races, the resistance of rice cultivars breaks often within a short span of time.

Some of the varieties, such as Co.4, Co.25, Co.26, TKM.1, ADT.23, Tetep, Zenith and Tadukan are found to have high degree of resistance to blast and are being used as parents in many breeding programs. Several promising blast resistant varieties have been evolved and are being cultivated. Co.25, Co.37, Co.43, ADT.36, ADT.40, IR.20 and Ponmani are resistant, while Co.45, ADT.37, ASD.18, IR.62, IR.64, Jaya and Vikas are moderately resistant to blast. IR.36, IR.50, TKM.9, J.13, HR.12, Ponni and White ponni are susceptible and should not be grown in blast prone areas and in seasons favorable for the disease occurrence.

Disease forecasting. Plenty of research work has been carried out to correlate the influence of various meteorological factors on the occurrence of blast disease, so as to develop a method for forecasting blast outbreaks. From data collected over a number of years, it has been found that a minimum night temperature of 20°- 26°C, very high relative humidity of 88 - 90 % lasting for a continuous period of 7- 10 days, cloudy, overcast weather and rainfall or dewfall during any of the vulnerable stages of the crop growth viz., seedling stage, active tillering stage or booting stage favor large scale incidence of the disease.

Maintaining trap nurseries planted with highly susceptible varieties ahead of the main crop and setting up of spore traps above the trap-crop canopy also help to forecast and forewarn the outbreak of the disease in the main crop at least 10-15 days in advance, so that suitable control measures can be taken up, as soon as the disease appears.

2. Brown spot of rice

Bipolaris oryzae / Helminthosporium oryzae
(Cochliobolus miyabeanus)

'Brown spot', **'Helminthosporiose'** or **'sesame leaf spot'** is a very serious disease of rice and is prevalent throughout the world, wherever rice is grown. In India, the disease was first recorded in 1919. Subsequently, it has become an endemic disease and occurs every year in a more or less severe form, sometimes in an epidemic form. All cultivated rice varieties are found to be susceptible to the disease to a more or less extent

The disease causes damage in different ways: **(i)** Germination of infected seeds is poor, resulting in loss of seeds. Infected seeds, even if they germinate, the coleoptile and later, the leaves of young seedlings are affected and they may die **(ii)** Leaf infection and development of spots results in the reduction of photosynthetic area of leaves, leading to marked weakening of plants **(iii)** Due to general weakening of plants, grain setting is adversely affected and the grains produced are shrivelled, chaffy and of poor quality **(iv)** Grains are also affected by the disease and such affected grains are undesirable for use as seeds or for consumption.

Symptoms. The symptoms appear on the coleoptile of germinating seeds, leaves, leaf sheaths, rachis and the glumes. On the coleoptile, the spots produced are small, circular to oval and brown in color and the affected seedlings may die in case of severe infection. On the leaves, the spots vary in size and shape, from minute dots to oval or eye-shaped or sesame seed-shaped spots and measure from 1.0 - 1.4 x 0.5 - 3.0 mm. They are distinct, isolated and mostly evenly distributed on the leaf surface. In highly susceptible varieties, the spots may be larger, up to 1.0 cm. in length. While the smaller spots are dark-brown or purplish brown, the larger spots are dark-brown at the margin and the center may be pale yellow, dirty-white, brown or gray. Sometimes, the spots are surrounded by a yellowish halo. Rarely the spots may coalesce and become irregular in shape. In severe infections, the entire leaf may turn brown and dry. Similar symptoms appear on the leaf sheaths also. If the plants are attacked at the early growth stage in a severe form, earheads may fail to emerge from the sheath and perish prematurely. Even if the earheads emerge under such conditions, they are distorted and there is no grain setting or very poor grain setting.

On the heads, lesions first appear on or near the lowest joints of the rachis. The heads do not break from the joints, but grain formation is very badly affected and most of the grains are chaffy. Spots appear on the glumes, which have a gray center and brown border. Entire glumes may be covered by several, small or one large spot on which a dark-brown velvety coating of conidiophores and conidia are present **(Fig.2).**

Causal organism. The mycelium of the fungus is profusely branched, septate, dark-brown or grayish-brown in color and prostrate. The hyphae are both inter- and intracellular. The conidiophores, which are more or less erect, arise as lateral branches from the hyphae. They are stout, rarely branched only at the base and lighter in color than the hyphae. When developing from the internal mycelium, conidiophores arise in tufts through the stomata. They possess characteristic knee joints at the points where conidia had been attached. The lowest conidium is the oldest and the youngest conidium is formed at the tip of the conidiophore. Conidia are 5 - 10 septate, thick-walled, brown, typically obclavate, with both the ends rounded and measure 56 - 104 x 15 - 20 mm. Conidia germinate characteristically from the two terminal cells. The germ tube develops an appressorium and attaches itself to the host surface. From the appressorium, infection hypha develops, penetrates the host epidermis, enters the host and in about 24 hours develops a new spot. When infection takes place through stomata, no appressoria are formed.

Perithecia are very rarely produced in nature. In cultures, perithecia are globose, pseudoparenchymatous and black, with an ostiolar beak and measure 560 - 950 x 368 - 377µ. Asci are cylindrical to long-fusiform, slightly curved and contain 4 to 6 ascospores. Ascospores are long, filiform, 6 - 15 septate and coiled in a close helix in the ascus. The pathogen produces a toxin, '**cochliobolin**', which adversely affects the growth of rice seedlings. Besides this, a few other '**proteolytic enzymes**' are also produced by this pathogen, which are capable of disintegrating the host cell walls and cause damage to the plants **(Fig.2).**

Mode of survival, spread and epidemiology. The disease is both soil- and seed-borne. The mycelium and conidia are carried over from season to season on the seed and diseased crop residues left over in the field. The mycelium is also carried inside the seed coat in a dormant state. Aeroscope studies have shown that conidia of the fungus are present over rice fields during the sowing seasons and this air-borne inoculum from some external source appears to cause primary infection in a locality. The source of this initial inoculum may be perennial grass hosts and early sown paddy crop. *Leersia hexandra, Setaria glauca, S. italica, Zea mays, Echinocloa frumentacea, E. colona, Sorghum vulgare, Panicum miliaceum, P. milara, Cynodan dactylon* and *Digitaria sanguinalis* are some of the collateral hosts of this pathogen.

Minimum temperature of 27.0° - 28.5°C, relative humidity 90 - 99 per cent, wind velocity of 4.0 - 8.8 km./ hour and low to moderate rainfall of 0.4 - 14.4 mm are most conducive for the dispersal of conidia in the atmosphere and for causing primary and secondary infection.

Cloudy weather favors disease development. The disease is common in tracts where rice is grown under irrigation. Deep sowing results in slow emergence of seedlings, which favors more seedling infection and high pre-emergence loss, where seed-borne inoculum plays an important role. Susceptibility to this pathogen increases with the age of the host. Nutritional deficiencies, especially those of nitrogen and potash predispose rice to infection by this pathogen. An optimum temperature of 21° - 26°C and a relative humidity of 92.5 % or more are favorable for conidial formation. For the germination of conidia, an optimum temperature of 25° - 30°C and 92 % or more of relative humidity is ideal. Lesions on aerial parts develop rapidly at high temperature and moisture in a cloudy weather.

Disease management

Agronomic practices. (i) In field, where the disease has occurred in an epidemic form, plant debris and stubble in the field should be removed and burnt after crop harvest (ii) The pathogen is capable of surviving in many other collateral hosts in the absence of rice crop. These collateral hosts should be destroyed as far as possible in and around the fields (iii) From heavily infected fields, irrigation water should not be allowed to pass to

Fig. 2: Brown spot of rice-*Cochliobolus miyabeanus (Bipolaris oryzae)*

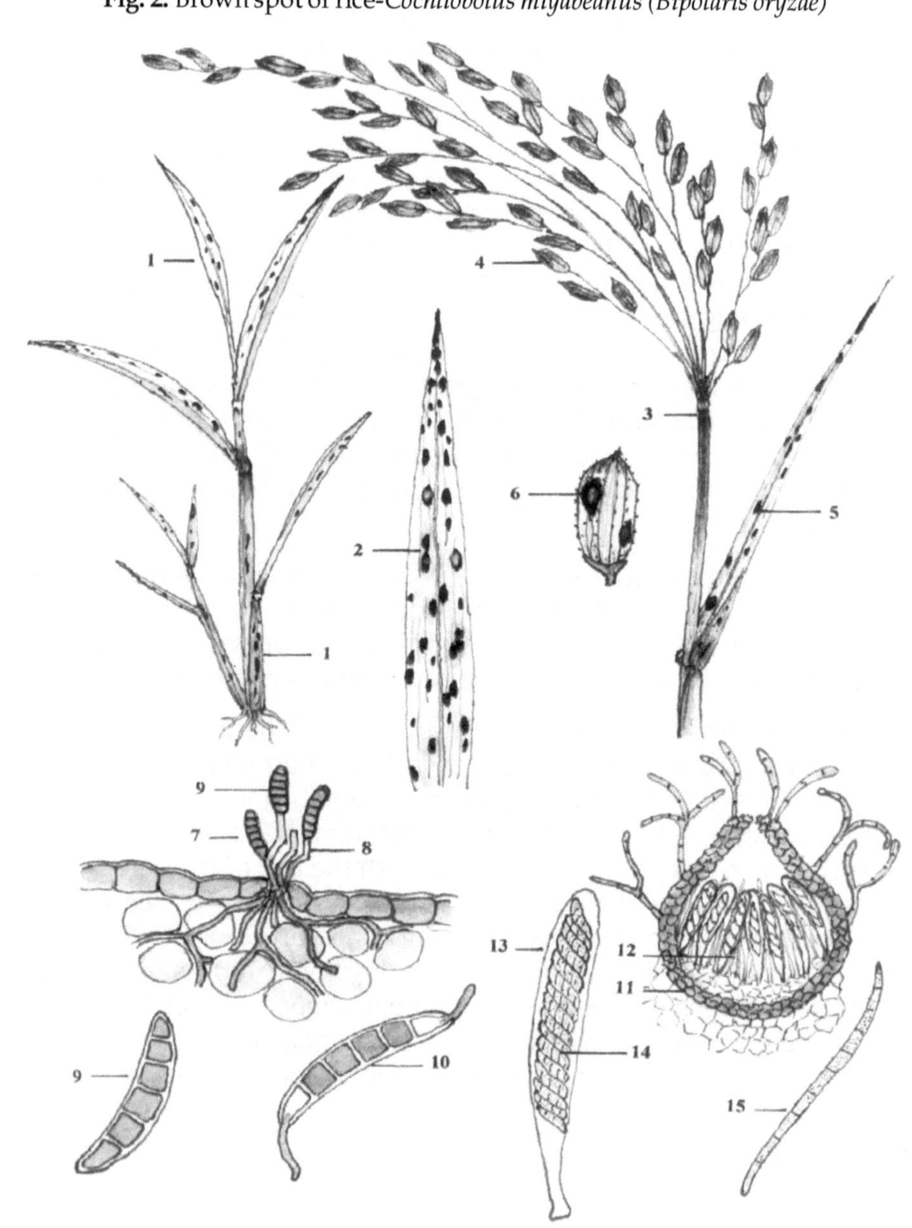

1. Spots on leaves and leaf sheaths of seedlings 2. Spots on leaf 3. Lesion on the rachis 4. Chaffy grains on the infected head 5. Spots on the boot leaf 6. Spots on the glumes 7. Conidiophores and conidia arising from the internal mycelium through the stomata 8. Conidiophore 9. Conidium 10. Germinating conidium 11. Perithecium 12. Hymenium 13. Ascus with coiled ascospores 14. Coiled ascospores 15. Ascospore

other fields (iv) More care should be bestowed in the selection of seeds. Seeds should be obtained from disease free crops and from disease free fields (v) Proper fertilizer and water management should be followed. Nitrogen and potash deficiencies are known to predispose rice plants to attack by this pathogen. So, recommended dose of these fertilizers should be applied. Use of slow release nitrogenous fertilizers reduces the incidence of this disease.

Seed treatment (i) As the disease is primarily seed-borne, seed treatment is effective in controlling the disease. Treating the seeds with thiram or captan at 4 gm./ kg.of seeds, 24 hours prior to sowing helps in eradicating the inoculum on the seed surface (ii) Hot water treatment of seeds at 55°C for a continuous period of 10 minutes is effective in eradicating the inoculum inside the seed (iii) Soaking the seeds in a solution of Aureofungin Sol. - 2 gm. + copper sulfate - 2 gm. in 100 liters of water for 24 hours, allowing the seeds to sprout and then sowing the sprouted seeds in the nursery beds also helps in eliminating the internal inoculum.

Chemical control (i) Secondary air-borne spread of the disease can be controlled by spraying, either edifenphos - 200 ml., mancozeb - 400 gm., captafol - 250 gm., copper oxychloride - 500 gm. or organo tin compound (Du-ter) - 250 gm. in 200 lit. of water per acre, twice or thrice, at fortnightly intervals.

3. Sheath blight of rice

Rhizoctonia solani

(*Thanatephorus cucumeris*)

The disease was first reported from Japan in 1910. The disease is prevalent in both tropical and temperate countries of the world, such as Brazil, Fiji, India, Iran, Japan, Nigeria, Sri Lanka, Taiwan, Venezuela, United States of America etc. In India, it occurs in Kerala, Tamil Nadu, Uttar Pradesh and Pondicherri states. In Tamil Nadu, the disease is found in some parts during some seasons in a very serious form and causes severe damage and loss.

Symptoms. Seedlings and grown up plants are vulnerable to attack by this pathogen. When the seedlings are affected, they dry and die. When older plants, especially plants in the booting stage are affected, the grain setting is very badly affected. High tillering, semi-dwarf varieties are more susceptible to the disease.

Initial symptoms appear as small, greenish-gray spots on the leaf sheaths and culms at the water level. The spots on the leaf sheath are ellipsoid or ovoid and about 10 mm.in length. The spots enlarge rapidly and may reach 2.0 - 3.0 cm. in length and become irregular in shape. They become grayish-white in the center, with brown or purplish margin. Many such spots coalesce and the leaves present a characteristic banded appearance. Hence, it is also called **'banded sclerotial disease'**. Such spots appear on the culm and leaf lamina also and when large areas are affected, the plants dry and die. The older leaves are affected first, followed by all the younger leaves. The affected plants turn yellow, dry and die. Even if ears are produced in the affected plants, the grains become chaffy. On the surface of the lesions and sometimes on the inner surface of the sheath and on the culm, brownish, silky, tufts of mycelium are seen. From these tufts of mycelium, brown or dark sclerotia are produced. Only during periods of high humidity and under water-logged conditions, mycelial growth and sclerotial formation is found **(Fig.3)**.

The causal organism. The mycelium of the fungus is found inside the host tissues, as well as on the surface of the host. The mycelium inside the host tissues is slender, hyaline and longer-celled, while the superficial mycelium found growing on the host surface is entirely different. This mycelium is dark colored, short-celled, much branched and stouter. On this mycelial mat, superficial sclerotia are formed. The sclerotia are spherical to irregular in shape. The sclerotia, which are whitish in the beginning, later become brown or black and hard and are up to 5.0 mm. in diameter. Many such sclerotia may join together and present an irregular, scabby appearance. There is no differentiation of the sclerotial tissue into a rind and internal medulla, but the outer cells are darker and thick-walled. The sclerotia are more or less covered with a web of mycelium, which is almost colorless. They fall off easily to the ground and remain in the soil. High humidity is essential for the mycelial growth and sclerotial production.

The basidial stage of the fungus is of little importance and is mostly saprophytic. The thallus appears as a flaky pellicle on the surface of the substrate or leaves near the ground level or on dead parts of the stem under high humid conditions. The basidia are borne on small, imperfectly symmetrical cymes. They are barrel-shaped or clavate. From the apex of the basidium, four, horn-shaped strigmata are produced, at the tips of which basidiospores are borne singly. Basidiospores are ellipsoid, flattened on one side, hyaline, thin-walled, truncate and measure 5.0 - 14.5 x 4.0 - 8.0µ in size. They germinate and produce secondary basidiospores similar to the mother basidiospores **(Fig.3)**.

Mode of survival, spread and epidemiology. The pathogen is primarily soil-borne in nature. It can survive as sclerotia and mycelium in diseased plant debris left over in the soil for about 20 months under dry conditions. However, in moist soils they are short-lived and remain viable only for 5 - 8 months. Exposure to sunlight inhibits infection, mycelial development and formation of sclerotia. The sclerotia, which remain in a viable state in the plant debris, are brought up to the soil surface during ploughing, puddling, leveling and other operations. They come in contact with newly planted seedlings and cause infection.

The pathogen is known to attack 188 plant species in 32 families. Sheath blight occurs on a number of crops, such as maize, sugarcane, sorghum, wheat etc, and many weed hosts, such as *Cynodon dactylon, Setaria glauca, Paspalum flavidum, Echinocloa crusgalli, Ergrostis pilosa, Panicum repens, Cyperus rotundus, Chloris* sp., *Digitaria* sp. etc. These weeds in and around rice fields and in irrigation channels may serve as sources of primary inoculum. Wild rice varieties are also susceptible to this disease.

Sheath blight is basically a disease of wet conditions. Under conditions of high relative humidity of 96 - 97 % and high temperatures of 30° - 32°C, the disease becomes very destructive. The sclerotia are dispersed through irrigation water, soil, farm implements, cattle, contaminated seeds etc. The disease usually appears, when the crop has passed the tillering phase. High tillering varieties, closer planting and heavy doses of nitrogenous fertilizer favor sheath blight incidence.

Disease management

Agronomic practices (i) Rice fields, irrigation channels and areas adjoining rice fields should be kept free from weeds, so as to reduce the inoculum potential (ii) Diseased crop debris, which may harbor the mycelium and sclerotia of the pathogen should be collected after the crop harvest and destroyed by burning (iii) Closer planting should be avoided, as it tends to increase the relative humidity within the crop canopy, which makes the plants more vulnerable to attack by the pathogen (iv) Application of excessive dose of nitrogenous fertilizer should be avoided. Balanced dose of fertilizers should be applied as per recommendations and they should be applied in split doses (v) Irrigation through infected fields to adjoining fields should be avoided, as the sclerotia from infected fields are likely to be carried to the other fields through irrigation water (vi) Deep summer ploughing should be encouraged, so as to bring the sclerotia in the soil to the surface of the soil to be exposed to the heat and light of the sun and get killed (vii) Soil amendment with neem cake at 1,000 kg./acre at the time of planting is found to be effective in controlling the disease.

Chemical control. Spraying the crop with carbendazim or EBP (Kitazin) or thiophanate (Topsin M) at 200 gm. in 200 lit. of water per acre, 2-3 times, at 10-15 days interval affords good control of the disease. Care should be taken to give a thorough coverage of the lower leaf sheaths and stem.

Seed treatment. Treating the seeds with carbendazim at 2 gm./ kg. of seed is also found to be effective in controlling the disease.

Biological control. Mycoparasites belonging to *Trichoderma* sp. are known to parasitize and kill the sclerotia in the soil. So, to encourage the growth of such organisms already present in the soil, organic manure should be applied to the fields.

4. Sheath rot of rice

Sarocladium (Acrocylindrium) oryzae

The disease was first reported from Taiwan in 1922. It is commonly found in India, Japan, United States of America and West Africa. In India, the disease is prevalent in all the rice growing states. In recent times the disease occurs in Tamil Nadu in all the districts and in some seasons causes very serious damage to the crop and yield loss.

Symptoms. The disease appears only during the booting stage and the flag-leaf sheath, which encloses the panicle, is affected. The symptoms appear as small, oblong or irregular brown spots, about 0.5 - 1.5 cm. in length. The lesions enlarge and become grayish-brown or gray with a brown margin. The lesions coalesce and may cover almost the entire leaf sheath. Due to the infection, the flag-leaf sheath fails to unfurl normally, as a result the young panicle is choked. The panicle may not emerge out from the flag-leaf sheath or may emerge only partially. The grains produced from such affected panicles become shrivelled, chaffy, discolored and are gray or black in color. The panicles, which fail to emerge out, rot completely. White or grayish-white fungal growth is seen inside the affected sheath and on the surface of the grains **(Fig.4).**

The causal organism. The mycelium of the fungus is hyaline, septate, slender and much branched. From the mycelial mat, conidiophores, which are slightly thicker than the ordinary hyphae arise. Conidiophores are hyaline, erect, branched once or twice and with 3 or 4 branches in a whorl. The

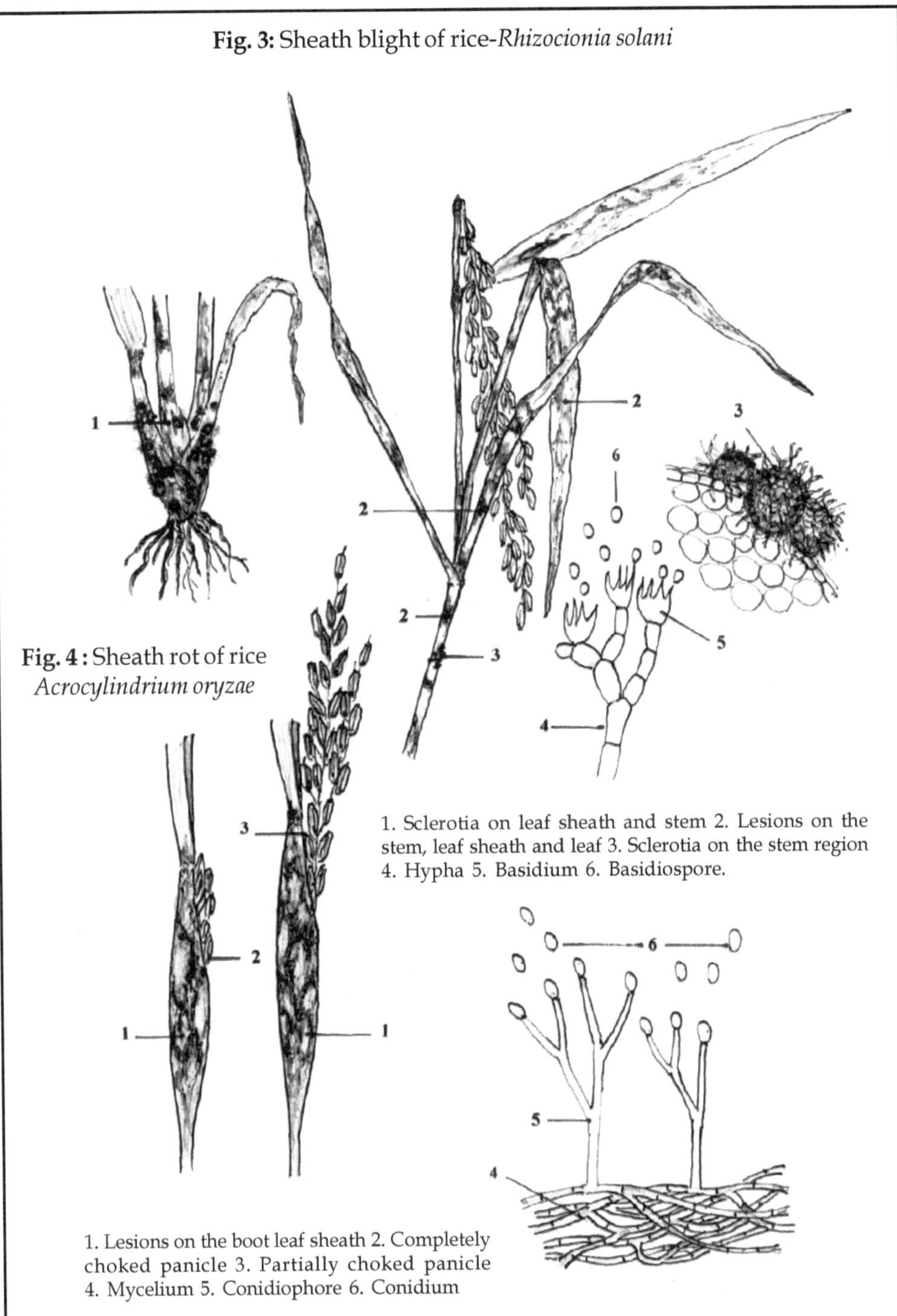

Fig. 3: Sheath blight of rice-*Rhizocionia solani*

1. Sclerotia on leaf sheath and stem 2. Lesions on the stem, leaf sheath and leaf 3. Sclerotia on the stem region 4. Hypha 5. Basidium 6. Basidiospore.

Fig. 4 : Sheath rot of rice *Acrocylindrium oryzae*

1. Lesions on the boot leaf sheath 2. Completely choked panicle 3. Partially choked panicle 4. Mycelium 5. Conidiophore 6. Conidium

branches taper towards the tip and conidia are borne consecutively from the tips of the branches one by one. The spores are single-celled, hyaline, smooth, cylindrical and measure 4.0 - 9.0 x 1.0 - 2.5µ in size. They germinate by producing a germ tube and cause fresh infection **(Fig.4).**

Mode of spread and epidemiology. Crops affected by sucking insect pests and crops affected by diseases, such as 'tungro' and 'yellow dwarf' are more vulnerable to attack by this disease. Dwarf and semi-dwarf, and high tillering varieties are more susceptible to this disease.

The conidia are dispersed by wind, rainsplash etc. Contaminated seeds may also spread the disease.

Temperature of 30° - 31°C and high humidity are favorable for the rapid spread of the disease. Low night temperature and high day temperature and high relative humidity during the booting stage favor large-scale disease occurrence. The disease does not occur at temperatures above 37°C and below 13°C.

Disease management

Agronomic practices (i) Balanced dose of fertilizers as per recommendations should be applied in split doses (ii) Closer planting should be avoided (iii) Soil amendment with gypsum at 200 kg./acre in two split doses, one at the time of planting and the second at the active tillering stage helps in reducing the disease incidence.

Chemical control. Spraying the crop with carbendazim - 200 gm. or edifenphos - 200 ml. or captafol - 250 gm. or mancozeb - 400 gm. or chlorothalonil - 250 gm. in 200 lit. of water per acre, twice, at the boot leaf stage and 15 days later gives effective control of the disease. Care should be taken to give a thorough covering of the boot-leaf sheath (ii) Spraying the crop with neem oil - 2 % (neem oil - 4.0 lit. + liquid soap - 200 ml). in 200 lit. of water or neem seed kernel extract - 5 % (neem seed kernel extract - 10 lit. + liquid soap - 200 ml.) in 200 lit. of water is also effective in controlling the disease.

5. Narrow brown leaf spot of rice

Cercospora janseana / C. oryzae
(*Sphaerulina oryzina*)

The disease was first reported from Japan in 1910. Now, the disease is prevalent in many of the rice growing countries of the world including India.

Symptoms. The disease attacks the leaves mainly, but in severe cases of infection, the leaf sheaths and the grains are also attacked. The symptoms appear as small, narrow, linear lesions, about 2.0 - 10.0 mm. long running along the veins. The lesions become longer and may reach several cm. long. They are dark- brown in the center, with a light-brown, dull margin. When environmental conditions are favorable for the pathogen, several such lesions develop on the same leaf. The spots coalesce and most of the leaf lamina is affected, as a result the leaves turn yellow, dry and eventually die. The lesions on the leaf sheaths and grains are slightly broader. The disease occurs only during some seasons sporadically. The disease appears mostly during the booting stage and progresses as the crops mature **(Fig.6).**

The causal organism. The mycelium of the fungus is both inter- and intracellular. From the intercellular hyphae, haustoria are produced. The mycelium is much branched, septate and olive-brown in color. From the internal mycelium, conidiophores, which are slightly thicker than the ordinary hyphae emerge in bunches through the stomata. They are erect, unbranched or with few branches from the base, septate and olive-brown in color. From the tips of the conidiophores, conidia are produced singly. The conidiophores are geniculate and a scar is discernible at each of the knee bends from where the conidia had dislodged. They measure 68.0 - 140.0 x 4.5μ in size. Conidia are cylindrical, slightly narrowing towards the tip, 3 - 10 septate, olive-brown in color and measure 206.0 x 5.0μ in size on an average.

The perfect stage of the pathogen has been identified as *Sphaerulina oryzina*, which is non-pathogenic. The perithecia are globose or sub-globose, black and 60 - 100μ in diameter. Asci are cylindrical or club-shaped, round at the top, stipitate and 50 - 60 x 10 - 13μ in size. Ascospores are spindle-shaped, hyaline, 3-septate, and are 20.0 - 23.0 x 4.0 - 5.0μ in size **(Fig.6).**

Mode of survival, spread and epidemiology. Air-borne conidia are mostly responsible for the spread of the disease. Infected seeds also serve to propagate the disease. The pathogen attacks *Panicum repens* also and the conidia produced from this host may provide the initial inoculum. Many pathogenic races of this fungus exist in nature.

The disease spreads rapidly under conditions of moderate temperature and high humidity.

Disease management

Agronomic practices (i) Seeds should be selected from disease-free crops (ii) The crop debris left over in the field after the harvest of the crop should be destroyed by burning (iii) Rice fields, bunds, irrigation channels and areas around the field should be kept free from weeds

Chemical control. Spraying the crop with copper oxychloride - 500 gm. or mancozeb - 400 gm. or edifenphos - 200 ml. or carbendazim - 200 gm. in 200 lit. of water per acre, twice, the first when the disease intensity has reached the economic threshold level, followed by another spraying 10 - 15 days later controls the disease

Seed treatment. The seeds should be treated with captan - 4 gm. or thiram - 4 gm. or carbendazim - 2 gm./ kg, of seeds, 24 hours prior to sowing.

6. Stem rot of rice

Sclerotium oryzae
Helminthosporium sigmoideum
(*Leptosphaeria salvinii*)

The disease was reported in India in 1911. It is prevalent in Bengal, Bihar, Tamil Nadu, Punjab and other major rice growing states of the country. The disease is also commonly found in Japan, Sri Lanka, Vietnam, United States of America, the Philippines and Italy.

Symptoms. The initial symptoms of the disease appear as small, black lesions on the outer leaf sheaths of the plants at the water level. The lesions enlarge and the stem at this region shows rotting and finally the entire plant is killed. The culm of the affected plant is covered with black discoloration near the soil, which later spreads to all the leaf sheaths. Due to the pressure exerted by the invading pathogen, the epidermis of the plant ruptures and the plant withers and dies. The base of the stem, at the lowest internode or the next one or two above is thus affected. Inside the culm, dark, grayish, weft of mycelium is found and the inner surface may be dotted with small, round, shining black sclerotia. Within the rotting tissues of the lower leaf sheaths also, mycelial growth and sclerotia are formed. The affected plant may produce light ears with unfilled grains, while green shoots with thin, small, pale and rigid leaves may arise from the basal nodes **(Fig.5).**

The causal organism. The conidial stage of the pathogen has been identified as *Helminthosporium sigmoideum* and the ascigerous stage as *Leptosphaeria (Magnaporthe) salvinii*. The mycelium of the fungus is present in the subepidermal layers of the culm and leaf sheaths. It consists of much branched, fluffy, creeping, septate hyphae, which are whitish in color. On the white mycelium, sclerotia are formed, sometimes in large numbers.

Fig. 5: Stem rot of rice-*Sclerotium oryzae*

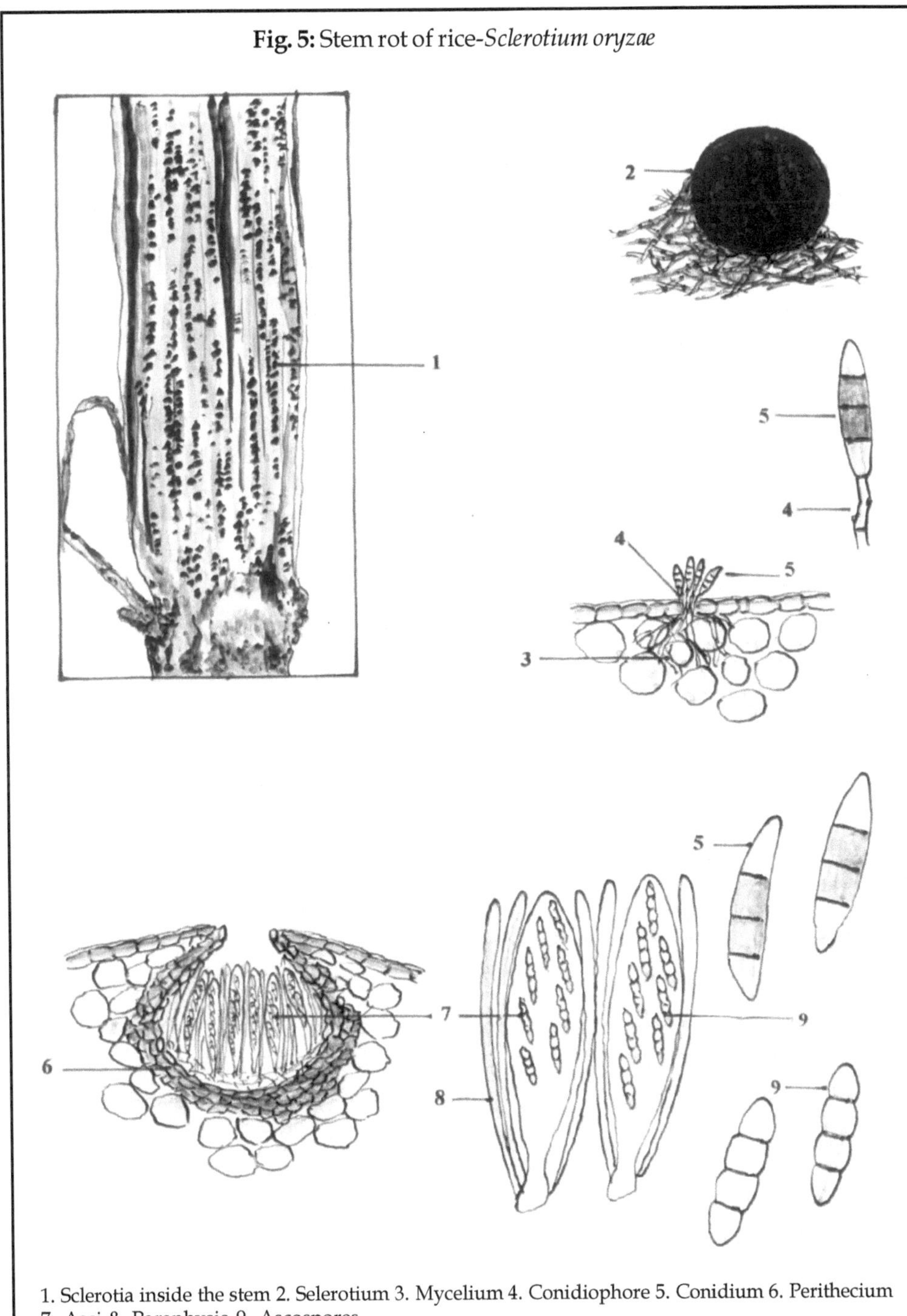

1. Sclerotia inside the stem 2. Selerotium 3. Mycelium 4. Conidiophore 5. Conidium 6. Perithecium 7. Asci 8. Paraphysis 9. Ascospores

They are large, globose, of almost uniform size, white when immature, becoming dark-brown to black on maturity. Each sclerotium is differentiated into an outer melanized rind, a middle cortex and an innermost area of loosely arranged hyphae. The sclerotia measure 230 - 270μ in diameter. The sclerotial stage is common on rice and a number of wild grasses.

The conidial stage is not of common occurrence. The conidiophores are dark, 8 - 10-septate, simple or sparsely branched. The conidia are large, falcate-sigmoid, 3-septate and are borne singly at the tips of sterigmata on the conidiophores. The middle cells of the conidia are dark-brown in color, while the apical cells are light green or hyaline with obtuse ends. They measure 55 - 65 x 11 - 14μ.

The perfect stage is scarcely found in nature. The perithecia are black, globose, clustered in the parenchyma of the leaf sheath. They have short beaks, flush with the host epidermis and measure 350 - 400μ without the beak. The asci are clavate, stipitate, thin-walled, 120μ long and contain 8 ascospores, which are arranged in 2 - 3 series. Ascospores are oblong-fusiform, slightly curved, 3-septate, the central cells constricted at the septum and measure 60 x 90μ in size. The fungus is heterothallic **(Fig.5).**

Mode of survival, spread and epidemiology. The pathogen, which is mainly a wound parasite, survives during unfavorable seasons through sclerotia. The sclerotia, which are present in large numbers in old rice straw and stubble left over in the field after the harvest of the crop, serve as the primary inoculum. The sclerotia can survive in dry soils for a much longer period of time than in wet soils. The viability of the sclerotia is lost quickly, if they are exposed to direct sun on the soil surface.

The sclerotia present in the crop debris and top soil are disseminated with irrigation or rainwater. Sclerotia may also be transported along with the soil by cattle, agricultural implements and even by humans.

Close planting and stagnating more than the required quantity of water in the fields, especially in heavy clay soils favor the disease occurrence. Application of high doses of nitrogenous fertilizer increases the incidence of stem rot.

Disease management

Agronomic practices (i) The straw, stubble and other crop residues left over in the field after the harvest of the crop should be burnt *in situ*. (ii) Close planting, application of excess doses of nitrogenous fertilizer and stagnation of too much water in the fields should be avoided (iii) Seeds, which may be contaminated with the sclerotia should not be used for sowing (iv) Deep summer ploughing may be encouraged, so as to bury the sclerotia deep into the soil or to bring them to the surface of the soil to be exposed to direct sun (v) Varieties resistant to the disease may be grown.

Chemical control. No chemical treatment has been found to be effective in controlling the disease.

7. False smut of rice

Ustilaginoidea virens

(*Claviceps oryzae - sativae*)

The occurrence of the disease has been reported from almost all rice growing countries of the world including China, Japan, South East Asian countries, North and South America, Burma, Sri Lanka, Fiji, Africa and the Philippines. In India, the disease is found in Punjab, Uttar Pradesh, Bihar, and Tamil Nadu states. In Tamil Nadu, the disease was first reported from Tirunelveli district in 1878.

Symptoms. The symptoms appear only after flowering. The pathogen attacks the ovary of individual kernels and transforms them into large, velvety, green smut balls (pseudomorphs), which sometimes attain the size, twice the diameter of normal grains. The color of the ball is orange-yellow on the periphery and whitish in the center. The spore balls are small in the beginning and are visible in-between the glumes. They gradually increase in size enclosing all the floral parts and are smooth, yellow and covered by a membrane. As the ball enlarges in size, the membrane ruptures, exposing the ball. Usually only a few grains in a panicle are affected. However, in severe attacks, many such balls may aggregate together. Florets beside the smut balls may remain sterile **(Fig.9).**

The causal organism. The fructifications replacing the grains represent the conidial and sclerotial stage of the pathogen. They are yellow to orange in the beginning, later turning olive-green to black. When young, they are fleshy and become hard later. The center of the spore ball consists of compactly packed interwoven mycelium of thin, hyaline, septate hyphae. Three sporiferous layers surround this central hard core. The innermost layer is pale yellow, consisting of radiating mycelium and spores in the process of formation. The spores (conidia) are borne laterally on minute sterigmata on hyaline, septate hyphae. The next layer is orange-yellow and is composed of

Fig. 6: Narrow brown leaf spot of rice-*Cercospora oryzae*

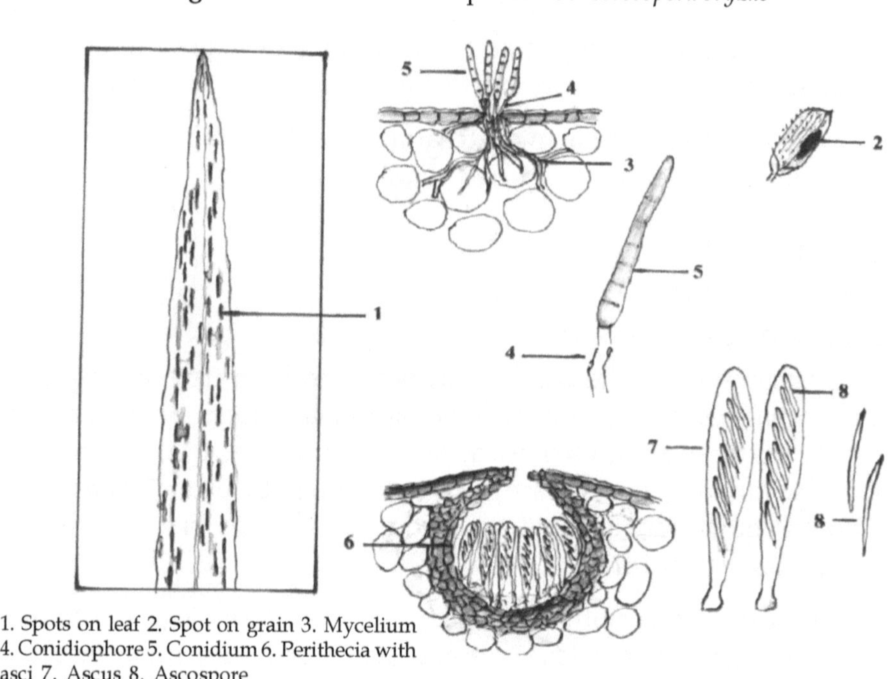

1. Spots on leaf 2. Spot on grain 3. Mycelium 4. Conidiophore 5. Conidium 6. Perithecia with asci 7. Ascus 8. Ascospore

Fig. 7: Bacterial leaf blight of rice *Xanthomonas campestris* pv. *oryzae*

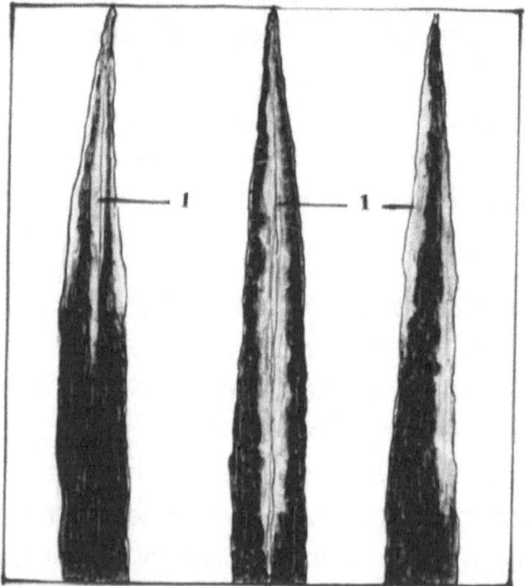

1. Blighted area on leaf

Fig. 8: Bacterial leaf streak of rice *Xanthomonas translucens* f.sp. *oryzicola*

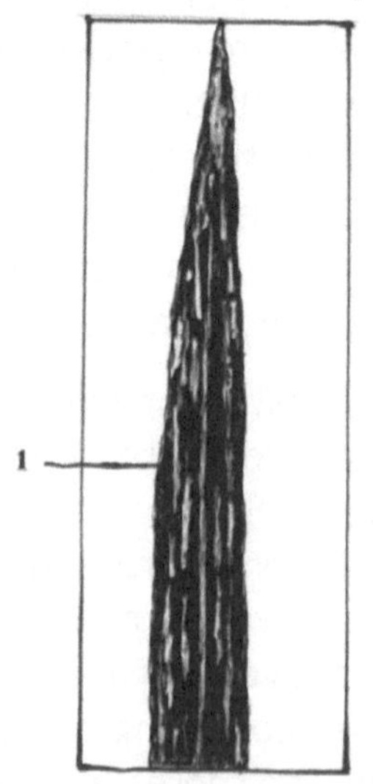

1. Streaks on leaf

mycelium and spores. The outermost layer is olive-green to black in color and consists of matured spores and fragments of mycelium. The surface is covered with powdery, dark green spores in mass. The conidia are almost spherical, echinulate, olivaceous and measure 4.0 - 6.0 x 3.0 - 5.0μ. Young spores are smaller, light-colored and smooth-walled. The conidia germinate in water by a short germ tube bearing 1 - 3 secondary conidia. The pathogen differs from other species of *Claviceps* in that, the conidial state in this fungus is dry instead of in a honey dew.

The sclerotial stage follows the conidial stage. In each of the green ball (pseudomorph), one or more, usually two sclerotia are formed. At first, they are buried inside the conidial mass and become exposed after the conidia are dispersed. These sclerotia are hard, variously shaped and are 5.0 - 13.0 x 2.0 - 5.0 mm. in size. They look black but are white inside and pseudoparenchymatous. The sclerotia germinate and produce several perithecial heads, borne on stipes 5.0 - 15.0 mm. in length. The stromatic heads are dark greenish-yellow to olive in color and 1.0 - 3.0 mm. in diameter. In each such head, several perithecia, which are ovate to pyriform, with prominent ostiole are formed. In each perithecium, many hyaline, cylindrical asci are formed. Each ascus has 8 hyaline, one-celled, filiform ascospores, measuring 50.0 - 80.0 x 0.5 - 1.0μ in size. These ascospores germinate by 1 - 2 germ tubes, which produce secondary ascospores **(Fig.9).**

Mode of survival, spread and epidemiology. The fungus perennates through sclerotia. Infection occurs through ascospores produced from overwintered or oversummered sclerotia and through conidia. Infection also occurs in the seedling stage, when germ tubes from conidia penetrate the cuticle of coleoptile and grow intercellularly causing systemic infection. The conidia can remain viable for a period of 2 months, while the sclerotia can remain viable for a period of 9 months in the soil under field conditions. *Oryza officinale* and *Chionachne koenigii* serve as alternate hosts of this pathogen in India. Conidia and ascospores produced from these hosts may serve as initial inoculum and infect rice flowers.

High humidity, cloudy and damp weather and moderate temperature at flowering time favor disease occurrence. Hyphal growth and conidial germination can occur within a wide range of temperatures from 12° - 34°C, but the optimum temperature is 28°C. Optimum relative humidity for germination of conidia is 98 %. The conidia do not germinate when the humidity is less than 92 %. Applying high doses of fertilizers, especially at the time of flowering, predisposes the plants to infection by the pathogen.

Disease management

Agronomic practices (i) Sclerotia-free seeds should be used for sowing (ii) At the time of harvest, diseased plants with scleroria should be harvested first, so as to prevent the sclerotia falling to the field (iii) Varieties, which show resistance to the disease may be grown. No chemical control measures have been advocated to control this disease.

8. Rice grain discoloration

Several fungal organisms are responsible in causing grain discoloration of rice. The grains may be affected before the harvest of the crop in the field or after harvest. Depending upon the pathogens, the grains may be partially or completely affected and spoiled. The discoloration may appear externally on the glumes or internally on the kernels or both. Grain discoloration results in reduction in quantity of the produce, quality of the grains and the market value. The disease is common in all rice growing countries.

Symptoms. Depending upon the pathogens, the grains may be affected after the milk or dough stage in the field or at the harvest time or during storage. Infection during the milk or dough stage results in total damage of the affected grains. The infection may be external, which may cause discoloration of the glumes or internal, which may cause discoloration of the kernels and make them powdery. On the surface of the grains, dark-brown or black spots appear that may sometimes cover the entire grain surface. Due to internal infection the kernels may become dirty white, yellowish, pink or red depending upon the causal organism, environmental conditions and varieties of rice.

The causal organism. Many fungi are responsible for causing grain discoloration in rice. Among them, *Helminthosporium oryzae, Curvularia lunata, Acrocylindrium oryzae* and *Trichoconis padwickii* are more important. *Alternaria tenuis, Drechslera tetramera, Fusarium moniliforme, Pyricularia oryzae* and *Nigrospora* sp. are also responsible for grain discoloration in rice to some extent. Some species of *Aspergillus* and *Penicillium* cause grain discoloration in rice during storage under humid conditions.

Mode of spread and epidemiology. Almost all the organisms, which cause grain discoloration produce large number of conidia, which are easily carried and dispersed by wind. Some of these fungi can also survive as conidia or mycelium on the surface of the grains or inside the grains.

High humidity and moderately high temperatures favor the growth, sporulation, spread and infection of most of these fungi.

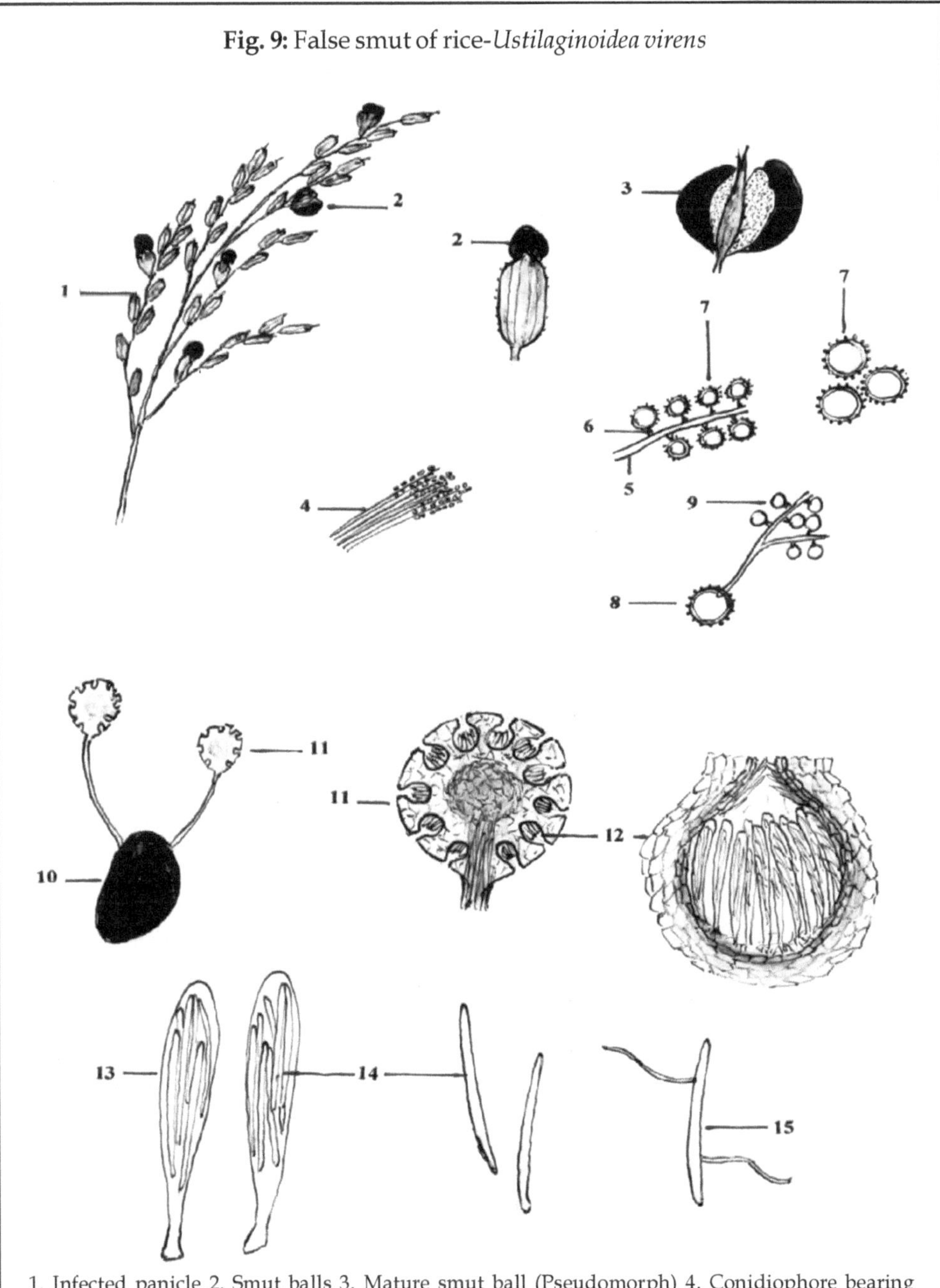

Fig. 9: False smut of rice-*Ustilaginoidea virens*

1. Infected panicle 2. Smut balls 3. Mature smut ball (Pseudomorph) 4. Conidiophore bearing conidia 5. Conidiophore 6. Sterigmata 7. Conidia 8. Germinating conidium 9. Secondary conidia 10. Germinating sclerotium 11. Perithecial head 12. Perithecium 13. Asci with ascospores 14. Ascospores 15. Germinating ascospore.

Disease management

Agronomic practices (i) The grains should be well dried to bring down the moisture content to 12 - 14 % before storage (ii) The grains should be stored in well-ventilated, dry store-houses to prevent build up of humidity (iii) Diseased and discolored grains should not be used for sowing.

Chemical control. Spraying the crop with carbendazim- 200 gm. or captafol - 250 gm. or mancozeb - 400 gm. in 200 lit. of water per acre at the time of panicle initiation, followed by another spraying, 10 days later controls grain discoloration.

9. Bacterial leaf blight of rice

Xanthomonas oryzae pv. *oryzae (X. campestris* pv. *oryzae)*

'Bacterial leaf blight' was first reported from Japan in 1884. The disease appeared in India in 1951 in the Maharashtra state. In Tamil Nadu, the disease appeared in 1961 in the Thanjavur district. With a plan to increase rice production, several high yielding varieties of rice, including Taichung Native-1 (TN.1) were introduced to India from foreign countries. Along with these varieties, this disease was also introduced to this country and had spread rapidly throughout the country, attacking many of the cultivated varieties and had caused heavy damage to the crops and heavy losses. In Tamil Nadu, it is prevalent in all the districts where rice is grown. The disease occurs in many major rice growing countries, such as Japan, China, the Philippines, Taiwan, Thailand, Vietnam, Korea, Malaysia, Bangladesh and Australia.

Symptoms. The symptoms of the disease vary depending upon the growth stage of the crop and weather conditions. Three types of symptom viz., **'leaf blight'**, **'Kresek'** or **'wilt'** and **'yellow leaf'** are found however, leaf blight is quite common. The disease is a typical vascular wilt, leaf blight being the mild phase resulting from secondary infection.

(i) Leaf blight phase

The disease may appear at any growth stage of the crop, but the leaf blight phase, which is more common is found mostly in crops, 4 - 6 weeks after transplanting. The disease is mostly confined to the leaves only. In the nursery, small, circular to elongate spots appear on the margins of the leaves. These spots enlarge in size and eventually, the leaves turn yellow, dry and die completely. In the transplanted crop, when the plants are about 4 - 6 weeks old, leaf blight symptoms appear. Small, water-soaked streaks appear from the tip of the leaf and extend downwards along the margins of the leaf or on either side of the midrib. Very rarely the streaks may appear on one side. As the disease advances, the streaks enlarge in size, extend towards the base of the leaf and become long streaks. Sometimes, more than two streaks may appear on either side of the mid rib, but the streaks always start from the tip of the leaf lamina and extend downwards. Sometimes, the streaks may extend into the leaf sheath also. The streaks turn yellow to straw-colored and become necrotic stripes. In due course, the streaks enlarge and large areas of the leaf lamina are affected, as a result they dry and die. In the diseased leaves, the inner necrotic areas show wavy margins, which is characteristic of the disease. The tips of the affected leaves become twisted and dried. Under humid and warm conditions, creamy white bacterial ooze comes out of the young lesions in the early morning hours in minute droplets, which on drying leaves a yellowish encrustation on the leaf surface. This bacterial encrustation is washed down in rainwater and leads to secondary infection. Blighting may extend to the leaf sheath and culms, resulting in the death of the affected tiller or the whole clump. The glumes of seeds and even kernels are also infected. The glumes become discolored and water-soaked spots develop on them **(Fig.7).**

Blighting of leaves may be due to several other causes, physiological or other diseases, such as 'tungro'. For proper diagnosis of the disease, the affected leaves should be examined microscopically. A small piece of the affected leaf is cut, mounted on a drop of water on a glass slide and examined under a microscope. Bacterial ooze can be seen as a whitish, cloudy mass coming out at the cut ends.

(ii) "Kresek' or wilt phase

This is the most destructive phase of the disease and is known by the name **'Kresek'** in Indonesia. During this phase, the young plants are affected. The outer leaves of the affected plants wither, become grayish - green or light brown and often float on the surface of water. Some of the younger leaves roll up and wither. The bacteria multiply in the vascular region, spread through the xylem vessels to the growing point of the young plant and infect other leaves. As a result, the entire plant dries and rots. Bacterial exudate, which comes out of the broken ends of leaves and sheaths get mixed up with water and cause fresh infection. The bacteria, which multiply in large numbers in the xylem vessels, block the vessels and prevent the flow of nutrients to the leaves, thereby the plants wilt and die.

(iii) Yellow leaf phase

This is another common symptom found on the plants in the tropics. Some of the youngest leaves in a clump become pale-yellow or whitish. These leaves later turn yellowish-brown, wither and die.

The causal organism. The bacteria causing this disease are rod-shaped, with rounded ends, 1.0 - 2.0 x 0.8 - 1.0μ in size and have monotrichous, polar flagellum, which measures 6.0 - 8.0μ in length. They are gram- negative and non-spore forming. The bacterial cells are surrounded by mucous capsules and joined together to form an aggregated mass. They are aerobic in nature. Several strains of bacterial leaf blight pathogen exist, which vary in their virulence. Most of the tropical strains are virulent forms, capable of causing the 'Kresek' phase and considerable damage to the crop during the early stages of growth.

Mode of survival, spread and epidemiology. The bacterial leaf blight is primarily a vascular disease. The primary inoculum for the perpetuation of the disease may come from infected seeds or stubble and other diseased plant materials left over in the field after harvest or from collateral hosts. Several workers have demonstrated seed infection and survival of the bacterium on the glumes or in the endosperm for varying lengths of time, sometimes up to the next growing season. In India, the pathogen is propagated from season to season through infected seeds. Infected straw left over in the field and stubble in the field after the harvest of the crop is another source of perpetuation of the pathogen. In areas where two or three crops of rice are grown consecutively in a year, this mode of carry over of the disease is quite possible, although the bacterium can survive in the soil for only a short period. Chaff dumped near threshing places also plays an important role in causing primary infection. The pathogen may also survive in the ratoon crops and in self-sown plants. Many grass hosts have been found to harbor the pathogen and serve as sources of primary inoculum for initiation of the disease. In India, several grass hosts, including *Cyperus rotundus, C. defformis, Leersia hexandra, Paspalum scrobiculatum, Panicum repens* etc. have been found to be collateral hosts of this pathogen. Diseased wild rice growing in ponds may also serve as source of primary inoculum.

The hibernating bacterial cells, on coming into contact with the rice seedling are activated and multiply and invade the seedling through wounds caused by root development at the basal part of the stem or through hydathodes on the leaf or through stomata. They multiply rapidly in the intercellular spaces of the parenchyma in the coleoptile and leaf sheath of young seedlings.

Secondary infection is through wounds caused, as a result of chafing and striking of leaves against one another in the wind, through the cut ends of seedling leaves, through broken roots while transplanting or through hydathodes. Once inside the leaf, the bacterium multiplies and moves in both directions. Bacterial exudates are produced on the surface of leaves and during rainstorms, these bacteria disperse and come in contact with other leaves or fall into the water. These bacteria are carried by irrigation water from one field to another.

Continuous rainfall over prolonged periods, damp and humid weather and moderate temperatures between 22°- 25°C favor the development of the disease. Temperatures below 20°C, severe heat and drought are unfavorable for disease development and spread. Heavy nitrogen fertilization, ill-drained and shaded conditions aggravate the disease severity. Though rice plant is susceptible to infection at all stages of growth, with age the susceptibility to vascular infection decreases.

Disease management

Agronomic practices (i) Diseased plant parts, straw, stubble etc. left over in the field after the harvest of the crop should be destroyed by burning (ii) Chaff dumped near threshing places should be destroyed by burning (iii) Paddy fields, irrigation channels and areas adjoining rice fields should be kept free from weeds (iv) Closer planting should be avoided (v) Stagnating more than the required amount of water in rice fields should be avoided (vi) Irrigation from or through infected fields to other fields should be avoided (vii) Seeds should be selected from fields, which have not been affected by the disease (viii) Clipping the tips of leaves of seedlings at the time of transplanting may be avoided (ix) Varieties highly susceptible to the disease should not be grown in disease prone areas.

Chemical treatment (i) Spraying the crop with the antibiotic Agrimycin (Containing 15 % Streptomycin and 1.5 % Oxytetracyclin) - 120 gm. in 200 lit. of water/ acre affords good control of the disease (ii) Spraying the crop with Agrimycin - 120 gm. + copper oxychloride - 500 gm. in 200 lit. of water/ acre is more effective than the former (iii) Spraying with the antibacterial substance Bactrinol - 60 gm. in 200 lit. of water is also found to be very effective. The spraying should commence, as soon as the initial symptoms appear on the leaves. Two to three sprayings should be given at 10 days interval.

Resistant varieties. Mashuri, Prasad, Ramakrishna, Salet-4, Sasyasree, CNM.540, IR.20, IR.22, IR.36, IR.42, IR.54, BAM.9, MTO.15 and N.22 are fairly resistant to the disease. IR.8, IR.24, Jaya, Padma, Pankaj, Jagannath, Bala, Krishna, Cauvery, Jamuna, Kanchi, Vijaya, Sona, Sabarmathi and ADT.38 are susceptible.

10. Bacterial leaf streak disease of rice

Xanthomonas translucens f.sp. *oryzae (X. translucens* f.sp. *oryzicola)*

The disease was first described as **'stripe disease'** in 1918 from the Philippines. Later, it was identified and described as **'leaf streak'** from China in 1957. The disease is prevalent in many of the Asian countries, including the Philippines, China, Thailand, Indonesia, Malaysia, Bangladesh and India. In India, it was reported in 1967 and it occurs in Uttar Pradesh, Madhya Pradesh, Maharashtra, Bihar, Karnataka, Haryana, Telungu Desam, West Bengal, Orissa and Tamil Nadu. The disease, which appeared in Tamil Nadu in 1968 occurs sporadically in some places during some seasons.

Symptoms. The initial symptom of this foliar disease appears as fine, linear, water-soaked to translucent streaks in-between the veins. The streaks may be 1.0 - 10.0 cm. in length and are restricted by the veins. Soon, the streaks turn yellow or orange-brown and minute, yellow or amber-colored droplets of bacterial exudate come out from them through the stomata. When these droplets dry, the leaf surface becomes rough along the streaks. When the disease occurs in a severe form, the streaks may coalesce to form large patches, which may cover the entire leaf surface. Finally, the leaves are blighted completely. In highly susceptible varieties, the streaks may be surrounded by a yellow halo. The disease may affect the leaf sheath and the seeds also **(Fig.8).**

The causal organism. The bacteria causing this disease are rod-shaped, 1.0 - 2.5 x 0.5 - 0.8µ in size, monotrichous with a polar flagellum, gram negative and aerobic in nature.

Mode of survival, spread and epidemiology. The pathogen can survive in infected seed from one season to the next, but cannot survive in the debris. The bacteria hibernate under the glumes of matured grains. At the time of germination, the plumule is infected from this source and from the first leaf, the bacteria are carried to the other parts. From the infected aerial parts of the plant, secondary infection occurs through wounds and stomata. The bacteria do not enter the vascular systems and are not systemic. They mainly infect the parenchyma cells and multiply in the intercellular spaces and come out through the stomatal openings to the leaf surface in minute droplets-like beads. Secondary spread of the bacteria is mostly by rainsplash and irrigation water. Young rice leaves are more susceptible to the disease, while older leaves develop some kind of resistance.

Rain, storm, high relative humidity of 83 - 93 % continuously for a few days or dew during morning hours and moderate temperatures of 26.0° - 30.5°C favor disease development. Temperatures below 22.4°C are unfavorable for the disease occurrence. Heavy doses of nitrogenous fertilizer, shade and close planting are conducive for the spread and development of the disease. Hairy and glabrous varieties are more prone to attack by the pathogen. Many wild rice varieties and a few grasses serve as collateral hosts for the pathogen.

Disease management

Agronomic practices (i) Seeds should be selected from disease free crops (ii) Wild rice varieties and grass hosts in and around rice fields should be removed and destroyed (iii) Application of high doses of nitrogenous fertilizer and close planting should be avoided

Seed treatment (i) Wet ceresan - 25 gm. and Streptocycline - 7 gm. are mixed in 25 lit. of water. Soaking 25 kg. of seeds in this solution for 8 - 10 hours prior to sowing eradicates the seed-borne bacteria (ii) Streptocycline - 7 gm. is mixed with 25 lit. of water. Soaking 25 kg. of seeds in this solution for 12 hours, followed by hot water treatment at 52 - 54°C for 30 minutes prior to sowing eradicates the seed-borne inoculum.

Chemical treatment. Spraying the crop with carboxin - 200 gm. or copper oxychloride - 500 gm. or captan - 250 gm. or Agrimycin - 120 gm. or Bactrinol - 60 gm. in 200 lit. of water per acre controls the disease effectively. The spraying should commence, as soon as initial symptoms appear, followed by 2 more sprayings at 10 days interval.

Resistant varieties. IR.20, Krishna and Jagannath are tolerant to this disease, while IR.8, Jaya and Padma are highly susceptible.

11. Rice 'tungro' virus disease

'Tungro', a word from the Philippine language means degenerated growth. Although the disease was known for a number of years, only in 1965, its viral origin was established. The disease is known by different names viz, **'tungro'**, **'mentek'**, **'yellow-orange leaf'**, **'leaf yellowing'** etc. in different countries of the world. The disease is more common in Indonesia, Thailand, Malaysia, the Philippines, Bangladesh and India. In India, the disease, which was known as leaf yellowing, is prevalent in Uttar Pradesh, West Bengal, Bihar, Orissa, Telungu Desam, Karnataka and Tamil Nadu. In Tamil Nadu, the disease occurs in an epidemic form once in 5 or 6 years and ravages rice crops, causing severe economic losses. The disease occurred in many parts of Tamil Nadu during 1977, 1984 and 1992 and caused serious damage to rice crops. The yield loss due to attack by this disease is found to be much more, when young plants are attacked.

Symptoms. Rice plants affected by 'tungro' disease are stunted in growth and produce lesser number of tillers than normal plants. Depending upon the rice varieties and the age of infection, the symptoms vary. The symptoms are manifested to a lesser extent, if grown-up plants are attacked and the severity of the disease decreases with the age of the plants. When 15 days old young plants are infected, the yield loss may go up to 68 %, while the loss is as low as 7 %, when 75 days old plants are infected.

Striking symptoms are seen on the leaves of affected plants. The color of lower leaves turns yellow or yellowish-orange from the leaf tip and margins and extends downwards and inwards. Part of the leaves or the entire leaves may be discolored. Irregular, dark-brown spots may also develop on the discolored leaves.

In young leaves, whitish-green or whitish streaks of various lengths appear in-between and parallel to the veins. Mottling of young leaves may also occur. In some varieties, the newly emerging leaves become yellow, slightly rolled outward, spirally twisted and are not fully exerted because of limited elongation of leaf sheaths. If grown-up plants are affected, such symptoms are not pronounced clearly, but in the ratoon crops, the symptoms of 'tungro' appear quite distinctly. Leaf discoloration varies with the varieties and environmental conditions. The root development is also very poor. Plants infected early may be killed, but most of the plants live until maturity. Flowering is unduly delayed in infected plants and the panicles are small in size and most of the grains become sterile. The grains, which are formed do not mature properly and are lighter in weight and discolored. The symptoms may disappear after a few good showers or after top dressing with fertilizers, but reappear again after a few days.

The causal organism. *Rice tungro viruses* cause rice tungro disease. Two distinctly separate viruses are involved. They are the *Rice tungro bacilliform virus (RTBV)* and *Rice tungro spherical virus (RTSV)*. *RTBV* is a double stranded, non-enveloped, rod-shaped, DNA particle of size 150 - 300 nm. in length and 30 - 35 nm. in width and belongs to the Genus - *Badnavirus*. *RTSV* is a single stranded RNA isometric particle of size 30 - 34 nm. in diameter and belongs to the Genus - *Waikavirus*. Both the viruses multiply independently. *RTSV* is restricted to the phloem cells only, while *RTBV* is found both in the xylem and phloem cells. If the plants are infected by *RTBV* alone, mild symptoms of tungro appear and if infected by *RTSV* alone, no specific symptoms are produced. But, when both *RTBV* and *RTSV* are present in the host plant, then typical symptoms of tungro appear and the intensity of the disease is also severe.

Mode of spread and epidemiology. *Rice tungro viruses* are transmitted exclusively by the green leaf hoppers *Nephotettix virescens, N. malayanus, N. parvus* and *Recilia dorsalis,* of which *N. virescens* is the most efficient vector. The females are more efficient in transmitting the viruses than males and the nymphs. The minimum acquisition feeding time is 10 - 30 minutes and the minimum feeding time for virus transmission is 30 seconds. There is no incubation period in the vector and the insect is capable of transmitting the viruses immediately after acquisition feeding. However, the virus is non-persistent in the vector. After a period of 5 - 6 days or after moulting, the insect becomes non-viruliferous. However, the vector can again acquire the viruses by feeding on the infected plants. The viruses are not transmitted through rice grains from infected plants or through eggs of the vector. The viruses are also not sap transmissible.

Cool and humid weather conditions favor the multiplication of green leafhoppers and also results in the rapid spread of the disease. In regions where rice is grown all through the year, the viruses and the vector can thrive continuously throughout the year. Several grass hosts found in and around rice fields, such as *Echinocloa colonum, E. crusgalli, Leersia hexandra, Setaria glauca, Sorghum vulgare, Ischaemum rogosum, Oryza officinalis* etc. have been found to harbor the viruses and the vectors. The viruses and the vectors may also survive in several wild rice varieties and in rice ratoon crops throughout the year.

Detection of rice tungro virus disease. Nutrients, especially micronutrient deficiency may cause certain symptoms, which may resemble the symptoms caused by 'tungro' virus disease. In order to differentiate and ascertain whether the symptoms are due to 'tungro' virus disease, certain tests have to be conducted.

(i) Chemical test (Iodine test)

When a solution of iodine is added to sugars or starch, due to chemical reaction, the color changes into dark blue. Based on this principle, the iodine test is carried out to detect 'tungro' virus disease in a plant.

In the daytime, during photosynthetic activity of the green plants, the carbondioxide present in the atmosphere is converted into carbohydrates, such as sugars and starch and are stored in the leaf tissues.

During night time, when there is no photosynthetic activity, the carbohydrates stored in the leaf tissues are converted into proteins and are translocated to different parts of the plants. In plants affected by 'tungro'virus disease, such physiological activities are carried out to a much lesser extent. Further, in 'tungro' affected plants, the number of phloem cells in the leaves are very much reduced. So, translocation of manufactured food to different parts of the plants is hampered. As a result, the carbohydrates remain as such in the leaf tissues without being converted into proteins and translocated. So, when a cut end of a leaf affected by 'tungro' is dipped in iodine solution, the carbohydrates present in the leaf take the blue color, which is seen at the cut end of the leaf.

Potassium iodide - 6 gm. and iodine crystals - 2 gm. are dissolved in 100 ml. of water. The solution is taken in a glass test tube. Second or third leaf from the top of the suspected plant is collected early in the morning before sunrise. A 10 cm. long piece is cut from the tip of the leaf and the cut end is dipped into the iodine solution and kept as such for 15 - 30 minutes and then washed in plain water. Dark blue streaks or patches in the interveinal areas around the cut end indicate RTV infection, while leaves from healthy plants do not show such discoloration.

Certain precautions have to be taken while conducting this test. The test should be conducted in the morning hours before sunrise. The test should be conducted indoors or under shade. Leaves from suspected and healthy plants should be tested simultaneously.

This test can also be performed with tincture iodine available in medical shops. Tincture iodine is mixed with water in the ratio 1 : 15 and the test is conducted with this solution as mentioned before.

(ii) Biological test

The biological test can be done only in laboratories by trained personnel. Virus-free green leafhoppers, grown in cages are allowed to feed on the suspected 'tungro' affected plant for about 5 hours. Then, the insects are collected and allowed to feed on young, healthy plants for about 12 hours under protected conditions. Then, the seedlings are planted in small mud pots and kept in insect-proof cages and observed regularly. If the suspected plants are really affected by 'tungro' virus, the seedlings will develop 'tungro' symptoms in 7 - 10 days. However, by this method, it will take 7 - 10 days or more to ascertain whether a suspected plant is affected by RTV or not.

(iii) Physical method

This test can be done only in well-equipped laboratories. The sap of the suspected plant can be examined under electron microscope and the presence of the virus particles can be detected.

Disease management

Agronomic practices (i) Rice fields and surrounding areas should be kept clean and free from weeds and other grass hosts, which may harbor the viruses and the vectors (ii) Rice fields should be ploughed immediately after harvest, so as to avoid growing of rice ratoon crops (iii) As the green leafhoppers are positive phototrophic in nature, nurseries should not be raised near light posts and other sources of light (iv) Light traps should be set up to monitor the presence of green leafhoppers. Control measures to eliminate the green leafhoppers should be taken up, even when the population is one insect per hill in RTV prone areas (v) Close planting should be avoided (vi) Application of excessive amounts of nitrogenous fertilizer should be avoided, so as to keep down the vegetative growth

Chemical control. Green leafhoppers may spread the disease in the nursery itself. So, chemical control measures should be taken up even in the nursery stage. Application of carbofuran 3G - 1.4 kg. or phorate 10G - 400 gm. or quinalphos 5G - 800 gm., mixed with 1.0 kg. of sand uniformly over 8 cent nursery area 10 days after sowing and impounding water to a height of 2.5 cm. for 2 - 3 days is effective in eliminating the green leafhopper vectors.

If granular insecticides are not applied, spraying with monocrotophos - 40 ml. or phosalone - 25 ml. or phosphamidon - 10 ml. or fenthion - 25 ml. in 20 lit. of water per 8 cent nursery area may be done. The first spraying should be given 10 days after sowing, followed by another spraying 10 days later.

In the main field, spraying with monocrotophos - 400 ml. or phosphamidon - 200 ml. or fenthion - 250 ml. in 200 lit. of water per acre is found to be effective in eliminating the vector. The first spraying should be given 15 days after transplanting, followed by a second spraying 15 days later. Spraying with neem oil - 2 % (neem oil - 4 lit. + liquid soap - 200 ml. in 200 lit. of water per acre) is also effective in controlling the vector.

Resistant varieties. IR.36, IR.50, IR.56, PTB.18, CR.1009, ASD.8, PY.3 and Ratna are less susceptible to this disease. IR.20, ADT.36, TN.1, Co.37,

Co.40, TKM.9, Vaigai, Kanchi and IET.1722 are highly susceptible to the disease.

Table 1: Differences between rice 'tungro' virus disease and nutrient deficiency diseases

Rice 'tungro' virus disease	Nutrient deficiency diseases
1. Caused by *Rice tungro bacilliform virus (RTBV)* and *Rice tungro spherical virus (RTSV)*	Caused by macro and micro nutrient deficiency, especially 'nitrogen', 'zinc' and 'iron' deficiency
2. One or a few tillers in a hilt may be affected first and show typical symptoms of 'tungro'. Later on all the tillers may be affected and show the disease symptoms.	All the tillers in a hilt will show identical symptoms simultaneously.
3. The color of the lower leaves turn yellow or yellowish-orange from the tip and the sides and spreads downwards and inwards respectively.	All the leaves may become completely yellow or yellowish-white simultaneously.
4. Young leaves may show whitish-green or whitish streaks in between and parallel to the veins.	The entire area of all the leaves becomes yellowish-white in the plants in the field.
5. Symptoms may appear in some plants at random in the field especially in plants near the field bunds and irrigation channels.	All the plants in the field will show the symptoms uniformly at the same time.
6. Young leaves may roll outwards and mottling and crinkling of leaves are seen. In some varieties, rusty spots may appear in the older leaves.	No mottling or crinkling is seen.
7. Depending upon the population of green leafhoppers in the field, the intensity and spread of the disease vary. Generally higher population of green leafhoppers results in higher incidence of the disease.	There is no correlation between green leafhopper population and disease incidence and severity.
8. By controlling green leafhoppers by timely application of suitable insecticides, the spread of the disease can be curtailed	Controlling green leafhoppers has no effect on the disease incidence.
9. Application of plant nutrients has no ameliorative effect on the incidence of the disease.	Application of plant nutrients will have definite ameliorative effect, thereby the symptoms may disappear and the plants will recover and become normal.

12. Grassy stunt disease of rice

The disease was first recorded in the Philippines in 1963 and its virus origin established. Later, it has been found to occur in India, Malaysia, Sri Lanka and Thailand.

Symptoms. The disease occurs at all growth stages of the crop however, it is more severe in the seedling stage. The infected plants are severely stunted, produce more number of tillers, the leaves pale yellow or pale green in color, short and narrow. The stem portion is very short and thin and the plant looks like a grass. Young leaves show mottling and crinkling. Rusty spots develop on older leaves. The infected plants may not die but live until maturity. Most often, the infected plants fail to produce panicles. Even if panicles are formed, they are deformed and the grains produced are discolored, ill-filled and powdery.

The causal organism. The disease is caused by *Oryzavirus*. It is an icosahedral, double-stranded RNA virus, with segmented genome and is encapsidated within an isometric particle. It has a very narrow host range.

Mode of spread. The brown plant hopper *(Nilaparvata lugens)* transmits the virus causing the disease and the spread of the disease depends on the presence of this vector. The disease is not transmitted through sap or seed. Once the vector acquires the virus, it persists in the insect for a long period. Apterous and brachypterous forms of both male and female sexes are capable of transmitting the disease. The acquisition feeding time is 30 minutes and after acquisition, the vector becomes viruliferous only after about 10 days. In the mean time, the virus multiplies within the vector in enormous numbers.

Disease management

Agronomic practices (i) In the absence of rice crop, several weed hosts, such as *Panicum repens, Cyperus rotundus* etc. harbor the vectors. So, weeds in and around rice fields should be removed and destroyed (ii) After the harvest of rice crop, the stubble should be burnt, otherwise the ratoon crop that may grow will harbor the vector and the virus (iii) Impounding excessive quantities of water in the fields should be avoided (iv) Light traps should be set up and the vector population monitored (v) Application of excess dose of nitrogenous fertilizer should be avoided. The required quantity of nitrogen should also be applied in split doses (vi) Diseased plants may be uprooted and destroyed, as and when they are detected.

Chemical control (i) When only a few brown plant hoppers are present, dusting with carbaryl 10 D at 10 kg./ acre may be done (ii) If the vector population tends to increase, spraying should be done with phosphamidon - 200 ml. or monocrotophos - 500 ml. or phosalone - 600 ml. in 200 lit. of water per acre. A higher concentration of insecticides is recommended for the control of brown plant hopper. The vector is mostly found on the stem region, just above the water level. So, spraying should be given to cover the stem portions thoroughly.

Resistant varieties. Varieties, such as IR.28, IR.29, IR.30, IR.32 and IR.34 have been reported to be resistant to brown plant hopper. Varieties PY.3, Co.42, ADT.36, Pankaj, Rohini and Jaya are tolerant to this disease

13. Rice yellow dwarf disease

'Yellow dwarf disease' was first reported from Japan in 1919. The disease is prevalent in Taiwan, Sri Lanka, the Philippines, China, India and many other Asian countries. In India, the disease was recorded in 1967. In Tamil Nadu, the disease occurs sporadically in parts of Pudukkottai, Madurai and Thanjavur districts.

Symptoms. The disease may occur at any growth stage of the crop however, when seedlings are infected the intensity of the disease and the resulting yield loss is much more. General chlorosis of the newly emerging leaves is a characteristic symptom of the disease. Young leaves become chlorotic, the color turning to yellowish-green or whitish-green. The infected plants show abnormal increase in the number of tillers. The plants remain stunted and weak, with rather soft and narrow leaves. Most often, the plants live until maturity, but the plants do not produce any panicles or may produce one or two degenerated panicles, with few ill-filled grains. If the disease occurs after the tillering stage, the symptoms are not marked. But the symptoms are well pronounced in the ratoon crop, which comes out after the harvest of the crop. The infected plants become highly susceptible to brown leaf spot and narrow brown leaf spot diseases.

The causal organism. The disease is caused by a *phytoplasma,* which is classified under the mollicutes

Mode of spread and epidemiology. The green leafhopper vectors, *Nephotettix cincticeps, N. virescens* and *N. nigropictus* transmit the *phytoplasma* in a persistent manner. The acquisition feeding time is 1 - 3 hours. After acquisition, the *phytoplasma* remains inside the vector and multiplies for about 23 - 30 days. Only after this period, the vector is capable of transmitting the pathogen. The feeding time for transmitting the *phytoplasma* is just about 1 - 3 minutes. The vector remains infective for the rest of its life. Both the nymphs and adults are capable of transmitting the *phytoplasma.*

The disease is found to occur in wild rice varieties, such as *Oryza cubensis* and other grass hosts, such as *Alopecuros aequalis* and *Glyceria acutifolia.*

Cool and humid weather conditions, favorable for the rapid multiplication of the vectors are also favorable for the occurrence of the disease.

Disease management

Agronomic practices (i) Rice fields, irrigation channels, field bunds and surrounding areas should be kept free from weeds and grasses (ii) After the harvest of rice crop, the field should be ploughed and the stubble removed and burnt, so as to prevent regeneration of a ratoon crop (iii) As the leafhoppers are positive phototrophic in nature, nurseries should not be raised near lamp posts or any other light sources (iv) Light traps should be set up and the population of green leafhoppers should be monitored regularly. When the population of the vector reaches 5 insects per hill, control measures should be taken up (v) Application of excessive doses of nitrogenous fertilizer should be avoided and the recommended dose of nitrogen should also be given in split doses, so as to curtail the vegetative growth (vi) Closer spacing should be avoided (vii) Diseased plants should be uprooted and destroyed.

Chemical control. Chemical control measures should be taken up from the nursery stage itself to prevent build up of green leafhopper population (i) In the nursery carbofuran 3G - 1.4 kg. or phorate 10G - 400 gm. or quinalphos 5G - 800 gm., mixed with 1.0 kg. of sand should be applied uniformly in 8 cent nursery area, 10 days after sowing and water should be impounded to a height of 2.5 cm. for 2 - 3 days (ii) If granular insecticides are not applied, spraying should be done. Monocrotophos - 40 ml. or phosalone - 60 ml. or phosphamidon - 10 ml. mixed with 20 lit. of water per 8 cent nursery area should be sprayed on the 10th and 20th day after sowing.

In the transplanted main field, spraying should be done with monocrotophos - 400 ml. or phophamidon - 200 ml. or fenthion - 250 ml. in 200 lit. of water on the 15th and 30th day after transplanting. Spraying with neem oil - 2 % (Neem oil - 4 lit. + liquid soap - 200 ml. in 200 lit. of water per acre) also gives effective control of the vector.

14. Zinc deficiency disease of rice

The disease or disorder also known as **'Khaira'** disease was reported from the Tarai region of Uttar Pradesh. The disease is also found in Karnataka, Haryana, Punjab and Tamil Nadu. The disease is caused because of non-availability of zinc to the rice plants, even though enough zinc may be present in the soil. Zinc forms an integral part of the enzymes involved in auxin synthesis and in oxidation of sugars. Non-availability of zinc to the plants causes certain specific symptoms in the plants. Zinc, readily available to the plants is more important than the quantity of zinc present in the soil. Soluble zinc is quickly leached from open sandy soils, while in clay soils with high humus or organic matter, it is partially fixed in the soil. Under these conditions, zinc may not be available to the plants readily, which results in the incidence of zinc deficiency disease. It is also known that soil microorganisms are also responsible for rendering zinc unavailable to the plants.

Symptoms. The disease usually appears 10 - 15 days after transplanting, a period, which coincides with the peak period of decomposition of last year's stubble in the flooded field. This decomposition of organic matter may indirectly affect the availability of zinc by fixing it in the colloidal clay particles. Leaves of affected plants show chlorosis, starting from the base of the leaves. Following this, a large number of small brown or bronze-colored, rusty spots appear on the leaf lamina, which coalesce to form bigger spots or patches and cover the whole leaf surface of the older leaves. The leaves then become brown or bronze- colored entirely and eventually dry and die. Such affected plants are stunted in growth. The root system is also affected adversely and is discolored. In severe cases, the plants neither grow further nor do they produce panicles. Sometimes, the plants may recover naturally six weeks after transplantation and produce ears with few grains.

Control measures

Agronomic practices. Sufficient time should be given for the decomposition of stubble and other organic matter added to the field before transplanting seedlings.

Chemical control. Spraying with zinc phosphate - 1.0 kg. + slaked lime - 500 gm. in 200 lit. of water per acre, once as soon as initial symptoms of the disease appear, followed by a second spraying, 10 days later controls the melody completely.

Diseases of minor importance. Besides the diseases discussed above in detail, rice crop is vulnerable to several other diseases, which may be of minor importance and may occur sporadically. However, when conditions become highly favorable for the occurrence and spread of the diseases, they may assume epidemic proportions and may cause considerable damage to the crop and economic loss.

'Stackburn disease' caused by *Alternaria (Trichoconis) padwickii* affects seedlings, leaves of grown-up plants and grains, and produces typical spots; 'Foot rot' caused by *Fusarium moniliforme (Gibberella fujikuroi)* causes 'Bakanae disease', resulting in abnormal elongation, yellowing and death of seedlings; 'Leaf smut' caused by *Entyloma oryzae* causes spots on leaves; 'Bunt' or 'Kernel smut' caused by *Tilletia barclayana (T. horrida)* affects the floral parts and produces mass of black, powdery, smut spores in the place of kernels; 'Udbathi' or 'Udbatta disease' caused by *Ephelis oryzae (Balansia oryzae)* affects the panicle; 'Dwarf disease' caused by *Rice dwarf virus* affects the plants leading to stunting; 'Ragged stunt' caused by *Ragged stunt virus*, results in stunted growth and deformation of leaves.

Wheat *(Triticum aestivum)*

1. Black rust or stem rust of wheat

Puccinia graminis tritici

'Black rust' is known to attack wheat crop from very early times. It is prevalent in all the wheat growing countries of the world and is responsible to cause severe crop losses. The disease is very common in the United States of America, Australia, Mexico and several other European countries and causes severe damage to wheat crop. In India, the disease occurs in all wheat growing states and in Tamil Nadu, it is found in the Nilgiris and Kodaikanal. The disease appears rather late in the season, mostly after the emergence of the inflorescence. Besides wheat and some other cereal crops, the pathogen is found to attack several wild grasses, including *Bromus japonicus*.

Symptoms. Early symptoms on wheat appear as elongated, brown pustules, parallel to the long axis of the stem, leaf stalks, leaves and even on the neck and glumes of the spikes. But, the part usually affected first and most severely is the stem, hence the name **'stem rust'** is often applied to black rust. The early pustules, which consist of uredosori, are about 0.6 cm. or more in length and frequently join together in stripes. Due to pressure exerted by the uredospores from below, the epidermis is broken and dense masses of brown, powdery uredospores are released.

Later on, the pustules assume a darker color and from the same pustules teliospores develop or telia develop independently and produce teliospores. The telia are darker in color than the uredia. The telia, which are formed in large numbers on the stem region, present a black appearance to the plants and hence the disease is called **'black rust'**. The teliospores break through the epidermis. When large areas of the plant parts are affected, the affected plants look sickly and fail to produce normal ears. The grains formed are all shrivelled and are of poor quality.

Attack due to black rust, results in marked physiological changes in the host plant. Large amount of water is lost by transpiration and translocation of carbohydrates is hindered. The rate of respiration is also increased markedly. All these factors are responsible for affecting the vitality of rusted plants.

The causal organism. The pathogen causing black rust is an obligate parasite. It is a macrocyclic, heteroecious rust .The uredial and telial stages are found in wheat, which is the primary or main host, while the pycnial and aecial stages are found in the alternate or secondary hosts, such as species of Berberis and Mahonia.

The aeciospore produced from an alternate host falls on the leaf of a wheat plant (primary host), germinates, enters the host leaf through the stomata and develops into a dikaryotic, intercellular mycelium. Small, round or branched haustoria arising from the hyphae are sent into the host cells to absorb nourishment. As the disease advances, a mass of hyphae gathers beneath the host epidermis and develops into an uredium or uredosorus. From the base of the sorus, numerous, short, erect stalks arise and at the end of each stalk, a single uredospore is formed. As the spores develop, pressure is exerted on the epidermis, which ruptures and the spores are set free into the air. Uredospores are single-celled, oval in shape, brown-colored, thick-walled and provided with tiny spines on the surface. They measure 26-30 x 17-20μ. Each spore has got usually four germ pores, situated at equal distances in an equatorial band. Uredospores are the repeating spores that cause secondary infection on wheat crop. The spores falling on the host surface germinate, when environmental conditions, such as temperature and humidity are favorable. The uredospore germinates in a film of water by issuing a germ tube through a germ pore. Sometimes, two germ tubes are produced from two different germ pores. The germ tube enters the host through the stomata and its tip swells into an elongated appressorium. From the appressorium, a narrow infection hypha emerges, passes through the stomatal opening, enters the sub-stomatal cavity and forms a vesicle. From this sub-stomatal vesicle, intercellular mycelium is produced. The haustoria arising from the mycelium, invade the host cells and establish organic relationship between the host and the parasite, thus causing infection. The mycelium continues its growth and produces rust pustules within 10-15 days. From each pustule, hundreds of uredospores are produced, each one capable of initiating a new sorus.

Late in the season, telia arise from the same or similar dikaryotic mycelium. Frequently uredospores and teliospores are found in the same sorus. The dark-colored teliospores are more firmly attached to the beds than the uredospores and their stalks are more rigid and thicker. They are composed of two, super-imposed cells and have a thick and smooth wall. The apex is rounded or pointed. They measure 40 - 56 x 15 - 20 μ. Each cell is provided with a germ pore, the upper germ pore at the apex and that of the lower at the side just below the septum. The wall is much thicker at the apex. Unlike the uredospore, the teliospore is not capable of immediate germination. It requires a resting period of several months before germination.

On germination, the teliospore gives rise to a four-celled promycelium from each of the two cells. From each cell of the promycelium, a short sterigma grows out and swells into a rounded, small basidiospore (sporidium) at the tip. The basidiospores are unicellular, uninucleate and haploid. Of the four basidiospores formed, two are of one sex (+) and the other two of the opposite sex (-). These spores dislodge easily and are carried by wind. They are capable of immediate germination in the presence of water, but cannot infect the cereal host. They can infect only the alternate host species.

When the monokaryotic basidiospore reaches a suitable barberry host, it sends out a germ tube, which directly penetrates the epidermis. The leaves, stems, spines, petioles, sepals and even the berries of the barberry may be attacked by the sporidia and on all these parts, red spots develop. Inside the barberry tissue, the infection hypha develops into a uninucleate, septate mycelium. The mycelium produces pycnia (spermagonia) on the upper surface of the leaf. The flask-shaped pycnia are provided with an ostiole, which protrudes out through the cuticle. The pycnia are lined with very narrow spermatial hyphae, which converge towards the narrow ostiole. These slender hyphae cut off numerous spermatia (pycniospores) that are exuded at the ostiole in drops of sweet nectar or honeydew. The honeydew attracts insects, which help in the dissemination of the spermatia. The spermatia are unicellular and uninucleate. These spores, as well as the mycelium produced by them are all monokaryotic and correspond to the sex of the basidiospores from which they are produced. The initiation of a binucleate condition in the mycelium constitutes the first stage in the sexual reproduction of these fungi. This is effected by the fertilizing process of the spermatia through receptive hypha of the opposite sex produced by the spermagonium.

The receptive hyphae are connected to the mycelium forming the aecial primordium. In this way, the male nuclei from the spermatia pass into the

Fig. 10: Black rust or stem rust of wheat-*Pucciia graminis tritici*

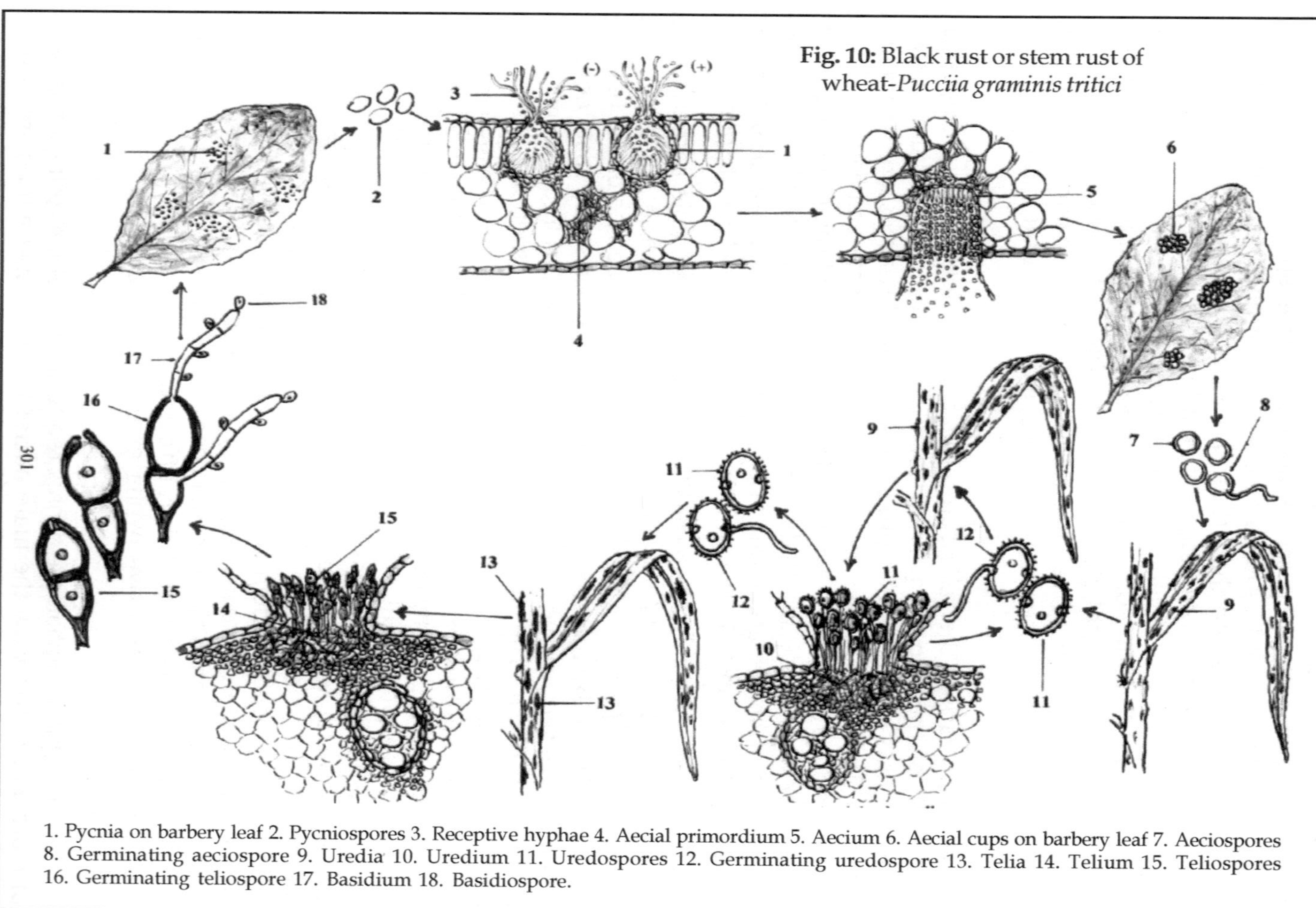

1. Pycnia on barbery leaf 2. Pycniospores 3. Receptive hyphae 4. Aecial primordium 5. Aecium 6. Aecial cups on barbery leaf 7. Aeciospores 8. Germinating aeciospore 9. Uredia 10. Uredium 11. Uredospores 12. Germinating uredospore 13. Telia 14. Telium 15. Teliospores 16. Germinating teliospore 17. Basidium 18. Basidiospore.

female cells of the receptive hyphae and viceversa within the aecial primordium, rendering them binucleate. In cells with such paired nuclei, one nucleus belongs to that of a **(+)** sporidium and its mate to that of a **(-)** receptive hypha. A number of such fertile cells eventually form a well-defined group of binucleate cells, all in close lateral contact. They divide transversely. Meanwhile, their nuclei also divide conjugately and each cuts off a binucleate aeciospore mother cell, which in turn divides to form a small, flat interstitial cell and a larger cell, which is the true aeciospore. Both these cells are binucleate. Each fertile cell repeats these divisions several times to form a chain of aeciospores and interstitial cells. By such simultaneous development of all the fertile cells at the base of the aecidium, a cup-shaped aecidium, with mass of spores and interstitial cells are formed in chains in the host leaf, ready to break out at the lower epidermis. The outermost spore chains forming the periphery and top of the aecial cup are sterile and form a protective cover (peridium) to the spore chains. Due to the pressure exerted by the developing aeciospores, the lower epidermis of the host leaf is ruptured and subsequently the peridium is also broken forming a fringe around the aecial cup. The whole thing appears as a bell-shaped structure. The orange-colored aeciospores are exposed and dispersed by wind.

While inside the aecial cup, the aeciospores are hexagonal in shape due to mutual pressure of the spore chains. But when set free, the aeciospores are spherical. Matured aeciospores are binucleate, 14 - 26 µ in diameter and have about six germ pores in the wall. These spores serve to return the rust to the graminaceous host, but cannot infect the host (barberry), which produced them. The aeciospores germinate on the wheat in the same manner as uredospores, producing germ tubes and gaining access to the host tissue through stomata. These binucleate spores establish a dikaryotic mycelium within the cereal host.

Later, this mycelium gives rise to uredosori of binucleate uredospores and teliosori of teliospores. The teliospores are binucleate when young and as they mature, the nuclei in each cell fuse to form a diploid nucleus, thus forming the diploid phase in the rust life cycle **(Fig.10)**.

Mode of survival, spread and epidemiology. In the black rust of wheat, about 50,000 - 4,00,000 uredospores are produced from each uredosorus. These spores alone are quite sufficient to infect and cause disease in one acre of wheat crop. Each uredospore can infect a plant and in about 10 - 15 days it produces new uredosori and uredospores. Thus, the uredospores are mainly responsible for spreading the disease very rapidly over large areas.

The uredospores are mostly carried by wind over long distances and spread the disease on a wide scale. The wind-borne spores can travel even from one continent to another by crossing natural barriers, such as oceans, mountains etc. and spread the disease. Depending upon the velocity of wind, these spores can be carried to a distance of 50 - 350 km. in a single day.

Besides wheat crop, black rust can infect several grass hosts, such as *Bromus japonicus, Briza minor, Brachypodium sylvaticum, Avena fatua* etc. The uredospores produced from these grasses throughout the year can infect the wheat crop, as and when it is cultivated. So, the absence of the alternate hosts viz., the barberry or mahonia makes little difference in the initiation of the disease through aeciospores.

In Tamil Nadu, where wheat is grown in the hilly regions of the Nilgiris, Kodaikanal, Palani etc. throughout the year, the pathogen survives as uredospores and cause infection. In the plains of North India, wheat is sown during October - November and the crops come to harvest during March - April. In these crops, in the first 2 - 3 months there is no incidence of rust, as there are not enough spores to initiate the disease. As such, the disease appears only later in the season. But, in the hilly regions of Himachal Pradesh, the disease appears much earlier, probably through aeciospores produced from the alternate hosts. The uredospores produced from the hilly regions are carried by wind and reach the plains and cause disease later in the season.

A number of factors influence the incidence and development of epidemics of black rust. The first and most important factor is that there must be an abundant supply of spores for infection. Second important factor is the weather conditions. Black rust is severe in seasons of abundant moisture, when the temperature is 17° - 18°C. This combination is often experienced during wet spells, alternating with bright, warm periods. Continuous cool weather, as well as dry-cool or dry-hot periods are not favorable for large scale incidence of rust. However, infection with rust spores is possible, only when the host plants are wet and even a saturated atmosphere is not enough for the germination of uredospores. The germ tubes require a film of water for penetration of the host plant.

Because of the existence of numerous physiological races of the rust fungus, most of the cultivated varieties of wheat are vulnerable for attack by one or more races of the pathogen. But, efforts are being made on a permanent basis to evolve new varieties of wheat resistant to black rust throughout the world.

Disease management

Agronomic practices (i) The alternate and collateral hosts of the pathogen should be destroyed as far as possible, so as to prevent build up of inoculum (ii) Use of excessive dose of nitrogenous fertilizer should be avoided. Balanced fertilizer application helps to reduce the incidence of rust (iii) The crop should be raised early in the season, so that it can escape the onslaught of the disease.

Seed treatment. Seed dressing with carboxin or oxycarboxin at 2 gm./ kg. of seeds, 24 hours prior to sowing protects the seedlings from rust infection for about 7 weeks.

Chemical control (i) Dusting the crop with sulfur dust at 10 kg./ ac., 3 - 4 times, at 10 - 14 days interval is effective in controlling the disease (ii) Spraying the crop with wettable sulfur at 800 gm. or dithiocarbamate at 400 gm. in 200 lit. of water per acre, 3 - 4 times, at 10 - 14 days interval is also found to be effective. The treatment should commence as soon as first symptoms of the disease are noticed.

Resistant varieties. The presence of numerous physiological races of the fungus makes it very difficult to evolve varieties resistant to all the races. Sonora-64, Lerma Rojo, Safed Lerma and Sonalika are some of the varieties resistant to black rust

2. Yellow rust or stripe rust of wheat

Puccinia striiformis (P. glumarum)

'Yellow rust' or **'stripe rust'** occurs in Britain, Europe, the United States of America, Canada, Mexico, Argentina, South Central Asia, Russia, India, China and parts of Africa. Throughout Europe, it is the most destructive of all cereal rusts. Yellow rust appears earlier than black rust and in some years, it causes serious losses due to destruction of the foliage and production of badly shrivelled grains.

Symptoms. The uredosori appear as bright-yellow pustules, chiefly on the leaves, but may also appear on leaf sheaths, stem, spikelets and even on the grains. Following the infection, the green color of the leaves fades in long streaks and later on, small pustules, consisting of uredosori develop along the streaks imparting a distinct, yellow, striped appearance to the leaves and hence it is called **'stripe rust'**. The oval-shaped sori are arranged end to end in a series and each sorus is quite distinct from one another. In severe attacks, the serial arrangement may be lost and the pustules appear as large patches of uredosori. On the glumes, the uredosori are in rows parallel with the veins.

Later, towards the maturity of the host, the teliosori appear abundantly on the leaf sheaths and glumes, but rarely on the leaf blades. The teliosori may also appear along the edges of old uredosori. But uredospores and teliospores are rarely produced in the same sorus, as often found in black rust. The teliosori are compact, dull black spots, arranged in rows like the uredosori. However, as in black rust they do not break through the epidermis at all, but remain as flat, black crusts.

The causal organism. The uredosori are sub-epidermal and remain covered for a much longer time than in the other rusts. When they finally break through, the yellow uredospores are shed, dispersed by wind and infect wheat crop. The uredospores are spherical to ovate, binucleate, minutely echinulated and have 6 - 16 germ pores. They vary much in size and measure 23 - 35 x 20 - 35μ. The uredospores germinate on the wet surface of a leaf by a germ tube, which forms an appressorium over a stoma and swells up in the sub-stomatal cavity to form a thick-walled vesicle. From the vesicle, one or more infection hyphae arise and proceed into the mesophyll. From these hyphae, club-shaped or sometimes branched haustoria are formed and enter into the host cells. Other hyphae, arising from the sub-stomatal vesicles run in a longitudinal manner under the epidermis from one stomatal cavity to another like runner hyphae, resulting in a continuous band of infection, which may extend the full length of the leaf. The hyphae are non-septate in the beginning, but later become septate and finally to a binucleate condition. These binucleate hyphae, which collect beneath the stomata produce the uredosori finally.

The teliospores are dark-brown in color and flattened at the top, because of their contact with the host epidermis. They are two-celled and measure 35 - 63 x 12 - 20μ. They are interspersed with brown, unicellular paraphyses. When released, the teliospores are capable of immediate germination without any dormancy, but the fate of the basidiospores they produce is not known in the absence of any alternate host. The pycnial and aecial stages of this fungus have not been discovered so far **(Fig.11).**

More than 40 physiological races of this pathogen have been identified

Mode of survival, spread and epidemiology. The fungus overwinters in the form of uredospores or as mycelium in the host leaf. The pathogen also attacks several wild grass hosts. Some of the grass hosts may harbor the races of the fungus, which can attack wheat.

Plants affected severely by yellow rust have a very poor root system. Because of heavy leaf infection, the roots, which are dependent upon direct translocation of elaborated food from the leaves suffer very badly. Further, abnormally high rate of transpiration of water from the diseased leaves takes a heavy toll on the root system. As the absorbing capacity of the roots cannot cope up with such heavy demand for water, the root system is adversely affected.

Puccinia striiformis is intolerant to high temperatures and both spores and mycelium grow best at about 11°C. The disease is practically absent in areas, which experience low rainfall and high summer temperatures. In general, heavy rainfall and low temperatures are essential for the spread of the disease. No infection occurs at temperatures above 23°C and below 2°C.

Disease management

Agronomic practices (i) As the disease appears even in very young plants, affected plants may be removed and destroyed (ii) Wild grasses in and around wheat fields should be removed and destroyed (iii) Wheat is adversely affected by yellow rust due to unbalanced manurial application. Excess of nitrogen favor disease occurrence. Potash and phosphorus confer high degree of resistance to this disease. Application of balanced dose of fertilizers helps to reduce the incidence of the disease.

Chemical control. Spraying the crop with wettable sulfur - 1000 gm. or mancozeb - 500 gm. or caboxin - 250 gm. or oxycarboxin - 250 gm. in 250 lit. of water per acre affords good control of the disease. Spraying should commence, as soon as the symptoms are manifested and 3 - 4 sprayings, at 10 - 14 days interval should be given.

3. Brown rust or orange rust or leaf rust of wheat

Puccinia triticina (P. recondita)

'Brown rust' occurs in all wheat growing countries of the world. The disease is mainly confined to the leaves and hence, it is also known as **'leaf rust'**. However, rarely the rust pustules appear on the stems and heads. In India, the disease occurs in the states of Punjab, Bihar and Uttar Predesh and causes more extensive damage than the other rusts. In Tamil Nadu, the disease is found in the hilly tracts, as well as in the plains.

Symptoms. The disease appears earlier than the other two rusts, by the time the crop is 5 - 6 weeks old. Heavy infection of leaves greatly retards photosynthetic activity of the leaves and when the disease appears

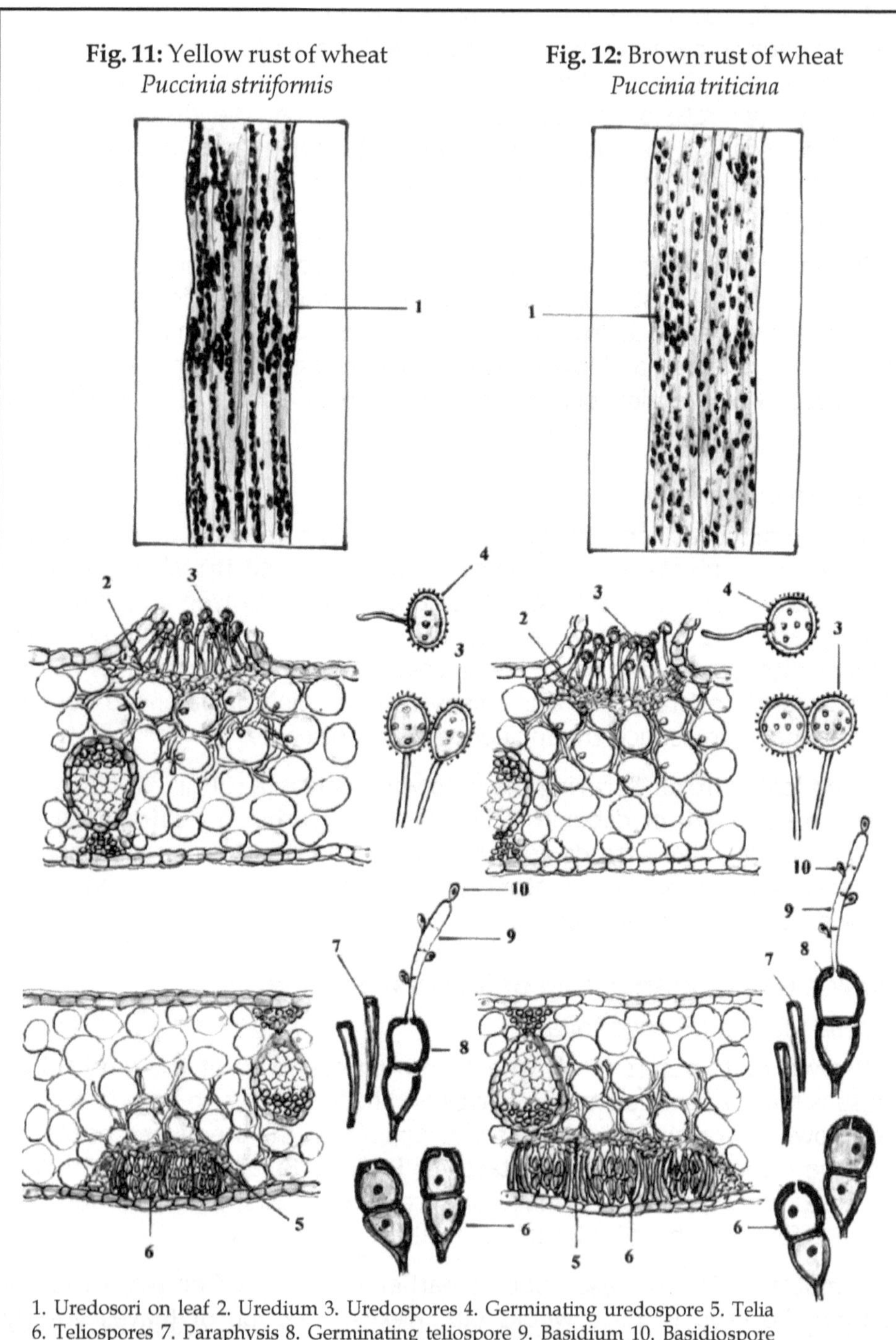

Fig. 11: Yellow rust of wheat *Puccinia striiformis*

Fig. 12: Brown rust of wheat *Puccinia triticina*

1. Uredosori on leaf 2. Uredium 3. Uredospores 4. Germinating uredospore 5. Telia 6. Teliospores 7. Paraphysis 8. Germinating teliospore 9. Basidium 10. Basidiospore

early, entire leaves turn yellow and wither within a short time. Increased transpiration, as a result of infection, cause the plants to take a much longer time to head and mature. Heavy rusting of the foliage, retards normal maturity and leads to reduction in yield and quality of grain and straw. The root system is also affected badly. The uredosori break out from under the epidermis, as orange or brown specks on the upper leaf surface and hence, it is also called as **'orange rust'**. They do not arise in rows or stripes as in yellow rust, but are grouped in small clusters or irregularly scattered. The pustules are slightly larger in size than those of yellow rust. The sori burst early and shed the uredospores.

P. triticina rarely forms teliospores. However, in some years they are produced in abundance. The telial sori are scattered chiefly on the under surface of the leaves and on the sheaths. They are flattened at the top and dull black in color. Like those of yellow rust, they do not break through the epidermis.

The causal organism. *Puccinia triticina*, a heteroecious, macrocyclic rust is an obligate parasite. The uredospores are orange to brown in color and spherical, 16 - 28μ in diameter, the wall minutely echinulated and provided with 7 - 10 germ pores. Infection by uredospores occurs through the stomata on either side of the leaf. On germination, the germ tube forms an appressorium and the infection hypha arising from it penetrates into the host through the open stoma. Within the sub-stomatal cavity, the invading hypha expands to form a vesicle, from which branching hyphae develop and invade the leaf. Intercellular mycelium produces haustoria, which penetrate the mesophyll cells.

The teliospores resemble those of yellow rust in size and shape, but the teliosorus is divided into small groups or compartments of teliospores by unicellular paraphyses. The teliospores are smooth, brown and two-celled and may remain viable for up to 2 years.

The pycnial and aecial stages of this fungus are found on *Thalictrum flavum* and several other species of *Thalictrum*. But, aecial formation is not very common. On the leaves of *Thalictrum*, the establishment of intercellular mycelium in the mesophyll follows sporidial infection. Small, unbranched haustoria produced from the mycelium enter into the host cells. Spermagonia appear on both surfaces of the leaf. Aecia are formed on the under surface of the leaf **(Fig.12)**.

Table 2 : Differences between the wheat rusts

Black rust or stem rust Puccinia graminis Tritici	**Yellow rust or stripe rust Puccinia Striiformis**	**Brown rust or Orange rust Puccinia Recondita**
1.Affects mostly the stems and leaf sheaths	Affects mostly the leaf blades	Affects mostly the leaves and rarely the leaf sheaths
2.Uredosori are large, elongated and frequently join together in stripes. The epidermis is broken early. The sori are reddish-brown in color.	Uredosori small, oval and are arranged end to end in a series in-between the veins, imparting a distinct yellow, striped appearance. The son remain covered for a much longer time. The sori are distinctly yellow in color.	Uredosori small, oval or round and are grouped in small clusters or irregularly scattered on the upper leaf surface. The son burst early. The sori appear bright orange or brown in color.
3.Uredospores oval, brown in color, provided with tiny spines on the sin-face and are 25-30 x 17-20μ in size, with 4 germ pores situated equidistant in an equatorial plane.	Uredospores spherical to ovate, yellow colored, minutely echinulated and are 23-35 x 20-35μ in size, with 6-16 scattered germ pores.	Uredospores spherical, brown colored, minutely echinulated, in diameter, with 7-10 scattered germ pores.
4.Newly formed teliosori are black in color, large, elongated and scattered on the stems and leaf sheaths. The telia burst soon.	Teliosori appear on the leaf sheaths and glumes, rarely on the leaf blades. They are dull black and arranged in rows. The telia do not break through the epidermis at all.	Teliosori appear chiefly on the under surface of the leaves. They are flattened, dull black and do not break through the epidermis.
5.Teliospores are dark-brown, thick and smooth-walled, 2-Celled, the apex rounded or pointed and measure 42-56 x 15 -20μ in size. The wall is much thicker at the apex.	Teliospores are dark-brown in color, thick and smooth-Walled, 2-celled, flattened at the top in contact with the epidermis and measure 35-63 x 12-20μ in size.	Teliospores are brown in color, thick and smooth-walled, 2-celled and with broader and rounded apex.
6.Paraphyses are not present	The teliospores are interspersed with brown, unicellular paraphyses.	The teliosorus is divided into small groups of teliospores by unicellular paraphyses.

Mode of survival, spread and epidemiology. In the absence of an aecial host in nature, *P. triticina,* like *P. striiformis* depends upon the survival of the uredospores to carry over the disease from season to season. As the spores can tolerate low temperatures, they can survive on host debris in the field, as well as on volunteer plants. The fungus can also overwinter as mycelium in the host leaf. The uredospores produced from such sources are spread through wind over long distances to start new infections in the ensuing season. In India, the disease appears first in the hilly tracts of the Himalayas in Northern India and in the hilly regions of Kodaikanal and Ooty in Southern India and from these localities, the disease spreads to the plains by wind-borne uredospores.

P. triticina embraces over 120 physiologic races, some of which have probably evolved as a result of mutation. More than 14 such races have been identified in India.

Weather factors play a vital role in the perpetuation of the disease. The environmental conditions favorable for black rust incidence also favors the incidence of brown rust. However, brown rust can develop over a wider range of temperatures, especially at temperatures below 15°C. Susceptibility to brown rust increases with increase in temperature, soil moisture and light intensity.

Disease management. The control measures recommended for the control of black rust of wheat may be followed (Page 47)

4. Loose smut of wheat

Ustilago segetum var. *tritici (Ustilago tritici)*

'Loose smut' of wheat occurs worldwide, but is more abundant and serious in humid and subhumid regions. In India, the disease is found in almost all the states where wheat is grown, but the incidence is generally more in the comparatively cooler and moist parts of Northern India than Southern India. In cases of severe incidence, the yield loss may go up to 30 %.

Symptoms. There are no signs of the disease in the crop until the heading stage is reached, when infected plants are recognized by the black, smutted appearance of the ears. The smutted heads consist of deformed spikelets, filled with black, dry, powdery masses of spores. The spore masses, which are covered by a thin, gray, papery membrane, ruptures most often before the head emerges from the sheath and form dense, aggregation of dark, olive-brown spores. All the floral parts and glumes are entirely replaced by the spore masses, except the ends of the awns that usually escape transformation. After a gust of wind or heavy rain, the spores are blown away or washed down the smutted ear, leaving behind only the naked stalk. Most often, all the heads on a plant are affected. However, in some varieties only a few tillers show smutted ears, while others are free from the disease. Smutted heads emerge from the leaf sheath, usually at the same time or mostly in advance of the heads of healthy plants. In dry weather, the spores are blown in clouds throughout the crop. When the heads are in bloom and when spikelets are open for pollination, infection of the flowers takes place. The smut spores falling on the stigma of a healthy flower, infect the developing grain. Such infected grains cannot be differentiated from the healthy grains, but when sown, systemic infection takes place, resulting in the production of smutted ears from such infected plants **(Fig.13).**

The causal organism. The smut spores are olive-brown in color, spherical or occasionally oval in shape, finely echinulated and are 5.0 - 9.0μ in diameter. The spores falling on a sterigma, germinate in water to produce a germ tube, which develops into a promycelium (basidium) with four, uninucleate cells. However, there is no formation of sporidia (basidiospores). Diplodisation takes place between the cells of the basidium by fusion of nuclei of opposite sexes. From the fusion cell, a dikaryotic mycelium is formed. From such dikaryotic mycelium, infection hyphae arise, which enter the style, grow forward intercellularly and enter the ovary by penetrating the ovary wall. Generally, the integument of the ovary becomes cutinized and impermeable in about 10 days after the normal time of fertilization and so, successful penetration of the ovary usually takes place between the 7th and 10th day. The hyphae cross the cells of the pericarp, enter the testa (seed coat) and then proceed towards the ovule. By the 10th day the hyphae reach the ovule, where they branch profusely and the mycelium becomes well ramified in the ovule and also in the endosperm, scutellum and embryo. In about a further period of 4 weeks, the mycelium establishes itself in all parts of the embryo, except the root. The scutellum is also found to harbor a considerable amount of mycelium. The invasion of the various parts of the grain occurs intercellularly at all times, but no haustoria are formed. Despite this infection, the swelling of the grain is in no way impeded and the tissues remain unharmed. The mycelium becomes somewhat thicker-walled, and remains dormant in the seed and infected grains cannot be distinguished from sound grains at the time of harvest.

The fungus revives when the infected grain is sown along with the good seeds and the hyphae present just behind the growing point of the embryo, keep pace with the apical growth of the plant. The fungus invades mostly the stem, the leaf sheath to a lesser extent and rarely the lamina. Finally, when the heading stage is reached, the fungus becomes active within the spikelets and in place of normal grains, dense masses of spores are produced. At the time of earhead formation, the hyphae accumulate in great quantity in the floral parts, which are completely destroyed. The hyphal segments become swollen, round off, the walls become thick and they separate as individual spores. The spores are called **'chlamydospores'** or **'teliospores'** or **'smut spores'**. On maturity of the spore, the two nuclei in the dikaryotic spore fuse to form a single, diploid nucleus. The spores remain viable for a period of 5 - 6 months under normal conditions **(Fig.13).**

Mode of survival, spread and epidemiology. The causal organism being systemic in nature, the expression of disease intensity depends to a great extent on the environmental conditions in which the plant is grown from the time of sowing until heading. However, some others are of the opinion that the incidence of the disease is unaffected by conditions of temperature, rainfall, date of sowing etc., but depends only upon the amount of infection in the seed.

Secondary infection of healthy flowers is through air-borne smut spores.

High humidity of 65 - 85 % and a temperature of 23°C favor high degree of disease development. Plants grown at 29°C are found to be free from disease incidence, even if they are grown from infected seeds.

Many physiologic races of this fungus exist in nature in different countries. But, the races found in one country differ from those occurring in other countries. In India, two physiologic races have been found.

Disease management

Agronomic practices (i) Healthy seeds should be selected from disease-free fields for sowing (ii) Smutted heads should be removed and destroyed, as soon as they emerge out and before the spores are dispersed.

Heat treatment. Jensen (1909), a Danish Scientist advocated the hot water treatment to destroy the dormant mycelium within the seed. According to this method, the seeds are soaked in water at a temperature of 26° - 30°C for 4 - 5 hours, which induces the dormant mycelium to revive. Then the seeds are quickly transferred to hot water at 54°C for 10 minutes, which kills the revived mycelium completely. Then the seeds are dried and stored for sowing. The thermal death point of the fungal mycelium and the embryo are very close and hence extreme care should be taken to keep the temperature of water at 54°C throughout the period of treatment. During hot water treatment, the seeds respire anaerobically and in this process **'alcohol'** and other **'catabolic substances'** are released, which are lethal to the fungus. Subsequently, several modified heat treatment methods have been advocated. In 1934, **Luthra** and **Sattar** recommended the solar energy treatment. This simple and economic method consists of soaking the seeds in water for 4 hours in the forenoon, followed by sun-drying the seeds for 4 hours in the noon at a temperature of 42° - 44°C, which is very common in Northern India during the summer months. Just soaking the infected seeds in water for 23 - 29 hours at 33°C or for 41 - 48 hours at 25°C has also been found to be effective in eliminating the mycelium present inside the seeds. Another method, known as anaerobic treatment is followed in the United States of America, which is found to be very effective. In this method, the infected seeds are soaked in water at 16° - 21°C for 2 - 4 hours and then kept in airtight containers for 65 - 70 hours and finally the seeds are dried.and stored for sowing.

Seed treatment. Seed dressing with systemic fungicides, such as carboxin or benomyl at 2.0 - 2.5 gm./ kg. of seeds, 24 hours prior to sowing is found to control loose smut of wheat effectively. Seed dressing with a mixture of carboxin - 2 gm. + thiram - 4 gm./ kg. of seed affords better control of the disease.

Resistant varieties. Kalyansona, PV.18, WG.307, C.302, NP.710, NP.718, NP.761, NP.770 and Bansi 224 have shown resistance to this disease.

5. Bunt of wheat

Tilletia caries (T. tritici) and *Tilletia foetida (T. laevis)*

'Bunt' or **'covered smut'** or **'stinking smut'** of wheat is of common occurrence in all wheat growing countries of the world and is much more serious than the loose smut of wheat. There are three kinds of bunts in wheat, caused by related, but different fungi, **'the common bunt'** or **'hill bunt'** caused by *Tilletia caries* and *T. foetida* and the **'Karnal smut'** caused by *Neovossia (Tilletia) indica*. Bunt caused by *T. caries* is also known as **'low smut'**, as the culms of the affected plants are shorter than the normal plants. A **'dwarf bunt'** caused by a variety of *T.caries*, which is now known as *T. controversa* has also been reported. Bunt caused by *T. foetida* is known as **'high smut'**, as the culms are mostly as high as the normal ones. The disease is also found to occur on many wild and cultivated grasses, such as species of *Triticum, Agropyron* and *Lolium*. In India, the disease is confined mostly to the cooler regions, such as Kashmir, Himachal Pradesh, parts of Punjab and Western Uttar Pradesh.

Symptoms. Bunt of wheat is a systemic disease resulting from seedling infection, however the symptoms of the disease are not usually manifested until the heading stage. Symptoms of the disease are seen only in the inflorescence. At maturity, the ears of the infected plants remain erect, dark or olive-green and with open pales and glumes, within which the black bunt ball is seen. Bearded varieties of wheat have deformed awns or the awns may be absent. Bunted grains are plumper and shorter than healthy grains and are grayish-brown, rather than the normal golden-yellow or red color. In a diseased stool, some ears may escape infection. Sometimes, a few grains in an ear may not be affected. A bunt ball may contain 1 - 4 million spores, united into a greasy mass that hardens on drying and emanates a foul smell of rotten fish due to the presence of **'trimethylamine'**. It is because of this bad smell, the disease is called **'stinking smut'**. The bunted grain may remain unbroken during harvesting, but at threshing is often broken and contaminate the sound grains at any stage, from harvest to sowing. There is no spread of the disease in the standing crop and infection occurs only in the soil, when contaminated grains germinate. The infected plants show greater susceptibility to some other diseases including yellow rust.

The causal organism. Bunt of wheat is caused by *Tilletia. caries* and *T. foetida,* individually or jointly. The life cycle of the two species is quite similar. However, the two species differ, chiefly in the morphological characters of the spores. In *T. caries,* the spores are rounded, 15 - 21µ in diameter and the walls are reticulate. So, it is also called **'rough spored bunt'**. In *T. foetida,* the spores are more irregular in shape, 16 - 25µ in diameter and the walls are smooth. So, it is also called **'smooth spored bunt'**. In both the species, the spores are mixed with sterile cells, which are round, thin-walled, hyaline and slightly smaller than the true spores. The color of the spores is of various shades of brown and appears black in masses. The spores of both the species germinate in the same way by producing a stout germ tube (promycelium), which bears a cluster of up to 24 elongated filiform, hyaline primary sporidia at its tip. The nuclei inside the promycelium move into the sporidia. The primary sporidia become binucleate by fusion in pairs *in situ* to form characteristic **'H'**-shaped structures. After fusion, the binucleate sporidium gives rise to a binucleate germ tube (fusion tube), which enters the host plant and causes infection. Both *T. caries* and *T. foetida* are heterothallic and only two mycelia from two sporidia of sexually compatible strains can fuse and produce a binucleate mycelium, which only can carry on and complete the parasitic life cycle of the fungus. The binucleate mycelium penetrates into the host plant and causes systemic infection.

After systemic development through the plant, the mycelium again forms teliospores in the kernels. Fusion of paired nuclei and formation of a single diploid nucleus takes place just prior to maturation of the smut spores. Several physiologic races of the two species have been identified in India.

Mode of survival, spread and epidemiology. The bunt pathogens overwinter as teliospores on contaminated wheat grains and in the soil. The spores are generally short-lived in moist soils, but can retain their viability for up to 2 years under dry conditions. The spores being sticky, adhere easily to the grains during threshing or to the soil particles, if shed in the fields. When contaminated seeds or healthy seeds are sown in bunt infected fields, the conditions that favor germination of seeds, favor the germination of the bunt spores also. As the wheat seeds germinate, the teliospores also germinate and produce basidia, primary sporidia and secondary sporidia. Then the secondary sporidia also germinate and produce a dikaryotic mycelium. This dikaryotic hyphae form appressoria in contact with the cuticle of young seedlings. The infection tube produced from the appressorium invades the host tissues intercellularly. The intercellular mycelium invades the developing leaves and the meristematic tissue at the growing point of the plant. When the plant forms the head, the mycelium invades all parts of it, even before the head emerges. The mycelium consumes all the parts of the kernel, except the pericarp, which forms a sturdy covering over the smut spore mass. At the same time most of the hyphal cells are transformed into teliospores

Environmental conditions, such as temperature, moisture, soil type, soil fertility etc, influence spore germination and disease development. The optimum temperature for spore germination is 18° - 20°C. Low soil temperature and high soil moisture are conducive for disease development.

Disease management

Agronomic practices (i) Straw and other crop residues in the infected fields should be burnt after the harvest of the crop (ii) Bunt-free seeds should be used for sowing (iii) Shallow planting reduces the disease incidence (iv) Diseased plants should be removed and burnt (v) Whenever possible, crop rotation should be followed in bunt affected fields

Seed treatment. Seed dressing with carboxin - 2 gm. or benomyl - 3 gm. or chloranil - 2 gm. or thiobendazole - 2 gm. or thiram - 4 gm./ kg. of seed affords good control of the disease.

Resistant varieties. Kalyansona, S.227, PV.18, HD.2012, HD.4513 and HD.4519 have been found to be resistant.

Fig. 13: Loose smut of wheat
Ustilago segetum var. *tritici*

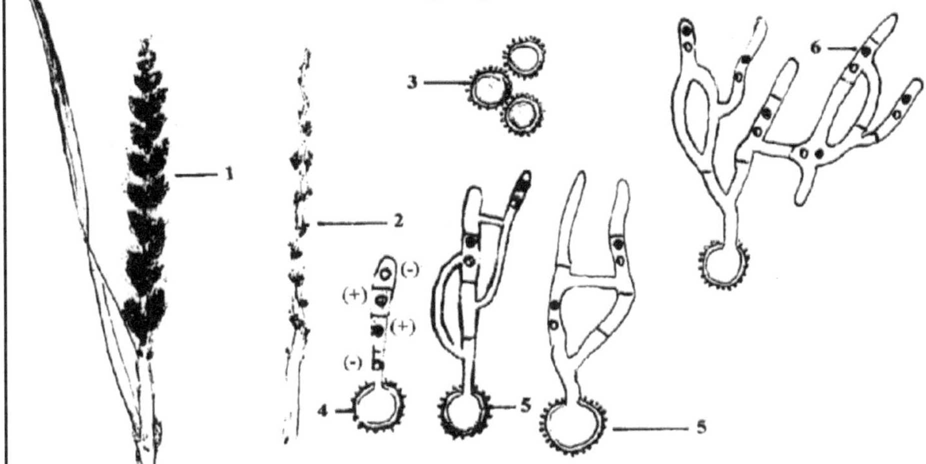

1. Smutted earhead 2. Bare rachis after the spores have been blownaway by wind 3. Teliospores 4. Germination of teliospore to form promycelium 5. Cell fusion between the cells of the promycelium 6. Extension of binucleate mycelium from the fusion cells and repeated cell fusion.

Fig. 14: Bunt of wheat
Tilletio curies and *T. foetida*

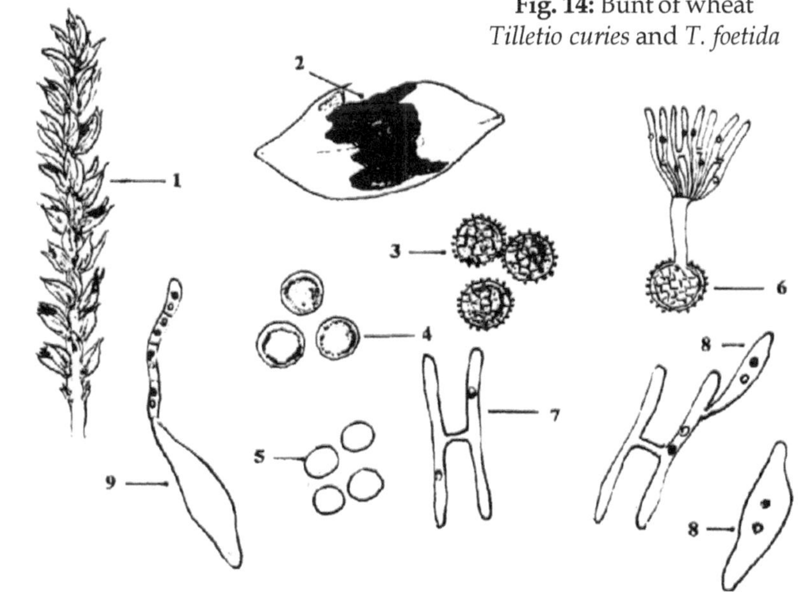

1. Smutted wheat head 2. Smutted kernel full of teliospores 3. Teliospores of *T. caries* 4. Treliospores of *T. foetida* 5. Sterile spores 6. Germinating spore producing primary sporidia 7. Fusion of sporidia 8. Binucleate secondary sporidia 9. Germination of secondary sporidia to form binucleate mycelium.

6. Karnal bunt of wheat

Neovossia (*Tilletia*) *indica*

'Karnal bunt' or **'new bunt'** or **'partial bunt'** has become a dreaded disease during the recent past and is more prevalent than the other bunt diseases of wheat. The disease, which was first reported in India from Karnal (Haryana) in 1931, now occurs in Delhi, Rajasthan, Haryana, Punjab, Himachal Pradesh, Jammu and Kashmir, Bihar and Uttar Pradesh. Most of the cultivated dwarf varieties of wheat are susceptible to this disease. This disease adversely affects both the yield and the quality of grains.

Symptoms. The symptoms caused by this pathogen are totally different from the hill bunts and this bunt becomes evident only when the grains have developed. In a stool, all the ears are not infected and in the ears, all the grains are not infected. Only some grains are partially, rarely completely converted into black powdery mass of spores, which are enclosed by the pericarp. The smutted grains are irregularly scattered indicating air-borne, localized infection. As the grains attain maturity, the outer glumes spread out slightly and the inner glumes of the spikelet expand. The spore mass remains covered by the pericarp for sometime and rupture later on, exposing the black, powdery spores. As in the case of covered bunt, here also the foul, stinking smell due to **'trimethylamine'** is present.

The causal organism. The teliospores or bunt spores or chlamydospores of this fungus are darker brown than those of *T. caries* and *T. foetida.* They are spherical to oval, with reticulations on the epispore that appear as curved spines and measure 22 - 49µ in diameter. A thin, hyaline membrane, the perisporium or sheath surrounds the spores. Numerous, large, sterile cells are present mixed with the spores. The spores are easily carried by wind. The spores germinate by producing a short, stout promycelium that bears a whorl of 60 - 185 sporidia at the apex. The sporidia are long and sickle-shaped. They do not fuse in pairs, but germinate by producing a germ tube, which acts as infection hypha and infects wheat floret. A long rest period is required prior to germination of the spores.

Mode of survival, spread and epidemiology. 'Karnal bunt' is mostly a soil-borne disease. The spores get into the soil during threshing or as external seed contaminant. Under suitable conditions of soil moisture and temperature, the spores germinate and produce sporidia. Secondary infection is by air-borne sporidia. The sporidia germinate and enter the developing grain through the ovary wall. The mycelium of the pathogen penetrates the embryonic tissues, but does not cause any damage to the embryo.

The optimum temperature for the germination of teliospores is 20° - 25°C. High relative humidity of about 100 %, frequent light rains, cloudy and damp weather and low temperature, particularly at the time of flowering, favor rapid and severe development of the disease. Excess irrigation and high doses of nitrogenous fertilizer make the plants more vulnerable to attack by the disease. The environmental factors favorable for the disease occurrence usually coincide with the flowering time of the crop, especially in Northern India. Many physiologic races of the fungus are present. The pathogen is known to infect several wild species of *Triticum*.

Disease management

Agronomic practices (i) Late sown crops are found to suffer more and hence, early sowing should be taken up (ii) Application of excessive doses of nitrogenous fertilizer and excessive irrigation should be avoided (iii) Field sanitation should be followed (iv) Growing of green manure crops prior to wheat crop results in reducing the disease incidence (v) Contaminated seeds should not be used for sowing (vi) Highly susceptible varieties, such as HD.1982, HD.2009, WL.711, WG.357, WG.377, UP.319 and UP.262 should not be grown in 'Karnal bunt' prone areas (vii) Deep summer ploughing should be practiced to bury the spores deep into the soil

Chemical control. Spraying with carbendazim - 250 gm. or propiconazole - 500 gm. in 250 lit. of water per acre, once at the boot leaf stage has been found to be very effective in controlling the disease.

Resistant varieties. Varieties HD.1907, HI.358, HP.743, L.176, L.191, M.137 A, MW.59-4 X 6A. UP.270, UP.368, HD.2222, HD.2227 and HD.2235 have shown resistance to this disease.

7. Powdery mildew of wheat

Erysiphe graminis f.sp. *tritici*

'Powdery mildew' of wheat is of worldwide occurrence. The species *Erysiphe graminis* is found on a wide range of cereals and grasses however, different races of this pathogen attack barley, oats, rye, wheat, *Agropyron, Bromus, Dactylis, Poa* and *Elymus*. The variety attacking wheat, does not attack barley or oats and vice versa. So, the variety that attacks wheat is known as *Erysiphe graminis* f.sp. *tritici* and the one that attacks barley is known as *Erysiphe graminis* f.sp. *hordei*. In India, the disease is prevalent in the hilly tracts of North and South India. Sporadic occurrence of the disease has been reported from parts of Uttar Pradesh, Punjab, Haryana, Rajasthan and Delhi. The disease is quite common in low-lying, ill-drained soils. Monocropping with wheat increases the chances of the disease occurrence to a larger extent. In some parts of the country, powdery mildew appears along with downy mildew and leaf rust.

Table 3 : Differences between the wheat bunts

Rough-Spored bunt or Low smut Tilletia tritici	Smooth-Spored Bunt or high Smut Tilletia Laevis	Partial bunt or Karnal bunt Neovossia indica
1. All spikelets in the earhead are infected completely.	All spikelets in the earhead are infected completely.	Few spikelets in the earhead are infected partially or rarely completely.
2. Systemic infection takes place in 6-10 days old seedlings.	Systemic infection takes Place in 6-10 days old seedlings.	Air-borne, localized infection of spikelets takes place.
3. Spores spherical, brown in color, rough-walled, reticulate and measure 15-21μ in diameter.	Spores irregular in shape, brown in color, smooth-walled and measure 11-25μ in diameter	Spores spherical to oval in shape, darker brown in color, reticulate and are 27 - 49μ in diameter. Reticulations appear as curved spines.
4. Germinating spores produce about 24, filiform primary sporidia that fuse in pairs to form 'H' shaped structures	Germinating spores produce about 24, filiform primary sporidia that fuse in pairs to form 'H' shaped structures.	Germinating spores produce 60 - 185, long, sickle-shaped primary sporidia that do not fuse.
5. The pathogen is soil-borne and externally seed-borne.	The pathogen is soil-borne and externally seed-borne.	Infection of florets is through air-borne sporidia produced from bunt spores in the soil.
6. Optimum temperature for germination of teliospores is 18-20°C.	Optimum temperature for germination of teliospores is 18-20°C.	Optimum temperature for germination of teliospores is 20-25°C.

Symptoms. The fungus develops most conspicuously on the leaves, usually on the upper surface, sometimes on the under surface also and on the leaf sheaths and stems. The disease may attack the seedlings as soon as the first leaves appear, sometimes killing the affected seedlings. But, mostly the disease is vigorous on grown-up plants. In case of severe incidence, the leaves become wrinkled, twisted or variously deformed. The tips of the shoot may droop and wither and ear development may be partially or completely checked. Transpiration and respiration are increased and photosynthesis is much reduced as the result of infection **(Fig.15).**

The causal organism. The mycelium is entirely superficial, forming a flocculent matted growth, white in the initial stages when the conidia are formed and changing into gray or reddish-brown color when cleistothecia are formed. Because of the coverage of vast areas of the leaf surface by the fungus, photosynthesis is adversely affected, leading to chlorosis and weakening of the plant. The superficial mycelium is anchored to the host by lobed haustoria that enter the epidermis and rarely into the mesophyll tissues. Hyphae are septate, interlaced to form a web, covering large areas of the leaf and the stem. Conidiophores are produced from hemispherical swellings on the prostrate mycelium at right angles to the leaf. Conidia are borne in chains of about 10 - 20 from each conidiophore, the oldest at the terminal end. They are elliptical, hyaline, uninucleate and 25.0 - 30.0 x 8.0 - 10.0μ in size. They fall off readily and are dispersed by wind.

At the end of the conidial stage, the superficial mycelial weft develops the perfect ascigerous stage in the form of small, dark, cleistothecia. The cleistothecia are globose-depressed, 160 - 192 x 120 - 130μ in diameter, black, partly immersed in the mycelial weft and furnished with simple or slightly branched, pale brown appendages. Each cleistothecium contains 9 - 30 cylindrical or ovoid, pedicellate asci, measuring 70 - 108 x 25 - 40μ in size. Each ascus contains 8 ascospores (rarely 4), that are elliptic, sub-hyaline to pale brown and measure 20 - 23 x 10 - 13μ in size **(Fig.15).**

Mode of survival, spread and epidemiology. Powdery mildew of wheat is soil-borne through cleistothecia, which can survive up to 13 years at low temperatures. The asci remain immature in the cleistothecia on fallen leaves and the ascospores take about 10 months to attain maturity under suitable conditions, such as alternate dry and wet soil conditions. Such conditions may be available in low-lying fields and hilly, temperate regions of North India. Initial infection may take place under such conditions. The disease may subsequently be introduced in the plains through air-borne conidia from the hilly regions. Secondary spread of the disease is through air-borne conidia.

Several physiologic races of this pathogen exist in nature.

Unlike most other powdery mildews, which are favored by dry weather, powdery mildew of cereals is favored by low temperature and thrives at low or high humidities. Conidia are produced most abundantly in a relatively cool and moist environment. At high temperatures, conidia do not retain their viability.

A temperature of 15° - 20°C and relative humidity of 100 % is optimum for conidial germination. Mycelium of the fungus grows best at 20° - 21°C. A temperature of 30°C or more has a deleterious effect on the disease.

Disease management

Agronomic practices. Application of excessive doses of nitrogenous fertilizer should be avoided. Potash and phosphorus application tends to reduce the severity of the disease.

Chemical control. Spraying with wettable sulfur - 1200 gm. or ethirimol - 300 gm. or diamethirimol - 300 gm. or benomyl - 300 gm. or tridemorph - 300 ml. in 300 lit. of water per acre is found to be effective in controlling the disease.

Resistant varieties. Sharbati Sona, Sonora 64, Chhoti Lerma, Kalyan Sona, HD.1980, NP.710, NP.718, K.53, E.750 and C.591 are resistant or moderately resistant to this disease.

8. Leaf blight of wheat

Alternaria triticina

The disease had been noticed in India since 1924. It is a common and destructive disease in various parts of India however, the disease has not been reported from any other country so far. It is found to occur in Bihar, Maharashtra, Uttar Pradesh, Punjab, West Bengal and Telungu Desam. Many of the recent wheat varieties, resistant to rust diseases have been found to be highly susceptible to this disease.

Symptoms. The disease does not affect young seedlings. Grown-up plants, which are 7-8 weeks old are vulnerable to attack by this disease. Initial symptoms appear as small, oval, discolored lesions, scattered on the leaves. Lowermost leaves and plants along the irrigation channels are affected first. The spots enlarge in size, become irregular in shape and turn brown to gray in color and are surrounded by a bright yellow halo. Several such lesions coalesce and cover large areas of the leaf, resulting in the death of the entire leaf. In some cases, the leaf starts drying downwards from the tip.

Similar symptoms may appear on leaf sheaths, ears and glumes. Under moist conditions, the lesions are covered with black powdery conidia **(Fig.16).**

The causal organism. The mycelium of the fungus is hyaline in the initial stages and becomes brownish-yellow later on. The hyphae are branched and septate. Conidiophores, which are olive-buff in color, are septate, mostly simple, but rarely branched, erect and emerge singly or in fascicles through the stomata. They measure 17.0 - 28.0 x 3.0 - 6.0µ. Conidia are smooth, irregularly oval, with both ends rounded or ellipsoid, gradually tapering into a beak and are borne singly or in chains of 2 - 4. They are light-brown to dark-olive-buff and become darker with age. They measure 15.0 - 89.0 x 7.0 - 30.0µ including the beak and have 1 - 10 transverse septa and 0 - 5 longitudinal septa. At least, 6 physiologic races of the fungus are known to exist in nature **(Fig.16).**

Mode of survival, spread and epidemiology. The pathogen is both externally and internally seed-borne, as well as soil-borne. In the case of internal seed infection, the mycelium of the pathogen is deep-seated in the embryo of the seed. The soil-borne inoculum is susceptible to extremes of temperature. However, the pathogen can oversummer in plant debris in the soil under normal environmental conditions and under shade and may perpetuate the disease. The possibility of seed-borne inoculum initiating the disease is definitely more. Secondary spread of the disease in the field is by air-borne conidia.

Optimum temperature for germination of conidia is 15° - 27°C and 100 % humidity is necessary for maximum germination of conidia. A maximum period of 48 hours of saturated atmosphere is required for successful infection. The most suitable temperature for development of the disease is 25°C.

Disease management

Agronomic practices (i) The stubble and other crop residues in the field should be destroyed by burning after the harvest of the crop (ii) Disease-free seeds should be obtained and used for sowing.

Chemical control (i) Spraying the crop with Cuman L - 600 ml. or mancozeb - 600 gm. or Dithane-M 45 - 600 gm. or Dithane-Z 78 - 600 gm. or carboxin - 300 gm. in 300 lit. of water per acre gives suitable control of the disease (ii) Spraying the crop with propiconazole (Tilt) at 300 gm. in 300 lit. of water per acre has been reported to be very effective in reducing the blight incidence, when applied twice at fortnightly intervals, commencing from the initiation of the disease.

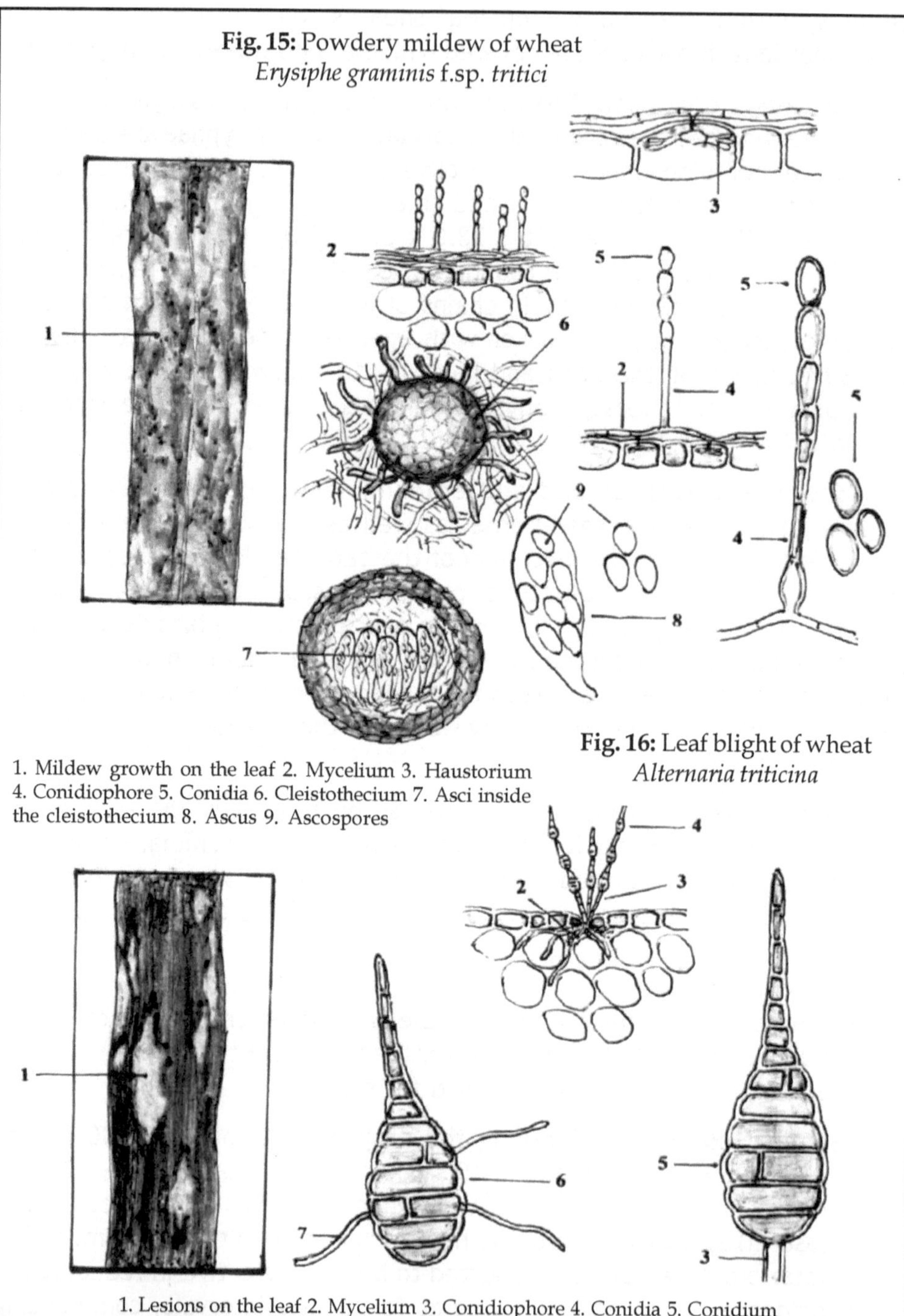

Fig. 15: Powdery mildew of wheat
Erysiphe graminis f.sp. *tritici*

1. Mildew growth on the leaf 2. Mycelium 3. Haustorium 4. Conidiophore 5. Conidia 6. Cleistothecium 7. Asci inside the cleistothecium 8. Ascus 9. Ascospores

Fig. 16: Leaf blight of wheat
Alternaria triticina

1. Lesions on the leaf 2. Mycelium 3. Conidiophore 4. Conidia 5. Conidium 6. Germinating conidium 7. Germ tube.

Resistant varieties. NP.4, NP.52, NP.200, NP.809, Arnautka, E.6-6160, K.7340, HP.1163, HD.1941, Janak, M.134. K.8904, K.8908 and UP.262 have been reported to be resistant to this disease.

Diseases of minor importance. Wheat crop is subjected to attack by several other diseases, besides the diseases dealt with in detail above. However, they are of minor importance and occur less frequently and sporadically. 'Black point disease' caused by *Cochliobolus sativus (=Helminthosporium sativum)* affects the seedlings and grown-up plants and causes crown rot; 'Speckled leaf blotch' caused by *Leptosphaeria tritici (=Septoria tritici)* affects the leaves and causes blotches; 'Foot rot of seedlings' caused by *Pythium graminicolum* and *P. arrhenomanes* affects the seedlings; 'Downy mildew' caused by *Sclerophthora macrospora* affects the leaves, leaf sheaths and stems; 'Take-all and white-head disease' caused by *Ophiobolus graminis* affects the ears; 'Eyespot disease' caused by *Cercosporella herpotrichoides* produces eyespots on leaves, leaf sheaths and stems; 'Sooty mould' or 'Black mould' caused by *Cladosporium herbarum* affects the leaves, leaf sheaths, stems and ears; 'Brown foot rot' and 'Ear blight' caused by *Fusarium avenaceum* and *F. culmorum* affect the stems and ears; 'Scab disease' caused by *Gibberella zeae (=Fusarium graminearum)* affects the ears; 'Yellow ear rot' or 'Tundu disease' caused by *Corynebacterium tritici* affects the ears; 'Mosaic streak disease' caused by *Chirkavirus* affects the leaves and leaf sheaths.

Sorghum, Jowar or Cholam *(Sorghum vulgare)*

1. Downy mildew or leaf-shredding disease of sorghum

Sclerospora (Peronosclerospora) sorghi

'Downy mildew'of sorghum is of common occurrence in many Asian and African countries and in Italy. In India, it is prevalent in the states of Tamil Nadu, Telungu Desam, Karnataka, Madhya Pradesh, Maharashtra and Uttar Pradesh. Fodder sorghum is more vulnerable to this disease than grain sorghum. The pathogen also attacks maize and teosinte.

Symptoms. The symptoms of the disease vary depending upon the growth stage of the crop at which infection occurs. When systemic infection is caused by oospores present in the soil, symptoms develop on the seedlings as a whitish, cottony, soft, downy growth, mostly on the under surface of leaves, consisting of sporangiophores and sporangia. The corresponding upper surface of the leaves turns yellowish. Such plants continue to grow, but eventually dry and die before ears are produced. The infected leaves turn brown and tear into strips, while the mid rib remains intact. Secondary infection may take place on 60 - 70 days old plants by sporangia produced by primary infection. Such infected leaves also turn whitish or yellowish and later become brown stripes. In such lesions also, sporangiophores and sporangia may develop on the under surface as downy growth under conditions favorable for the pathogen. In these leaves also, partial or complete shredding may occur. Only under conditions of high humidity and low day temperatures secondary infection takes place and fructifications are formed. Oospores, which are produced in abundance along the vascular bundles of the leaves, make the tissues weak, causing the leaves to tear apart and producing the characteristic leaf-shredding symptom. Leaf shredding is also attributed to the production of **'pectolytic enzymes'** by the pathogen. Grain setting and quality of the seeds are adversely affected.

The causal organism. The causal organism is an obligate parasite. It grows systemically in the host plant. The mycelium is coenocytic, intercellular and is present in the paraenchymatous tissues of the roots and stem and the mesophyll tissues of the leaves. The hyphae produce finger-like haustoria.

Sporangiophores emerge singly or in clusters through the stomata. They are stout, and have one or two transverse septa at the basal portion. They branch dichotomously at the tips and the branches end in a series of pointed sterigmata. The sporangiophores measure about 200µ in length. The sterigmata bearing the sporangia spread out in a hemispherical plane. Sporangia are produced singly at the tips of the sterigmata. They are single-celled, thin-walled, hyaline, globose, measuring 15 - 29µ in diameter and lack apical papilla. The sporangia are smaller than those produced by *Sclerospora graminicola* and they germinate by issuing one or more germ tubes. They never produce zoospores. Oospores are produced in large numbers in parallel bands in the vascular bundles, which weaken the tissues, making the leaf blades to tear into stripes along the brown streaks. They are thick-walled, spherical, dark-brown in color and fill the oogonium completely. They measure 25 - 43µ in diameter. Matured oospores germinate by means of one or more germ tubes.

Mode of survival, spread and epidemiology. The pathogen is both soil-borne and seed-borne. The oospores can retain their viability for 3 - 4 years. Plant debris and haystacks harbor the oospores. The oospores germinate, when they come into contact with the germinating seedlings and cause primary systemic infection. Secondary spread is by means of sporangia dispersed by wind, rainwater, direct contact and by some insects mechanically. In the seeds, the fungus is carried as dormant mycelium and when the seeds germinate, the mycelium revives, starts growing systemically and cause infection of the seedlings.

Maize, teosinte and some weeds serve as alternate hosts of this fungus, which may serve to provide the initial inoculum. This fungus does not infect pearl millet and finger millet.

Optimum temperature for sporulation and sporangial germination is 21° - 23°C. Formation of sporangia is suppressed at temperatures below 15°C. High relative humidity and a damp, saturated atmosphere, with a film of water on the leaf surface and a temperature of 21°C is highly favorable for severe disease occurrence.

Disease management

Agronomic practices (i) Deep ploughing and burying the crop debris and the oospores present in the soil deep into the soil helps to reduce soil-borne infection (ii) Systematic roguing of affected plants before they produce spores helps to reduce secondary spread of the disease (iii) Long term crop rotation with crops like pulses and oilseeds helps to eliminate the soil-borne inoculum (iv) Healthy seeds obtained from disease-free areas should be used for sowing.

Chemical control. Foliar spraying with Cuman L - 600 ml. or mancozeb - 600 gm. or zineb - 600 gm. metalaxyl - 300 gm. in 300 lit. of water per acre is effective in controlling the disease. Spraying with a mixture of metalaxyl - 300 gm. + mancozeb - 600 gm. in 300 lit. of water per acre has been reported to give very good control of the disease.

Seed treatment. Seed dressing with metalaxyl at 2 gm./ kg. of seeds is found to be effective in controlling seed-borne infection.

2. Grain smut of sorghum

Sporisorium (Sphacelotheca) sorghi

'Grain smut', **'kernel smut'**, **'short smut'** or **'covered smut'** as the disease is known, is present in Asia, Australia, the United States of America, Sri Lanka and Venezuela. In India, it is one of the most serious diseases of the crop in the states of Tamil Nadu, Telungu Desam, Karnataka, Uttar Pradesh, Madhya Pradesh and Maharashtra. The disease affects both rainfed and irrigated crops.

Symptoms. The majority of the grains in an earhead are mostly converted into smut sori. In the place of normal grains, oval or cylindrical, gray, sac-like sori, 5.0 - 12.0 x 3.0 - 5.0 mm. in size are formed in-between the glumes. The sac-like sorus is covered by a tough, whitish to light-brown membrane (peridium) that remains intact without rupturing till the time of threshing. The interior of the sorus, except a slender, sometimes curved central column of hard tissues called **'columella'** is completely filled with black or dark-brown, powdery spores. The columella is bulbous at the base and extends up to the apex. It is composed of the host tissues consisting of parenchyma, traversed by fibro vascular bundles **(Fig.17)**.

The causal organism. The smut spores present in the sorus in-between the peridium and columella are dark-brown to black in mass. The spores are globose to sub-globose, thick-walled, olive-brown in color, apparently smooth-walled and measure 5.0 - 9.0µ in diameter. Though the spores are capable of immediate germination, they remain in a viable state for a number of years. Under favorable environmental conditions, the spore germinates to produce a 3 septate basidium and from each of the 4 basidial cells, a single-celled, uninucleate sporidium is formed. The sporidia are spindle-shaped and they multiply by repeated budding till the protoplasm is depleted. They measure 10.0 - 12.5 x 2.0 - 3.0µ in size. Sometimes, instead of sporidia, germ tubes are produced. The sporidia or germ tubes of compatible sexes fuse to form a dikaryotic mycelium, which causes infection **(Fig.17)**.

Mode of survival, spread and epidemiology. The disease is externally seed-borne. At the time of threshing, the smut sori break and the seeds get contaminated with the spores. When such contaminated seeds are sown, the spores germinate and attack the seedlings from the time they emerge out of the seeds and before they appear above the soil. The fungus penetrates the seedlings through the radicle or mesocotyl region and establishes systemic infection. The mycelium grows intercellularly and keeps pace with the growing point of the shoot. No external symptoms are visible till the plants produce earheads. During earhead formation, the hyphae accumulate in the immature ovary forming a dense mass, the outer part of which becomes the peridium of the sorus, while the inner portion is converted into a mass of smut spores or teliospores or chlamydospores.

A temperature of about 25°C and medium to low soil moisture favor infection and development of the disease. The smut infects all cultivated varieties of sorghum, besides *Sorghum halepense* and *S. seudanense*. Several physiologic races of the fungus have been identified.

Disease management

Agronomic practices (i) Affected earheads should be collected before harvest and destroyed (ii) Disease-free seeds should be used for sowing (iii) Crops sown early in the season i.e. by early June usually escape disease incidence.

Seed treatment (i) Solar heat treatment. The seeds are soaked in cold water for 4 hours. Then the seeds are dried under direct sunlight. The spores, which have initiated the germination process, are killed by direct exposure to sunlight (ii) Seed dressing with sulfur dust or thiram or captan at 4 gm./ kg. of seeds, 24 hours prior to sowing, eliminates the externally seed-borne spores.

Reistant varieties. CSH.5, CSH.7-R, CSH.9, RSV.1-R, SDM.9, SPV.102, SPV.115, SPV.138, SPV.245, SPV.297, T.29/1, PJ.7 K, PJ.23 K, Nandyal and Bilichigan are resistant to this disease.

3. Loose smut of sorghum

Sporisorium (Sphacelotheca) cruenta

'Loose smut of sorghum', which is distributed in many sorghum growing countries of the world, such as Asia, Europe and the United States of America, is less common than grain smut of sorghum. In India, it occurs in the states of Telungu Desam, Maharashtra, Karnataka, Madhya Pradesh and Tamil Nadu.

Symptoms. The infected plants can be detected before the emergence of earheads. They are shorter than the healthy plants by about a foot, have thinner stalks and show increased tillering. The plants usually come to flowering prematurely, about 2 weeks earlier than normal plants. The ears are looser and bushy compared to normal ears and show hypertrophy of glumes and proliferation of spikelets. Usually, all the florets of infected ears are smutted and sori may be formed on the rachis and branches of the inflorescence. The peridium of the sorus usually ruptures prior to head emergence. The sori measure 3.0 - 18.0 mm. in length. Late tillers, as well as ratoon crops show higher incidence of the disease **(Fig.18)**.

The causal organism. The spores are formed in the ovaries and floral bracts. The thin, membranous peridium covering the sorus is made up of loosely joined, rounded, gray fungal cells that are about twice the diameter of the spores. In the center of each sorus, there is a long, conical **'columella'** as in the grain smut, but they are longer and more curved. The well-developed, unbranched, black, pointed columella, extending to the tip of the sorus is made up of host tissues. The columella is surrounded by mass of smut spores, which appear dark-brown or black. The spores are globose to subglobose, light yellowish-brown in color, finely echinulated and measure 6.0 - 10.0µ in diameter. They germinate to form a 4-celled promycelium and from each cell of the promycelium, a basidiospore is borne laterally. The basidium may develop into branched or unbranched hyphae instead of producing sporidia at higher temperatures. **(Fig.18)**.

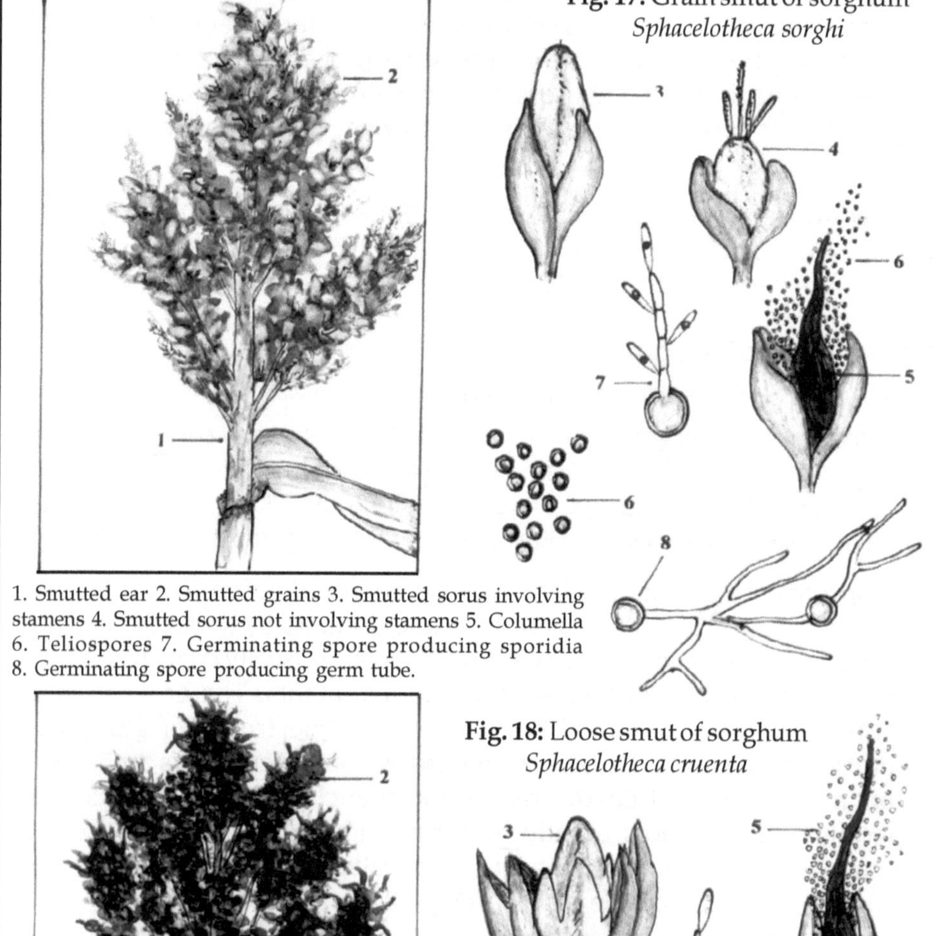

Fig. 17: Grain smut of sorghum
Sphacelotheca sorghi

1. Smutted ear 2. Smutted grains 3. Smutted sorus involving stamens 4. Smutted sorus not involving stamens 5. Columella 6. Teliospores 7. Germinating spore producing sporidia 8. Germinating spore producing germ tube.

Fig. 18: Loose smut of sorghum
Sphacelotheca cruenta

1. Smutted head 2. Smutted grains 3. Smut sorus 4. Columella 5. Teliospores 6. Germinating spore producing sporidia 7. Budding sporidia 8. Germinating spore producing germ tube.

Hybridization has been recorded between *Sphacelotheca cruenta*, as well as *S. reiliana* and *S. sorghi*. Only the dikaryotic mycelium is capable of producing the disease

Mode of survival, spread and epidemiology. The disease is mainly externally seed-borne however, in dry regions it may also be soil-borne. The spores remain viable for 4 years. Infection of seedlings occurs at the time of germination of seeds. As soon as the seedling emerges, the fungus enters the seedling through the radicle, mesocotyl or epicotyl. Further development follows as in other systemic smuts, such as grain smut of sorghum. Floral infection and spread of the disease within the crop through air-borne sporidia has been reported. Three physiologic races of the fungus have been identified

The optimum temperature for germination of spores is 18°-32°C. Low temperature and low soil moisture favor infection. Deep sowing of seeds predispose the seedlings to infection by this pathogen.

Disease management

The control measures recommended for the control of grain smut of sorghum is applicable for controlling this disease also (Page 70).

4. Head smut of sorghum

Sporisorium (Sphacelotheca) reiliana

'**Head smut**' is found in many countries, such as Southern Europe, Africa, Asia and the United States of America. In India, the disease occurs in Kashmir, Telungu Desam, Tamil Nadu, Karnataka, Gujarat, Madhya Pradesh, Maharashtra, Uttar Pradesh and Punjab.

Symptoms. Generally, only a few plants in a field are affected and there is no noticeable difference between the affected and healthy plants, either in size or growth habitat. In the infected plant, the entire inflorescence is converted into a big smut sorus, about 10 - 13 cm. in length and 4 - 6 cm. in width. The large sorus replaces the tassel and ear entirely. The sorus consists of the conductive elements of the floral parts, surrounded by the spore mass. The fragile peridium made up of fungal tissue that covers the sorus, ruptures before the emergence of the sorus from the boot leaf and exposes a mass of brownish-black, powdery, smut spores. After the spores are blown away by wind, the vascular elements remain as long, thin, dark-colored filaments. Sometimes, only portions of the head are destroyed completely. Head smut can be easily differentiated from the other smuts in which only individual florets are infected. Sometimes, sori may develop on the leaves and the basal part of the rachis **(Fig.19)**.

The causal organism. The sorus is composed of loosely united smut spores and the conductive tissues of the inflorescence. The spore mass is powdery, dark-brown and is quickly dispersed to expose a tangled mass of vascular strands of the host or sometimes a single central columella. The spores are reddish-brown to black in color, finely echinulated, spherical to irregular in shape and 9.0 - 14.0μ in diameter. The spores germinate to produce a 4-celled promycelium and sporidia **(Fig.19)**.

Mode of survival, spread and epidemiology. The disease, which is primarily externally seed-borne, is soil-borne to some extent. The spores remain viable for about 2 years. They germinate under favorable environmental conditions producing stout promycelium and sporidia. The sporidia fuse to form a dikaryoyic mycelium, which infects the young seedlings before they emerge from the soil. The mycelium grows systemically keeping pace with the host and affects the entire inflorescence.

A soil temperature of 21°-28°C and soil moisture of 15-25 % is favorable for infection of the seedlings. The disease incidence is more at relatively high soil temperature and low soil moisture, than in wet and cool soils. Frequent irrigation after sowing has been found to reduce the disease incidence.

Disease management

Agronomic practices (i) Rouging and destroying smutted plants from the field before dispersal of the spores reduce infection of the succeeding crop (ii) Frequent irrigation to the young crops for 18 - 21 days after sowing helps to reduce infection (iii) Crop rotation and growing resistant varieties may be followed in infected fields.

Seed treatment. The seed treatments recommended for the control of covered smut of sorghum may be followed (Page 70).

5. Long smut of sorghum

Tolyposporium ehrenbergii

The disease is prevalent in most of the sorghum growing Asian countries and is widespread in Africa. In India, it occurs in Tamil nadu, Karnataka, Telungu Desam, Maharashtra, Madhya Pradesh and Uttar Pradesh. It occurs only sporadically and is restricted to a few grains scattered on the head.

Symptoms. The sori are cylindrical, elongate, slightly curved with tapering end, up to 4.0 cm. long, 0.5 - 1.0 cm. wide and covered by a relatively thick, creamy-brown membrane of fungal tissue. The peridium normally splits from the apex, revealing black mass of spores and among the spores are found several, dark-brown filaments, which represent the vascular bundles of the infected ovary. The sori are much longer and wider than those of covered smut **(Fig.20)**.

The causal organism. The spores are united into solid, granular balls, which are black in color and intermixed with shreds of host tissue. The spore balls contain many spores, which are rather permanently united, irregular in shape and vary in size. They are globose to elongated, dark-brown in mass and measure 45 - 200μ in diameter. The spores found in the interior of the balls are pale in color, smooth, globose to subglobose or angular in shape, measuring 10.0 - 15.0μ in diameter. The spores on the surface of the balls are darker in color and have flattened echinulations on the exposed surface **(Fig.20)**.

Mode of survival, spread and epidemiology. The smut spores found in the soil germinate by the formation of an elongated promycelium, frequently branching. Clusters of sporidia are produced singly in chains. The wind-borne sporidia fall on the florets and initiate a systemic mycelium, which converts the ovary into smut sorus within 12 - 15 days in the same season. Secondary infection may take place in the same season through the spores released from the smutted ears in the field, as the spores do not have a dormancy period.

The optimum temperature for germination of spores is 28°C. A relative humidity of 70 - 80 % is favorable for the occurrence of the disease.

Disease management

Agronomic practices (i) The smutted plants should be removed and destroyed by burning (ii) Use of disease-free seeds, early sowing and crop rotation are other methods, which can help to reduce the occurrence of the disease.

Seed treatment. Since the spores of the pathogen are air-borne, seed treatment is of no use.

Resistant varieties. In Tamil Nadu, 'Irungu cholam' is found to be free from infection, as the glumes cover the floral parts completely.

Table 4: Differences between the four smuts of sorghum

Grain smut *Sphacelotheca sorghi*	Loose smut *Sphacelotheca cruenta*	Head smut *Sphacelotheca reiliana*	Long smut *Tolyposporium ehrenbergii*
1. Infected plants appear normal till the time of flowering.	Infected plants are stunted, thinner, produce more tillers, flower prematurely and the glumes are longer.	Infected plants appear normal till the time of flowering.	Infected plants appear normal till the time of flowering.
2. Majority of the grains in an ear are converted into smut sori. The smut sori are oval or cylindrical, gray, sac-like, 5-12 x 3-5 mm. in size and covered by a peridium that remains without rupturing till the time of threshing.	All the grains in the infected ear are smutted. The smut sori are small, oval and 3-18 x 2-4 mm. in size. The peridium covering the sorus ruptures prior to the emergence of the head.	The entire inflorescence is converted into a big smut sorus, 10-13 x 4-6cm. in size. The fragile peridium covering the sorus ruptures before earhead emergence.	Only a few individual grains in an ear are transformed into smut sori. The sori are very conspicuous, long, cylindrical, slightly curved, with tapering end and are up to 4.0 cm. long and 0.5-1.0 cm. wide. The peridium covering the sorus splits from the apex.
3. Columella slender, sometimes curved, with a bulbous base and extends up to the apex.	Columella well-developed, unbranched, black, pointed, extending to the tip of the sorus and are longer than that of grain smut.	Usually no columella is present, but a tangled mass of vascular strands is seen. Sometimes a single central columella may be present.	Columella not present, instead several dark-brown filaments consisting of vascular strands are found.
4. Teliospores are single, globose to sub-globose, thick-walled, olive-brown in color, smooth and are 5-9μ in diameter.	Teliospores are single, globose to sub-globose, light yellowish-brown in color, finely echinulated and are 6-10μ in diameter.	Teliospores are loosely bounded into balls. The spores are spherical to irregular in shape, reddish-brown in color, finely echinulated and are 9-14μ in diameter.	Teliospores are rather permanently united into solid, granular spore balls, which are 45-200μ in diameter. Each spore ball contains many spores. The spores in the interior of the balls are pale in color, smooth, globose or irregular in shape and are 10-15μ in diameter.
5. The disease is externally seed-borne.	The disease is mainly externally seed-borne, but soil-borne to some extent.	The disease is primarily externally seed-borne, but soil-borne to some extent.	The disease is air-borne.

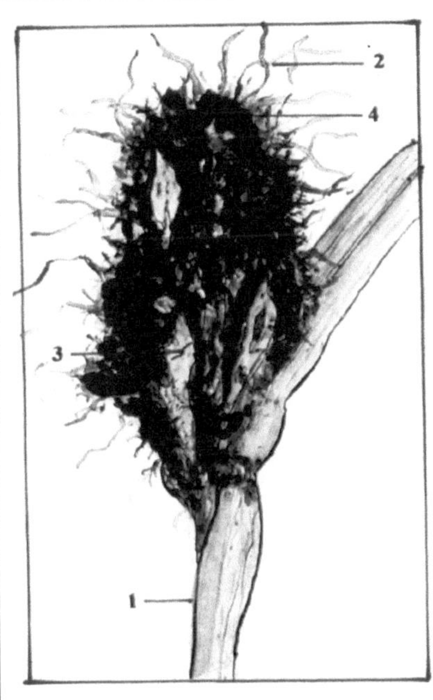

Fig. 19: Head smut of sorghum
Sphacelotheca reiliana

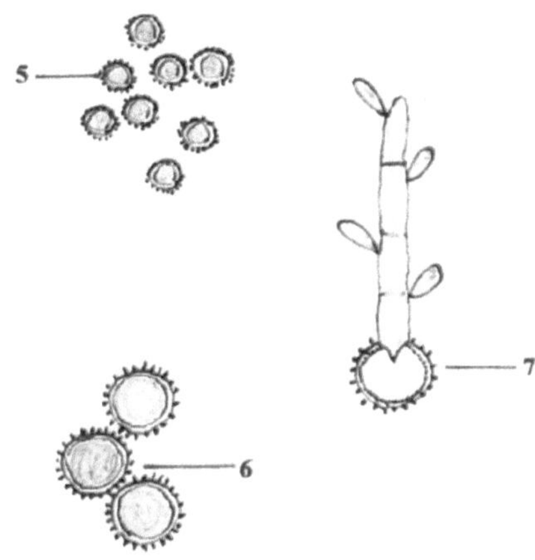

1. Smutted head 2. Vascular strands of the inflorescence 3. Pieces of the ruptured peridium 4. Spore masses 5. Spores 6. Spores showing fine echinulations 7. Germinating spore producing sporidia.

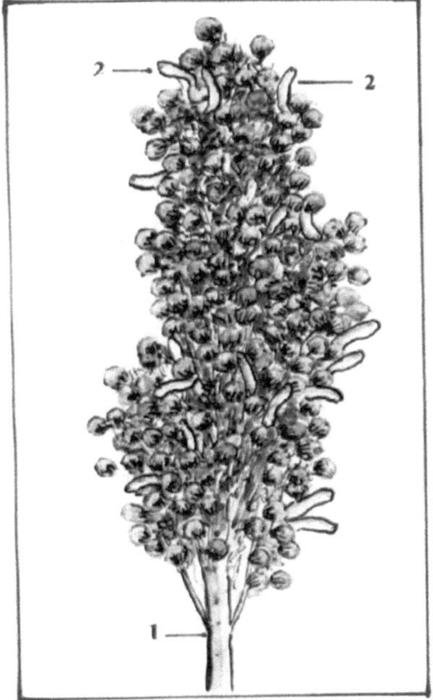

Fig. 20: Long smut of sorghum
Tolyposporium chrenbergii

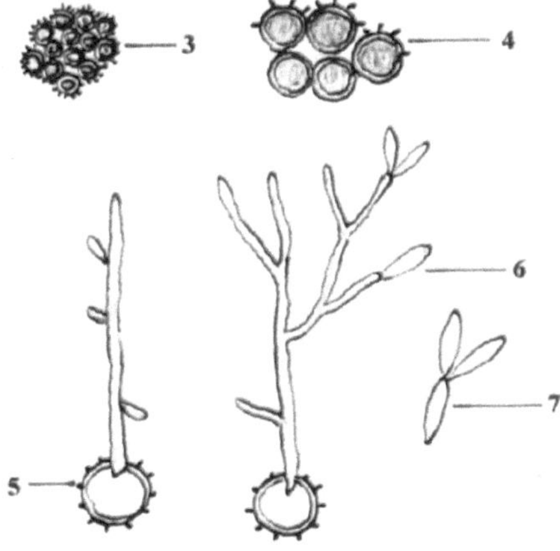

1. Smutted ear 2. Smutted grains 3. Spore ball 4. Spores on the surface of the spore ball showing echinulations 5. Germinating spore 6. Germinating spore producing sporidia 7. Budding sporidia

6. Anthracnose and red rot of sorghum

Colletotrichum graminicola

The disease occurs in all the countries, where sorghum is grown. In India, the disease is prevalent in Telungu Desam, Karnataka, Madhya Pradesh, Rajasthan, Tamil Nadu and Uttar Pradesh. The leaf spot phase is more common in India.

Symptoms. The pathogen causes a leaf spot disease, **'anthracnose'** and a stalk rot **'red rot'**. The symptoms on the leaves appear as small, sesame-seed-shaped or elliptical spots, measuring 2.0 - 4.0 x 1.0 - 2.0 mm. in size. The spots have straw colored center and wide purple or reddish-brown, irregular margin. The center of the spots appears slightly depressed. Usually the lower leaves are subjected to attack by the pathogen. When the incidence is severe, many lesions may develop close together and coalesce, covering large areas of the leaf, eventually killing the leaf. Few or numerous pinhead-like, black dots are seen on the surface of the center of the lesions, representing the fruiting bodies (acervuli) of the pathogen. Often, lesions are formed on the midribs of the leaves also, which appear as elongate-elliptical, red or purple lesions, on which the black acervuli can be seen clearly. If the disease attacks young plants, all the leaves may dry and the plants also die.

The red rot phase may occur in the stalks or in the inflorescence or in both, as a result small, circular cankers may develop on the outside, particularly in the inflorescence. Infected stems when split open, show discoloration, which may be continuous over a large area or most often discontinuous, giving the stem a marbled appearance. Nodal tissues are rarely discolored. Stalk rot is usually preceded by leaf anthracnose **(Fig.21)**.

The causal organism. The mycelium is hyaline, septate, inter- and intracellular and is localized around the point of infection. Stromata develop below the epidermal layer and numerous, short, hyaline, unbranched conidiophores arise by rupturing the epidermis. These are the fruiting bodies (acervuli) of the pathogen. The acervulus appears as an inverted saucer. The conidiophores, arranged like palisade cells, produce conidia one after another in succession. The conidia are single-celled, hyaline, thin-walled, slightly curved and measure 21.0 - 32.0 x 3.0 - 7.0μ in size. Intermingled with the conidiophores and along the margin of the acervulus are seen long, rigid, pointed, black, septate, setae that measure 175.0 x 4.0 - 6.0μ in size. The setae help to rupture the host epidermis. The conidia germinate by means of a germ tube from either ends and form an appressorium at the tip **(Fig.21)**.

Mode of survival, spread and epidemiology. The mycelium and conidia can survive in the crop debris for a long time and initiate fresh infection. The pathogen is also carried through seeds. Weed hosts, such as *Dactyloctenium aegyptum, Digitaria sanguinalis* and *Echinochloa colonum* serve as alternate hosts of this pathogen and may provide the primary inoculum. Secondary spread of the disease is through wind-borne conidia.

A temperature of 28° - 30°C, high relative humidity and intermittent rainfall favor the occurrence and spread of the disease.

Disease management

Agronomic practices (i) The disease usually attacks the lower, older leaves. Severely affected leaves may be cut and destroyed to minimize the inoculum (ii) Removal and destruction of alternate hosts in and around the fields, field sanitation, selection of healthy seeds etc. help to prevent the disease occurrence.

Chemical control. Foliar spraying with zineb - 500 gm. or mancozeb - 500 gm. in 250 lit. of water per acre, 2 or 3 times, at 15 days intervals, commencing from 45 days after sowing controls the disease.

Seed treatment. Seed dressing with captan or thiram at 4.0 gm./ kg. of seed, 24 hours prior to sowing protects the crop from seed-borne infection.

7. Leaf blight of sorghum

Exerohilum turcicum / Helminthosporium turcicum

(*Trichometasphaeria turcica*)

The disease is found in Africa, Argentina, Italy, Rumania, America and India. In India, it has been reported from Telungu Desam, Karnataka, Madhya Pradesh, Rajasthan and Tamil Nadu.

Symptoms. The pathogen causes **'seed rot'** and **'seedling blight'** of sorghum. On older plants, especially on the lower leaves, typical symptoms are produced. The symptoms appear as long, elliptical necrotic lesions, with straw colored center and dark-brownish margin. The lesions may be several centimeters long and 1.0 - 2.0 cm. wide. Under humid conditions, a faint gray bloom is seen on the lesions. This bloom consists of the conidiophores and conidia. Many lesions may develop and coalesce, as a result large areas of the leaf tissues are destroyed, giving the crop a burnt or blasted appearance **(Fig.22)**.

The causal organism. The mycelium of the fungus is inter- and intracellular and is localized around the point of infection. Before fructifications are formed, the fungus forms stromata below the substomatal cavities. From the stromata, conidiophores arise and emerge through the stomata in clusters of 3 - 6. They are simple, olive-brown in color, septate, slightly thicker than the hyphae and measure 150.0 - 250.0 x 7.0 - 9.0μ in size. From the tips of the conidiophores, conidia are produced acrogenously. After one spore is formed, the tip of the conidiophore continues to grow, pushing aside the conidia already formed and produces another conidia. So, at each of the point where a spore is formed, there is a slight bend in the conidiophore and at the point from where a matured conidium had fallen down, a scar is also seen. In this manner, several spores may be produced from each conidiophore. The conidia are long, cylindrical, thick-walled, slightly curved, olive-brown in color, 3 - 8 septate and measure 45 - 132 x 15 - 28μ in size. They germinate by producing a germ tube from either of the terminal cells **(Fig.22)**.

Very rarely, the fungus produces the ascigerous stage. On the surface of the lesions, black, ellipsoidal to globose ascocarps measuring 359 - 721 x 345 - 497μ in size are produced. The asci are clavate, bitunicate and are 176 - 249 x 24 - 31μ in size. The ascospores, which are 8 in number in each of the ascus are hyaline, straight or slightly curved, typically 3-septate and are 42 - 78 x 13 - 17μ in size.

Mode of survival, spread and epidemiology. The mycelium and conidia of the pathogen can survive in the soil in the infected plant debris for a long period of time. The pathogen is also seed-borne and this seed-borne inoculum causes seed rot and seedling blight. Secondary spread is mostly through wind-borne conidia.

Moderate temperature, high atmospheric moisture and intermittent rainfall favor the disease occurrence and spread. The pathogen attacks maize and Johnson grass.

Disease management

Agronomic practices (i) Disease-free seeds should be used for sowing (ii) Field sanitation helps to reduce the inoculum potential in the soil.

Chemical treatment. Foliar spraying with mancozeb - 500 gm. or captafol - 325 gm. in 250 lit. of water per acre controls the disease effectively.

Seed treatment. Seed dressing with captan or thiram at 4 gm./ kg. of seeds helps to eliminate the seed-borne inoculum.

8. Gray leaf spot or rectangular leaf spot of sorghum

Cercospora sorghi

The disease is prevalent in China, Italy, Africa and India. In India, it was first reported from Tamil Nadu in 1931. Now, the disease is very common in the states of Telungu Desam, Karnataka, Madhya Pradesh, Rajasthan and Uttar Pradesh besides Tamil Nadu.

Symptoms. The symptoms begin as small leaf spots, which enlarge to become roughly rectangular lesions that may reach 5.0 - 15.0 mm. long and 2.0 - 5.0 mm. wide. The lesions are typically dark-red to purplish, with somewhat light-colored or straw-colored center and are slightly sunken. The lesions occur on the leaf blades and leaf sheaths and are mostly isolated. They are mostly confined between the veins. In case of severe incidence, the lesions become contiguous to form long stripes. Banding may occur within the lesions. Under humid conditions, a grayish-white bloom is produced on the lesions, when the fungus produces conidiophores and conidia. The older leaves are more vulnerable to attack by the disease.

The causal organism. The mycelium of the fungus is olive-brown in color, septate, inter- and intracellular. Before fructifications are formed, the mycelium forms stromata below the substomatal cavities, from which clusters of conidiophores, numbering 3 - 16 emerge through the stomata from both surfaces of the leaves. The conidiophores are light-brown, simple or rarely branched and have 0 - 5 septa. They are produced one after the other acrogenously from the tips of the conidiophores. The conidia are hyaline, thin-walled, 1 - 12 septate, long and filiform, with the base slightly broader than the tip and measure 30.0 - 132.0 x 3.0 - 8.0µ in size. Matured conidia dislodge and germinate readily by producing germ tubes from one or more cells **(Fig.23)**.

Mode of survival, spread and epidemiology. The mycelium and conidia can survive in the soil in infected plant debris for up to 5 months. The pathogen attacks the grass host *Amphilophis pertusa*, which may provide primary inoculum. Secondary infection takes place by means of air-borne conidia.

Cool, moist weather, high relative humidity of about 90 % and intermittent rainfall are favorable for the incidence of the disease.

Fig. 21: Anthracnose of sorghum
Colletotrichum graminicola

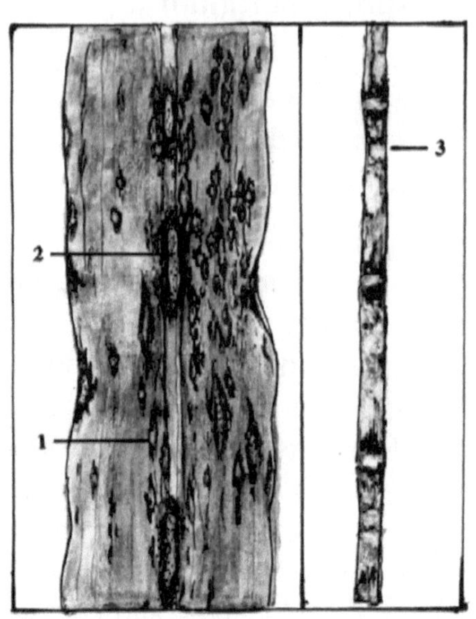

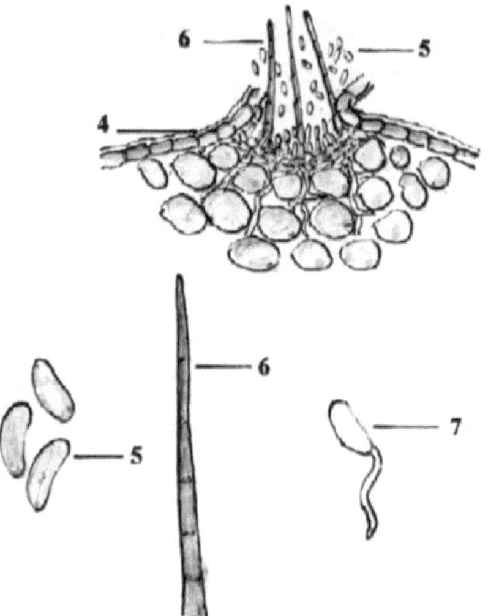

1. Leaf spots on the leaf 2. Lesions on the leaf mid-rib showing acervuli 3. Stalk rot and the marbled appearance in the inner tissues of the stem 4. Acervuli 5. Conidia 6. Seta 7. Germinating conidium

Fig. 22: Leaf blight of sorghum
Helminthosporium turcicum

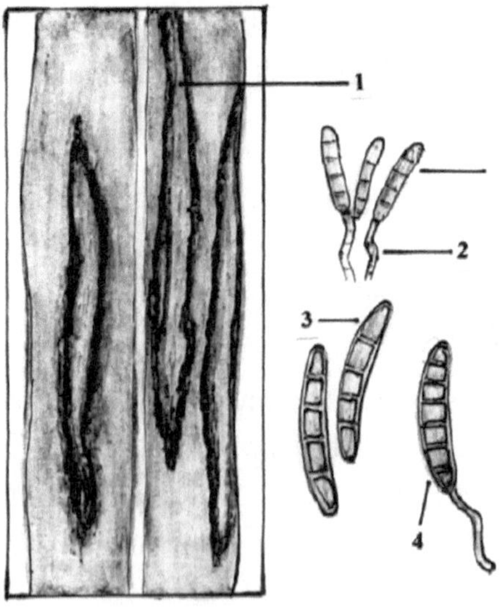

1. Leaf spots on the leaf 2. Conidiophore 3. Conidium 4. Germinating conidium

Fig. 23: Rectangular leaf spot of sorghum-*Cercospora sorghi*

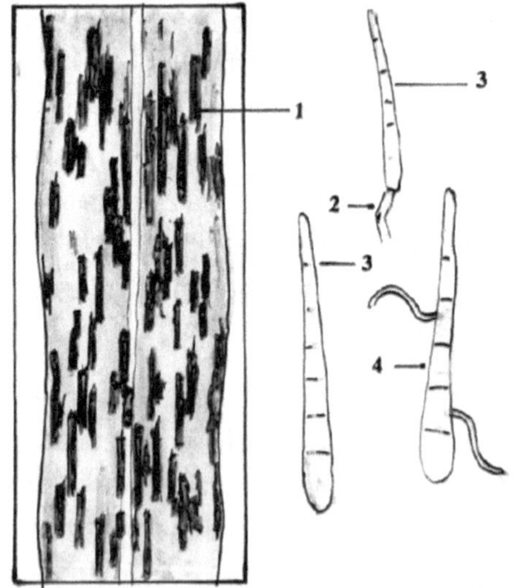

1. Leaf spots on the leaf 2. Conidiophore 3. Conidium 4. Germinating conidium

Disease management

Agronomic practices. The measures suggested for the control of other sorghum leaf diseases may be followed.

Chemical control. Foliar spraying with ziram - 500 gm. or Cuman L - 500 ml. in 250 lit. of water per acre gives adequate control of the disease.

9. Head moulds or grain moulds

'Head moulds' occur in all sorghum growing countries of the world. In India, it is quite common in all the states, where sorghum is cultivated when there is rain during the flowering and grain filling stages.

Symptoms. The symptoms appear as mycelial growth and fructifications of the organisms on the surface of the inflorescence and grains, which may be of different colors and textures depending upon the causal organisms. Sometimes, the fungal growth may be superficial, while in other cases the infection may go deeper into the tissues.

The causal organisms. More than 30 Genera of fungi have been found to be associated with grain moulds of sorghum. However, the most frequently occurring Genera are *Fusarium, Curvularia, Aspergillus, Penicillium, Alternaria* and *Phoma.* The most commonly occurring *Fusarium* species are *Fusarium moniliforme* and *F. semitectum.* Grains infected with these fungi develop a fluffy white or pinkish coloration. *Curvularia lunata* is also frequently found and this fungus produces black coloration. The infection is deep seated within the grains and the attack may result in poor germination. Species of *Aspergillus* produce green or black mouldy growth on the surface of grains. Species of *Penicillium* also produce green or blue mouldy growth. All the cultivated varieties of sorghum are vulnerable to attack by the mould fungi.

Mode of spread and epidemiology. Most of the mould producing organisms are either saprophytes or semi-parasites and are found everywhere throughout the year and their spores are present in the atmosphere all through the year. These air-borne spores can cause infection whenever the host is present under conditions favorable for their existence.

Prolonged wet weather during flowering and grain filling stages and moderate temperature favor mould development.

Disease management

Agronomic practices. The time of sowing should be so adjusted, that the flowering and grain filling stages may not coincide with the monsoon period.

Chemical control. Spraying with mancozeb - 500 gm. or captafol - 350 gm. or captan - 350 gm. in 250 lit. of water per acre, once at the time of earhead emergence, followed by a second spraying, 7 days later will control mould attack. The spraying should be taken up, when there is a break in the rainfall.

Minor diseases. Besides the above diseases discussed in detail, sorghum is subjected to several other diseases, which may not be of much economical importance normally. However, these diseases may assume serious proportions and cause serious losses, if conditions become favorable for the pathogens. 'Charcoal rot' caused by *Macrophomina phaseolina,* causes rotting of the lower stem portion of plants, leading to lodging and death of infected plants; 'Crazy top downy mildew' caused by *Sclerophthora macrospora* affects young plants, as well as old plants and causes a downy mildew disease. In this downy mildew disease, no leaf splitting occurs; 'Rust disease' caused by *Puccinia purpurea,* a heteroecious, macrocyclic rust, infects the leaves; 'Sugary disease' or 'Ergot' caused by *Claviceps (Sphacelia) sorghi* affects the florets; 'Zonate leaf spot' caused by *Gloeocercospora sorghi* produces characteristic zonate leaf spot lesions on the leaves **(Fig.24)**; 'Rough leaf spot' caused by *Ascochyta sorghina* produces lesions on the leaves; 'Oval leaf spot' caused by *Ramulispora sorghicola* produces roughly circular lesions on the leaves **(Fig.25)**; 'Tar spot' caused by *Phyllachora sorghi* produces black, roughly circular, raised spots on the leaves **(Fig.26)**; 'Head blight' caused by *Fusarium moniliforme* invades the tissues of the inflorescence; 'Bacterial leaf streak' caused by *Xanthomonas holcicola* and 'Bacterial leaf stripe' caused by *Pseudomonas andropogoni* affect the leaves **(Fig.27)**; Virus diseases, such as 'Maize dwarf mosaic', 'Sugarcane mosaic' and 'Yellow sorghum stunt' also infect sorghum; 'Witch weed' *(Striga* sp.*),* a phanerogamic parasite, parasitizes sorghum plants.

Pearl millet, Bajra or Cumbu *(Pennisetum typhoides)*

1. Green ear disease or downy mildew of pearl millet

Sclerospora graminicola

'Green ear' disease is commonly found to affect pearl millet and occurs in several countries, including India, Iran, Israel, China, Fiji, Japan, the United States of America and many other European and African countries. However, so far the disease has not been reported from South America and Australia. The disease, which was reported as a sporadic one in India during 1910, has become very serious and occurs regularly in many parts of the country, causing heavy losses, especially in the high yielding hybrid varieties.

The loss caused by the disease in India is estimated to be up to 27 - 30 per cent. Besides bajra, the pathogen attacks maize, sugarcane and a number of other hosts belonging to Family - Gramineae.

Symptoms. The symptoms appear in two stages: **(i)** the **'downy mildew stage' (ii)** the **'green ear stage'.** The downy mildew stage occurs quite early in the growth stage, even when the plants are very young. At this stage, the leaves are predominantly affected.and the diseased plants remain stunted, become pale yellow and present a sickly appearance. The leaves of such plants show chlorotic streaks on the upper surface. On the under surface of these streaks, a fine, cottony, downy growth of the fungus is seen. Soon afterwards, the chlorotic areas turn brown and in advanced stages, shredding of the leaves along the veins occurs. Very often, the nodal buds are stimulated to develop into lateral shoots, giving an abnormal bunchy appearance to the plant.

The second, green ear stage is the main characteristic symptom that is produced in the inflorescence. The visual deformities of the ear are characterized by the transformation of the floral organs into twisted, leafy structures. This gives the ear an appearance of green, leafy mass, hence it is known as the **'green ear disease'**. In some cases, the entire length of the ear is converted into a leafy mass and in some other cases, only the lower portion is converted into a leafy mass, while the upper portion bears grains normally. Yet in some other cases, the development of the inflorescence is altogether suppressed and only a bunch of small, leafy structures are formed.

The bristles of the spikelets become hypertrophied and variously contorted. They may grow to a length of about 2.5 cm. and are flattened, round or angular, the upper part being often twisted or screw-like. Increase in the number of spikelets and florets in the spikelets are also common. The stamens become leaf-like and elongated or stumpy. The anther may also become elongated and contorted. The pistil rarely develops, but becomes leafy outgrowth. Generally, the downy mildew stage on the leaves is not so conspicuous as the green ear stage on the inflorescence. However, in susceptible varieties, both the stages are commonly seen. While the severely affected young plants dry and die within a short time, the grown up plants may not die, but may serve as source of inoculum for secondary spread. In older plants, the newly emerging young leaves are more severely affected and are more completely whitened **(Fig.29).**

Fig. 24: Zonate leaf spot of sorghum
Gloeocercospora sorghi

Fig. 25: Oval leaf spot of sorghum
Ramulispora sorghicola

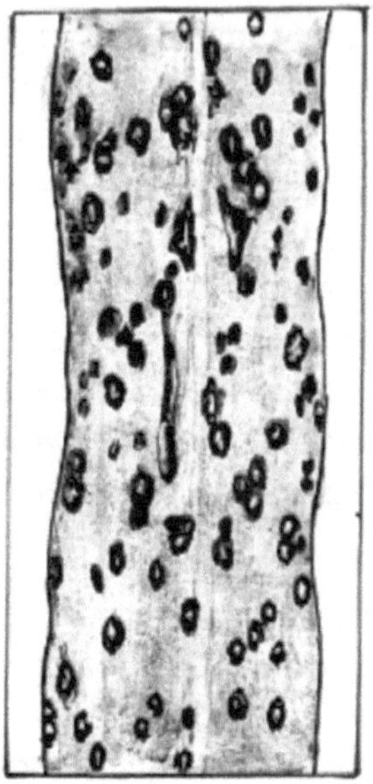

Fig. 26: Tar spot of sorghum
Phyllachora sorghi

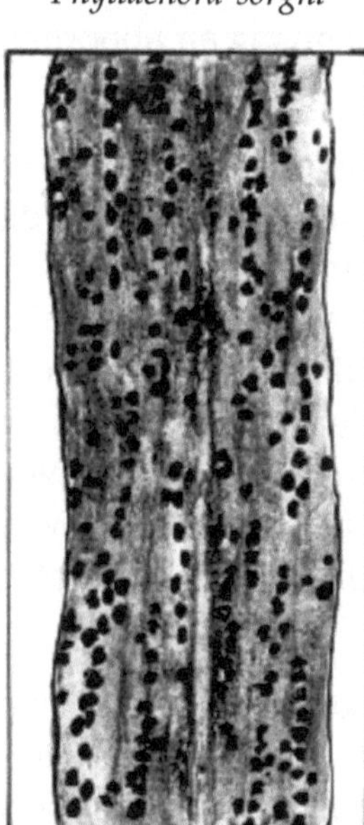

Fig. 28: Leaf blast of pearl millet
Pyricularia setariae

Fig. 27: Bacterial leaf stripe of sorghum
Pseudomonas andropogoni

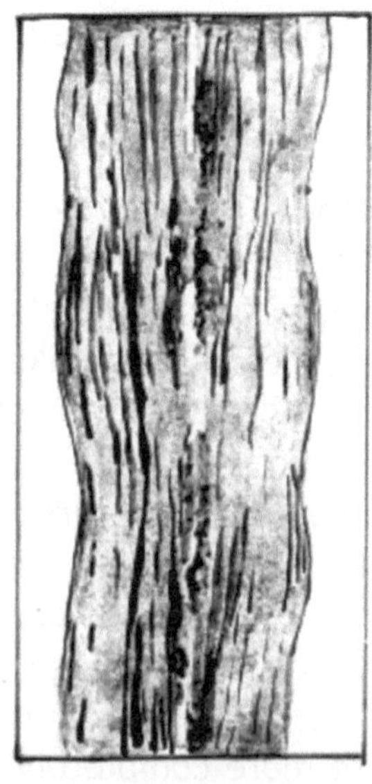

The causal organism. *Sclerospora graminicola*, the causal organism of this disease is an obligate parasite. The mycelium is found in all parts of the systemically affected plants. The hyphae are non-septate (coenocytic), intercellular, with small, bulbous haustoria and branch freely in-between the cells. In fully infected leaves, the hyphae collect chiefly between the cells of the mesophyll adjoining the bundles and also in the inner layers between them. Before formation of sporangiophores, tufts of hyphae reach the air space beneath the stomata, which are arranged in parallel rows, one on either side of the vein. These spore-producing structures absorb more food from the living cells of the host, as a result the cells collapse and die leading to shredding of leaves along the veins. Prior to this, the chloroplasts of the leaf cells are wholly or partially destroyed, giving a chlorotic appearance to the leaves, depicting the downy mildew stage of the fungus.

The sporangiophores emerge in clusters through the stomata. The sporangiophores are specialized hyphae, which are broad and short. They measure about 100 μ in length and 12-15μ in width. They are unbranched in the lower part, but usually give rise to a few thick, short branches at the top. The sporangiophores expand to a diameter of about 30μ towards the tip, where branching occurs. The final branches or the sterigmata bear sporangia, which are formed by swelling of the tip of the sterigmata and cut off by a septum. The sporangia are hyaline, broadly elliptical, papillate at the free end, with a smooth wall and measure 13-34 x 12-23μ. Matured sporangia fall off soon and germinate in the presence of a film of water by producing zoospores. Each sporangium may release 3-13 zoospores. The zoospores are irregularly reniform, biflagellate and swim about for some time before coming to rest. The resting zoospores measure 9.0-12.0μ in diameter. They germinate by a germ tube that exhibits a positive chemotrophic response to roots of bajra.

The sexual stage of the fungus is more common and more conspicuous. The oospores develop within the host tissues, mostly in leaves and malformed spikelets. The oospore consists of a thin outer layer (exospore) and a thicker inner layer (endospore). The original oogonial wall shrinks to touch the new oosporial wall at many points, giving the entire oospore an elliptical, angular or irregular shape, but the oospore proper is usually perfectly spherical. The matured oospores measure 34-52μ in diameter. The oospores do not germinate readily, but have a prolonged resting period. They germinate after the resting period by means of 1-4 germ tubes **(Fig.29)**.

Fig. 29: Green ear disease of pearl millet-*Scierospora graminicola*

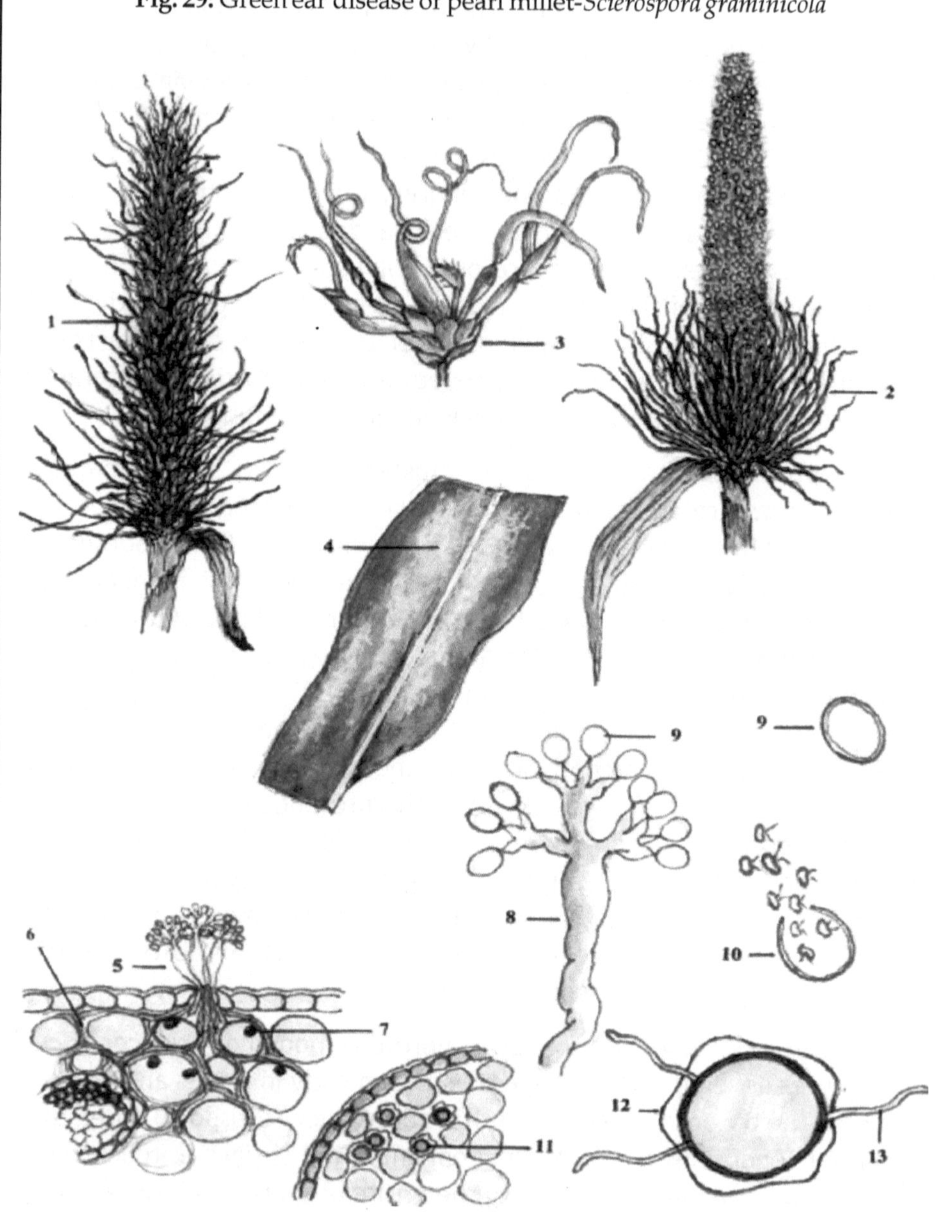

1. Entire ear converted into a leafy mass 2. Lower part of the ear alone converted into a leafy mass 3. Hypertrophied floret 4. Downy growth of the fungus on the under surfacre of leaf 5. Sporangiophores emerging through the stomata 6. Intercellular mycelium 7. Haustoria 8. Sporangiophore with sporangia 9. Sporangium 10. Germinating sporangium releasing zoospores 11. Oospores inside the malformed part of floret 12. Germinating oospore 13. Germ tube.

Mode of survival, spread and epidemiology. The disease is primarily soil-borne. The oospores, abundantly present in the diseased leaves and malformed spikelets, which fall down to the ground, perennate in the soil debris during winter. They remain viable in the soil for about 10 years and give rise to primary infection however, one year old oospores cause more infection than fresh or more than one year old oospores. Further, the oospores are found to remain viable even after passing through the alimentary canal of cattle or even after composting the infected plant materials. Contaminated seeds also serve as an important source of inoculum. Oospores may be carried with the seed during harvesting and threshing. Oospores carried along with seed and oospores present in the soil play the main role in establishment and spread of the disease in the field. The mycelium of the fungus may also be present inside the seed and may cause disease. At least two physiological races of the pathogen are known to exist.

When conditions are favorable, the oospores present in the soil germinate and cause infection of the underground parts of the seedling and the infection spreads systemically upwards. The susceptibility of the underground part of the stem and of the roots decreases with advancing age of the host. Abundant air supply in the soil, low soil moisture content and a temperature between 20° - 25°C favor germination of oospores. Dry soils are more conducive for infection.

Secondary inoculum of the fungus consists of sporangia, which are disseminated by wind, water and insects. Formation of sporangia is favored by a temperature range of 15°-25°C. Temperatures of 12.5°-29.0°C and a saturated atmosphere, with a film of water on the leaves and presence of dew are conducive for the germination of sporangia and liberation of zoospores. Downy mildew phase of the disease affects young seedlings up to the age of 15 days more rapidly than older plants.

Disease management

Agronomic practices (i) Increasing the seed rate slightly and then roguing the diseased plants within a month of sowing helps to minimize the disease incidence (ii) In the grown-up crop, periodical roguing of plants showing symptoms of the disease should be followed (iii) Deep ploughing during the summer months helps to bury the oospores deep into the soil (iv) Infected leaves and green ears should not be fed to cattle or used for preparing compost, as the oospores, which remain viable in the faeces of cattle may reach the FYM or compost (v) Crop rotation can be advocated in severely infected fields (vi) In irrigated areas, transplanting with disease-free and healthy seedlings can be taken up (vii) Good seeds, obtained from disease free crops should be used for sowing.

Seed treatment. Seed treatment with metalaxyl (Ridomil) - 4 gm. or thiram - 4 gm./ kg. of seed gives protection against seed-borne infection.

Chemical treatment. Spraying the crop with mancozeb - 400 gm. or captafol - 250 gm. or metalaxyl - 200 gm.in 200 liters of water per acre, 20 days after sowing gives good control of the disease. Foliar application with metalaxyl - 200 gm. + mancozeb - 400 gm. is found to be more effective in controlling the disease, besides other foliar diseases.

Resistant varieties. Resistant varieties, such as Co.6, Co.7, X.5 and WCC.75 may be grown in the place of susceptible varieties. In the recent past, many varieties, such as ICMS.7703, ICTP.8203, ICMV.155, ICMV.221, RCB-IC.9 and hybrids, such as ICMH.356, ICMH.423, ICMH.501, HHB.68, Pusa 23, Pusa 322 and RHB.30 have been released in India as downy mildew resistant varieties.

2. Sugary disease or 'Ergot' of pearl millet

Claviceps fusiformis (C. microcephala)

'Ergot' of pearl millet is considered as a disease of International importance. The disease has been reported from India, Pakistan and many African countries. In India, the disease was first reported from south India. It appeared in an epiphytotic form in Maharashtra in 1956. Severe epiphytotics of ergot were reported in several states of India, such as Delhi, Rajasthan, Maharashtra, Karnataka and Tamil Nadu in 1967. Subsequently, the disease has been reported from other bajra growing states also. The disease has assumed great importance recently due to its occurrence on new, high-yielding bajra hybrids, such as HB.1 and HB.2. In Tamil Nadu, the disease occurs on a large scale in several districts, such as Coimbatore, Trichirapalli, South Arcot, Salem and Tirunelveli.

The disease gained all the more importance because of the presence of alkaloids 'ergotoxin', 'ergotamine' and 'ergometrine' in the matured sclerotia, which are poisonous to human beings and cattle. Assimilation of the toxins in the blood stream of animals and human beings causes the disease 'Ergotism'.

Symptoms. The symptoms of the disease are noticed only at the time of flowering. The pathogen infects the florets at the protogyny stage and develops in the ovaries. As a result of infection, the infected spikelets start exuding copious creamy, pinkish or reddish, sweet, sticky fluid called **'honeydew'**, which carries conidia produced by the pathogen. Drops of the honeydew containing the spores fall down the inflorescence on to the upper leaves making them sticky and often pollen and anther sacs adhere to the honeydew. Subsequently, from the infected florets long, dark-colored, hard structures, the sclerotia or ergot develop. The sclerotia, which appear creamy in the early stages of their formation, turn black as they mature. The sclerotia measure 3.6 - 6.1 x 1.3 - 1.8 mm. in size. In case of severe incidence, all the florets are infected, with the large, black sclerotia sticking out over the entire surface of the earhead **(Fig.30)**.

The causal organism. Initial infection takes place through air-borne ascospores. The ascospore falling on the stigma of the floret, germinates by producing a germ tube, penetrates the style, grows through it and reaches the ovary or sometimes, the germ tube penetrates the ovary wall directly and enters the ovary before fertilization has taken place. Infection by the pathogen is not possible once fertilization is completed. The mycelium developing from the germ tube, ramifies into the tissues and occupies the ovary completely. The ovary is destroyed and occupied by the fungal mycelium, which is hyaline and septate. As a result of infection of the ovary, honeydew is secreted from the floret. Soon, the mycelium starts growing outside the ovary also. From this mycelium, conidiophores arise. The conidiophores are slightly different from the ordinary hyphae and are septate, with few branches. From the tips of the branches, conidia are produced in large numbers and are mixed with the honeydew secretion. Within the anthesis period, 2 - 3 generations of conidia are produced. The conidia are hyaline, single-celled, and broadly falcate and measure 13.0 - 18.0 x 3.0 - 4.0µ in size. They may remain viable for 13 months. However, they germinate readily by producing germ tubes, which bear secondary and sometimes tertiary conidia. The honeydew attracts various insects, which act as carriers of conidia along with the honeydew, thus causing secondary infection of fresh ovaries in the same season. This asexual type of life cycle goes on till the ovary is completely exhausted and is replaced by repeated intertwining hyphal mass, which becomes a compact, hard structure, that is the **'sclerotium'**. The sclerotia, which are soft and light-colored in the beginning, soon become hard, black or violet and slightly curved like a horn. The interior consists of simple pseudoparenchymatous cells. The horny sclerotium grows in-between the glumes, attains a large size and is quite conspicuous because of its size, color and shape.

Matured sclerotia fall down on the soil or are carried with the seed, only to return to the soil at the time of sowing. In the ensuing season, when the crop starts growing, the sclerotia also germinate to produce the ascigerous or perfect stage of the fungus. Before germination, a number of swellings appear on the surface of the sclerotium. Through these swellings the internal loose mass of hyphae from the sclerotium grows out into vertical stromatic stalks or stipes. The fully developed stipe, which is 2.5 cm. or more in length develops a spherical, orange or pink-colored, small head or the stroma proper at its tip. This head bears numerous spine-like outgrowths representing the neck and ostiole of perithecia embedded in the stroma. The flask-shaped perithecia are arranged side by side in the peripheral tissue of the stromatic head. Each perithecium consists of a number of cylindrical or clavate asci, which project upward converging towards the ostiole. The asci are intermingled with numerous paraphyses. Each ascus contains a bundle of eight, slender, filiform, hyaline, septate ascospores, measuring 18.0 - 23.0 x 1.0μ in size. The cap of the ascus bursts open and the ascospores are ejected out with force above the ground level. The ascospores are caught in the air current and are disseminated **(Fig.30)**.

Mode of survival, spread and epidemiology. The disease is carried over from one season to the next through sclerotia present in the soil or through contaminated seeds. The sclerotia remain viable in the soil for 6 - 8 months. Under favorable environmental conditions, the sclerotia take about 30 - 45 days to germinate. This coincides with the time the host plants come to flowering and receive the air-borne ascospores, resulting in primary infection. Under favorable weather conditions, the fungus produces large number of conidia, which are carried in the honeydew and the disease spreads very rapidly. The conidia along with the honeydew are picked up by various insects or dispersed by rain.

The pathogen infects several other hosts, such as *Pennisetum alopecuros, P. hohenackeri, P. polystachyon, P. purpureum, P. ruppelii, Cenchrus ciliaris, C. setigerus* and *Panicum miliaceum*. The conidia produced on these hosts also serve to provide initial inoculum that can infect pearl millet crop.

High relative humidity, ranging from 85 - 95 % during the morning hours and 60 - 90 % during the evening hours, cloudy and damp weather, with overcast sky, low sunshine and daily showers during the flowering period are highly conducive for the occurrence and spread of the disease.

Disease management

Agronomic practices (i) The sowing time should be so adjusted that the flowering time should not coincide with the rainy season (ii) Sclerotia-free seeds should be used for sowing (iii) The field and surrounding areas should be kept clean and free from weed hosts (iv) Disease affected earheads should be removed and destroyed (v) Crop rotation, preferably with legumes helps to reduce the soil inoculum (vi) Deep summer ploughing and exposure of the sclerotia to the sun destroys the sclerotia (vii) Application of heavy doses of nitrogen increases the disease incidence.

Chemical control (i) Spraying with ziram - 500 gm. or a mixture of copper oxychloride - 625 gm. + zineb - 500 gm. in 250 lit. of water per acre, 2 - 3 times, at 5 - 7 days intervals starting from the time of earhead emergence gives satisfactory control of the disease (ii) Carbendazim - 250 gm. or benomyl - 250 gm. or Brestan - 325 gm. in 250 lit. of water per acre has also been found to be effective.

Seed treatment. The seeds are steeped in 20 % common salt solution. The sclerotia and fragments of sclerotia, which float on the salt solution are removed, the seeds thoroughly washed in clean water till the salt is completely removed, dried, seed treated with fungicide and then used for sowing.

Biological control. *Fusarium roseum, F. sambucinum* and *Dactylium fusarioides* are mycoparasites of sclerotia of this fungus and have been found to reduce the incidence of ergot.

3. Smut of pearl millet

Tolyposporium penicillariae

'Bajra smut' occurs in Africa, India, Pakistan and the United States of America. The disease was first reported in India in 1933 and is more prevalent in the states of Tamil Nadu, Telungu Desam, Gujarat, Haryana, Maharashtra, Punjab and Rajasthan. The yield loss due to this disease may be as high as 30 %.

Symptoms. Only individual grains, scattered in an ear are infected, while majority of the grains remains normal. The affected grains may be single or sometimes in groups and often confine to one side of the ear or near about the base of the ear. The smut sori are pear or oval-shaped and project beyond the glumes. They are ½ - 2 times bigger than the normal grains, and are 3.0 - 4.0 mm. long and 2.0 - 3.0 mm. broad. The top portion is broader and the apex rounded to conical in shape. A membrane made up of host tissue, which appears bright-green or dark-brown in the early stages, becoming black on maturity, covers the sorus. This tough membrane ruptures at maturity exposing masses of black spores **(Fig.31)**.

The causal organism. The pathogen is mostly confined to the sorus, the wall of which is made up of a tough membrane made up of a combination of host tissue and fungal hyphae that surrounds the powdery spore mass. The spore mass is granular and black. The spores are held together in compact balls that remain persistent even in water. The spore balls measure

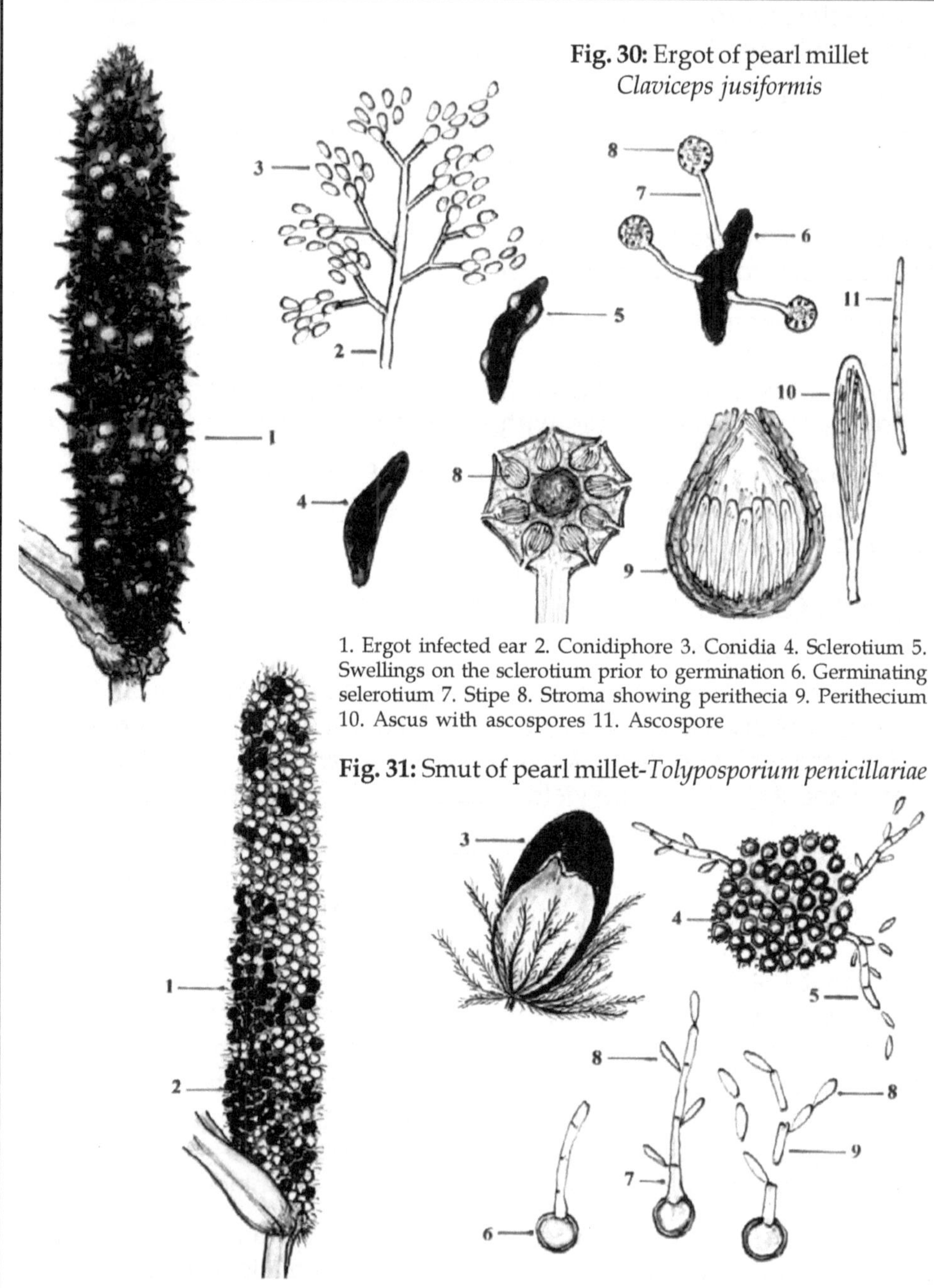

Fig. 30: Ergot of pearl millet
Claviceps jusiformis

1. Ergot infected ear 2. Conidiphore 3. Conidia 4. Sclerotium 5. Swellings on the sclerotium prior to germination 6. Germinating selerotium 7. Stipe 8. Stroma showing perithecia 9. Perithecium 10. Ascus with ascospores 11. Ascospore

Fig. 31: Smut of pearl millet-*Tolyposporium penicillariae*

1. Smutted ear 2. Smut sori 3. Smut sorus 4. Spore ball 5. Germination of spores from the spore ball 6. Germinating spore producing promycelium 7. Promycelium 8. Sporidium 9. Cells of promycelium producing sporidia by budding.

100 - 150μ in diameter and are not wind-borne. Individual spores are globose to irregular or angular, light-brown in color, 6.0 - 10.0μ in diameter and have a slightly roughened wall. They germinate, while still held in the balls and produce 4-celled promycelium. Each cell of the promycelium separates, falls down and buds out sporidia **(Fig.31)**.

Mode of survival, spread and epidemiology. The disease is primarily soil-borne. The smut balls remain in the soil and germinate in the next season. Floral infection occurs initially by air-borne sporidia produced from spores in the soil and later by spores from early infected plants. The fungus invades the ovule through the nucellus and mesophyll. Spikelets are most susceptible before the anthers and stigmas emerge. Once the flowers are pollinated, infection cannot take place. Smut sorus develops in about 15 days after successful infection. No resting period is necessary for the spores. Secondary infection of late sown crops or of ears appearing late in the season, also occurs by spores produced in the same season. The disease may also spread through seeds, which become contaminated with the spores at the time of harvesting and threshing.

Infection is higher under humid conditions, accompanied by frequent rains at the time of flowering and much less in the drier months. Successive cropping with pearl millet increases the intensity of infection.

Disease management

Agronomic practices (i) Smutted ears should be removed and destroyed by burning (ii) Disease-free seeds should be used for sowing (iii) Deep summer ploughing should be followed, so as to expose and destroy the spore balls (iv) Crop rotation practices should be adopted (v) Sowing should be so adjusted that the flowering time should not coincide with the rainy season

Chemical control. As the spore balls remain viable in the soil for a long time and successive infection of the florets is through air-borne sporidia, it is very difficult to control the disease by chemical treatments.

4. Rust of Pearl millet

Puccinia penniseti

'Rust' occurs in almost all the bajra growing countries of the world, such as the United States of America, Africa, Sudan, Kenya, Nigeria, Rhodesia, Tanganyika, Uganda, India and Pakistan. It is found all over India and causes extensive damage to both dry and irrigated crops, especially when the infection occurs at the early stages of crop growth.

Symptoms. Pearl millet rust is a heteroecious and macrocyclic rust. The pycnial and aecial stages are found in brinjal (eggplant), while the uredial and telial stages occur in pearl millet. On brinjal leaves, hypertrophied, yellowish-green, circular patches develop. These patches are concave on the upper surface, while the corresponding lower surface is convex and appear as a hemispherical bulge. Several, minute, orange-yellow dots, which are the pycnia of the fungus develop on the upper surface. On the lower surface of the patches, orange colored aecia develop. In pearl millet, the disease appears first on the lower, older leaves as typical erumpent pustules containing a reddish-brown powder consisting of uredospores. Uredosori appear on both the surfaces of the leaf, but are more numerous on the upper surface. Highly susceptible varieties develop large pustules, densely grouped on leaf blades and on leaf sheaths. The teliosori, which develop later, are darker in color than the uredosori **(Fig.32)**.

The causal organism. The pathogen is an obligate parasite. The uredosori are sub-epidermal and at maturity, the epidermis is ruptured, exposing reddish-brown, powdery uredospores. The uredospores are oval, elliptical or pyriform, yellowish-brown in color, pedicellate, very finely or sparsely echinulate, with 3 - 5 germ pores arranged equidistantly in an equatorial orbit and are 33 - 48 x 23 - 31μ in size. Marginal, flexuous or incurved, sub-hyaline paraphyses, with swollen tips are present. Teliospores are produced, either in the uredosori or in new teliosori that are formed at a later stage. Teliospores are dark-brown, 2-celled, elliptical, cylindrical or clavate, constricted at the septum, obtuse or truncate at the top, thick at the apex, smooth - walled and pedicellate, the pedicels brownish and about 20μ long and measure 40 - 80 x 22 - 27μ in size. The teliospores have two germ pores. The teliosori are divided into compartments by sterile cells.

The pycnia, which appear on the upper surface of eggplant leaves, are sub-epidermal, flask-shaped, with an ostiole that protrudes out through the cuticle and measure 100 - 140μ in diameter. The pycnia are lined with slender spermatial hyphae, which cut off numerous pycniospores (spermatia). The spermatia are exuded through the ostiole in minute droplets of honeydew. The honeydew attracts several insects that help in the dissemination of the spermatia. The spermatia are unicellular, uninucleate, haploid, hyaline, oblong or elliptic and measure 4.0 - 8.0 x 2.0 - 3.0μ in size. From the ostiole of the pycnium, several receptive hyphae are also produced. Fertilization takes place by the fusion of the nucleus of the spermatium and that of the receptive hyphae of the opposite sex. The spermatia carried by insects fall on the receptive hyphae of other pycnia and as a result of fertilization, a binucleate condition is initiated. The binucleate mycelium thus formed, passes down to the under surface of the leaf and produces aecia. The cup-shaped aecia are produced in clusters. The aecium, which appears orange in color, cuts off aeciospores in chains from mother cells, situated at the base of the aecium like palisade cells. The aeciospores are dikaryoyic, globose to angular, yellowish-orange, single-celled, thin-walled, and smooth and measure about 20μ in diameter. The aeciospores infect pearl millet and produce the uredial stage. Only the uredospores are capable of reinfecting pearl millet crop and produce the uredial and telial stages **(Fig.32)**.

Mode of survival, spread and epidemiology. The uredospores can retain their viability only for 8 days at 41°C and as such, they cannot survive the summer, when the temperature is very high.

Besides pearl millet, the pathogen is known to attack several other collateral hosts, such as *Pennisetum purpureum*, *P. spicatum* and *P. leonis.* The uredospores produced from such collateral hosts can infect bajra. The uredospores produced from bajra crop can also reinfect bajra crop. Further, the pathogen has many pycnial alternate hosts besides eggplant, such as *Solanum panduriforme*, *S. pubescens*, *S. torvum* and *S. xanthocarpum.* These hosts, which grow all round the year, produce aeciospores that can infect pearl millet crop. Secondary infection is through air-borne uredospores or aeciospores.

Moderate temperature and high humidity favor the occurrence and spread of the disease.

Two distinct physiologic races of the pathogen have been identified.

Disease management

Agronomic practices (i) Growing of brinjal crop near about bajra fields should be avoided (ii) Solanaceous weeds, which serve as alternate hosts, as well as collateral hosts around bajra fields should be removed and destroyed (iii) Closer spacing should be avoided.

Chemical control (i) Foliar spraying with wettable sulfur - 1000 gm. or mancozeb - 500 gm. or captafol - 325 gm. or carboxin - 250 gm. in 250 lit. of water per acre is effective in controlling the disease.

Resistant varieties. Varieties, such as IP.537-B, P.159, P.1564, P.1577, P.1581, P.2880, P.2890 and P.2084-1 are resistant to this disease.

Biological control. *Trichoderma koningii*, *Chaetomium globosum*, *Aspergillus japonicus* and *Fusarium oxysporum* are mycoparasites of this fungus.

Fig. 32: Rust of pearl millet
Puccinia penniseti

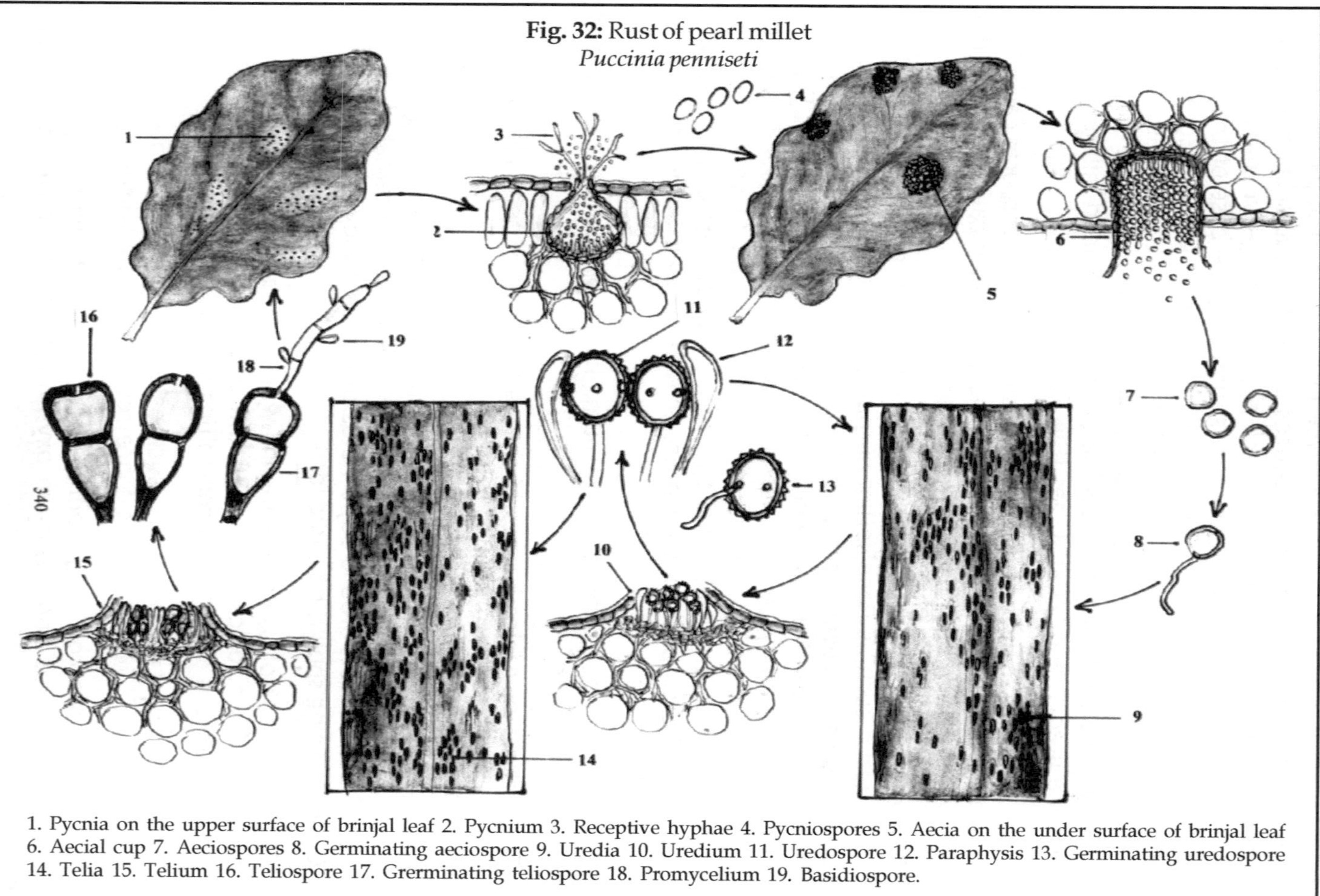

1. Pycnia on the upper surface of brinjal leaf 2. Pycnium 3. Receptive hyphae 4. Pycniospores 5. Aecia on the under surface of brinjal leaf 6. Aecial cup 7. Aeciospores 8. Germinating aeciospore 9. Uredia 10. Uredium 11. Uredospore 12. Paraphysis 13. Germinating uredospore 14. Telia 15. Telium 16. Teliospore 17. Grerminating teliospore 18. Promycelium 19. Basidiospore.

Diseases of minor importance. Besides the diseases discussed above in detail, bajra is affected by many other diseases, which are considered to be economically less important and occur only sporadically. 'Leaf blast' caused by *Pyricularia setariae* produces roughly diamond-shaped to circular spots on the leaves **(Fig.28)**; 'Top rot', 'Twisted top' or 'Pokah Boeng' caused by *Gibberella fugikuroi (=Fusarium moniliforme)* is characterized by deformed, twisted and discolored leaves near the top of the plant; 'Zonate leaf spot' caused by *Gloeocercospora sorghi* produces roughly circular lesions, with alternating roughly concentric bands of straw and brown coloration on the leaf blades; 'Leaf spot' caused by *Curvularia penniseti* var. *poonensis* produces oval leaf spots on the leaves and also causes grain discoloration; 'Bacterial leaf spot' caused by *Xanthomonas penniseti* produces water-soaked, linear, reddish-brown spots on the leaves; 'Leaf blotch' caused by *Xanthomonas annamalaiensis* produces narrow, minute, reddish-brown streaks on the leaves.

Maize or Corn *(Zea mays)*

1. Brownspot of maize

Physoderma zea maydis

'Brownspot' of maize or corn was first reported in India in 1910. Subsequently, it has been reported from other countries of the world, such as China, Japan and the United States of America. All cultivated varieties of maize are found to be susceptible to the disease. It is a sporadic and minor disease, but may become severe, when environmental conditions are favorable for disease occurrence.

Symptoms. The disease attacks the leaf blades, leaf sheaths and culms of young plants. The initial symptoms appear as small, slightly bleached or yellowish spots, about 1.0 mm. in diameter. These spots become darker and finally turn reddish-brown, with a lighter margin. Numerous spots may develop in a small area giving the leaf blade or sheath a rusty-brown appearance. The spots may coalesce and form larger spots. Spots on the mid rib and leaf sheath are irregular in shape and larger than the spots on the leaf lamina. The browning of the host tissue is due to necrosis of the affected cells and accumulation of large number of spores of the fungus. Eventually, the affected leaves dry and die prematurely. The disease attacks the lower portions of the plants more than the upper portions. On the culms, spots are produced mostly on the nodal region, as a result the plants become weak and may lodge.

The causal organism. The pathogen causing this disease is an obligate parasite. The older lesions contain a large number of thick-walled resting spores or resting sporangia. They are smooth, brown and flattened on one side, where a cap-like lid is present. The resting sporangia survive in the diseased plant debris or in the soil and carry over the disease from one season to the next season. Under favorable conditions, the resting sporangia germinate by opening the lid-like structure and a thin-walled vesicle comes out through the opening. The vesicle ruptures at the apex and liberates uniflagellate zoospores. The zoospores, when they come in contact with the host leaves, germinate and infect the host cells and form a slipper-shaped structure called **'apophysis'**, which is anchored to the host surface by means of rhizoids. From the apophysis, secondary sporangia develop, which on maturity produce about 300 secondary zoospores. These secondary zoospores function as planogametes and fuse in pairs and become zygotes. From the zygote, infection hypha arises and penetrates the host epidermal cell and the mycelium becomes endobiotic. Inside the host cell, the diploid, intramatrical mycelium gives rise to more, elongated cells called **'turbinate cells'**. These are directly converted into resting sporangia or produce special hyphae called **'rhizomycelium'**, which bears terminal resting spore. When sporangial formation is complete, the mycelium disappears entirely and in the infected host cells only the thick-walled resting sporangia remain.

Mode of survival, spread and epidemiology. The resting sporangia can survive in the crop remains and debris for up to 4 years in the field. Primary infection occurs through the soil-borne resting sporangia and secondary dissemination is by spore dust carried by wind. The disease is common in low-lying and ill-drained fields. High temperatures of 28° - 29°C and abundant moisture during the early phase of crop growth are favorable for the occurrence of the disease.

Disease management

Agronomic practices. Field sanitation and crop rotation reduce the intensity of the disease. No chemical control measures have been suggested to control the disease.

2. Corn smut

Ustilago maydis

The disease is prevalent in most of the maize growing countries of the world, except Australia and New Zealand. It is one of the worst diseases of corn in the United States of America. In India, it occurs in the Himalayas, the Kashmir valley and West Bengal.

Symptoms. The fungus produces galls on almost all aboveground parts of the maize plant, such as ears, axillary buds, tassels, stalks and sometimes on the leaves. The initial symptoms appear as galls of different sizes. The galls enlarge and may reach 10.0 cm. or more in diameter. When galls are formed on the cob, extensive damage is caused. Mostly galls are found, wherever embryonic tissues are present. The mycelium developing between the thin-walled cells of vigorously growing embryonic tissues, induces hyperplasia and hypertrophy and excessive development of phloem elements of the vascular bundles. The galls, as they enlarge appear almost white, while the inner tissues become dark due to spore formation. On maturity of the spores inside the gall, the outer membrane, which is actually the host epidermis ruptures and black spore mass is exposed. From each point of infection, a gall is produced. Infected female flowers give rise to galls instead of grains. Galls on the leaf are small and usually they become hard and warty and contain only small number of spores. If seedlings are affected, they remain stunted, weak and may die.

The causal organism. The smut spores are spherical to ellipsoid, black, finely echinulated and 8 - 11μ in diameter. They germinate to produce a three-septate basidium and from each of the basidial cell, a basidiospore (sporidium) is formed. The basidiospores are ovate, hyaline, single-celled and uninucleate. These sporidia may multiply by repeated budding. Fusion of sporidia of opposite sexes occurs commonly. Infection of the host takes place through wounds or natural openings and air-borne spores and sporidia cause such infection. The sporidia germinate and produce a uninucleate mycelium that infects the host. Dikaryotization takes place within the host by means of somatogamy (cell fusion) between sexually compatible haploid hyphae and the pathogenic dikaryotic mycelium is formed. This dikaryotic mycelium further invades the tissues and forms galls. Later, most of the cells of the dikaryotic mycelium are converted into smut spores.

Mode of survival and spread. Manure heaps and crop refuse are the main sources of primary infection.

Disease management

Agronomic practices. Field sanitation and crop rotation can help in checking the disease outbreak.

3. Downy mildews of maize

Sclerospora (Peronosclerospora) sorghi, *Sclerophthora rayssiae* var. *zeae*, *Sclerospora (Peronosclerospora) sacchari* and *Sclerospora (Peronosclerospora) philippinensis*.

Maize crop is subjected to attack by several species of downy mildew fungi. Of these, the **'sorghum downy mildew'** caused by *Sclerospora (Peronosclerospora) sorghi*, the **'brown stripe downy mildew'** caused by *Sclerophthora rayssiae* var. *zeae*, the **'sugarcane downy mildew of maize'** caused by *Sclerospora (Peronosclerospora) sacchari* and the **'Philippine downy mildew of maize'** caused by *Sclerospora (Peronosclerospora) philippinensis* are very important and are responsible for causing severe damage to the crop and yield loss under conditions favorable for the pathogens. These downy mildews are found in different parts of South East Asian countries to a larger extent. In India, the diseases occur in Himachal Pradesh, Punjab, Uttar Pradesh, Haryana, Sikkim, Delhi and West Bengal. The symptoms produced by the different downy mildew pathogens and their morphological characters vary. The species of *Sclerospora*, which show typical conidial germination, are included in the **Genus - *Peronosclerospora*.**

(i) Sorghum downy mildew of maize

Sclerospora (Peronosclerospora) sorghi

Symptoms. The symptoms appear as long, rather broad, chlorotic stripes on the leaf, sometimes along the entire length of the leaf. A number of such streaks fuse and form large, irregular patches. The stripes, which are yellow in color initially, turn brown finally. On both the surfaces of younger leaves, fine, white, cottony mildew growth of the fungus is seen. The downy growth is more prominent on younger leaves than on the older leaves. After the leaves turn brown, leaf-shredding symptoms are noticed. Downy growth is also formed on the bracts of green, unopened male flowers in the tassel. Plants affected at the early stage of growth do not produce cobs. Even if cobs are formed, grain formation is drastically affected.

The causal organism. The pathogen causing this disease is an obligate parasite. The mycelium is non-septate, intercellular and grows systemically in the host tissues. From the mycelium, finger-shaped haustoria are produced, which penetrate the cells of the mesophyll tissues. Sporangiophores emerge through the stomata singly or in groups of two to three. They are stout, wedge-shaped, short and dichotomously branched at the apex. They are about 185µ in length, thin-walled and have 1 - 2 septa near the basal portion. Sporangia are produced singly from the tips of the branches. They are thin-walled, hyaline, spherical or oval, without apical papillae and measure 15 - 29µ in diameter. They germinate by producing one or more germ tubes and not by producing zoospores. Oospores are produced in large numbers in parallel bands, between fibrovascular strands, which makes the tissues weak, leading to leaf shredding. They are round, thin-walled, reddish-brown in color, completely filled with the oogonium and measure 25 - 43µ in diameter. The oospore germinates by producing a germ tube and infects the underground parts of the host.

Mode of survival, spread and epidemiology. The pathogen perennates through oospores present in the soil and plant debris. They retain their viability for 3 - 4 years. The pathogen has many alternate hosts, such as *Saccharum spontaneum, Sorghum bicolor, S. helepense* etc. It is also internally seed-borne and remains inside the seed as dormant mycelium, which can initiate systemic infection in the seedling. Secondary infection takes place through sporangia carried by wind or rain splash.

The optimum temperature for sporangial formation and germination of sporangia is 21° - 23°C. Maximum sporulation occurs at 100 % relative humidity. Cool, damp weather, accompanied by light drizzling is conducive for disease development and rapid spread of the disease.

Disease management

Agronomic practices (i) Diseased crop debris should be destroyed by burning (ii) Fields and surrounding areas should be kept free from weeds and alternate hosts (iii) Deep ploughing helps to bury the oospores deep into the soil (iv) Long term rotation, including pulse and oilseed crops should be followed (v) Young plants showing symptoms of the disease should be uprooted and destroyed.

Chemical control (i) Spraying the crop with Dithane - M 45 or mancozeb - 600 gm. or Brestan - 375 gm. in 300 lit. of water per acre controls the disease effectively (ii) Spraying with a mixture of metalaxyl - 300 gm. + mancozeb - 600 gm. in 300 lit. of water per acre is very effective in controlling the disease.

Seed treatment. Seed dressing with metalaxyl or Apron at 4 gm./ kg. of seeds is claimed to give 100 % control of the disease.

Resistant varieties. Ph.DMR.1 is resistant to this disease.

(ii) Brown stripe downy mildew of maize
Sclerophthora rayssiae var. *zeae*

Symptoms. The symptoms appear as narrow, chlorotic spots, about 3.0 - 7.0 mm. in width, with well-defined margins. The spots extend in a parallel manner in-between the veins of the leaf blade. The stripes later turn reddish-brown and become necrotic. If the disease occurs prior to flowering, there is no development of grains and the plants die prematurely.

The causal organism. Sporangiophores arise from a stroma in the sub-stomatal space. They are determinate, unbranched and short. They produce large sized sporangia sympodially in groups of 2 - 6. The sporangia are hyaline, ovate, obclavate, cylindrical or elliptical and smooth, with a rounded apex, measuring 29.0 - 69.0 x 18.5 - 26.0μ. They germinate by issuing 4 - 8 zoospores. Oospores are produced in abundance in the leaf mesophyll or under the stomata. They are spherical or sub-spherical, slightly smaller than oospores produced by *Sclerospora*, smooth and have glistening, hyaline walls. They measure 29.5 - 37.0μ in diameter. The oogonial wall is confluent with the wall of the oospore.

Mode of survival, spread and epidemiology. The pathogen survives through oospores in the soil and crop debris for more than 3 years. Once primary infection is established, secondary spread of the disease is brought about by sporangia produced abundantly on the lower surface of infected leaves under conditions of high humidity and moderate temperature. The sporangia are dispersed by wind, rainwater and by direct contact with infected leaves and by some insects. The disease may also be transmitted through seeds of diseased plants as dormant mycelium in the embryo of seeds. *Digitaria bicornis, D. sanguinalis* and *Zea mexicana* are some of the alternate hosts of this pathogen.

The disease is favored by high relative humidity of above 80 % and fairly high temperature of 29° - 32°C. Optimum temperature for sporangial production is 22° - 25°C and for zoospore production 20° - 22°C, while oospores are produced at higher temperatures. A film of water on the host surface is essential for the zoospores to germinate and initiate infection.

Disease management

Agronomic practices (i) Rouging of diseased young plants, 20 days after sowing or at the time of thinning out of seedlings helps to prevent the spread of the disease through sporangia (ii) Field sanitation and crop rotation also help to check the disease occurrence.

Chemical control (i) Spraying the crop with Dithane-M 45 - 600 gm. or Dithane-Z 78 - 600 gm. or Brestan - 375 gm. in 300 lit. of water per acre gives adequate control of the disease (ii) A combination of metalaxyl - 300 gm. + mancozeb - 600 gm. in 300 lit. of water per acre is highly effective in controlling the disease.

Seed treatment. Seed dressing with metalaxyl at 4 gm./ kg. of seeds has been reported to give 100 % control of the disease.

(iii) Sugarcane downy mildew of maize

Sclerospora (Peronosclerospora) sacchari

Symptoms. The symptoms of the disease are almost similar to that produced by *Sclerospora sorghi*. The most characteristic symptom is the development of long, rather broad, chlorotic stripes along almost the entire length of the leaf. In the case of secondary infection, the stripes appear short and narrow. The color of the stripes, which is whitish or light-yellowish in the initial stages, changes to dark or light-brownish-yellow in the later stages. A number of such stripes coalesce to form irregular elongated patches. Leaf shredding is not commonly found. In the striped areas of the leaf, puckering may occur. The downy growth of the fungus appears on the stripes on both the surfaces of the leaf. Downy growth is more prominent on young leaves than on old leaves. On the bracts of green, unopened male flowers in the tassel also, such downy growth appears. Young unfolding leaves usually become uniformly yellow.

The causal organism. The sporangiophores emerge through the stomata singly or in groups of two, rarely three. They are massive, wedge-shaped, short and stout structures, widening gradually towards the upper portion and dichotomously branched at the top 2-3 times, rarely 4 times. They measure 160-170µ in length. They are thin and smooth-walled, one or rarely two-septate near the base, the basal cell narrowing down towards the base. The branches are stout and swollen in the middle. The sterigmata taper to a long, fine tip, which bears a single sporangium. The sporangia are hyaline, thin-walled, elliptical, cylindrical or ovate-oblong, with rounded apex and measure 24.0-46.6 x 12.0-20.0µ in size. Oospores are round, thin-walled and measure 22-30µ in diameter. Sporangia and oospores germinate by producing hyaline, slender, non-septate germ tubes. The sporangia germinate by issuing germ tubes either from the ends or from the lateral side. No zoospores are formed.

Mode of survival, spread and epidemiology. The pathogen perennates mostly through oospores present in the soil and crop debris. Sugarcane and teosinte are alternate hosts of this pathogen and the disease may alternate between maize and sugarcane. The fungus is also found to be seed-borne, but it is quickly destroyed when seeds are dried.

Temperature, humidity and light greatly influence the production and germination of sporangia. The optimum temperature for production and germination of sporangia is about 25°c. Sporangial production is completely suppressed at temperatures below 13°c and above 31°c. Sporangia are mostly produced during the night time. Very high humidity near saturation is necessary for the production of sporangia and a film of water on the host surface is essential for the germination of the sporangia.

Disease management

The control measures suggested for the control of other downy mildews of maize are adequate for the control of this disease also.

(iv) Philippine downy mildew of maize

Sclerospora (Peronosclerospora) philippinensis

Symptoms. The symptoms produced by this fungus are almost similar to that of *Sclerospora sacchari.*

The leaves become pale-yellow or chlorotic initially. The under surface of the leaves bear a woolly, white downy growth, followed by yellowing, browning and necrosis of the leaf blades. Infected plants become stunted and have shorter internodes. They may not produce any cobs and even if cobs are produced, grains are not formed.

The causal organism. *Sclerospora philippinensis* and *S. sacchari* are morphologically quite similar, but in *S. philippinensis,* oospore production is suppressed. Sporangiophores emerge through the stomata in groups of 2-4. They are 140-410μ in length. Sterigmata are produced at the tips of the branches of sporangiophores. From each sporangiophore 6-30 sterigmata are formed. They are 14-15μ long and bear sporangia singly at their tips. Sporangia are usually barrel-shaped and measure 17-57 x 11-27μ in size. The sporangium germinates by issuing a germ tube that penetrates the host through the stomatal opening.

Mode of survival, spread and epidemiology. The fungus has many alternate hosts, such as *Saccharum spontaneum* (kansgrass), *Sorghum helapense, S. bicolor, S. propinaquum* etc., that may serve as source of primary inoculum. Secondary infection may occur through sporangia, which are dispersed by wind, rainwater, direct contact and by some insects.

High relative humidity is essential for sporangial formation and their germination. The sporangia are produced only during the night time. Optimum temperature for germination of sporangia is 19°-20°c.

Disease management. The control measures suggested for the control of other downy mildews can be followed to control this disease also.

4. Charcoal rot of corn

Macrophomina phaseolina

Rhizoctonia bataticola

'Charcoal rot' is prevalent in many states of India, such as Tamil Nadu, Telungu Desam, Haryana, Delhi, Karnataka, Madhya Pradesh, Punjab and Uttar Pradesh.

Symptoms. The disease usually attacks the plants just prior to emergence of the earheads. The affected plants start wilting and eventually dry and die. The stalks of the infected plants become weak and the collar region turns charcoal-black and is shrunken. On the rind, large number of sclerotia are formed. The pith region becomes shredded and large number of sclerotia are formed on the vascular region. Sometimes, gray or whitish mycelial growth is also seen. The roots also become shredded and black sclerotia are formed in the disintegrated tissues. Sometimes, on the surface of the stem, pycnidia are also produced, which appear as minute, black dots.

The causal organism. The pycnidial stage of the fungus is *Macrophomina phaseolina* and its sterile stage is *Rhizoctonia bataticola.* The pycnidia are black, globose to flask-shaped, subepidermal, ostiolate and measure 100 - 200µ in diameter. Through the ostiole, large number of pycnidiospores are exuded. The pycnidiospores are hyaline, single-celled, and oval to cylindrical in shape and measure 14.0-30.0 x 5.0-10.0µ in size. The sclerotia are black in color, hard, and globular to irregularly globular in shape **(Fig.43).**

Mode of survival, spread and epidemiology. The fungus is mainly soil-borne. The sclerotia can survive in the plant debris in the soil for a long period of time. The fungus is also capable of continuing a saprophytic life in the organic matter present in the soil. Whenever pycnidia are formed, the pycnidiospores are spread through wind. Further, the pathogen has a wide range of hosts, such as pigeon pea, black gram, green gram, cowpea, soybean, gingelly, sunflower, groundnut, cotton, tomato, bean, maize, sweet potato etc. So, whenever suitable hosts and favorable environmental conditions occur, the fungus becomes pathogenic. The fungus gains entry through the roots and then proceeds upward into the stem region. Injuries caused by insects, nematodes or by mechanical means make the plants more vulnerable to attack by the fungus.

High temperatures of 36°-40°c and low soil moisture are favorable for the occurrence of the disease.

Disease management

Agronomic practices (i) Balanced fertilization and timely irrigation help to minimize the incidence of the disease (ii) Affected plants may be removed and destroyed by burning.

Chemical control. No chemical control measures have been advocated to control this disease.

Resistant varieties. CM.103, CM.104, CM.107, CM.111, CM.112, CM.202, CM.300, CM.600, Ganga 5, Ganga 101, Ambar and Jawahar have shown resistance to this disease.

Diseases of minor importance. Besides the diseases dealt with in detail, maize is subjected to attack by many fungal, bacterial and virus diseases, which may not be of much economic importance. 'Rust disease' caused by *Puccinia sorghi* produces erumpent pustules on the leaves; 'Head smut' caused by *Sporisorium (Sphacelotheca) reiliana* affects a portion or the entire inflorescence, converting them into smut sori; '*Diplodia* stalk rot' caused by *Diplodia maydis* causes sudden wilting and death of affected plants; 'Wilt disease' caused by *Cephalosporium maydis* and *Fusarium moniliforme* affect the vascular tissues, leading to the death of affected plants; 'Leaf blight' caused by *Drechslera (Helminthosporium) maydis* produces linear lesions on the leaves; 'Bacterial stalk rot' caused by *Erwinia chrysanthemi* pv. *zeae* causes rotting of lower nodes and stem region and ultimate death of the infected plants; 'Mosaic' caused by *Sugarcane mosaic virus* and 'maize mosaic' caused by *Maize stripe virus* are the important virus diseases that affect maize.

Finger millet or Ragi *(Eleusine coracana)*

1. Blast of finger millet

Pyricularia setariae / Pyricularia grisea

(Magnaporthe grisea)

'Blast disease' of finger millet or ragi occurs in India, Malaysia and some of the African countries. In India, the disease is found in the states of Tamil Nadu, Pondichery, Karnataka, Telungu Desam and Maharashtra. In Tamil Nadu, the disease appears in epiphytotic form during some years and causes severe damage. When the disease appears in a severe form, the yield loss may go up to 90 per cent.

Symptoms. Plants of all ages from the nursery stage to maturity may be affected. However, seedlings and 40-50 days old plants are highly vulnerable to attack by this disease. When the seedlings in the nursery and young transplanted plants are affected, all the leaves dry and the affected plants may die. In grown up plants, the symptoms appear as small, scattered, long spindle-shaped or diamond-shaped or eye-shaped spots. The spots are grayish-green in the centre with yellowish-brown margin. The spots enlarge and become 1.0 - 2.0 x 0.3 - 0.5 cm. in size and the centre of the spots turn grayish-white, which disintegrate eventually. Under humid conditions, grayish growth of the fungus, consisting of conidiophores and conidia appear on the upper surface of the spots. Several spots may coalesce covering large areas of the leaves, which dry and give the crop a burnt appearance. Similar spots may appear on the mid rib of leaves and in such cases, the leaves break from the spots and hang down. The pathogen may attack the nodal region of the stems at the vegetative and earhead stages. When the nodes are attacked, the nodal region, to a length of 0.5 - 1.0 cm. up and down the nodes turns black, shrink, the tissues disintegrate and the stem may break at the infected nodes. At the maturity stage, the pathogen may also attack the node of the rachis. In such cases the nodal region of the rachis turns black up and down the node to a length of 0.5 - 1.0 cm., shrinks, becomes weak, breaks at the node and hangs down. Most of the grains in such affected ears become chaffy.

Even if grains are formed, they are shrivelled, discolored and of poor quality. Sometimes, individual spikes are affected at the base and the grains in such fingers are destroyed.

The causal organism. The mycelium of the fungus is septate, hyaline in the early stages and turns olive-brown later on. The hyphae are not much branched and are localized around the point of infection. Prior to production of fructifications, the hyphae aggregate below the epidermis to form a stroma. From the stroma, conidiophores are produced, which emerge through the stomata in groups. They are erect, simple, septate, dark-colored at the base and lighter above. From the conidiophores, conidia are produced acrogenously, one after another by sympodial growth. The conidiophores are geniculate and at each point from where a matured conidium has dislodged, a scar is seen. The conidia are obpyriform or pear-shaped, hyaline, thin-walled, mostly 3-celled, the middle cell broader and light-colored, and measure 19 - 31 x 10 - 15µ in size. They germinate by producing germ tubes from the terminal cells, which penetrate the host through the stomata or through the epidermal cells and cause fresh infection.

The perfect stage of this fungus, *Magnaporthe grisea* has not been found to occur in nature.

Mode of survival, spread and epidemiology. The pathogen can survive in the stubble left over in the field after the harvest of the crop. The fungus is also found to be present as conidia in the pericarp and endosperm of the shrivelled grains. These seed-borne conidia can cause primary infection. Secondary spread is through air-borne conidia. The fungus has many collateral hosts, such as foxtail millet, pearl millet, wheat, barley, oats, maize, *Eleusine indica* and *Dactyloctenium aegypticum.* This fungus does not attack rice and *Digitaria marginata.*

A temperature range of 25°-30°c, high relative humidity of 90% and above, cloudy and damp weather and intermittent rainfall are highly conducive for the occurrence and spread of the disease. A film of water on the leaf surface is necessary for the germination of conidia. Application of excess doses of nitrogenous fertilizer makes the plants more susceptible to attack by this pathogen.

Disease management

Agronomic practices (i) Disease-free seeds should be used for sowing (ii) The fields and surrounding areas should be kept free from weed hosts (iii) Application of excessive doses of nitrogenous fertilizer should be avoided (iv) Usually, a large number of lesions are found near the tip portion of leaves of seedlings. So, before transplanting, the tip portion of the leaves of the seedlings may be clipped and removed. Then the seedlings are dipped in Bordeaux mixture - 1% solution or copper oxychloride solution (2.5 gm. of copper oxychloride in 1 litre of water) and then transplanted.

Chemical treatment in the nursery. Spraying the nursery with edifenphos - 20 ml. or carbendazim - 20 gm. or thiophanate methyl - 20 gm. or Kitazin - 20 ml. or tricyclazole - 20 gm. in 20 lit. of water per 8 cent nursery area, 10-15 days after sowing controls blast disease in the seedling stage.

Chemical treatment in the main field. Foliar spraying with edifenphos - 200 ml. or carbendazim - 200 gm. or Kitazin - 200 ml. in 200 lit. of water per acre, once when initial symptoms are noticed and the second at the time of flowering gives good control of the disease.

Dry seed treatment. Dry seed dressing with thiram - 4 gm. or captan - 4 gm. or carbendazim - 2 gm./ kg. of seeds, 24 hours prior to sowing is effective in checking seed-borne infection.

Seed pelleting. Instead of dry seed treatment, seed pelleting may be done. Carbendazim - 2.5 gm. and gum - 1.25 gm. are dissolved in 100 ml. of water and mixed well. In the fungicide-gum mixture, 2.0 kg. of seeds required to cover one acre of field is put and mixed thoroughly, so as to give a uniform coating on the seed surface. Then, the seeds are air-dried in shade and used for sowing after 24 hours. Both externally and internally seed-borne infections can be checked by this method.

Resistant varieties. Co.10, Co.11, Co.12, Co.13, HR.374, PR.202, K.3 and Mozambique 359 are resistant to this disease.

2. Seedling blight and leaf blight of finger millet

Helminthosporium nodulosum / Drechslera nodulosum (Cochliobolus nodulosus)

The disease is very common in Africa, Japan, the Philippines, the United States of America and India and is responsible for causing heavy loss in yield. In India, it is prevalent in the states of Telungu Desam, Karnataka, Tamil Nadu, Maharashtra, Uttar Pradesh, Bihar and Pondichery.

Symptoms. The disease attacks seedlings, as well as grown-up plants. All parts of the plants viz., leaves, leaf sheaths, roots, collar region, culms, neck of panicles and fingers are affected. Germinating seeds may be killed before emergence of seedlings above the soil. After emergence, minute, light-brown, oval or sesame seed-like, scattered spots develop in large numbers on the leaves. As the seedlings grow, the lesions enlarge and become dark-brown, linear spots, about 10 cm. long and 0.1 - 0.2 cm. wide and parallel to the veins. The spots coalesce and become large necrotic patches, as a result the leaves wither and die. The affected seedlings may die within 15 days. The pathogen attacks the roots and collar region of the shoots and causes root rot and collar rot, resulting in the death of infected young plants.

Later on, lesions are produced on the leaf blades and leaf sheaths of grown-up plants. The spots on the leaves are oblong and dark-brown in color. Roughly oblong, dark-brown spots with faded margins develop on the leaf sheaths. Under humid conditions, grayish-brown, fine, cottony growth of the fungus consisting of mycelium, conidiophores and conidia appear at the central portion of the lesions. Many spots coalesce to form bigger patches and heavily infected leaves wither and die prematurely. The nodes of culms and the nodal region of the ears may also be affected. The nodal region, up and down the nodes turns dark-brown in color, shrivels, becomes weak and as a result of disintegration of the affected tissues, breaks easily and may hang down. As a result of neck infection, grain formation may be completely affected. Even if grains are formed, they are poorly developed, shrivelled and discolored. Individual fingers in an ear may also be affected.

The causal organism. The mycelium of the fungus is septate, light-brown in color and inter- and intracellular. The mycelium is mostly localized around the points of infection. Conidiophores arising from the stromata beneath the host epidermis emerge through the stomata or directly by penetrating the epidermis. Conidiophores may also arise from the superficial mycelium that grows on the surface of the lesions. They are long, erect, septate, and hyaline at first and later becoming dark-brown at the base and lighter at the top, geniculate and measure 80.0 - 250.0 x 5.0 - 7.0µ in size. Conidia are produced at the tip of the conidiophores, singly one after another. Conidia are thick-walled, light-brown in color, sub-cylindrical or obclavate, straight or slightly curved, 3 - 10 septate and measure 40 - 110 x 11 - 21µ in size. At each point from where a matured conidium has dislodged, a scar is seen in the conidiophore. Conidia germinate by issuing a germ tube from any one of the terminal cells and enter the host, either through the stoma or directly by penetrating the host epidermis and causes fresh infection. The ascigerous stage of this fungus, *Cochliobolus nodulosus* is not pathogenic.

Mode of survival, spread and epidemiology. The disease is mostly seed-borne. The spores, which are found on the seeds, cause primary infection of the seedlings. The spores produced from such affected seedlings are carried by wind and cause secondary infection.

The stubble left over in the field after the harvest of the crop and the crop debris in the soil also harbor the mycelium and conidia, which also may cause primary infection. Further, the disease attacks several other crops and weed hosts, such as maize, pearl millet, foxtail millet, sugarcane, few other lesser millets, *Eleusine indica* and *Dactyloctenium aegypticum*. The spores produced from such hosts provide inoculum that can infect finger millet crop.

High day temperature, high relative humidity and intermittent rainfall favor the occurrence and spread of the disease.

Disease management

Agronomic practices (i) The field should be ploughed after the harvest of the crop and the stubble and crop debris removed and destroyed by burning (ii) Field sanitation and selection of disease-free seeds are other practices that should be followed.

Chemical control. As soon as the symptoms of the disease appear, either in the nursery or in the main field, fungicidal spraying should be taken up. Spraying with copper oxychloride at 2.5 gm. or mancozeb at 2.0 gm. or edifenphos at 1.0 ml. per litre of water controls the disease.

Seed treatment. Seed dressing with captan or thiram at 4.0 gm./ kg. of seeds, 24 hours prior to sowing eradicates the seed-borne inoculum.

3. Virus diseases of ragi

(i) Mottle streak virus disease

The disease is prevalent in Tamil Nadu and Karnataka states of India. In Tamil Nadu, it is found in the districts of Coimbatore, South Arcot and Madurai.

Symptoms. The symptoms of the disease appear in young plants, 30 - 45 days after sowing. Initial symptoms usually appear in the newly emerging leaves as small, chlorotic, mosaic spots of irregular shapes. Later, such spots elongate and become intermittent, short streaks parallel to the veins. Mottling and yellowing of young leaves follow this. The infected plants are stunted in growth. Severely affected plants fail to produce panicles. Even if panicles are formed, the grains are poorly developed and are chaffy. Early infection leads to severe loss in yield.

The causal organism. *Ragi mottle streak virus* causes the disease.

Mode of spread. The leafhopper *Cicadulina bipunctella* and *C. chinai* transmit the causal virus. The acquisition feeding time for the vectors is 48 hours. An incubation period of 9 - 16 days in the body of the vectors is required before the vectors can become viruliferous. After inoculation of the virus into the healthy plants by the vector, it may take about 20 - 30 days for symptom expression in the host plant. Variety AKP.2 is tolerant to *Mottle streak virus*, while C0.7, Co.10 and Co.11 are highly susceptible.

(ii) Streak virus disease

The disease occurs in Coimbatore and other major ragi growing districts of Tamil Nadu.

Symptoms. The symptoms appear only 30 - 45 days after sowing. Thin, chlorotic streaks, parallel to the veins appear initially on the young leaves. Later, these streaks develop into long streaks, running parallel to the veins all along the entire length of the leaves. The affected leaves dry completely. The affected plants are stunted in growth and produce more tillers, but they rarely produce panicles. Even if panicles are formed, the grains are poorly developed, chaffy and discolored.

Mode of spread. The virus causing this disease is transmitted by the leafhopper, *Cicadulina bipunctella*. The acquisition feeding time for the vector is 6 hours. The vectors become viruliferous immediately after acquisition and no incubation period is required inside the body of the vectors. After inoculation of the virus by the vector, it takes about 20 days for symptom expression. The virus infects foxtail millet, pearl millet and sorghum. AKP.2 is tolerant to streak virus also.

Disease management of streak virus diseases

Chemical control. Because leafhopper vectors transmit the diseases, control of the vectors helps to minimize the spread of the diseases. In the nursery, spraying with monocrotophos - 20 ml. or dimethoate - 40 ml. or phosphamidon - 10 ml. in 20 liters of water per 8 cent nursery, 15 days after sowing, followed by spraying in the main field with monocrotophos - 200 ml. or dimethoate - 400 ml. or phosphamidon - 100 ml. in 200 lit. of water per acre, 15 days after planting controls the vector.

(iii) Eleusine mosaic virus disease

The disease is widespread and occurs in all the finger millet growing districts of Tamil Nadu.

Symptoms. The symptoms usually appear when the crop is 30 - 40 days old. The disease is characterized by chlorosis, mosaic, mottling of young leaves and severe stunting of infected plants. The diseased plants fail to produce spikelets. Even if spikelets are formed rarely, the grain setting is very poor and the grains are chaffy and discolored.

The causal organism. *Eleusine mosaic virus* causes the disease.

Mode of spread. *Aphis maydis, Myzus persicae, Rhopalosiphum maidis* and *Tetraneura rufiabdominalis* transmit the virus in a non-persistent manner. The virus attacks foxtail millet, maize and sorghum and a few grasses.

Disease management

Chemical control. In the nursery, spraying with monocrotophos - 20 ml. or dimethoate - 40 ml. or phosphamidon - 10 ml. in 20 lit. of water per 8 cent nursery, 15 days after sowing, followed by spraying in the main field with monocrotophos - 200 ml. or dimethoate - 400 ml. or phosphamidon - 100 ml. in 200 lit. of water per acre, 15 days after planting controls the vector transmitting the virus.

Minor diseases. Besides the diseases mentioned above, some other diseases, which are considered to be minor diseases, attack ragi. 'Wilt' or 'Foot rot' caused by *Sclerotium rolfsii* attacks the basal stem region, resulting in rotting of the collar region and death of the affected plants; 'Green ear',

'Downy mildew' or 'Crazy top' caused by *Sclerophthora macrospora* attacks the leaves and ear heads. The spikelets become abnormal and exhibit proliferation, as a result of hyperplasia and hypertrophy of the cells due to the infection; 'Smut' caused by *Melanopsichium eleusinis* produces enlarged, globose, smut sori in the place of normal grains; 'Tar spot' caused by *Phyllachora eleusinis* produces small, oval, black, elevated spots on the leaves; *Cercospora fusimaculans* and *Helminthosporium tetramera* produce characteristic leaf spots.

Foxtail millet, Italian millet, Siberian millet or Tenai *(Setaria italica)*

1. Smut of tenai

Ustilago crameri

The disease occurs in India, China and South Eastern Europe. The disease was first reported in India in 1921 and is prevalent in the districts of Tamil Nadu, Karnataka, Maharashtra and Telungu Desam.

Symptoms. In the affected earhead, normal grains are replaced by smut sori, which are covered by a grayish-white membrane, enclosing black mass of smut spores. The smut sori are larger than the normal grains, 2.0 - 4.0 mm. in size and spherical or pear-shaped. Later, the membrane covering the sorus bursts open exposing the black, powdery spores.

The causal organism. The smut sori usually rupture at the time of harvesting and threshing. The teliospores are spherical to angular in shape, dark-brown in color, smooth-walled and are 7.0 - 10.0µ in diameter. The spores germinate by producing promycelium. The promycelium rarely produces sporidia, but produces a slender hypha from the basidial cells, which infect the host.

Mode of spread and epidemiology. The disease is externally seed-borne. The seeds get contaminated with the smut sori or the smut spores at the time of threshing. The spores and the seeds of the host germinate simultaneously. The pathogen gains entry into the host through the coleoptile and moves into the embryonic growing point. Once inside the host, the fungus grows systemically, keeping pace with the growth of the plant. At the time of flowering, spores are produced by segmentation of the dikaryotic hyphae, which replace the ovary, converting it into a smut sorus. Secondary spread may take place through wind-borne smut spores.

Seedling infection is influenced to a large extent by soil temperature and soil moisture. Moderate soil temperature and high soil moisture is conducive for disease occurrence.

Disease management

Seed treatment. Seed dressing with thiram or captan at 4 gm./ kg. of seeds, 24 hours prior to sowing is effective in controlling the disease.

2. Rust of tenai

Uromyces setariae - italicae

The disease is found in Africa, Europe and several Asian countries, including India. In India, it occurs in Tamil Nadu, Telungu Desam, Karnataka, Maharashtra, Bihar, Uttar Pradesh and West Bengal. In Tamil Nadu, the disease is prevalent in all major tenai growing districts.

Symptoms. The disease attacks tenai at all stages of growth. However, the damage is more severe if infection occurs prior to the flowering stage. If the infection occurs after grain setting, the loss in yield is negligible. Initial symptoms appear as minute, brown, sesame-seed-shaped, erumpent pustules consisting of uredia on both the surfaces of leaves. The pustules are often arranged in rows parallel to the veins. Lesions also occur on the leaf sheaths and culms. On maturity of the uredia, the host epidermis is ruptured and the uredospores are exposed as powdery mass. Later on, when the crop is nearing maturity, telia are formed in large numbers on the leaves, leaf sheaths and culms. The telia are slightly bigger than the uredia and are black-colored. The epidermis covers them for a longer time. Formation of numerous lesions on the leaves, results in premature drying and death of leaves and in case of severe infection, affected plants may dry and die before flowering. Even if the plants come to heading, only few grains are formed.

The causal organism. The uredospores are pedicellate, single-celled, globose or oval in shape, finely echinulated, yellowish-brown in color, with hyaline walls. They have 3 - 4 scattered germ pores and measure 26 x 20μ in size. The teliospores are pedicellate, oblong, with broad, rounded bottom and blunt, tapering and thickened apex. They are smooth-walled, with one germ pore at the apex, yellowish-brown in color and measure 22 x 17μ in size. The pycnial and aecial stages of the fungus are not known.

Mode of survival, spread and epidemiology. The disease usually appears, when the plants are 20 - 25 days old and the disease intensity increases, as the plants grow older. The disease perpetuates as uredospores. When the crop is grown throughout the year, the disease spreads through wind-borne uredospores. The pathogen attacks *Eriochloa procera, Setaria viridis, S. glauca, S. verticillata* and *S. intermedia.* The uredospores produced from these hosts may serve to provide primary inoculum that can infect tenai. Secondary spread in the field may occur through wind-borne uredospores.

Moist and humid conditions and moderate temperature, which prevails during June - July and October - January is highly favorable for the occurrence and spread of the disease.

Disease management

Chemical control. Spraying with wettable sulfur - 800 gm. or mancozeb - 400 gm. or carboxin (Vitavax) - 200 gm. in 200 lit. of water per acre is effective in controlling the disease.

3. Blast of tenai

Pyricularia setariae

The disease is found in Japan, South Africa and India. In India, it was first recorded in 1919 and is prevalent in the states of Tamil Nadu and Maharashtra.

Symptoms. Lesions on leaf blades are roughly diamond-shaped to circular, 2.0 - 5.0 mm. in length, with dark- brown margin and light-brown center. The spots are scattered on the leaf surface, but are found in large numbers near the top portion. Lesions have chlorotic-yellow haloes around them. Under humid conditions, a light-gray bloom, consisting of conidiophores and conidia develop on the central portion of the lesions. Neck and nodal infections are very rare.

The causal organism. The conidiophores emerge singly or in small groups through the stomata or by penetrating the host epidermis. From each conidiophore, several conidia are produced one after another. The conidia are obpyriform, sub-hyaline and 3 - celled. They germinate by issuing germ tubes from the terminal cells.

Mode of survival, spread and epidemiology. The fungus survives on the stubble left over in the field after the harvest of the crop or on crop debris in the soil. The fungus is also carried through conidia on the grains. Besides tenai, the pathogen attacks pearl millet, finger millet, wheat and *Dactyloctenium aegypticum*. The primary inoculum may come from any of these sources. Secondary spread is through air-borne conidia.

Low night temperature, high relative humidity of 90 % and above, cloudy and damp weather and intermittent rainfall are favorable for the occurrence and spread of the disease. The disease is more common during the cooler months of October - December.

Disease management

Chemical control. Foliar spraying with edifenphos - 200 ml. or iprobenphos (Kitazin) - 200 ml. or carbendazim - 200 gm. in 200 lit. of water per acre controls the disease effectively.

Seed treatment. Seed dressing with thiram - 4 gm. or captan - 4 gm. or carbendazim - 2 gm./ kg. of seeds, 24 hours prior to sowing, helps to eradicate the seed-borne inoculum.

Diseases of minor importance. Several other diseases, which are not considered to be of much economic importance also, attack tenai crop. 'Leaf spot' or 'Leaf blotch' caused by *Helminthosporium setariae* affects the leaves; 'Downy mildew' and 'Green ear' caused by *Sclerospora graminicola* affects the spikelets and converts the florets into leafy structures; 'Leaf spot' is caused by *Colletotrichum graminicolum; Helminthosporium nodulosum* causes 'seedling rot' and 'Leaf spot'; 'False smut' caused by *Ustilaginoidea setariae* and 'Udbatta disease' caused by *Ephelis oryzae* affect the inflorescence; 'Bacterial brown stripe' caused by *Pseudomonas setariae* affects the leaves.

Japanese barnyard millet or Kudiraivali *(Echinochloa frumentacea)*

Most of the diseases that occur on other millets, occur on kudiraivali also. However, as the crop is mostly raised under rainfed conditions and as it is not very remunerative, not much importance is given to the diseases and their control. *Helminthosporium frumentacei* affects the leaves and causes 'Leaf spot' and 'Leaf stripe'; *Helminthosporium monoceros (crusgalli)* affects the leaves and causes 'Leaf blight'; *Ustilago crusgalli* causes 'Head smut' that affects the entire earhead; *Ustilago paradoxa* causes 'Kernel smut' that affects a few spikelets in the inflorescence.

Kodo millet or Varagu *(Paspalum scrobiculatum)*

'Head smut' caused by *Sorosporium paspali-thumbergii* affects the inflorescence and converts the entire inflorescence into a smut sorus; 'Rust' caused by *Puccinia substriata* affects the leaf blades and leaf sheaths and produces rust pustules; 'Sugary disease' or 'Ergot' caused by *Claviceps paspali* affects the spikelets and convert them into ergot; 'Tar spot' caused by *Phyllachora winkleri* affects the leaves and produce dull-black, elevated spots; 'Leaf blight' caused by *Helminthosporium turcicum* affects the leaves; 'Udbatta disease' caused by *Ephelis oryzae* affects the inflorescence; *Xanthomonas oryzae pv. oryzae* affects the leaves and causes 'Bacterial leaf blight'

Rye *(Secale cereali)*

'Sugary disease' or 'Ergot' caused by *Claviceps purpuria* affects the earhead and converts some of the spikelets into ergot. *Erysiphe graminis-secalis* causes 'Powdery mildew'; *Colletotrichum graminicola* causes 'Leaf spot' and 'Anthracnose'.

Proso millet or Hog millet or Samai *(Panicum miliaceum)*

Sclerospora graminicola causes 'Downy mildew' that affects the leaves; 'Head smut' caused by *Sphacelotheca destruens* affects the inflorescence and converts the entire inflorescence into a smut sorus; *Pyricularia grisea* causes 'Blast disease' that produces typical spots on the leaves; *Colletotrichum graminicolum* causes 'Leaf spot' and 'Anthracnose'; 'Bacterial leaf stripe' caused by *Xanthomonas panici* produces streaks on the leaf blades, leaf sheaths and occasionally on the culms.

Small millet or Panivaragu *(Panicum miliare)*

'Rust disease' caused by *Uromyces linearis* produces minute, narrow, erumpent pustules on the leaves; *Ephelis oryzae* causes 'Udbatta disease'; *Helminthosporium oryzae* affects the leaves and produces 'leaf blight'.

PULSES

Red gram or Pigeon pea *(Cajanus cajan)*

1. *Fusarium* wilt of red gram

Fusarium udum

The disease is found in India, Bangladesh, several African countries and Thailand. It was first reported in India from Bihar in 1931. Though the disease occurs in the entire red gram growing states of India, it is more prevalent in Bihar, Madhya Pradesh, Rajasthan, Maharashtra, Uttar Pradesh and Tami Nadu.

Symptoms. The disease attacks the crop at all growing stages however, it is more prevalent in grown up plants, especially during the flowering and pod formation stages after heavy rains. The disease incidence is much less during the first two months of growth. The characteristic symptoms of the disease are sudden wilting of young and grown up plants, even when there is enough moisture in the soil. The leaves become yellow, droop and dry, followed by drying of some of the branches or the entire plant. Such wilted plants appear scattered in the field. Then, the plants around the wilted plants start wilting, resulting in patches of wilted plants in the field.

Black, dry patches are seen on the basal stem portion and on the primary root portions. Dark streaks are visible under the bark of such affected plants. The streaks may extend up to a height of several feet in the stem and below in the roots. If the stem of such wilted plant is split open longitudinally, brown to black, vascular discoloration is seen extending up to the top of the plant and down to the roots. Rotting is also seen in the root portions, but no exudates are given out. Sometimes, black streaks are seen only on one side of the stem and in that case, partial wilting may occur and branches on that side alone wilt and dry. If the streaks appear all round the stem, then all the branches and the entire plant may wilt and dry. Similarly, infection starting and extending up from a few lateral roots, results in partial wilting.

In advanced stages of infection, aerial growth of the fungus may appear on the basal portion of the stem near the ground level as fine, white, cottony mycelial growth, along with slimy masses of pink spores. The dried leaves on the wilted plants remain attached to the plants without dropping off for a long period of time. Recovery of the affected plants is very rare **(Fig.33)**.

The causal organism. The fungus causing the disease is host specific and attacks mostly pigeon pea. The mycelium is both inter- and intracellular and is mostly confined to the vascular tissues. The hyphae are hyaline and septate. After gaining entry into the host, the mycelium runs across the cells and grows rapidly along the inside of the larger conducting vessels, partially plugging them with matted coils of hyphae.

A gummy substance produced by the pathogen is also responsible for plugging of the conducting vessels. This plugging of xylem vessels, interferes with free flow of water and mineral nutrients to the aboveground, green parts of the plant and causes drooping and wilting of the leaves and eventually death of the plant. The pathogen produces some toxins and enzymes that disintegrate the cell walls. As a result of this, the cells become dark in color and the conducting vessels appear as dark streaks along the stem and root regions. The fungus also produces a toxin, **'fusaric acid'** in the cells of affected plants and in the soil around the roots, which may also cause such wilting of the plants.

The pathogen produces three types of spores in its asexual phase. They are the macroconidia, microconidia and chlamydospores. Macroconidia are produced from minute, pointed, simple or branched conidiophores, that arise from cushion-like, stromata **(sporodochia)**. The macroconidia are linear, curved or fusoid, pointed at both the ends, knotched at the base, thin-walled, septate, with 3 - 4 septa and measure 15.0 - 50.0 x 3.0 - 5.0μ in size. The microconidia are produced in enormous numbers from the tips of hyphae. They are small, elliptical or curved, thin-walled, single-celled or with one or two septa and measure 5.0 - 15.0 x 2.0 - 4.0μ in size and are found even inside the cells of the conducting vessels. The chlamydospores, which are produced in the hyphae are spherical to oval, thick-walled, single or in chains of two to three and are terminal or intercalary. Sometimes, any cell of the macroconidia may also become a chlamydospore **(Fig.33)**.

The perfect stage of the fungus has been reported as *Glomerella indica.* The perithecia are superficial, globose, sessile, dark-violet in color and measure 350 - 550μ in diameter. The asci are sub-cylindrical and slightly broader in the middle. Ascospores, which are 8 in number in each of the ascus, are ellipsoidal to ovate, hyaline and 2-celled, rarely 3 - 4-celled. The perfect stage is not pathogenic.

Mode of survival, spread and epidemiology. The fungus is a facultative saprophyte. It can survive in the stubble of the diseased plants for a long time, in the absence of the living host plant. The macroconidia can also remain viable in the soil or in the plant debris for a number of days. The thick-walled chlamydospores can remain in the soil in a dormant state for many years. So, the fungus is a typical soil-borne organism. The spores may be carried on the surface of grains also.

Infection occurs through the fine rootlets, which are penetrated by the hyphae. As a result of infection, the rootlets become blackened and shrivelled. From the lateral roots, the fungus passes on into the larger roots. The larger roots cannot be penetrated directly by the fungus, unless they are injured by soil-inhabiting insects or nematodes or by cultural operations. The pathogen spreads within the field or from one field to another through irrigation water and through implements.

Wilt is common in areas, where pigeon pea is cultivated continuously year after year. Long and medium duration varieties are less prone to attack by the disease than short duration varieties. Ratooning predisposes the crop to attack by the pathogen. The pathogen can tolerate a wide range of environmental conditions, such as a pH range of 4.6 - 9.0 and high soil temperature of 35°C. Temperatures between 17° - 29°C are conducive for the development of the disease. Sandy soils favor the occurrence of the disease.

Fig. 33: Wilt of red gram-*Fusarium udum*

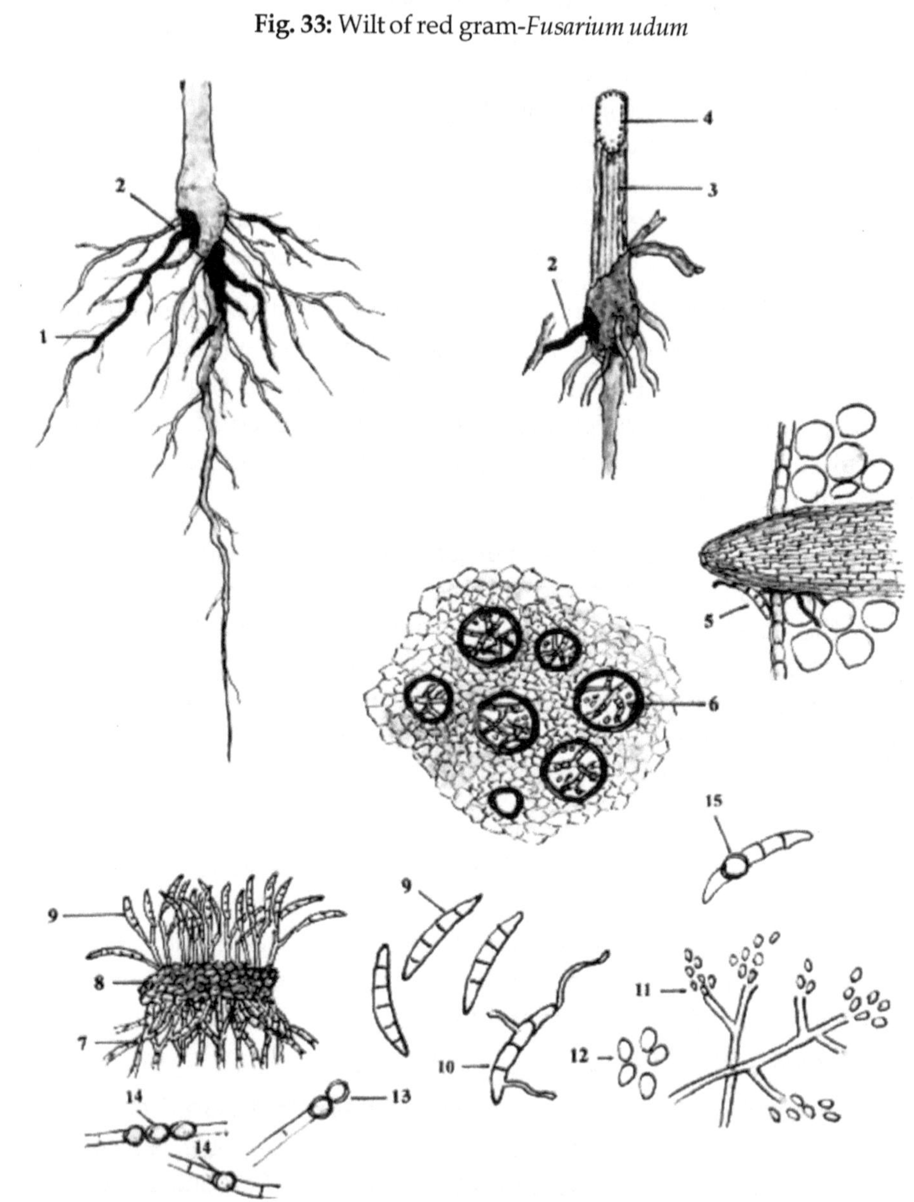

1. Infected lateral root becoming dark 2. Infection spreading to the stem 3. Streaks beneath the rind of the stem 4. Ring of discolored xylem vessels 5. Penetration through cracks formed by emerging lateral root 6. Mycelium, microconidia and chlamydospores inside the larger xylem vessels 7. Mycelium 8. Stroma 9. Macroconidia 10. Germinating macroconidium 11. Microconidia from hyphal tips 12. Microconidia 13. Terminal chlamydospores 14. Intercalary chlamydospores 15. Chlamydospores from macroconidium.

Disease management

Agronomic practices (i) Diseased plant parts left over in the field after harvest of the crop and crop debris should be removed and destroyed by burning (ii) Crop rotation is the best way of eliminating soil-borne infection. A long, 4 - 5 year rotation, preferably with sorghum, pearl millet or tobacco eradicates the soil-borne inoculum completely (iii) Deep summer ploughing helps to bring out the spores to the soil surface and get killed by direct exposure to sunlight (iv) Application of farm yard manure or compost or green manure at 5.0 tons per acre helps to increase the population of soil microbes, such as *Bacillus subtilis* and several species of Actinomycetes, which are present in the soil naturally. The bacteria *Bacillus subtilis* produces an antibiotic, **'Bulbiformin'** and the Actinomycetes also produce several other antibiotic substances, which are antagonistic to this pathogen (v) Soil amendment with cotton seed cake or groundnut cake at 100 kg. per acre has also been found to increase the soil microbial population, which are antagonistic to this pathogen.

Seed treatment (i) Seed treatment with captan - 4 gm. or thiram - 4 gm. or carbendazim - 2 gm./ kg. of seeds, 24 hours prior to sowing, eradicates the spores adhering to the seeds (ii) Seed treatment with *Trichoderma viride* at 4 gm./ kg. of seeds, prior to sowing helps to combat the pathogen in the soil.

Resistant varieties. ICP.4769, ICP.7118, ICP.7182, ICP.8863, ICP.9168, ICP.10958, and ICP.11299 are cultivars that are found to have resistance to this disease.

2. Powdery mildew of pigeon pea

Oidiopsis (Leveillula) taurica

The disease occurs in several red gram-growing countries, such as Ethiopia, Kenya, Tanzania, Uganda, Zambia and India. In India, the disease is found in the states of Telungu Desam, Tamil Nadu, Bihar, Gujarat, Karnataka, Maharashtra, Orissa and Rajasthan.

Symptoms. The disease usually attacks the older leaves. In case of severe infection, white, powdery growth of the fungus is seen in patches on the upper surface of the leaves. Infection on the petioles and stems may also occur. The affected leaves become chlorotic in patches initially and then turn yellow. Later the entire leaves turn yellow and drop off prematurely. When the incidence of the disease is severe, premature leaf shedding occurs on a wide scale.

The causal organism. The mycelium of the fungus is endophytic. The hyphae penetrate into the leaf through the stomata and spread in-between the mesophyll cells and send in globular haustoria into the cells. From the endophytic mycelium, conidiophores grow out of the stomata singly or in fascicles. The conidiophores are often branched, septate and up to 280μ long. Conidia are produced singly at the end of the conidiophores, on the upper surface of the leaves. The conidia fall off, as soon as they are formed and no conidial chains are formed. The conidia are single-celled, sub-cylindrical, and sub-acuminate and measure 47.0 - 70.8 x 10.0 - 21.9μ in size.

The cleistothecia are symmetrical and spherical in shape and are similar in structure. They have mycelium-like, indefinite appendages, which resemble somatic hyphae. Several asci are formed inside each cleistothecium. Each ascus contains 8 ascospores.

Mode of survival, spread and epidemiology. The pathogen survives in the infected plant parts as cleistothecia. The ascospores, which are released from the asci after disintegration of the cleistothecial wall, infect the lower most leaves of the host near the ground level and cause primary infection. The pathogen can continue its life on perennial pigeon pea, volunteer plants and on the ratoon growth from harvested stubble. It survives as dormant mycelium on infected plant parts, such as axillary buds. Secondary infection occurs by air-borne conidia.

The conidia can germinate at a wide temperature range between 14° - 35°C. They do not require free water for germination and may germinate at very low humidity levels. A temperature of 25°C is optimum for the development of the disease however, the disease occurs at temperatures between 20° - 35°C. In general, a warm and humid climate is favorable for infection, development and spread of the disease. Varieties with thin and succulent leaves are more vulnerable to attack by the disease.

Disease management

Chemical control. Spraying with wettable sulfur - 1200 gm. or carboxin - 300 gm. or dinacap (Karathane) - 375 gm. in 300 lit. of water per acre, once at the onset of the disease, followed by a second spraying 15 days later is effective in controlling the disease.

3. Red gram sterility mosaic

'Sterility mosaic', which is the most important disease of red gram, is more prevalent in Sri Lanka, Burma, Thailand, Bangladesh and India. In

India, the disease was first reported from Bihar in 1931 and it occurs in the states of Maharashtra, Gujarat, Punjab, Uttar Pradesh and Tamil Nadu. If the disease attacks the crop within 45 days of sowing, the plants fail to produce flowers and pods and the yield loss may go up to 95 %.

Symptoms. Diseased plants can be easily noticed in the field from a distance, because of their pale green color and bushy appearance. The leaves of the infected plants are very much reduced in size, deformed and show mottling. The diseased plants are stunted, with short internodes and increased number of secondary and tertiary branches arising from the leaf axils. The leaves are closely packed and the floral parts grow abnormally into twisted, leafy structures and the whole plant appears as a bunch of diminutive leaves. Young leaves show vein clearing symptoms and older leaves show mosaic symptoms. The flowers become sterile. In the case of late infection, partial sterility may result.

The causal organism. *Red gram sterility mosaic virus* causes the disease.

Mode of spread. The disease is transmitted by the eriophyid mite, *Aceria cajani*. These mites, which are very minute, bury themselves deep into the leaf surface and are mostly found on the lower surface of the leaves in dense groups. They are pale crimson to orange in color. Both young and adult mites are capable of transmitting the virus. The acquisition feeding time is about 5 minutes and the time required for transmission of the virus into a new host is 20 minutes of feeding time. The mites remain viruliferous for about 48 hours after acquisition of the virus. The mites are carried from one place to another through wind or physically by some insects. The disease can also be transmitted by grafting. But the disease is not transmitted either through sap or seed. A few nematode species have also been reported to transmit the virus. Ratooning results in increased disease incidence.

Disease management

Agronomic practices (i) Self sown and volunteer pigeon pea plants should be removed and destroyed before taking up sowing of the regular crop (ii) Species of *Attilocia,* which harbor the mites during the off season should be destroyed (iii) Perennial red gram plants, which harbor the mites permanently throughout the year should be destroyed (iv) Plants showing symptoms of the disease in the early stages of growth should be uprooted and destroyed.

Chemical control. By eliminating the vectors transmitting the virus, the spread of the disease can be checked effectively. (i) Spraying the crop with wettable sulfur - 1200 gm. or dicofol (Kelthane) - 450 ml. or phosalone

- 600 ml. or monocrotophos - 450 ml. in 300 lit. of water per acre, twice, 15 and 30 days after sowing controls the vector (ii) Soil application of carbofuran 3G - 12 kg. or phorate 10G - 4 kg., mixed with 10 kg. of sand around the stem portion of the plants uniformly and irrigating the field soon after is also effective in controlling the mites.

Diseases of minor importance. Red gram is subjected to attack by several other diseases, which are mostly of minor importance. 'Dry root rot' caused by *Macrophomina phaseolina*, affects the basal stem and roots of both young and grown up plants, leading to their eventual death; 'Leaf spot' caused by *Cercospora cajani* and *C. indica* produce irregular spots on the leaves, leading to premature leaf fall; 'Leaf blight' caused by *Alternaria tenuissina* produces characteristic spots, which appear as dark and light-brown concentric rings; 'Frog's eye spot' caused by *Phyllosticta cajani* produces round, necrotic spots with dark- brown margin and gray center like a frog's eye; 'Anthracnose' caused by *Colletotrichum cajani* and *C. graminicola* cause sunken lesions on the stems and branches; 'Bacterial leaf spot and stem canker' caused by *Xanthomonas campestris* pv. *cajani* produces irregular or angular, water-soaked lesions on the leaves and petioles, as well as dark-brown cankers on the stems and branches; *Diplodia cajani* and *Phoma cajani* cause stem cankers; *Mung bean yellow mosaic virus* causes 'Yellow mosaic disease'.

Chick pea or Bengal gram *(Cicer arietinum)*

1. Blight of chickpea or Bengal gram

Ascochyta rabiei

(Mycosphaerella rabiei)

'Blight of Bengal gram', commonly known as **'*Ascochyta* blight'** is one of the most serious diseases of this crop and is prevalent in West Asia, Northern Africa and Southern Europe. It is a very common and serious disease in Punjab and Uttar Pradesh

Symptoms. All aboveground parts of the plants are affected. Circular spots appear on leaves and pods and elongated spots on petioles and stems. On leaflets, round or elongated spots develop, which become irregularly depressed, brown spots and are surrounded by a brownish-red margin. The spots may coalesce and the entire leaf turns brown, presenting a scorched appearance. On the green pods, the lesions are usually circular, with dark margin and pinhead-like dark dots representing the pycnidia are seen in concentric circles. Seeds within the pods may show brown lesions. On the stems and petioles, the lesions are elongated, 3 - 4 cm. in length, brown and bear black pinhead-like dots consisting of pycnidia. These lesions may gradually girdle the affected region, as a result the parts above the lesions droop and wilt. If the collar region is girdled, the whole plant dies. If moist, wet weather prevails, the spread of the disease is very rapid **(Fig.34).**

The causal organism. The mycelium of this pathogen is hyaline to brownish and septate. The pycnidia are immersed in the host tissue and become erumpent. They are globose, dark-brown and 140-200μ in diameter. The wall is composed of one to two layers of elongated, pseudoparaenchymatous cells and the ostiole is very prominent. Conidiophores are hyaline. The conidia (pycnidiospores) are hyaline, oval to oblong, straight or slightly curved, single-septate, some non-septate, slightly or not constricted at the septum, rounded at both the ends and measure 10.0-16.0 x 3.5μ in size. When the pycnidia get moistened, they absorb water, swell and a slimy mass of conidia oozes out through the ostiole. The conidia germinate by long germ tubes. Within the pycnidia, the conidia can remain viable up to a year or more.

The perfect stage is not found in India, but has been reported from Eastern Europe. The pseudothecia are produced on the plant debris, especially the pods. They are dark-brown or black, globose, with inconspicuous ostiole and measure 120-250 x 75-152μ. The asci are cylindrical to clavate, slightly curved, pedicellate and measure 48.0-70.0 x 9.0-13.7μ. The ascospores are single-septate, with one cell larger than the other, prominently constricted at the septum and measure 12.5-19.0 x 6.7-7.6μ in size.

In India, two physiological races of the pathogen have been reported.

Mode of survival, spread and epidemiology. The fungus survives as pycnidia on diseased plant debris left over in the fields and also on seeds. Mycelium may be present in the seed coat. If deep lesions are formed on the seed, pycnidia may be produced on the seed coat. Conidia disseminated from soil or seed by raindrop splashes, wind, insects and other agents cause primary infection on the hypocotyl, epicotyl and base of the stem, which spreads to the aerial parts

The rapidity of disease build-up and spread is more in wet weather, accompanied by strong winds and when the temperature is around 22-26°c. Years of high rainfall during the crop season favor epidemics of the disease.

Fig. 34: Blight of chickpea-*Mycosphaerella rabiel (Ascochyta rabiei)*

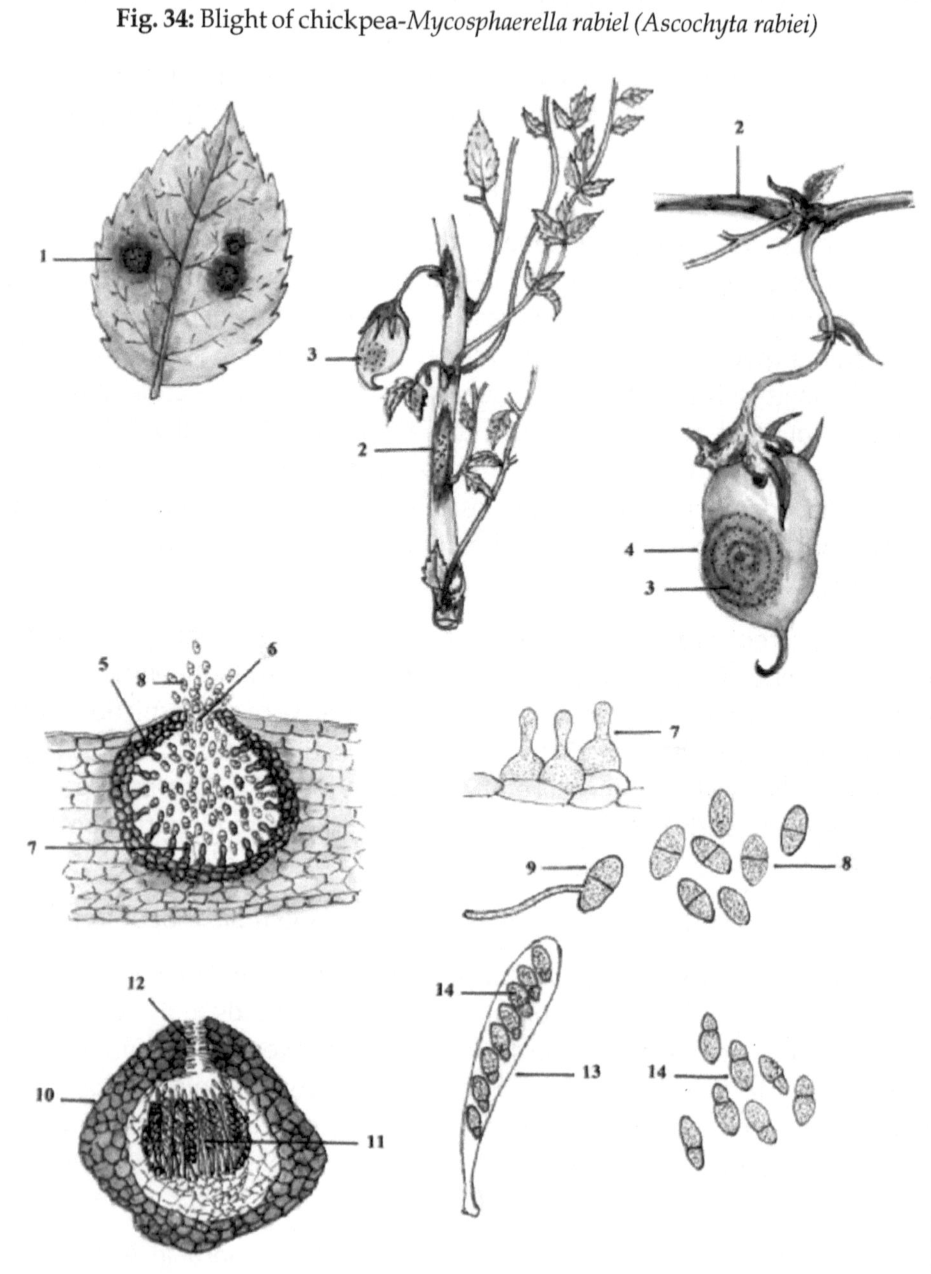

1. Lesions on the leaf 2. Lesions on the stem 3. Lesions on the pod 4. Pycnidia 5. Pycnidium 6. Ostiole 7. Conidiophore 8. Conidia 9. Germinating conidium 10. Pseudothecium 11. Hymenium (Asci and paraphyses) 12. Periphyses 13. Ascus 14. Ascospores.

Disease management

Agronomic practices (i) Field sanitation and cultural practices, such as removal and destruction of dead plant debris, deep ploughing to bury the debris deep into the soil, deep owing, crop rotation, selection of disease free seeds etc. will go a long way in minimizing the chances of occurrence of the disease.

Seed treatment. Since the disease is externally and internally seed-borne, seed treatment is an important means of controlling the disease. Seed treatment with tridemorph (Calixin M) or thiabendazole at 2 gm./ kg. of seeds affords effective control of the disease from seed-borne infection.

Chemical control. Foliar spraying with chlorothalonil (Bravo, Daconil or Kavach) - 300 ml or tridemorph (Calixin M) - 250 ml. in 250 lit. of water per acre, twice, at fortnightly interval is effective in controlling the disease.

Resistant varieties. Use of resistant varieties is the best and the cheapest method of preventing the occurrence of the disease. ILC.72, ILC.191, ILC.3279, ILC.3856, ICC.3634, ICC.4200, ICC.4248, ICC.4368, ICC.5181 and ICC.6981 have been found to be resistant to this disease.

2. Fusarium wilt of chickpea

Fusarium oxysporum f.sp. *ciceri*

The disease has been reported from Myanmar, Iran, Malawi, Mexico, Peru, Syria, the United States of America, Nepal, Pakistan, Spain, Tunisia, Bangladesh, Ethiopia and India. In India, it is a severe disease and occurs in all the states where chickpea is grown

Symptoms. The characteristic symptom of the disease is wilting of the leaves and sudden death of plants at all stages of growth. The leaves of 3 - 5 weeks old, young plants, turn dull-greenish-yellow, the plants collapse, fall to the ground and die. The collar region, up and down the soil level to a length of 2.5 cm. or more are shrunken, but no rotting of the tissues are seen. When split open longitudinally, the internal tissues show dark-brown to black, continuous streaks from the roots upward into the stem region.

Grown up plants also show wilting symptoms downwards from the upper part of the plants. The leaves droop, turn yellow and later become brown and dry. Dried leaves remain attached to the diseased plant and are not shed. Initially, affected plants show no rotting or drying. If the stem of such infected plant is split longitudinally, black discoloration of the xylem and pith are seen as long, continuous streaks, extending from the roots upward to a height of several centimeters in the stem. Sometimes, only a few branches of a plant are affected and in such cases, partial wilting occurs. Under favorable conditions, aerial mycelium of the fungus is seen as white to gray, fine, cottony mat on the collar region of the stem near the ground level and masses of slimy, pinkish spores are also seen.

The causal organism. The mycelium of the fungus is both inter- and intracellular, hyaline and septate. Microconidia are produced in large numbers from the tips of hyphal branches. They are oval to elliptical, unicellular, hyaline and measure 5.0 - 11.0 x 2.5 - 3.5μ in size. Conidiophores are formed in sporodochia or sometimes directly from the mycelium. They are single or branched and macroconidia are produced from their tips. Macroconidia are straight, fusoid or sickle-shaped, pointed at both the ends, hyaline, 3 - 5 septate and measure 25.0 - 65.0 x 3.5 - 4.5μ in size. Terminal or intercalary chlamydospores are also produced singly or in chains. They are thick-walled and smooth.

Mode of survival, spread and epidemiology. The fungus is mainly soil-borne and can lead a saprophytic life in the soil on the stubble of diseased plants for a number of years. The chlamydospores also can remain in the soil in a dormant state for many years. The disease is also seed-borne to some extent. As in the case of wilt of pigeon pea, infection occurs through the root system. Entry of the pathogen into the host plants is facilitated by injuries caused by soil-inhabiting nematodes and by mechanical means. Wilting occurs due to plugging of the xylem vessels by the mycelium, chlamydospores, microconidia and the gummy substance produced by the fungus. Production of some toxic substances in the vascular system, as well as in the root zone of the affected plants is also another reason for wilting. The fungus produces certain enzymes that cause disintegration of the cells.

The disease develops at a wide range of temperatures between 17° - 29°C. It can also tolerate a wide soil pH range of 4.6 - 9.0. The pathogen can attack pigeon pea and pea. Four physiologic races of the fungus have been reported.

Disease management

Agronomic practices. Field sanitation, selection of seeds, long term crop rotation and mixed cropping help to reduce the wilt incidence.

Seed treatment. Seed treatment with carbendazim - 2 gm. + thiram - 4 gm./ kg. of seeds, 24 hours prior to sowing is effective in checking seed-borne infection.

Resistant varieties. PPK.1 and PPK.2 are found to be resistant to this disease.

Diseases of minor importance. Several other diseases also occur on Chickpea. 'Dry root rot' caused by *Macrophomina phaseoli* attacks the roots of plants leading to death of the plants; 'Wet root rot' caused by *Rhizoctonia solani* is another root disease that kills the infected plants; *Sclerotium rolfsii* causes 'Collar rot' that affects the collar region leading to eventual death of the infected plants; 'Powdery mildew' caused by *Leveillula taurica* affects the leaves, stems and pods; 'Rust' caused by *Uromyces ciceris-arietine* produces rust pustules on the leaves, stems and pods; *Alternaria alternata* causes 'Leaf blight'; *Pellicularia filamentosa* causes 'Root rot'; *Pea leaf roll virus* causes 'Stunt disease'; Iron deficiency causes 'Chlorosis and stunting of plants'.

Pea *(Pisum sativum)*

1. Downy mildew of pea

Peronospora pisi

The disease occurs in all parts of the world, however it is more prevalent in Europe and North America. In India, it occurs in all the states, where pea is grown but is more common in the Indo Gangetic plains.

Symptoms. The disease makes its appearance, when the plants are young, with 3 or 4 leaves. Scattered, irregular, yellowish or yellowish-brown patches appear on the upper surface of the leaves and stipules. These patches may be small or sometimes large occupying a major portion of the leaf lamina. On the corresponding under surface of these patches, a grayish-violet, fine, cottony mildew growth is seen. This mildew growth consists of sporangiophores and sporangia. The infected tissues soon die and turn brown. The disease first appears on the lower leaves and then spreads to the upper ones. In the case of systemic infection, the plants become distorted and very much stunted. On the pods, the disease is noticed at the flat-pod stage as pale green, ellipsoidal or irregular lesions. These blotches gradually become bright- dark-brown in color. Seeds inside the infected pods, especially those beneath the lesions are aborted and much smaller in size. However, no fungal growth is seen on the surface of the pods **(Fig.35).**

The causal organism. The fungus causing this disease is an obligate parasite. The mycelium of the fungus is coenocytic, hyaline, endophytic, profusely branched and intercellular. Finger-shaped or branched haustoria produced from the hyphae, penetrate into the host cells. Sporangiophores develop from the internal hyphae and emerge in clusters through the stomata on the under surface of the leaves. Sporangiophores are hyaline, non-septate, long and thin, and are 400 - 700 μ in length. They are unbranched for two-thirds or more of their length and then branch dichotomously 2 - 7 times at the top. The ultimate branches, the sterigmata are at right angles or at obtuse angles with each other. Sporangia are borne singly at the tips of the sterigmata. The sporangia are hyaline, thin-walled, and elliptical and measure 22 - 27 x 15 - 19μ in size. They are short-lived and germinate readily by means of a germ tube.

Oospores are formed in the tissues of old and withered leaves or in the tissues of infected pods. They are almost spherical, greenish-yellow and measure 28 - 32μ in diameter. They have a thick epispore, marked with large, raised, reticulations and are formed within a larger, thin-walled oogonium, which disappears after some time. The oospores germinate after a rest period by issuing germ tubes **(Fig.35)**.

Mode of survival, spread and epidemiology. The disease is primarily soil-borne. The oospores present in the crop debris lying in the soil, cause primary and systemic infection of young plants. The oospores carried on the seeds along with fragments of diseased pods may also cause primary infection. Secondary local infection is caused by wind-borne sporangia. The pathogen also attacks *Pisum arvense, Vicia hirsuta, Lathyrus sativus, Trigonella polycerata* and *Lens esculentus*.

Moist and cool weather is favorable for the development of the disease, while a warm, dry weather retards its development. Relatively low night temperatures and dewfall are conducive for germination of the sporangia and infection.

Disease management

Agronomic practices (i) Diseased crop debris lying in the soil should be removed and destroyed by burning (ii) Steps should be taken to avoid growth of alternate hosts in the vicinity of pea fields (iii) Disease-free seeds should be used for sowing (iv) Crop rotation of 2 - 3 years may be followed, wherever possible.

Chemical control. As the disease is primarily soil-borne and as there are several alternate hosts for this pathogen, it is very difficult to control the disease by fungicidal applications. Spraying with mancozeb - 600 gm. in 300 lit. of water per acre, as soon as initial symptoms of the disease appear, helps to prevent further spread of the disease to some extent. While spraying care should be taken to cover the under surface of the leaves thoroughly. Spraying a mixture of matalaxyl (Ridomil) - 300 gm. + mancozeb - 600 gm. in 300 lit. of water per acre has been found to be very effective in controlling the disease.

2. Powdery mildew of pea

Erysiphe polygoni

The disease is distributed throughout the world and is more prevalent in Australia, Canada, England, India, Peru, South Africa and Western United States of America. In India, it occurs very commonly in Punjab and Himachal Pradesh.

Symptoms. Initial symptoms appear as small, irregular, powdery patches on the upper surface of leaves. These powdery patches enlarge in size and may cover the entire leaf area. When the plants are in the flowering and pod stage, the disease may assume epiphytotic proportions. Soon, the powdery, white patches completely cover the leaves, petioles, stems and even the pods, and the plants appear grayish-white from a distance. The leaves turn yellow, wither and are finally shed. The white, powdery coating, consists of conidiophores and conidia of the fungus, which are produced abundantly. Due to the infection, the epidermal cells and the mesophyll tissues penetrated by the pathogen, become necrotic and collapse. Sometimes, the fungus grows through the pods and infects the young seeds, which adhere to the pods and become brown.

Later in the season, when the production of conidia is almost over, the powdery mildew growth is interspersed with numerous, small, black, cleistothecia, which are the sexual fruiting bodies. The infected plants become very much stunted due to excessive transpiration brought about by the demands of the mycelial webs on the leaves. Further, in the infected plants the rate of respiration is increased and photosynthesis is decreased **(Fig.36)**.

The causal organism. The fungus is an obligate parasite. The mycelium is ectophytic, septate and forms a fine net work of white, superficial hyphae on the host surface. The hyphae are attached to the host surface by means of small, round, haustoria that penetrate the epidermal cells or may rarely penetrate deeper into some of the mesophyll cells. From the superficial mycelium, erect, unbranched, septate, conidiophores arise. Conidia are cut off at the ends of the conidiophores, one at a time in succession from below. Rarely they are formed in chains. They are hyaline, elliptical, single-celled and are 31 - 38 x 17 - 21µ in size. The cleistothecia, which are scattered on the surface of the white mycelium, mostly in the diseased plant debris in the soil are black, globose and are 85-126μ in diameter. The appendages on the cleistothecia are mycelioid, variable in number (10-30) and length, distinct or more or less densely interwoven with the superficial mycelium. The cleistothecia contain 4-8 asci usually, which are ovate to broadly ovate or subglobose, nearly sessile and measure 46-72 x 30-45μ in size. Ascospores are 3-5, rarely 6 in each ascus. They are hyaline, elliptical, single-celled and are 19.0-25.0 x 9.0-14.0μ in size. The cleistothecia persist in the soil until the following season, when the wall disintegrates and ascospores are liberated **(Fig.36)**.

Mode of survival, spread and epidemiology. The disease perennates through cleistothecia in the plant debris in the soil. The ascospores liberated from them, infect the lowermost leaves near the soil first. Secondary spread is carried out by means of conidia produced abundantly on the primary infections. The disease may also perpetuate as dormant mycelium in the pea seed. Pea is grown throughout the year in Punjab and Himachal Pradesh, as such the pathogen is present in the conidial stage all through the year. The primary inoculum in the form of conidia comes from these places by wind currents and infects the pea crop in the plains.

The disease usually appears in an epidemic form almost every year, when the plants are in the pod stage during January - February. Unlike the downy mildew, which is favored by moist and humid weather, powdery mildew affects the crop more, when conditions are dry and does more damage than the downy mildew. When the attack is severe, the yield loss may go up to 26-47%. It attacks several other leguminous crops, such as *Vicia, Lupinus, Lens, Trifolium* and *Medicago*. It also attacks cabbage, sugarbeet, turnip, clover and lentil.

Production and dispersal of conidia is more during the hours of daylight than in the darkness. Cool nights and warm days are favorable for the disease. Temperatures between 20°-24°c and relative humidity up to 70% favor conidial germination. Temperatures between 16°-27°c are favorable for the growth of the fungus. The conidia contain large quantities of water, which enable them to germinate in relatively dry conditions.

Disease management

Agronomic practices (i) Diseased plant debris should be collected and destroyed by burning (ii) Crop rotation may be followed (iii) Early pea varieties, which may escape disease occurrence may be raised.

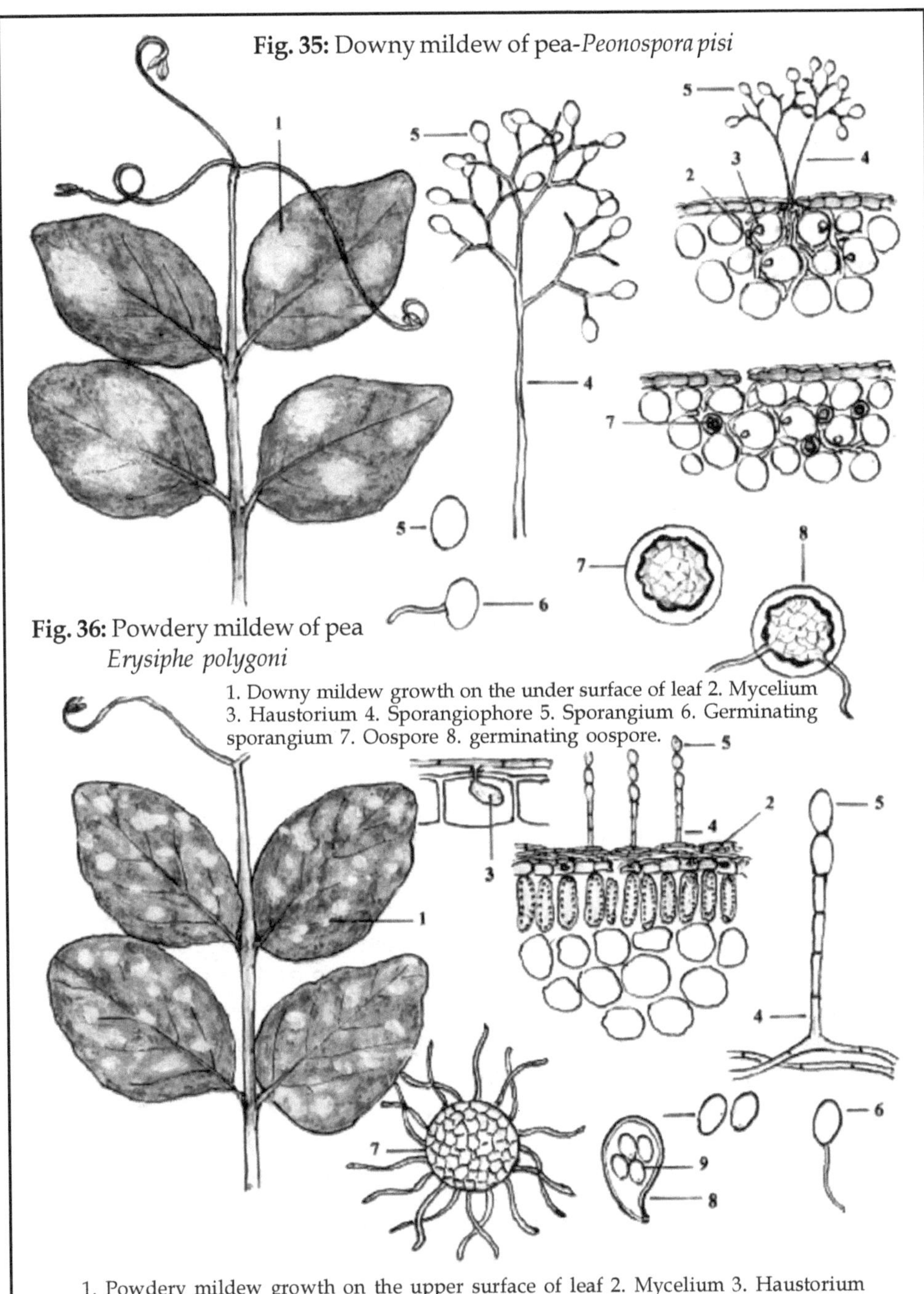

Fig. 35: Downy mildew of pea-*Peonospora pisi*

1. Downy mildew growth on the under surface of leaf 2. Mycelium 3. Haustorium 4. Sporangiophore 5. Sporangium 6. Germinating sporangium 7. Oospore 8. germinating oospore.

Fig. 36: Powdery mildew of pea *Erysiphe polygoni*

1. Powdery mildew growth on the upper surface of leaf 2. Mycelium 3. Haustorium 4. Conidiophore 5. Conidium 6. Germinating conidium 7. Cleistothecium 8. Ascus with ascospores 9. Ascospores

Chemical control. Spraying with wettable sulfur - 1200 gm. or dinacap (Karathane) - 375 gm. or tridemorph (Calixin) - 300 ml. in 300 lit. of water per acre controls the disease.

Seed treatment. Treating the seeds with captan or thiram at 4 gm./ kg. of seeds, 24 hours prior to sowing helps to eliminate the seed-borne inoculum.

Resistant varieties. The genotypes viz., P.185, P.383, P.388, P.6583, P.6587, P.6588, T.10, T.56, DPS.3, DRS.6, HUP.2, DMR.9 and Rachna show resistance to this disease.

3. Leaf, stem and pod spot, and foot rot of pea

(i) Ascochyta pinodes

(ii) Ascochyta pinodella

These closely related, but morphologically different species of *Ascochyta* attack pea plants at different stages of growth and cause leaf, stem and pod spots, besides causing foot rot. The disease is distributed in all the countries where pea is grown and causes severe damage to the crop and yield loss, when the conditions are favorable for the pathogens.

(i) Ascochyta pinodes

(Mycosphaerella pinodes)

Symptoms. The organism causes lesions on the leaves, stems and pods of pea, besides attacking the basal part of the stem below the ground level and causes foot rot. The initial symptoms on the foliage leaves, stems and pods appear as small, brown to purplish spots, which are not so delimited from the green portions. The spots increase in size, become circular in shape and turn brown to black. On the stems, the early lesions are black to purple and take the form of streaks, which are more evident at the nodal regions and may later enlarge into purplish, irregular areas over the entire stem for a length of 15 - 25 cms. above the highest point of insertion of the roots.

The underground stem just above the cotyledons (epicotyl) is attacked first and then the disease may extend a little below into the hypocotyl and passes into the main root, but the roots are invariably not attacked. Around the epicotyl, the entire stem portion of the seedling becomes a continuous, dark-brown lesion, extending up to the base of the taproot. Later, this region may become completely girdled by a dense blackening of the tissues and the shoot may become severed from the roots. This type of symptom, which destroys the basal part of the stem below the ground level, is termed **'foot rot' (Fig.37)**.

The causal organism. The pycnidia are dark, distinctly black when matured, scattered over the whole lesion, but occur more densely near the edge. The pycnidiospores are exuded through the ostiole along with a light-buff to flesh-colored exudate. The pycnidia are minute, sub-globose and are 75 - 175 x 60 - 160µ in size. The pycnidiospores are hyaline, oblong with rounded ends, straight or curved, mostly 2-celled, slightly constricted at the septum and measure 8.0 - 18.0 x 2.5 - 5.5µ in size. They contain numerous oil globules. They germinate by producing 1 - 3 germ tubes. The perithecia are formed on the necrotic lesions of leaves and stems in abundance after a prolonged wet weather. Perithecia are round, brownish in color, 100 - 300µ in diameter, with a short neck and are sunken in the host tissue. The asci are long, clavate, 60 - 80 x 12 - 16µ in size, thickened at the apex and provided with a pore at the apex at maturity. Each ascus contains 8 ascospores. They are hyaline, ovate to elliptical, 2-celled, with constriction at the septum and are 12.0 - 15.0 x 6.0 - 8.0µ in size **(Fig.37)**.

Mode of survival, spread and epidemiology. Pycnidia and perithecia are capable of surviving on diseased plant debris. The pycnidiospores are disseminated by splashing of rain to other plants in the immediate neighborhood and attack the lower leaves initially. The ascospores are ejected forcibly into the air and are borne by wind. The wind-borne ascospores have a wider range of infection. The fungus is also seed-borne. The mycelium and the fruiting bodies, which are present on the seeds, may also cause primary infection.

The optimum temperature for sporulation is 20° - 25°C and the optimum temperature for growth and development of the disease is 20° - 28°C. High relative humidity of 90 % and more and frequent rains during the crop growth favor disease development and spread.

(i) Ascochyta pinodella

Symptoms. This organism also attacks leaves, stems and pods and causes lesions, besides causing foot rot. The damage due to foot rot caused by this pathogen, is often much more severe than foot rot caused by *Ascochyta pinodes* in some localities. The fungus causes comparatively lesser number of spots on the leaves than *A. pinodes*. The spots tend to be more regularly circular and zonate. On the pods, the spots are more distinctive. They are very small, brown to black and scattered obliquely to the long axis of the pod, which is characteristic of the disease. On the stem and the basal part below the ground level, the lesions vary in size, purplish to dark-brown in color and resemble closely the symptoms caused by *A. pinodes*. The pathogen forms a dark-brown triangular lesion on the epicotyl and from it, infection spreads up and down through the cortical tissues even down to the tap root, but root infection is very much limited. Eventually, the epicotyl becomes completely girdled and when the vascular system is invaded, the roots become severed from the shoot. *A. pinodella* is more destructive to the tissues of the basal parts of the stem than *A. pinodes*. After the attack by *A. pinodella*, the stem shrinks very much and only strands of sclrenchyma are left at the foot.

Mode of survival, spread and epidemiology. Primary infection occurs through spores produced in pycnidia and perithecia, which survive on diseased plant debris in the soil. The spores are disseminated by rainsplash and by wind. The fungus is also seed-borne.

The optimum temperature for sporulation and spore germination is 20° - 30°C.

Strains of *A. pinodella* attack broad bean, lentil, vetch, soybean and lupine besides peas, but all the strains are more pathogenic to pea.

Disease management

Agronomic practices (i) Diseased plant debris in the field should be collected and destroyed by burning (ii) Disease-free seeds should be procured and used for sowing (iii) Long term crop rotation should be followed.

Seed treatment. Seeds should be treated with captan or thiram at 4 gm./ kg. of seeds, 24 hours prior to sowing, so as to eliminate the seed-borne inoculum.

4. Leaf, stem and pod spot of pea

Ascochyta pisi

The disease is common in all pea-growing countries including Britain, the United States of America, Ireland, Canada, Denmark, Germany, Holland, Australia, Ukraine and India.

Symptoms. The disease attacks mostly leaves, stems, stipules, pods and seeds. It does not usually affect the roots or the part of the stem below the ground. On leaves and pods, the lesions are round, up to 9.0 mm. in diameter, yellowish - brown to dark -brown and sunken. The spots are surrounded by a black and thickened margin. The spots on the stems and

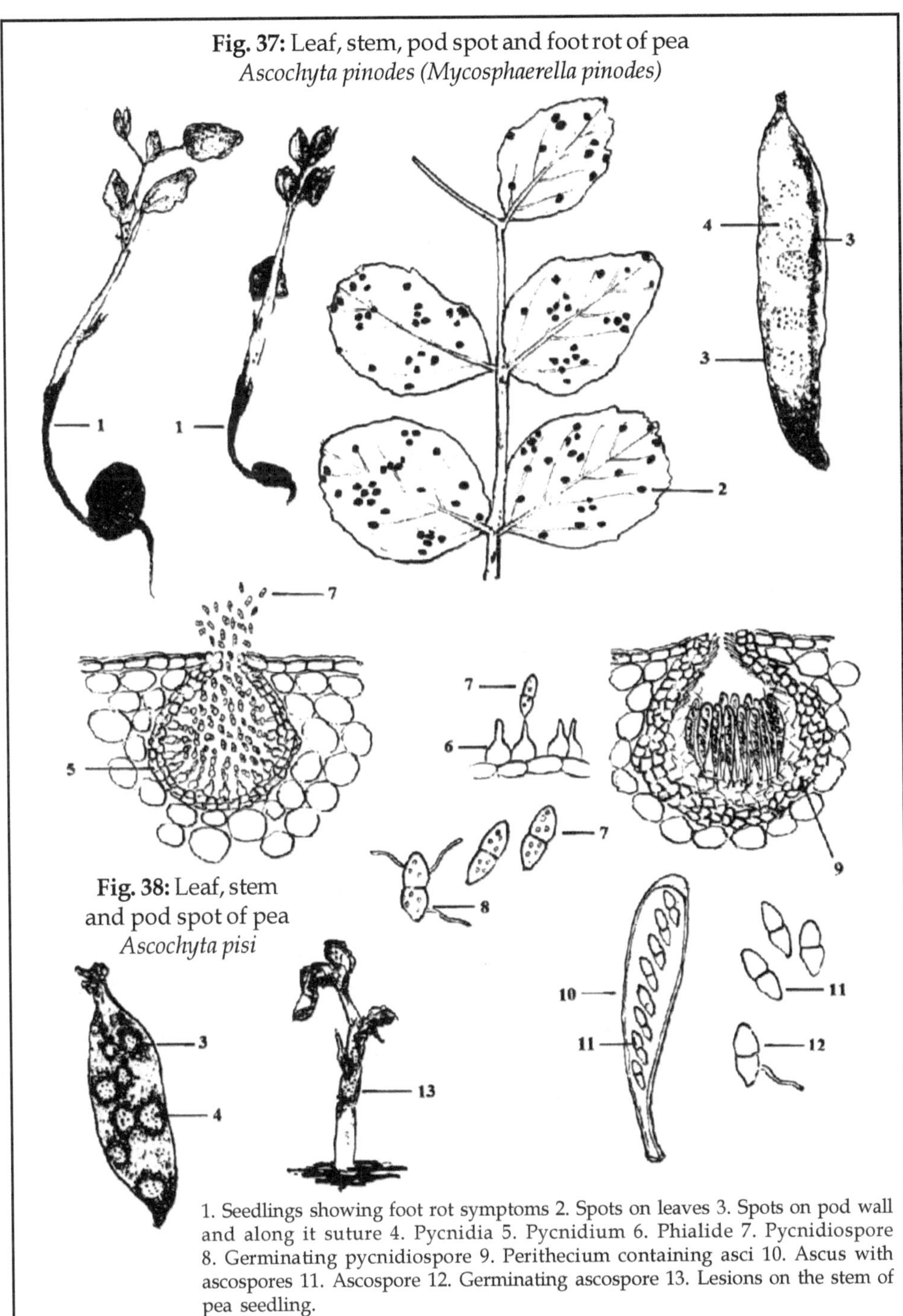

Fig. 37: Leaf, stem, pod spot and foot rot of pea
Ascochyta pinodes (Mycosphaerella pinodes)

Fig. 38: Leaf, stem and pod spot of pea
Ascochyta pisi

1. Seedlings showing foot rot symptoms 2. Spots on leaves 3. Spots on pod wall and along it suture 4. Pycnidia 5. Pycnidium 6. Phialide 7. Pycnidiospore 8. Germinating pycnidiospore 9. Perithecium containing asci 10. Ascus with ascospores 11. Ascospore 12. Germinating ascospore 13. Lesions on the stem of pea seedling.

petioles are more elongated and are yellowish-brown in color with a dark-brown margin. In case of severe incidence, the lesions on the stem may girdle the stem and the shoots above the lesions may wither. When seedlings, just emerging from the seeds are attacked, the tips of the shoots wither and the seedlings die. In the later stages of crop growth, the pods are affected and the seeds, which are in contact with the pod lesions, are also affected. Under humid conditions, the seeds show ill-defined, yellowish to grayish-brown or black, wrinkled spots. Sometimes, the infection spreads through the testa to the cotyledons and the embryo **(Fig.38)**.

The causal organism. The mycelium is hyaline, septate and inter- and intracellular. The fungus produces only pycnidia. The pycnidia are present on lesions on the leaves, stems or pods. They are yellowish-brown, round, somewhat flattened and are sunken in the necrotic tissues. They are produced in large numbers and are found in groups at the center of the lesions or in concentric circular zones, especially on the matured pods. They measure 75 - 225 x 75 - 205µ in size. The pycnidiospores, which are hyaline appear pink or reddish in mass and are exuded through the ostiole in a mucilaginous matrix. They are oblong, obtuse at the ends, 1-septate, rarely 2 - 3 septate, each with a conspicuous oil globule and measure 10.0 - 18.0 x 2.5 - 5.5µ in size. The spores germinating on moist leaf or pod, put forth one or more germ tubes and the young hyphae may fuse before the cuticle is penetrated. The hyphae do not enter the host through the stomata. Spots usually appear on the leaves in about 6 - 8 days after successful penetration and infection.

Mode of survival, spread and epidemiology. The disease is primarily seed-borne and the fungus remains viable inside the infected seeds for at least 6 years. Plants growing from infected seeds are usually not prevented from emerging from the ground and the primary lesions occur on the leaves and stem at the base of the plant. Infection may also occur from old, infected pods present in the soil. Secondary spread is through pycnidiospores, which are dispersed by splashing rain or by some insects.

Optimum temperature for sporulation is 18° - 23°C and for growth of the fungus is 15° - 23°C. Maximum infection of leaves and pods occurs during wet weather.

Disease management

Agronomic practices (i) Removal and destruction of diseased plant debris from the field (ii) Use of disease-free seeds.

Table 5: Differences between *Ascochyta* diseases of peas

Ascochyta pinodes	*Ascochyta pinodella*	*Ascochyta pisi*
1. Causes leaf, stem and pod spots and foot rot	Causes leaf, stem and pod spots and foot rot	Causes leaf, stem and pod spots only and not usually root or foot rot
2. Spots on leaves, stems and pods brown to purplish, not well-delimited. Stem spots mostly on nodes and appear as black to purplish streaks	Spots on leaves not so distinctive. Spots on pods smaller, brown or black and arranged obliquely. Stem lesions mostly on nodes and appear as black to purplish streaks	Spots on leaves and pods yellowish-brown to brown, round, sunken and with black, thickened edge. Lesions on stems and petioles elongated and yellowish-brown in color
3. Ascigerous stage, *Mycosphaerella pinodes*	No ascigerous stage found	No ascigerous stage found
4. Pycnidia scattered on the lesions, but denser near the edge of the lesions	Pycnidia scattered on the lesions. They are very indistinct and sunken in the lesions	Pycnidia found in a group at the center of the lesions
5. Spore exudate light buff or flesh-colored	No spore exudate	Spore exudate red in color
6. Pycnidiospores mostly 1-septate	Pycnidiospores mostly non septate, rarely 1-septate	Pycnidiospores mostly-1-septate

5. Bacterial blight of pea

Pseudomonas syringae pv. *pisi*

The disease is worldwide in occurrence and is quite common in Australia, England, Greece, Hungary, New Zealand, United States of America and India.

Symptoms. The disease attacks all the aboveground parts of pea plants. Symptoms on leaves appear as small, round, oval or slightly irregular spots, of about 2.5 cm. in diameter. The reddish-brown spots have a translucent center and dark-brown margin. Several such spots may coalesce to form big patches and the leaves appear blighted. On the stems and petioles, the lesions appear as dark-brown, linear streaks. In case of severe incidence, the entire stem becomes dark-brown and shrivelled and the plant may eventually die. Lesions on tender pods are dark-brown and the affected pods become malformed and shrivelled. The lesions tend to become large in older pods and the pods may dry. The seeds are also affected and they become discolored and shrivelled. The bacterial ooze coming out of the lesions dries on the surface of the affected areas and gives a glossy appearance.

The causal organism. The bacteria causing this disease are rod-shaped, with rounded ends, gram- negative and measure 1.5-1.75 x 0.5-0.75µ in size. They are motile, with 1-4 polar flagella.

Mode of survival, spread and epidemiology. The bacteria causing the disease is seed-borne and soil-borne. They can survive in the diseased plant debris and in the affected seeds during the off season. On germination of the diseased seeds, the bacteria present in the seeds, infect the emerging seedlings. Secondary infection may occur through the stomata or through wounds caused by insects or by mechanical means. The bacteria can enter the host through wounds caused by soil particles, which are blown by strong winds hitting the leaves and also by rubbing of leaves against one another. Once the bacterium enters the host, it multiplies very rapidly inside the host tissues and produces the symptoms. The bacteria can also be disseminated by irrigation water, rainsplash and by strong winds.

Moderate temperatures of about 28°-30°C, high humidity and free water on the host surface are favorable for infection and spread of the disease. The bacteria also attack cowpea and hyacinth bean.

Disease management

Agronomic practices. Selection disease-free seeds, destruction of diseased plant debris in the field, crop rotation and good drainage help to minimize the occurrence of the disease to a large extent.

Seed treatment. Streptomycin - 1 gm. is dissolved in 1 lit. of water and the seeds are thoroughly mixed with this solution, shade dried and then used for sowing.

Minor diseases. Several other diseases have also been reported to attack peas. *Fusarium oxysporum* f.sp. *pisi* causes 'Wilt disease' that eventually kills the affected plants; *Fusarium solani* f.sp. *pisi*, *Rhizoctonia solani*, *Pythium aphanidermatum*, *P. debaryanum* and *P. ultimum* cause 'Root rot' in young and grown-up plants; 'Rust disease' is caused by *Uromyces fabae; Colletotrichum pisi* causes 'Anthracnose' that produces lesions on leaves, stems and pods; *Pea leaf roll virus* causes 'Pea seed-borne mosaic'; *Pea enation mosaic virus* causes 'Enation mosaic disease'.

Cow pea *(Vigna sinensis)*

Most of the diseases that attack other pea crops are found to attack cowpea also. *Macrophomina phaseolina* causes 'Charcoal rot' or 'Dry root rot' and 'Stem blight'; *Fusarium oxysporum* f.sp. *tracheiphilum* causes 'Wilt disease'; *Erysiphe polygoni* causes 'Powdery mildew'; 'Anthracnose' is caused by *Colletotrichum lindemuthianum; Uromyces phaseoli* f.sp. *vignae* causes 'Rust disease'; *Cercospora cruenta* causes 'Leaf spot disease'; *Pythium aphanidermatum* causes 'Damping off of seedlings' and 'Stem rot'; *Xanthomonas vignicola* causes 'Bacterial blight' and 'Canker'; 'Cow pea aphid-borne mosaic' is caused by *Cow pea aphid-borne mosaic virus,* which also infects *Phaseolus vulgaris* and *Crotalaria juncea.*

Lima bean, Double bean or French bean *(Phaseolus vulgaris)*

1. Anthracnose of bean

Colletotrichum lindemuthianum

(Glomerella lindemuthianum)

The disease is of great importance and is prevalent in many countries, including India. Besides bean, the pathogen attacks cowpea, broad bean etc.

Symptoms. Plants at all stages of growth are vulnerable to attack by this pathogen however, susceptibility increases with age of the plant. The fungus is often present in or on the seeds produced in infected pods. Infected seeds may show yellowish to brown, sunken lesions. When infected seeds are sown, many of the germinating seedlings are killed before emergence. Dark-brown, sunken lesions, with pink masses of spores in the center of the lesions are often found on the cotyledons of emerging seedlings. The pathogen may destroy one or both the cotyledons. The spores spread and infect the leaves of seedlings, the stem, the bracts of inflorescence, stalks and sepals of flowers, but produce its most striking effects on the pods. The foliage leaves are not usually badly affected and the roots are not affected at all. Characteristic symptoms consist of black, somewhat sunken spots, surrounded by reddish or yellow, slightly raised margin. The spots, which begin as small, round, brownish or purplish specks, later become darker in color and enlarge in size to about 6.0 - 8.0 mm. in diameter. As the infected tissues dry and collapse, a depression is formed at the center of the lesion, which under moist conditions appear pink and oily due to the liberation of a large quantity of spores. The spots, especially on the stem region may coalesce and form long lesions, several centimeters in length. Young stems may collapse due to cracking and rotting of the tissues. The spots are quite conspicuous on the pods. The pathogen penetrates down to the seed coat and into the seed. On the foliage leaves, the lesions are mostly found on the veins and under surface of leaf stalks **(Fig.39).**

The causal organism. The imperfect stage of the pathogen, which was described as *Gloeosporium lindemuthianum*, was later changed as *Colletotrichum lindemuthianum* because of the presence of bristles or setae in the acervuli. Acervuli are produced on stromatic layers between the cuticle and epidermis. From the stroma, short, densely crowded, conidiophores arise, which abstrict conidia singly. The conidia exude through the broken cuticle and are deposited on the surface, as pink masses of spores, held together by mucilage. Conidiophores are cylindrical, hyaline, unbranched and non-septate. Conidia are hyaline, single-celled, and cylindrical to elongate-oval, sometimes slightly curved and with rounded ends. They measure 15.0-19.0 x 3.5-5.5µ. Dark-brown, stiff, unbranched and septate setae may be present among the conidiophores. They measure 30-90µ in length. In a single lesion, as many as 50 or more acervuli may be present.

The perfect stage, which is very rare, has been recorded as *Glomerella lindemuthianum*. The perithecia produced are dark and globose, with a rather long neck. Periphyses are present inside the ostiolar opening. Paraphyses are present during the early stages of development of the perithecium, but at maturity they may be sparse. Asci are sub-clavate and each ascus contains 8 ascospores. Ascospores are hyaline, unicellular, oblong, ellipsoid or curved with pointed tips **(Fig.39).**

Mode of survival, spread and epidemiology. Seeds taken from diseased pods give rise to diseased seedlings. The pathogen may be found in a viable state in the seed coat, in the fleshy cotyledons and even in the plumule. The fungus, which is largely confined to the surface tissues, breaks out and sporulates on the young stem and the cotyledons. These early-formed spores are disseminated by rain or watersplash and lead to further spread of the disease. The spores become attached to the cuticle because of the mucilaginous coat, germinate, form an appressorium and then directly penetrate the cuticle and enter the host. The fungus can also survive in the plant debris in the soil for long periods.

Cool, moist weather during summer affords the best conditions for the prevalence of this disease and high atmospheric humidity is essential for successful infection and dissemination of spores.

Fig. 39: Anthracnose of bean-*Glomerella limdemuthianum* (*Colletotrichum lindemuthianum*)

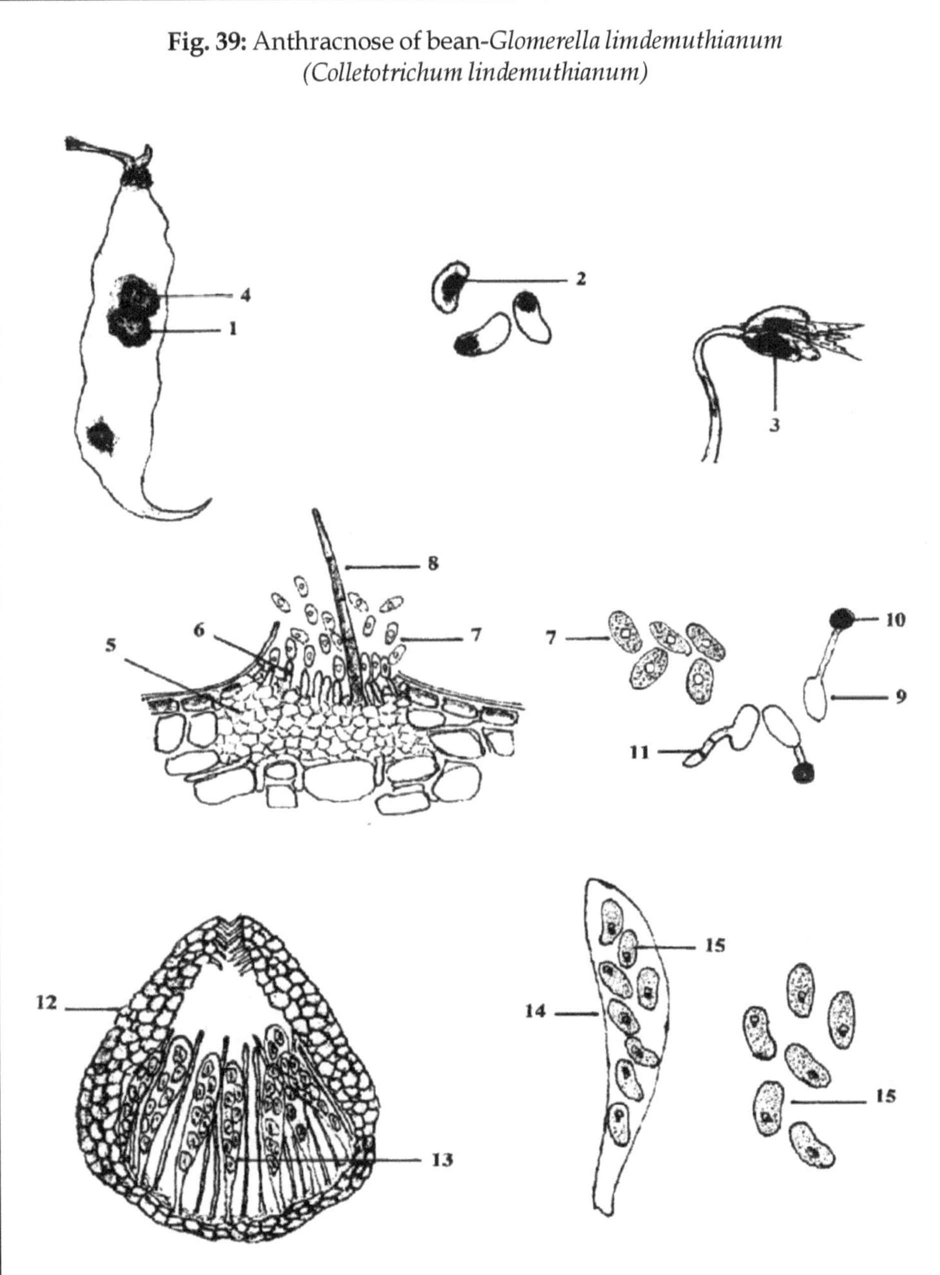

1. Lesions on the pod 2. Lesions on the seeds 3. Lesions on the cotyledons 4. Acervuli 5. Acervulus 6. Conidiophores 7. Conidia 8. Seta 9. Germinating conidia forming appressoria or septa 10. Appressorium 11. Septum 12. Perithecium 13. Hymenium (Asci and paraphyses) 14. Ascus 15. Ascospores.

Disease management

Agronomic practices (i) Priority should be given to obtain disease-free seeds for sowing. Seeds should be collected from healthy pods, which show no symptoms of the disease (ii) The plant debris should be removed and burnt or ploughed and buried deep into the soil (iii) A two to three year crop rotation should be followed to avoid inoculum build-up in the soil.

Chemical control. Foliar spraying with carbemdazim - 250 gm. or mancozeb - 500 gm.in 200 lit. of water/ acre, soon after the appearance of the disease is found to be effective.

Seed treatment. The seeds should be treated with carbendazim (Bavistin) or carboxin (Vitavax) at 2 gm./ kg. of seeds.

2. Rust disease of bean

Uromyces phaseoli typica

The disease is worldwide in occurrence and attacks different types of beans. In India, it occurs in all parts of the country, where beans are cultivated. The disease may prove to be very destructive during some seasons by causing severe defoliation.

Symptoms. The disease mostly attacks the leaves, rarely the stems and petioles. Sometimes, the pods may also be infected. The disease is more common on the under surface of leaves, but may appear on the upper surface also in some cases. The initial symptoms appear as minute, whitish, slightly raised pustules. These rust pustules may enlarge up to 2.0 mm. in diameter and become reddish-brown. In some cases the sori are surrounded by a yellow halo. These sori contain uredospores in large numbers. A ring of secondary sori may be formed around the primary sori. When the sori rupture by breaking the host epidermis, mass of powdery uredospores are released. Too many of these sori and the uredospores on the leaf surface give a rusty appearance to the leaves. Telia appear quite late in the season. With the formation of teliospores in the same sori, they turn dark-brown or black. Severely affected leaves turn yellow, dry and fall off prematurely leading to acute defoliation **(Fig.40).**

The causal organism. The fungus was earlier identified as *Uromyces appendiculatus*. The pathogen is an autoecious, macrocyclic fungus and all the spore stages occur on bean plants, but the pycnial and aecial stages are rarely found. The brownish, powdery uredospores give the uredia a reddish-brown color. The uredospores are pedicellate, globoid or ellipsoid, single-celled, echinulate and are 20 - 33 x 16 - 23µ in size. The walls are golden-brown in color. They have 2 equatorial or super equatorial germ pores. Teliospores are globoid or broadly ellipsoid, pedicellate, single-celled, smooth or with few verrucose marks and measure 24 - 32 x 20 - 26μ in size. The wall of the teliospore is chestnut-brown in color. Teliospores have an apical germ pore, with a hyaline papilla over it. Pycnia appear as yellowish dots on the upper surface of leaves. Pycniospores are very small, ovoid, hyaline single-celled, and thin-walled. On the under surface of leaves, around the opposite side of the pycnia, orange-colored aecia are formed. The aecia are inverted cup-shaped. Aeciospores, which are formed in chains from the basal cells of the aecium are ellipsoid, single-celled, thick-walled, hyaline, and minutely verrucose and measure 20 - 26 x 10 - 20μ in size **(Fig.40).**

Mode of survival, spread and epidemiology. The pathogen is not seed-borne. The uredospores, which are wind-borne, are the main source for the perpetuation of the disease. The fungus survives from season to season through different bean crops grown throughout the year continuously. In cooler regions, where the teliospores can survive and cause primary infection through sporidia, the disease may spread by wind-borne aeciospores also. Secondary local spread is by wind-borne uredospores.

High atmospheric humidity and fairly low temperature are conducive for disease development. When conditions are favorable for the pathogen, it can complete the infection cycle within 5 days and a new crop of uredospores is produced within 5 -10 days. Cloudy, damp and humid day and dew during the early morning hour are favorable for spore germination and infection. The uredospores, which are the repeating spores, germinate best at 15° - 24°C, while the resting spores (teliospores) found in the crop debris in the soil germinate best at 10° - 15°C. When the daily temperature reaches 34°C, there is no disease development. During periods of long-day hours, the disease spreads rapidly.

Disease management

Agronomic practices (i) Infected crop debris lying in the soil should be removed and destroyed (ii) Proper spacing should be given between plants and weeds in the field should be removed and the field kept clean, so as to avoid build-up of humidity (iii) Crop rotation may be followed.

Chemical control. Spraying the crop thoroughly with wettable sulfur - 1200 gm. or mancozeb - 600 gm. or chlorothalonil - 375 gm. or triadimefon - 300 gm. in 300 lit. of water per acre, 2 - 3 times, at 10 - 15 days interval controls the disease.

Resistant varieties. A number of resistant varieties are available and these varieties may be grown in rust prone areas.

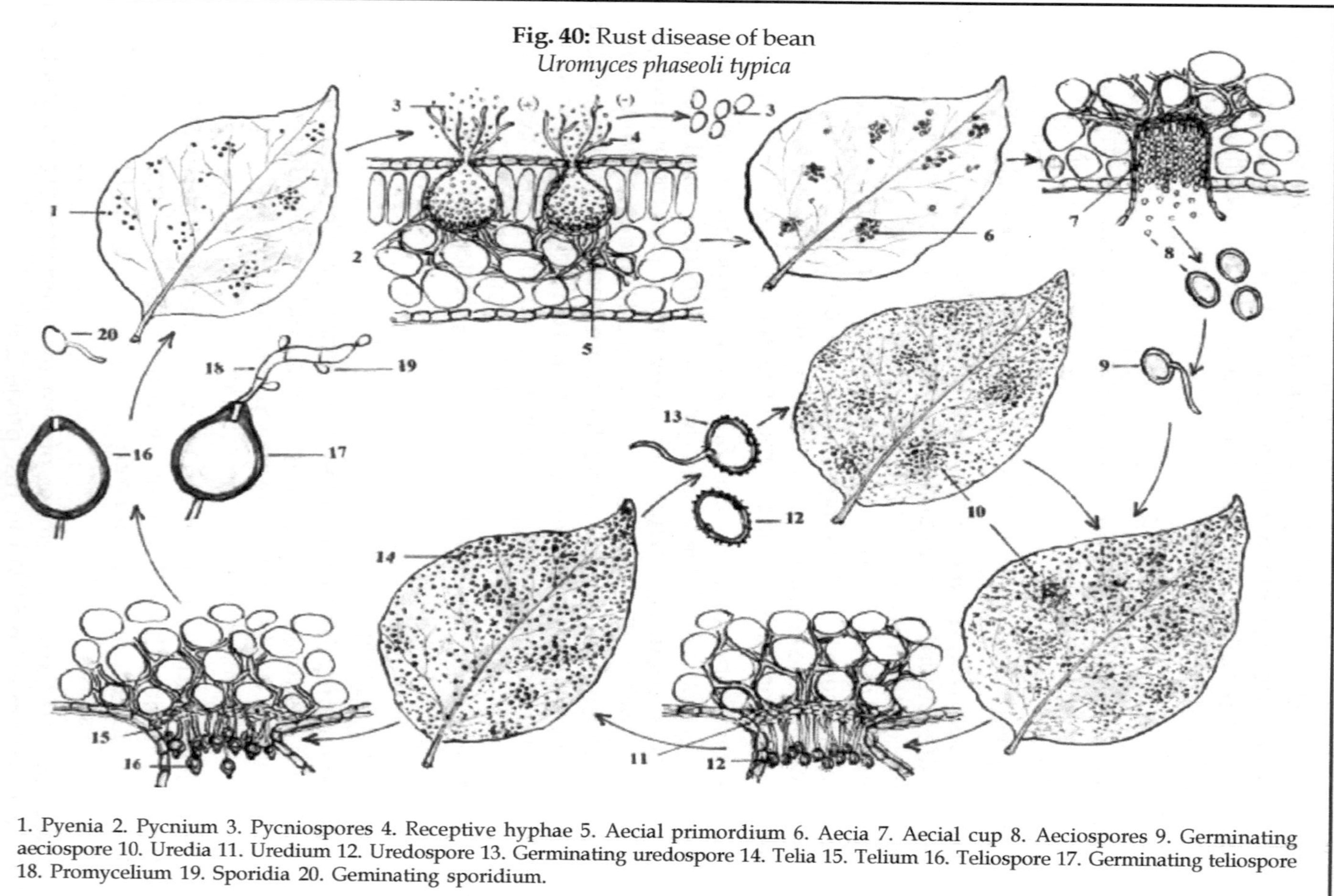

Fig. 40: Rust disease of bean
Uromyces phaseoli typica

1. Pyenia 2. Pycnium 3. Pycniospores 4. Receptive hyphae 5. Aecial primordium 6. Aecia 7. Aecial cup 8. Aeciospores 9. Germinating aeciospore 10. Uredia 11. Uredium 12. Uredospore 13. Germinating uredospore 14. Telia 15. Telium 16. Teliospore 17. Germinating teliospore 18. Promycelium 19. Sporidia 20. Geminating sporidium.

3. Foot rot of French bean

Fusarium solani f.sp. *phaseoli*

'Foot rot' or **'dry root rot'** affects all commercial varieties of French bean. The disease, which was first reported from England in 1929, is now prevalent in the entire pea growing countries.

Symptoms. No visible symptoms of infection, except a slight reduction in growth are seen during the first 5 - 6 weeks of growth. The infection, which starts at the tips of the main and lower lateral roots, causes a gradual drying of the tissues and killing them. The infection moves up until the parts near the soil surface are reached. However, the infection does not spread to the stem portion above the ground level. Usually, the symptoms are more pronounced, when the plants start producing the first pods. At that time, the lower- most leaves of affected plants begin to turn yellow, droop and wither at the edges and the pods formed become shrivelled. The diseased plants usually come to maturity earlier than the healthy plants, because of reduced rooting system and the drying effect of the disease on the tissues. The roots are affected very badly. The primary roots may be covered completely with a reddish discoloration or may be in the form of streaks, which may extend even above the soil level. Later, the discolored roots turn brown and develop cracks, sometimes extending as deep as the cortex. New roots may develop above the injured roots and these roots also get affected. In case of severe incidence, the infection may extend to the foot or base of the stem and in such cases, even the mass of fibrous rootlets developed at the foot are also destroyed and the plant may die eventually **(Fig.41).**

The causal organism. The fungus produces three type of spores viz., macroconidia, microconidia and chlamydospores. But, microconidia are rarely produced. They are very small, single-celled, oval and hyaline. In dry weather, even macroconidia are rarely formed, but develop abundantly during wet periods in the root region. The macroconidia are curved, 3 - 5-septate, mostly 3-septate, measuring 44.5 x 5.1µ in size on an average and are usually pedicellate. Chlamydospores are terminal or intercalary, arising singly or in short chains and measure on an average 11.6µ in diameter. They are formed in the host tissues mostly in the cortex of the finer rootlets and fewer in the larger laterals and taproot **(Fig.41).**

Mode of survival and spread. The fungus is a facultative parasite. It can survive in the soil as spores or mycelium and can remain active in the soil for as long as 10 years in the absence of its host. The spores are produced on old host roots and other host debris in the soil. The fungus is not carried through seeds.

The fungus gains entry into the host rootlets by either direct penetration or through wounds. Once inside the host, the fungus travels almost exclusively in the intercellular spaces of the cortex. After disintegration of the infected host tissues, strands of mycelium may be seen inside and on the exterior of the roots. The fungus rarely extends above the soil level and may reach the lower parts of the stem and in such cases the affected stem portion is also stained red. The fungus, which is concentrated mostly in the cortex of the taproots, may sometimes penetrate the vascular system and a deposit of gummy substance is found in the tissues. All varieties of French beans and scarlet runner beans are susceptible to foot rot

Disease management

Agronomic practices (i) Deep summer ploughing should be followed, so as to bring the spores and mycelium to the soil surface and get killed by direct exposure to sunlight (ii) Diseased crop debris in the field should be collected and destroyed by burning.

4. Halo blight of bean

Pseudomonas syringae pv. *phaseolicola*

'Halo blight' is world wide in occurrence and is more prevalent in Europe, America, Holland, Germany, France, Spain, Switzerland, Britain, New Zealand, Australia and India. In India, it is found in the entire major bean growing states.

Symptoms. The disease is characterized by a wilting of a part or sometimes of the entire plant. The leaves of the infected plants collapse, turn brown and cling to the plant. In case the leaflets drop off, the petioles usually remain erect. When the attack is severe, pods wither and fail to produce seeds. The symptoms on leaves vary considerably depending upon the mode of infection viz., local or stomatal infection or systemic infection through seeds. In the case of local infection, the symptoms appear as small, brownish, water-soaked, necrotic areas, surrounded by a chlorotic halo, which may vary from 1.25 - 2.50 cm. in diameter. The halo is much more evident during the early, cooler part of the season. The spots, which are formed at a later period, when the temperature rises are smaller, irregular and more numerous, and instead of each spot having its own halo, the whole lamina presents a pallid appearance. In the case of systemic infection, the spots are of indefinite shape and are furnished with a narrow, yellowish border.

Characteristic symptoms are formed on the stem, leaf petioles, pods, seeds and on seedlings. The lesions on the stem appear as elongated, chlorotic streaks, with a narrow, reddish border. Similar lesions may also appear on the stems of seedlings. Initially, the lesions appear as water-soaked area, which later turn dark- green and finally, as the spots dry out, a reddish margin is developed like that of the lesions on the older stem. Early infection of the seedlings may be local through stomata found on the stems or systemic from within the seeds. In the case of systemic infection, the bacteria within the tissues of the stem, along with large quantities of slime exudate may break out to form long streaks. Sometimes, deep cracks are formed in the stem, through which the bacterial exudate comes out as white, milky masses and get deposited on the surface of the stem. On the pods, the spots are rather circular. Many such spots coalesce to form irregular lesions. Like the spots on the leaves, they also appear water-soaked initially, but turn reddish-brown as they dry out. The bacterial ooze, which exudes from the pod lesions, dries into a thin, silvery crust on the surface. Systemic infection may cause lesions on the pod suture also. The seeds inside the infected pods may get infected through the pericarp or micropyle or the seeds may be infected systemically. The infected seeds become small, wrinkled and are yellowish in color **(Fig.42)**.

The causal organism. The bacteria causing this disease are rod-shaped and occur singly, in pairs or in chains. They are non-spore forming, motile by a single polar flagellum or 1 - 3 flagella, gram-negative and measure 1.5 - 3.0 x 0.5 - 1.25µ in size **(Fig.42)**

Mode of survival, spread and epidemiology. The disease is primarily seed-borne. Whether the seeds get infected from an external source or systemically, the bacteria collect under the seed coat and form raised blisters and invade the tissues of the testa and cotyledons. Only seeds, which are slightly infected, germinate. Some of the seedlings may be killed before emergence or after putting forth their cotyledons. But, if the infection from the cotyledons reaches the hypocotyl, then the stem is affected. The first signs of infection are manifested, when the seedlings are about 25 - 30 cm. high. Within the tissues of the cotyledons, the bacteria multiply enormously and move into the young shoot and travel upward. The stem of the young seedling is invaded by the bacteria through the vascular strands or through intercellular spaces in the parenchyma and cause destruction of the various tissues. Then, from the stem region, the bacteria pass into the leaves and then into the pods. Rapidly multiplying bacteria are exuded through the stomata or through cracks developed, in the form of milky, mucilaginous masses. Tissue destruction and occupation of water-channels may interfere with the translocation process, resulting in wilting and death of seedlings.

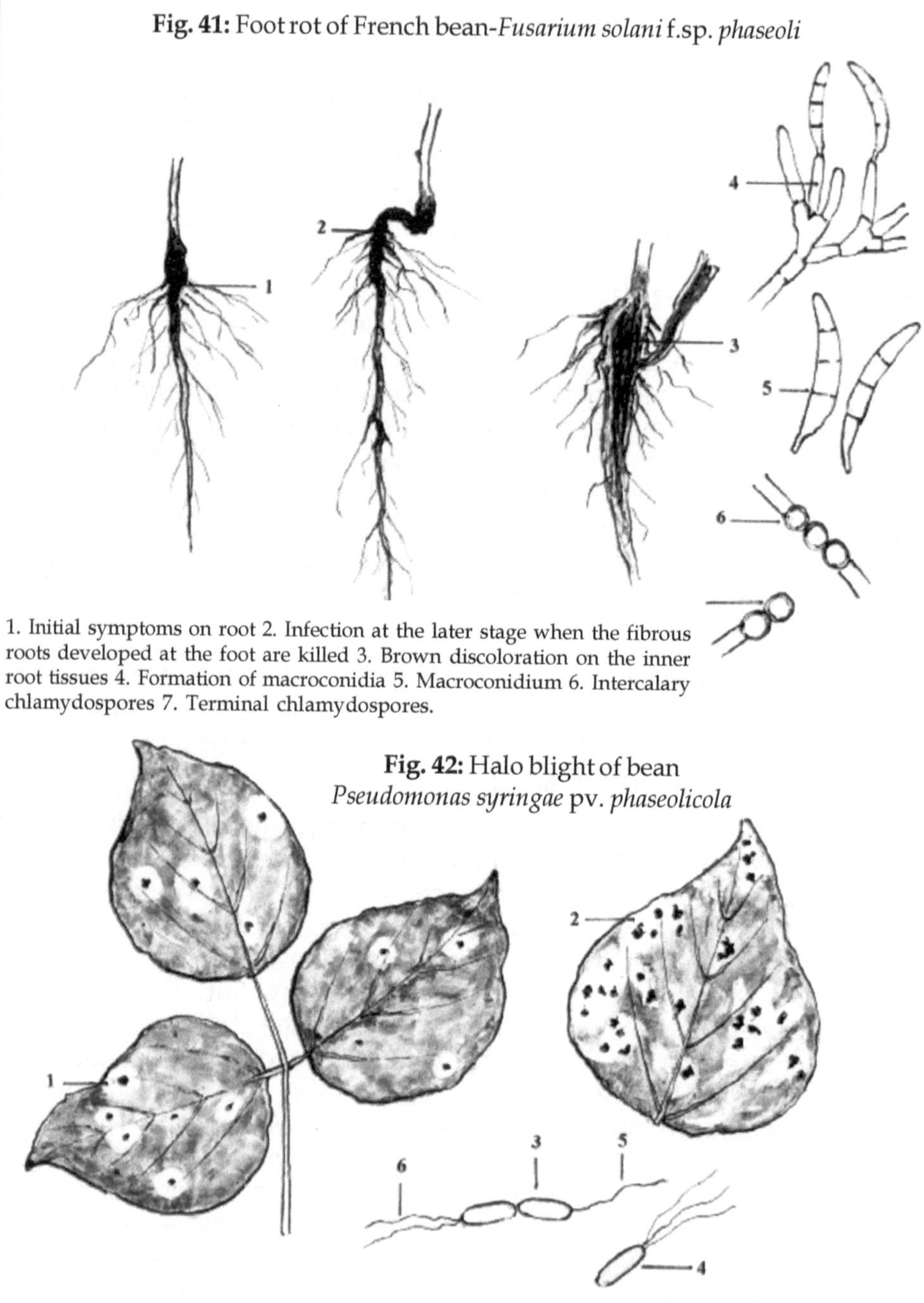

Fig. 41: Foot rot of French bean-*Fusarium solani* f.sp. *phaseoli*

1. Initial symptoms on root 2. Infection at the later stage when the fibrous roots developed at the foot are killed 3. Brown discoloration on the inner root tissues 4. Formation of macroconidia 5. Macroconidium 6. Intercalary chlamydospores 7. Terminal chlamydospores.

Fig. 42: Halo blight of bean *Pseudomonas syringae* pv. *phaseolicola*

1. Spots with chlorotic halo due to local infection . Haloed spots formed later when the temperature rises 3. Pair of bacteria 4. Single bacterium 5. Single flagellum 6. Flagella

Secondary infection is usually stomatal. Dew, splashing of rain, contact of infected plants with healthy plants and some insects are the sources of spread of the disease. The bacterial exudate washed down from the plant surface, which are carried through irrigation water may also cause secondary spread of the disease. In the stems, secondary infection mostly occurs from the infected leaves. The bacteria present in the infected plant debris in the soil may also cause primary infection in the seedlings.

Optimum temperature for the growth of the bacteria is 25°-30°C. Moist and humid conditions favor the occurrence of the disease.

Disease management

Agronomic practices. Selection of disease-free seeds, removal and destruction of infected plant debris from the field and crop rotation help to minimize the chances of occurrence of the disease.

Seed treatment. Steeping the seeds in hot water at 50°-55°C for a continuous period of 15-30 minutes, following a preliminary soaking for 12 hours in ordinary water eliminates the seed-borne bacteria.

Diseases of minor importance. Several other diseases have been reported to occur on bean. 'Charcoal rot' or 'Dry root rot' or 'Stem blight' caused by *Macrophomina phaseolina* invariably kills the affected plants; *Erysiphe polygoni* causes 'Powdery mildew'; *Fusarium solani* f.sp. *phaseoli*, *Rhizoctonia solani* and *Pythium aphanidermatum* cause 'Root rot' diseases; 'Bacterial brown spot' caused by *Pseudomonas syringae* pv. *syringae* produces necrotic spots on the leaves, stems and pods; *Bean common mosaic virus*, *Bean yellow mosaic virus* and *Cucumber mosaic virus* cause 'Common mosaic', 'Yellow mosaic' and 'Mosaic' disease respectively.

Soybean *(Glycine max)*

1. Charcoal rot or root rot of soybean

Macrophomina phaseolina

Rhizoctonia (Sclerotium) bataticola

The disease is world wide in distribution and is one of the most important diseases of soybean. The disease is otherwise called as **'summer wilt'** and **'dry root rot'**. In tropical countries, the fungus causes seedling blight also. The pathogen attacks several other crops, such as sorghum, maize, groundnut, cotton, potato, eggplant, sweet potato, flax, cucurbits, pulses etc.

Symptoms. The fungus attacks plants at all growth stages. When the seedlings emerging from the seeds are attacked, the tips of the seedlings become dark-brown and they die before coming out above the ground level. When the roots of slightly grown-up seedlings are attacked, the collar region and the stem portion above become dark-brown or black all around. The seedlings soon dry and die. If sufficient soil moisture is available and if the temperature is moderate, the seedlings may not die immediately. But, if dry conditions return, the disease intensity increases and the seedlings die eventually. This is the seedling blight phase of the disease.

When grown up plants are attacked by the pathogen, the tissues beneath the rind of the roots and basal part of the stems become light-brown in color. In the early stages of infection, no visible symptoms appear on the aboveground parts of the plants. But, when the disease intensity becomes severe, the leaves turn yellow and wilt. However, the wilted leaves remain attached to the plants and do not drop off. Superficial stem lesions are formed, which extend up from the collar region to the stem. If the outer rind of the stem is removed, large number of black, powdery, sclerotial bodies are seen on the surface of the tissues. If the stem and the main root is split longitudinally, a few black streaks are seen in the woody portion and black sclerotia are also seen in the pith region. In advanced stages of the disease, the outer rind of the stem and root disintegrates and becomes shredded. Affected plants produce very few pods with less number of seeds

The causal organism. The mycelium of the fungus is inter- and intracellular, hyaline or light-brown in color, much branched and septate. Fine, cottony mycelial growth may also be seen around the collar region of the affected plants. Mostly, the fungus produces sclerotia in large numbers. They are seen on the roots, stems, leaves and pods. They are jet-black in color, smooth and of widely varying sizes, ranging from 100 - 1000μ in diameter.

The pycnidial stage is found very rarely in nature. The pycnidia are produced singly or clustered together and are sunken in the leaves and stems. They are more or less globose, dark-brown to black in color, ostiolate and measure 100 - 200μ in diameter. Conidia are produced in succession from the tips of hyaline, short conidiophores (phialides), arranged in the inner wall of the pycnidium. The conidia are hyaline, ellipsoid to obovoid, single-celled and measure 14.0 - 30.0 x 5.0 - 10.0μ in size **(Fig.43).**

Mode of survival, spread and epidemiology. The disease is both soil-borne and seed-borne. The sclerotia can survive free in the soil for a long time and in the crop debris for a much longer time. The sclerotia can remain viable for a longer time in dry soils, whereas in wet soils, they remain viable for a much shorter time of 7 - 8 weeks only. The mycelium can survive in the soil for just about 7 days. The inoculum potential in the soil is very much increased, if the crop is grown continuously in the same field.

The seeds may carry the sclerotia on the seed coat. Infected seeds may fail to germinate or the seedlings emerging from such seeds are killed soon after emergence. Secondary spread may occur through wind-borne conidia, whenever pycnidia are formed.

Sclerotia germinate on the surface of roots and produce numerous germ tubes. Penetration of roots occurs from appressoria formed over the epidermal cells or natural openings. The fungal hyphae, which grow inter- and intracellularly through the xylem vessels, produce sclerotia in large numbers. The mechanical plugging of xylem vessels by the mycelium and sclerotia interferes with the translocation of nutrients and water to the upper parts of the plants, thereby causes wilting of the plants. The toxins and enzymes produced by the pathogen may also aggravate the disease.

Moderate temperature, ranging from 28° - 35°C and low soil moisture favor occurrence of the disease. Seedling blight incidence is more, when the soil temperature is 30°C or more at the time of sowing.

Disease management

Agronomic practices (i) Disease-free seeds should be used for sowing (ii) Closer spacing should be avoided, as crowding of seedlings reduces their vigor, thereby making them more vulnerable to attack by the pathogen (iii) In order to maintain the vigor of the plants, adequate quantities of fertilizers should be applied (iv) Whenever soil temperature rises, irrigation should be given (v) Wherever water is available in plenty, the fields should be kept flooded for about 2 - 3 weeks, so that the sclerotia present in the soil and in the plant debris may be destroyed (vi) Affected plants should be uprooted and destroyed (vii) Some of the microorganisms present naturally in the soil are capable of attacking and destroying the sclerotia. To encourage the growth of such organisms, farmyard manure, compost or green manure may be applied at the rate of 5 tons per acre (viii) Long-term rotation of crops in-between soybean may be adopted (ix) Irrigation through infected field to other fields should be avoided

Seed treatment. Seed treatment with captan or thiram at 4 gm./ kg. of seeds, 24 hours prior to sowing eradicates the seed-borne inoculum.

Fig. 43: Charcoal rot of soybean-*Macrophomina phaseolina* *Rhizoctonia (Sclerotium) bataticola*

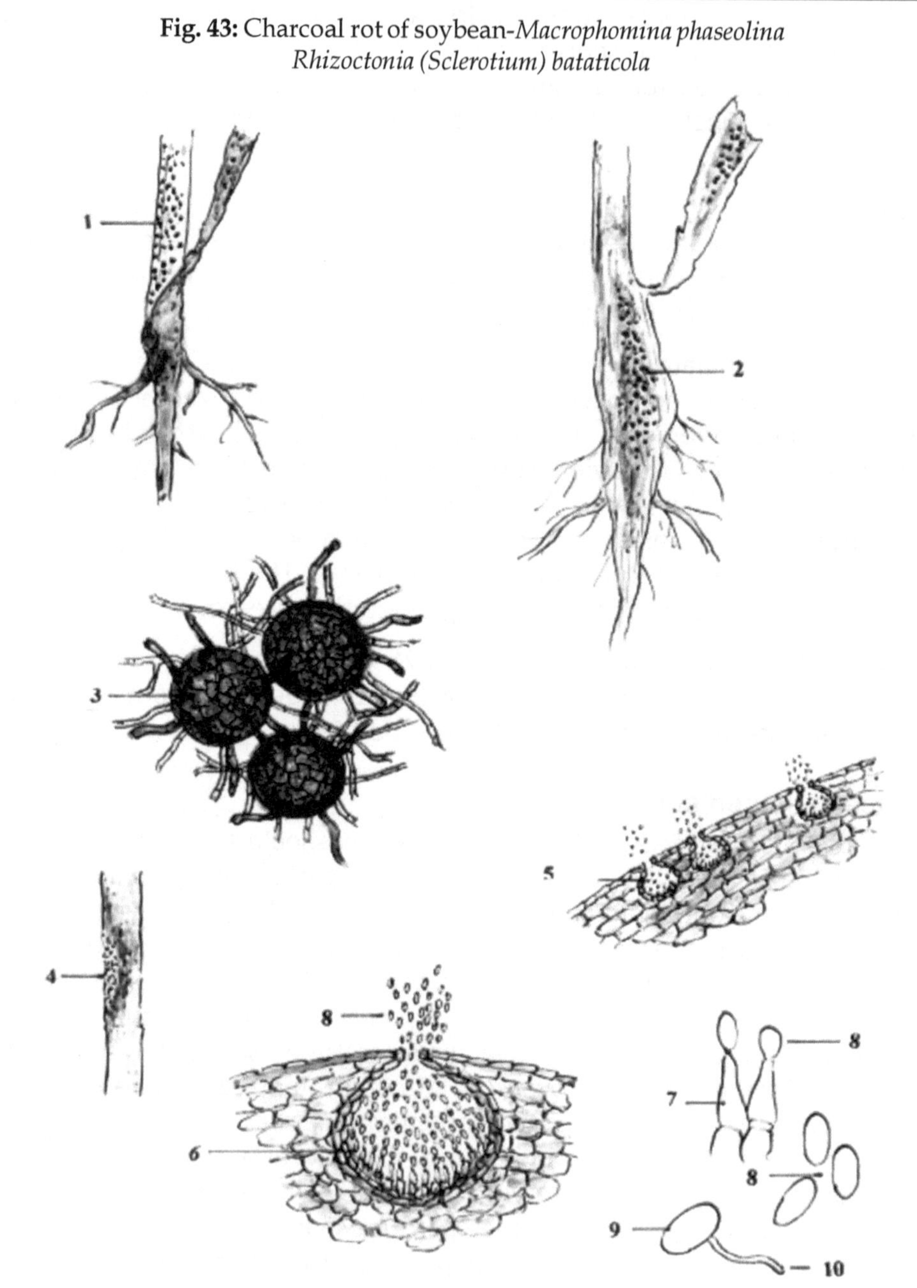

1. Sclerotia beneath the rind of stem 2. Split root and stem showing sclerotia in the pith region 3. Sclerotia 4. Pycnidia on thre stem portion 5. Pycnidia 6. Pycnidium 7. Phialide 8. Conidia 9. Genminating conidium 10. Germ tube.

2. Downy mildew of soybean

Peronospora manshuica (sojae)

The disease is prevalent throughout the world. In India, it was first reported from Kashmir in 1908 and now occurs in all parts of the country. It causes defoliation, reduction in yield and quality of grains.

Symptoms. Initial symptoms appear as irregular, yellowish-green areas on the upper surface of the leaves. In case of severe incidence, the entire leaf area becomes discolored. In advanced stages, these areas turn grayish-brown, with yellowish-green, faded margin. Under humid conditions, whitish to grayish, fine, cottony growth appears on the corresponding under surface of leaves. This growth consists of sporangiophores and sporangia of the fungus. Severely affected leaves drop off prematurely. Later, oospores are formed within the tissues of the fallen leaves. The fungus also grows within the pods and covers parts of the seeds with white, crusts of mycelium and oospores buried in it.

The causal organism. The fungus is an obligate parasite. The mycelium is coenocytic, endophytic, much branched, hyaline, intercellular and sends haustoria into the mesophyll cells. After a period of vegetative growth, numerous, erect, branched, sporangiophores emerge through the stomata, singly or in clusters on the under surface of the leaves. They are 240 - 980μ long and unbranched for a major portion of their length. Dichotomous branching occurs, 2 - 10 times at the top portion. The final branches, which are the sterigmata, are long, slender, more or less straight, pointed at the tips and are at acute angles with each other. At the tip of each of the sterigmata, a single sporangium is borne. Sporangia are sub-hyaline, single-celled, and broadly elliptical to sub-globose and are 19 - 24μ in diameter. No apical papilla is present. They readily fall off and germinate by a lateral germ tube.

The oospores, which are produced at a later period in the interior of the tissues of the fallen leaves and on the seeds, are globose, yellow or light-brown in color and measure 20 - 36μ in diameter. They germinate after a resting period by issuing a germ tube.

Mode of survival, spread and epidemiology. The pathogen perennates as oospores in the soil, in the crop debris and on seeds. They cause primary infection. Oospore contaminated seeds may cause systemic infection of seedlings under humid conditions. Secondary spread is through air-borne sporangia produced on the leaves. Young leaves are more prone to infection than older leaves, which develop some kind of resistance to localized infection.

Sporangial production occurs at temperatures between 10° - 25°C. High humidity and low temperatures of 20° - 22°C favor the occurrence and spread of the disease. Generally, cool, moist weather is highly favorable for the disease, while dry weather retards disease development.

Disease management

Agronomic practices (i) Disease-free seeds should be used for sowing (ii) Infected crop debris in the field should be removed and destroyed by burning or ploughed deep into the soil (iii) A 2 - 3 year crop rotation may be followed, as this period is enough to inactivate the oospores present in the soil.

Seed treatment. Seed treatment with captan or thiram at 4 gm./ kg. of seeds, 24 hours prior to sowing helps to eradicate the seed-borne inoculum and prevent systemic infection of seedlings.

3. Rust of soybean

Phakopsora pachyrrhizi

The disease is found in almost all the countries, where soybean is grown. It is more prevalent in Japan, Thailand, Taiwan and India.

Symptoms. The initial symptoms appear as small, chlorotic to gray-brown spots on the leaves. Small, round, rust pustules are produced in these spots on both the surfaces of leaves. The pustules increase in size and the host epidermis is ruptured and masses of reddish-brown, powdery uredospores are released. Rust pustules may be formed on the petioles and branches. Later in the season, teliosori are formed as dark- brown pustules. The disease causes premature leaf fall, reduction in pod formation and reduction in number and quality of seeds.

The causal organism. The fungus is hemicyclic. The uredia are sub-epidermal, erumpent and reddish to reddish-brown in color. Uredospores are short-pedicellate, yellowish-brown, subglobose, ovate or oblong, 15 - 21 x 15 - 28μ in size, with a finely echinulated wall. Paraphyses, which are present at the margin of the uredosorus are united at the base and forms a dome-like covering over the uredium. They are incurved, clavate or clavate-capitate, hyaline to straw colored, with a narrow lumen and are 7.5 - 15.0μ in length. The telia, which are formed later, are also sub-epidermal, but not erumpent. The teliospores are sessile, irregularly arranged in the sorus, not echinulated, single-celled and light-brown in color. The fate of the basidiospores is not known. The pycnial and aecial stages of the fungus have not been found.

Mode of survival, spread and epidemiology. The uredospores can remain in a viable state in the plant debris in the soil for 40 - 60 days. Because the pathogen has many alternate hosts, it can perennate as uredospores and cause primary infection. Once the pathogen gains entry into the host, it can produce a fresh crop of uredospores in about 10 days and the production of spores from each sorus continues for several weeks. The host range of this pathogen includes *Glycine clandestina, G. javanica, G. soja, G. tabacina, G. tomentella, Lupinus angustifolius, Phaseolus lathyroides, P. vulgaris, Vigna radiata* etc.

A temperature of 20°C., high humidity and free water on the host surface are favorable for the germination of uredospores and infection.

Disease management

Agronomic practices. Collection and destruction of diseased plant debris from the field, elimination of alternate hosts and growing resistant varieties go a long way in controlling the disease.

Chemical control. Spraying the crop with wettable sulfur - 1200 gm. or carboxin - 300 gm. in 300 lit. of water per acre, two or three times, at 10 - 15 days intervals, from the time initial symptoms of the disease are noticed controls the disease.

4. Soybean mosaic

Soybean mosaic virus

'Soybean mosaic' is one of the most important virus diseases of soybean and is found in all the countries, where the crop is cultivated.

Symptoms. Plants may be infected at different stages of growth. Primary infection occurs through seeds or by other means. Infected seeds produce diseased seedlings. These seedlings are spindly, with unifoliate leaves, curled downwards lengthwise, crinkled and are chlorotic. The subsequent trifoliate leaves produced are very much reduced in size, chlorotic, mottled and show rugosity. Plants infected in their early stages of growth become stunted, with shortened petioles and internodes. Leaves are diminutive in size, malformed, puckered and may have green enations along the veins and curl downward at the margin. The symptoms are more pronounced in new, vigorously growing leaves. Infected plants persist for a longer time than healthy plants. Symptoms may appear on the pods also. Such infected pods are reduced in size, flattened and are more acutely curved than healthy pods. They do not mature normally. The seeds from such pods are smaller in size, with poor viability. The affected plants produce lesser number of smaller root nodules.

The causal organism. *Soybean mosaic virus* or *Soja virus* -1 causes the disease

Mode of spread. The virus is mostly transmitted through seeds. The virus remains in a viable state in the seeds for at least 2 years. The virus is also transmitted by several species of aphid vectors, such as *Aphis craccivora, A. fabae, A. gossypii, Liphaphis erisimi, Myzus persicae* etc. in a non-persistent manner. The vectors become viruliferous by feeding on the leaves, stems or branches of infected plants. The acquisition feeding time is just 3 minutes and once they acquire the virus, they are capable of transmitting the virus for 25 - 30 minutes. The virus is also sap transmissible. The virus can infect a number of other hosts, such as *Cyamopsis tetragonoloba, Indigofera hirsuta, Lablab purpureus, Phaseolus vulgaris* etc. It can easily spread from these hosts to Soybean.

Disease management

Agronomic practices (i) Disease-free, healthy seeds, collected from disease-free fields should be used for sowing (ii) Plants, which are found to exhibit the symptoms in the early stages of growth should be removed and destroyed, so that secondary spread of the disease can be reduced (iii) Alternate hosts found in the fields and surrounding areas, which harbor the virus and the vestors should be removed and destroyed before taking up sowing of soybean crop.

Chemical control. Controlling the vectors, which transmit the virus, can control the disease to a large extent. Spraying with dimethoate - 600 ml. or methyl demeton - 600 ml. or monocrotophos - 300 ml. or phosphamidon - 150 ml. in 300 lit. of water per acre controls the vectors.

Diseases of minor importance. Many other diseases have also been reported on soybean *Glomerella glycines (=Colletotrichum truncatum)* causes 'Anthracnose', which is characterized by the formation of sunken lesions on the leaves, stems and pods; *Cercospora kikuchii* causes 'Purple stain' or 'Purple spot' that affects the leaves, stems, pods and seeds; *Diaporthe sojae (=Phomopsis sojae)* causes 'Pod and stem blight'; *Fusarium oxysporum* f.sp. *trachaephilum* causes 'Wilt disease'; *Pseudomonas syringae (glycines)* causes 'Bacterial blight'; 'Yellow mosaic virus disease' is caused by *Bean yellow mosaic virus.*

Broad bean *(Vicia faba)*

1. Rust of broad bean

Uromyces fabae

The disease occurs in several parts of India and causes damage by partial defoliation of the affected plants. This autoecious rust occurs very commonly on broad bean. Besides broad bean, it attacks peas, lentil *(Lens esculentus)*, *Lathyrus* sp. and *Vicia* sp.

Symptoms. Whitish spots, which soon turn brown, appear first on the leaves and later on the stems and other green parts. Reproductive spores are formed on these spots. Pycnia and aecia are rarely formed and the pycnia are not very conspicuous. The aecia are produced on the whitish spots singly or in round or elongated clusters, which are the initial, visible symptoms of the disease. The aecia are found scattered on the entire surface of the leaves and on the petioles, tendrils, and stems and even on the pods.

The uredosori are formed somewhat late in the season on both the surfaces of leaves, as well as on the stems and petioles. They are arranged in small circles and present a powdery, light-brown appearance. The teliosori occur in the uredosori and are developed from the same mycelium. They are formed on the leaves, but most commonly on stems and petioles. They are dark-brown or almost black in color **(Fig.44)**.

The causal organism. It is an autoecious, macrocyclic rust and all the spore stages occur on broad bean. The pycnia, which develop on all green parts of the plant, mostly on the under surface of leaves are flask-shaped and ostiolate. The pycniospores are very small. The aecia are also produced on the under surface of leaves. The peridium of the aecia is short, whitish, cup-shaped, with reflexed margin. The aeciospores are round to angular or elliptical, yellow, with fine warts and measure 14-22µ in diameter. The uredospores are round to ovate, light-brown, spiny, with 3-4 germ pores and measure 20-30 x 18-26µ in size. The teliospores are subglobose, ovate or elliptical with the apex rounded or flattened and the wall much thickened at this point. They are smooth, brown and measure 25-38 x 18-27µ in size. The stalk is persistent on the fallen spores, pale yellowish-brown, and thick and up to 90µ in length **(Fig.44)**.

Mode of survival, spread and epidemiology. The aeciospores play an important role in the dissemination of this rust. At lower temperatures between 17°-22°C, infection with aeciospores results in the formation of secondary aecia, while at 25°C, the infection causes development of uredia.

No infection by aeciospores occurs at 30°C or above. The viability of aeciospores lasts only for about 6 weeks and hence they may not survive during the off season and infect the next crop. The optimum temperature for the germination of uredospores has been reported to be 16°-22°C and beyond 28°C they fail to germinate. So, the uredospores may not survive the hot summer months to initiate fresh infection in the next crop. The teliospores have no dormancy period and can germinate readily at lower temperatures of 12°-22°C, but they can withstand the summer heat and remain viable for a long period of time. The teliospores, which are present in the diseased plant debris in the soil or mixed with the seeds as a contaminant, infect the new crop in the next season. So, the teliospores and the spores produced from the alternate hosts are mainly responsible for the perpetuation of the disease.

Disease management

Agronomic practices. Removal and destruction of diseased plant debris, crop rotation and destruction of alternate weed hosts help to minimize the occurrence of the disease.

Chemical control. Spraying with wettable sulfur - 1200 gm. or mancozeb - 600 gm. or tridemorph - 300 ml. or chlorothalonil - 375 gm. or triadimefon - 300 gm. in 300 lit. of water per acre controls the disease.

Seed treatment. Treating the seeds with captan or thiram at 4 gm./ kg. of seeds eradicates the seed-borne contaminants.

Lablab bean *(Dolichos lablab)*

Lablab is subjected to attack by many diseases, which occur commonly on other pulse crops. *Leveillula taurica* var. *macrospora* causes 'Powdery mildew'; *Uromyces appendiculatus (phaseoli)* causes 'Rust'; *Cercospora dolichi* causes 'Leaf spot'; *Xanthomonas phaseoli* causes 'Bacterial leaf spot'; 'Yellow mosaic' and 'Enation mosaic' virus diseases also attack Lablab.

Cluster bean *(Cyamopsis tetragonoloba)*

Cluster bean is also subjected to attack by several diseases and most of them occur in other pulse crops. *Rhizoctonia solani* causes 'Stem blight and Collar rot'; *Leveillula taurica* causes 'Powdery mildew'; *Colletotrichum capsici* causes 'Anthracnose'; *Alternaria cyamopsidis* causes '*Alternaria* blight'; *Xanthomonas campestris* pv. *cyamopsis* causes 'Bacterial leaf blight'.

Fig. 44: Rust of broad bean-*Uromyces fabae*

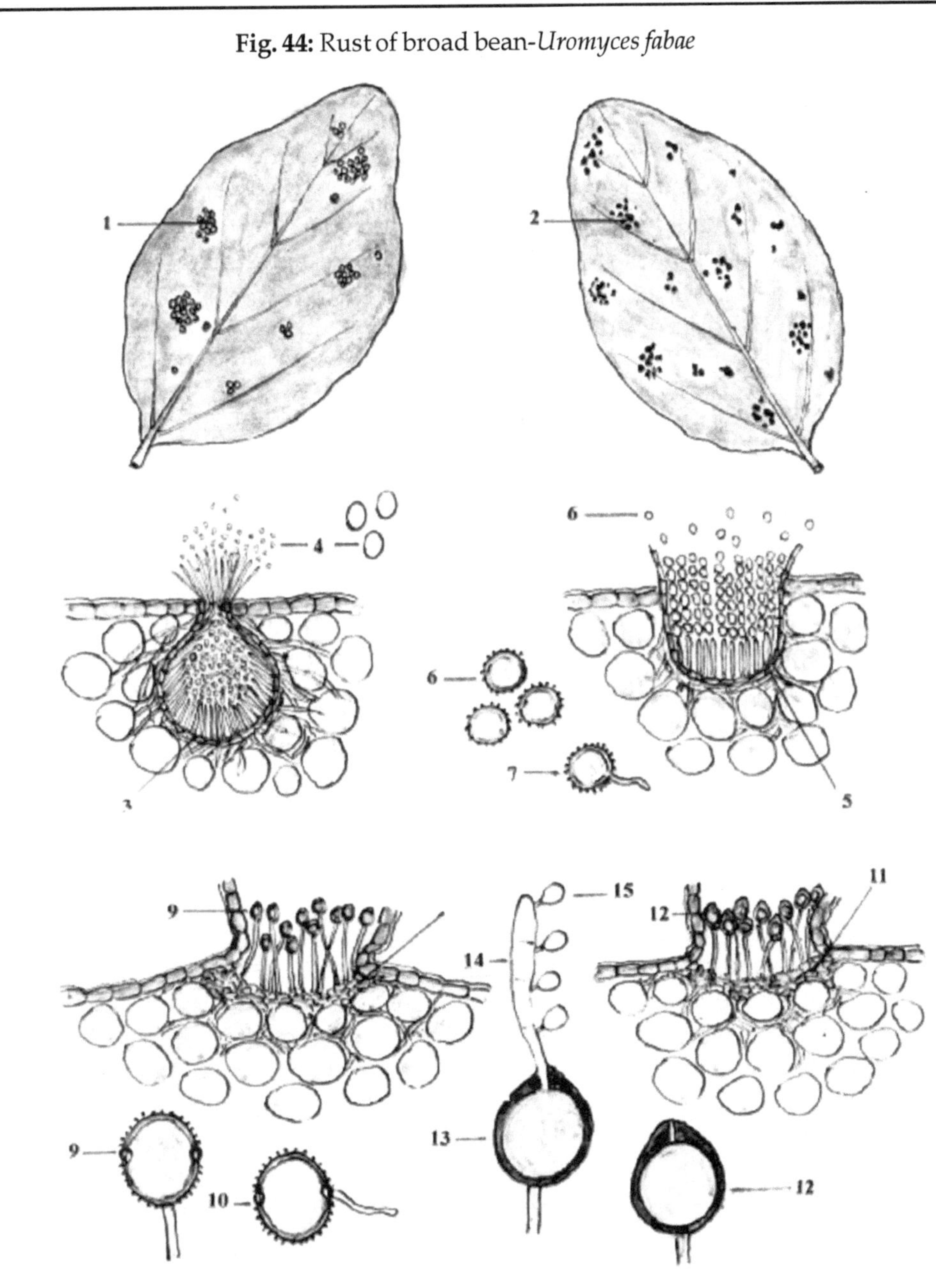

1. Aecial cups on the under surface of leaf 2. Uredosori on the under surface of leaf 3. Pycnium 4. pycniospores 5. Aecium 6. Aeciospores 7. Germinating aeciospore 8. Uredium 9. Uredospore 10. Germinating uredospore 11. Telium 13. Germinating teliospore 14. Promycelium 15. Basidiospore.

Green gram, Mungbean or Mung *(Phaseolus aureus)* and Black gram, Urdbean or Urd *(Phaseolus mungo)*

1. Yellow mosaic disease of black gram and green gram

Bean yellow mosaic virus

The disease was first reported in India in 1960 from Delhi. It occurs in all parts of the country and causes severe damage during the Kharif season. Besides black gram and green gram, it attacks soybean, pigeon pea and many other plants. Green gram is more susceptible to this disease than the other crops. If the infection occurs early in the season, there may be total loss of yield.

Symptoms. Initial symptoms appear, when the crop is about 30 days old. Small, yellow, mostly round spots appear scattered on the leaf lamina. The spots, which are diffused, expand rapidly to form irregular yellow patches, alternating with green areas. Soon, the green areas also turn yellow. Such discolored leaves gradually turn whitish and eventually become necrotic. The affected plants are quite conspicuous from a distance. The yellow discoloration increases in the newly formed leaves and sometimes, the apical leaves formed become completely yellow. The infected plants generally mature late and the number and size of the pods are very much reduced. The seeds are also reduced in size and sometimes discolored.

The causal organism. *Bean yellow mosaic virus (BYMV)* causes the disease. The virus belongs to *Potyvirus* group. The virus is a single, flexuous, rod-shaped particle, 680 - 900 nm. long and 12 nm. in diameter and contains a single, positive RNA.

Mode of survival, spread and epidemiology. The virus is not transmitted through sap, seed or soil, but only by the white fly vector, *Bemisia tabaci*. The acquisition feeding time is 15 minutes. However, after acquisition of the virus from a diseased leaf, an incubation period of 3 hours in the body of the vector is necessary before it can transmit the virus. The inoculation feeding time required is about 15 minutes. After acquisition, the virus is retained in the body of the vector for 10 days.and it continues to be viruliferous during the period. Primary and the first trifoliate leaves are most easily infected through the vector. Adult females are 3 times more efficient in transmitting the virus than the males. The vector is polyphagous and has several hosts and so, is found in nature all through the year. Transovarial transmission does not occur.

The virus has a wide host range of both cultivated and weed plants. These alternate hosts serve as a reservoir of the virus and are the source of primary inoculum. The alternate hosts include *Brachiaria ramosa, Cosmos bipinnatus, Cajanus cajan, Dolichos biflorus, Eclipta alba, Glycine max, Phaseolus acutifolius, P. lathyroides. P. vulgaris, Xanthium strumarium, Croton sparsiflorus, Acalypha indica* etc.

Temperatures between 30° - 35°C, which prevail during the kharif season are favorable for the outbreak of the vector and spread of the disease.

Disease management

Agronomic practices (i) Young plants up to 30 days, showing symptoms of the disease may be removed and destroyed to reduce the inoculum potential (ii) Weeds in and around the field, which may harbor the vector or the virus or both should be removed and destroyed (iii) Local varieties, which are highly susceptible should be replaced with resistant varieties.

Chemical control. Control of the disease by controlling the vector is effective to some extent. Because the vector can transmit the virus to several plants in a single day, insecticides, which can immobilize and kill the vectors quickly, have to be used. Spraying with methyl demeton (Metasystox) - 600 ml. or formothion (Anthio) - 600 ml. or monocrotophos - 300 ml. or neem oil 2 % (Neem oil 6.0 lit. + liquid soap - 300 ml. in 300 lit. of water) or neem seed kernel extract 5 % in 300 lit. of water per acre, 3 times, at 10 - 15 days interval controls the vector effectively.

Resistant varieties. In black gram, K 66-110, Mash - 1, NP.16, NP.21, Pant U.19, Pant U.26, Pant U.30, PLU.27, PLU.340, PLU.476, UG.135, UG.218, UPU.2, VBN.1 and VBN.2 are resistant. In green gram, Basanti, ML.1, ML.5, ML.267, MUM.2, Pant mung - 2, Pant mung - 3, PDM.11, PDM.54 and Pusa - 9072 are resistant.

Diseases of minor importance. Black gram and green gram are also affected by many other fungal, bacterial and viral diseases, most of them occur on other pulse crops also. 'Dry root rot' caused by *Macrophomina phaseolina* affects the crops at all growth stages, leading to death of the affected plants; *Cercospora canescens* and *C. cruenta* cause 'Leaf spots' that lead to premature leaf fall; *Erysiphe polygoni* causes 'Powdery mildew'; *Uromyces phaseoli-typica* causes 'Rust'; 'Anthracnose' caused by *Colletotrichum lindemuthianum*, produces sunken spots on leaves, stems and pods; *Xanthomonas campestris* pv. *phaseolicola* causes 'Halo blight'; 'Mosaic', 'Leaf crinkle' and' Leaf curl' virus diseases are caused by *Bean common mosaic virus, Urdbean leaf crinkle virus* and *Tomato spotted wilt virus* respectively.

Horse gram *(Dolichos biflorus)*

Many diseases, which occur on different pulse crops occur on horse gram also. *Macrophomina phaseolina* causes 'Dry root rot'; *Uromyces phaseoli-typica* causes 'Rust'; 'Anthracnose' caused by *Colletotrichum lindemuthianum* produces sunken lesions on the leaves, stems and pods; *Cercospora dolichi* causes 'Leaf spot'; 'Wilt' caused by *Fusarium oxysporum* results in the death of infected plants; 'Yellow mosaic' caused by *Horse gram yellow mosaic virus* infects many other crops, such as black gram, green gram, pigeon pea, soybean, bean etc. and a few weeds, such as *Indigofera hirsuta,* besides horsegram.

OILSEEDS

Groundnut *(Arachis hypogaea)*

1. Brown leaf spot or 'tikka' leaf spot

(i) Early 'tikka' leaf spot
Cercospora arachidicola
(Mycosphaerella arachidicola)

(ii) Late 'tikka' leaf spot
Phaeoisariopsis personata / Cercosporidium personatum / Cercospora personata)
(Mycosphaerella berkeleyii)

These diseases, which are also known as '*Cercospora* leaf spots' are the most important and destructive diseases and are found in almost all the countries where groundnut is grown, especially in America, Africa, the Philippines, Indonesia, Sri Lanka, China, Malaysia, Australia and India. In India, the diseases occur regularly in all groundnut-growing states and attack both rainfed and irrigated crops. Most of the cultivated varieties are susceptible to the diseases. The intensity of the diseases increases rapidly from the time of flowering up to the harvest stage. Severely affected leaves dry and fall off prematurely.

Symptoms. The time of appearance of the diseases, the symptoms produced by the pathogens and the morphology of the causal organisms differ. Leaf spots caused by *Cercospora arachidicola* appear earlier in the season, about 3-4 weeks after planting, while spots caused by *Cercospora personata* appear later after about 2-4 weeks. Hence the former is called 'early leaf spot' and the latter 'late leaf spot'. Under favorable conditions both the leaf spots may be produced on the same leaf side by side. In the initial stages, the spots produced by both the pathogens are not quite distinct and difficult to differentiate. The initial symptoms appear as slightly pale areas on the upper surface of leaves, especially the older leaves, while the corresponding lower surface shows collapse of epidermal cells and becomes brown. As the diseases progress, the two can be distinguished easily at the time of conidial production.

Cercospora arachidicola produces fewer, fairly larger and circular to irregular spots, about 1.0-10.0 mm. in diameter on the leaves, which tend to coalesce later. The spots on the upper surface are surrounded by a bright yellow halo from the very beginning. Such necrotic areas on the upper surface are reddish-brown to black in color, while they are light brown on the under surface and the yellow halo is indistinct or may not be present. Lesions may be formed on the petioles, stems and pegs.

Cercospora personata produces many, small, almost circular spots about 1.0-6.0 mm. in diameter on the leaves. The spots are dark brown to black in color on the upper surface of leaves, while they are carbon black on the under surface. The spots on the upper surface may be surrounded by a faint yellow halo in the later stages, but no halo is present in the early stages. Spots occur mostly on the leaf blades, however lesions are also produced on the petioles, stems, pegs and on the pod surface, and these lesions are oval to elongate in shape and have a more distinct margin **(Fig.45).**

Though late 'tikka' leaf spot caused by *C. personata* appear later than early 'tikka' leaf spot caused by *C. arachidicola,* it spreads much faster and due to excessive spotting and consequent necrosis, the leaves become weak, leading to severe premature leaf fall. This results in appreciable yield loss both in terms of quantity and quality.

The causal organism. The mycelium of *C. arachidicola* is intercellular in the beginning and becomes intracellular later on after the host cells are killed. No haustoria are produced. The killing of the host cells in advance of the mycelium may be the cause of formation of halo around the lesions. Conidial production is mostly confined to the upper leaf surface, but occasionally found on the lower surface also. They are sparse and not formed in concentric rings. Conidiophores are produced from stromata, which are thin and dark brown in color, 15.0-45.0 x 3.0-5.0µ in size, non-septate or 1-2 septate, unbranched and geniculate. Conidia are hyaline or pale yellow, obclavate, 38.0-108.0 x 3.0-6.0µ in size, 4-12 septate, with rounded ends or truncate with broader base narrowing towards the tip. The perfect stage of the fungus is *Mycosphaerella arachidicola.* The asci are clavate and contain eight ascospores arranged in a single row. Ascospores are two-celled, the apical cell larger than the lower cell, slightly curved, hyaline and measure 11.0 x 3.6µ on an average.

The mycelium of *C. personata* is septate, intercellular and sends in branched haustoria into the palisade and mesophyll cells, but the host cells

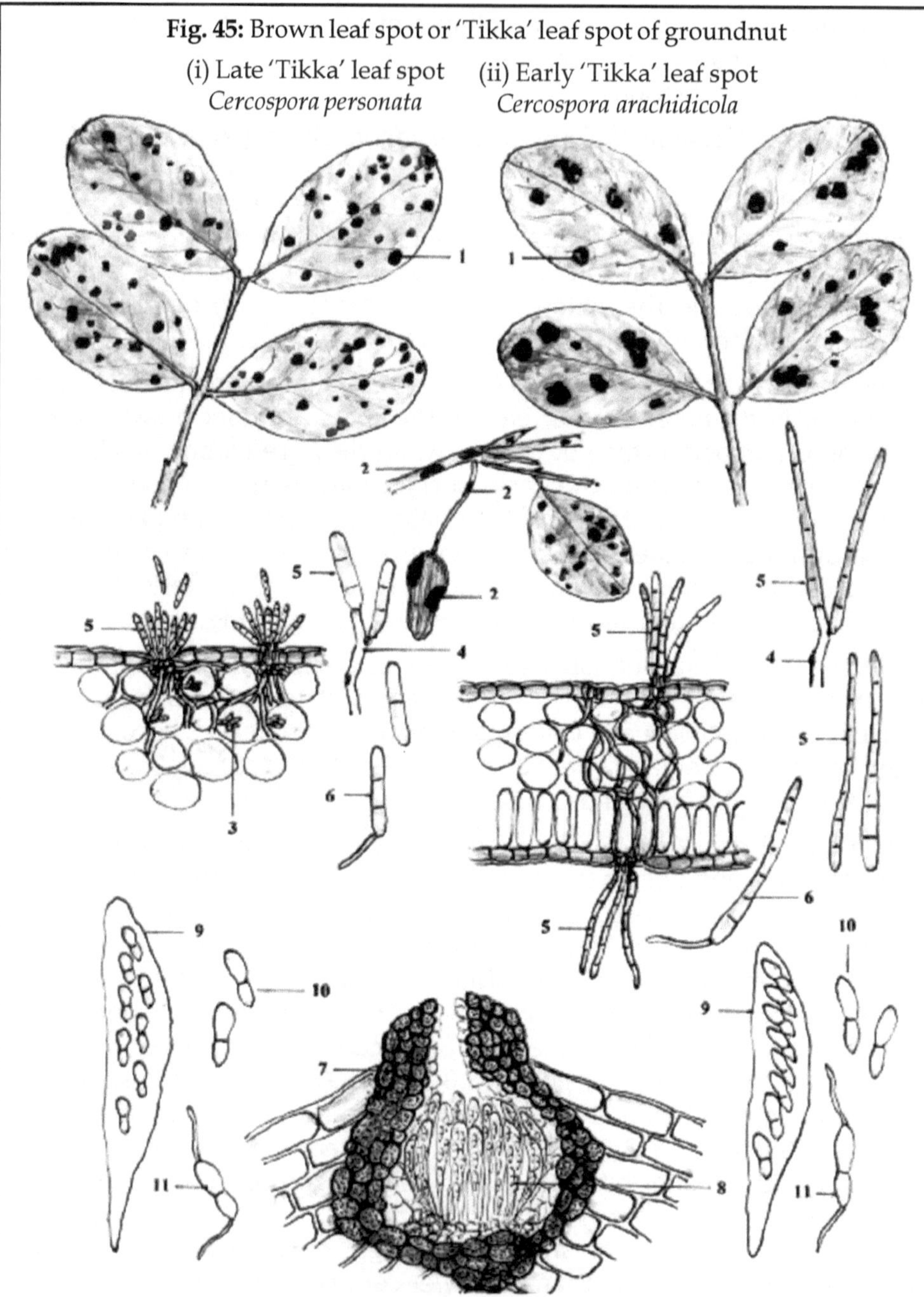

Fig. 45: Brown leaf spot or 'Tikka' leaf spot of groundnut

(i) Late 'Tikka' leaf spot *Cercospora personata* (ii) Early 'Tikka' leaf spot *Cercospora arachidicola*

1. Leafspots on the leaves 2. Lesions on the stem, peg and shell of pod 3. Haustorium 4. Conidiophore 5. Conidia 6. Germinating conidium 7. Ascocarp 8. Asci 9. Ascus with ascospores 10. Ascospores 11. Germinating ascospore 12. Germ tube.

are not killed in advance of the mycelium. Conidial production is restricted to the lower surface as cushions of conidiophores in concentric circles. Conidiophores are produced from dense, globose, brown to black stromata. They emerge by rupturing the epidermis in dense fascicles. The conidiophores are slightly thicker than the ordinary mycelium, olive-brown in color, non-septate or 1-2 septate, unbranched and geniculate. They measure 24.0-54.0 x 5.0-8.0 μ in size. Conidia are obclavate or cylindrical, light-colored, 18.0-60.0 x 6.0-11.0 μ in size, 1-7 septate and with bluntly rounded ends. The perithecial stage of the fungus is *Mycosphaerella berkeleyii*. The ascostromata are partially embedded in the leaf tissues and are broadly ovate to globose with papillate ostioles. Asci are cylindrical to clavate and contain eight ascospores. Ascospores are two-celled, constricted at the septum with the apical cell larger than the lower cell and measure 15.0 x 3.5μ on an average. The perfect stage is mostly found in dead leaves and during this stage the fungus does not cause any kind of damage to the host plants **(Fig.45)**.

Table 6 : Differences between early 'tikka' leaf spot and late 'tikka' leaf spot.

Early 'tikka' leaf spot	Late 'tikka' leaf spot
Causal organism - *Cercospora arachidicola*	Causal organism - *Cercospora personata*
Perfect stage - *Mycosphaerella arachidicola*	Perfect stage - *Mycosphaerella berkeleyii*
Symptoms - Symptoms appear about 3-4 weeks after planting	Symptoms - Symptoms appear about 2-4 weeks later than the symptom expression by *C. arachidicola*
Spots on leaves fewer, larger, 1-10 mm. in diameter, circular to irregular, which tend to coalesce. Spots reddish-brown to black on the upper surface and light brown on the under surface. Bright, yellow halo is present around the spots on the upper surface even from the very beginning.	Spots on leaves many, small, almost circular, 1.0-6.0 mm. in diameter. Spots dark brown to black on the upper surface and carbon black on the under surface. Faint yellow halo may be formed in the later stages on the upper surface around the spots, but no halo is formed in the early stages.
Mycelium intercellular in the beginning and becomes intracellular later on after the cells are killed. No haustoria are produced	Mycelium intercellular. Branched haustoria are produced and are present in the living cells.
Conidial production mostly confined to the upper surface of the spots, but may be produced on the under surface also. Conidia hyaline or pale yellow, obclavate or truncate with broader base narrowing towards the tip. They measure 38.0-108.0 x 3.0-6.0μ in size and are 4-12 septate.	Conidial production restricted to the lower surface of the spots. Conidia light colored, obclavate or cylindrical with bluntly rounded ends. They measure 18.0-60.0x6.0-11.0μ in size and are 1-7 septate.

Mode of survival, spread and epidemiology. The pathogens survive as mycelium and conidia in the soil on diseased crop debris or on the shells of pods harvested from diseased crop or on volunteer groundnut plants. The conidia on germination enter the host, either through the stomata or by direct penetration of the host epidermal cells, mostly the upper epidermis. Secondary spread of the disease is mostly by air-borne conidia and to some extent by rainsplash.

The optimum temperature range for the growth of the pathogen is 24° - 28°C. Maximum infection takes place, when the plants are about 30 days old and after 75 days the rate of infection is very low. In the infected plants, the disease progresses throughout the season and severe incidence leads to total defoliation of the plants. High relative humidity, prolonged low temperature and dew favor severe incidence of the disease. The disease is more severe in monocrop area, where groundnut is grown season after season and also in the rainfed crop. Application of heavy doses of nitrogen and phosphorus fertilizers increases the disease incidence, while application of gypsum and balanced dose of NPK fertilizers decreases the incidence of the disease. The spreading types of groundnut are comparatively less prone to attack by the pathogens than the bunch types. Micronutrient deficiency, especially magnesium deficiency makes the plants more vulnerable to attack by the pathogens. Two pathogenic races of *Cercospora personata* have been identified in India.

Disease management

Agronomic practices (i) Infected crop debris should be collected from the fields after harvest and destroyed by burning or should be buried deep in the soil (ii) Volunteer groundnut plants, that may grow in the fields should be removed and destroyed (iii) To avoid seed-borne inoculum, seeds should be selected from disease-free crops as far as possible (iv) Summer ploughing and burying of infected crop debris may be adopted (v) Monocropping of groundnut leads to inoculum build-up and severity of the diseases. So, crop rotation may be followed (vi) The crop should be sown sufficiently in advance, so that the plants may escape initial infection.

Chemical control. Chemical control measures should be taken up, as soon as the symptoms start appearing. Spraying with chlorothalonil (Daconil or Kavach)-500 gm. or Du-ter-500 gm. or wettable sulfur-1000 gm. or carbendazim-250 gm. + mancozeb-500 gm. in 250 lit. of water per acre gives adequate control of the diseases. Two to three sprayings may be given, at 2-3 weeks interval.

Biological control. *Verticillium lacanii* is found to parasitize the pathogens.

2. Groundnut rust

Puccinia arachidis

The disease was first reported from Russia in 1910 and later from China in 1937. In India, the disease was first reported in 1971. The disease occurs in a serious form in America, West Indies, Venezuela, China and Russia. Now, the disease has spread to almost all the countries, where groundnut is grown. In India, the disease is more prevalent in the states of Punjab, Telungu Desam, West Bengal and Tamil Nadu. In recent times, the disease has been very serious in Coimbatore, Salem, Trichirapalli, Pudukkottai, Madurai and Dindigul districts of Tamil Nadu.

Symptoms. The symptoms usually appear when the crop is about 6 weeks old. On the under surface of leaves, small, mostly circular, brown to dark-brown or orange pustules, 0.5 - 1.4 mm. in diameter appear. These are the uredinia. On maturity, the urediniospores rupture the epidermis of the host and come out. When large number of pustules are formed, the urediniospores come out in masses and are deposited on the leaf surface and appear as if dusted with a brownish dust. This gives the leaf a rusted appearance. On the upper surface of the leaves, brown spots develop opposite to the rust pustules. Production of large number of pustules leads to loss of chlorophyll area of the leaves, necrosis and death of entire leaves. The disease attacks most of the aerial parts viz., stem, petioles and shells of developing pods. Older leaves are more vulnerable to attack than younger leaves. The pods from infected plants become small and shrunken. The kernels from such pods are underdeveloped and the oil content is also less **(Fig.46).**

The causal organism. The fungus causing this disease produces only the uredinial stage. The telial stage has been reported only in South America. The pycnial and aecidial stages have not been found so far. The urediniospores are broadly ellipsoid or obovoid, 16 - 22 x 23 - 29µ in size, light- to dark-brown in color and the outer surface is finely echinulated. Three to four germ pores are present at equal distances along the equatorial band. They are borne singly at the tips of short, hyaline pedicels.

The telia produced later are 2.0 - 3.0 mm. in diameter, scattered and are dark-brown in color, turning to almost black on maturity. The teliospores, when matured, rupture the host epidermis and become exposed. The teliospores are oblong or ellipsoidal, single-septate, rarely 2 - 3 septate and constricted at the septum. The apical cell is rounded or tapering and thickened at the apex. They are dark-brown in color, smooth-walled, 38-42 x 14-16µ in size, with two germ pores, one at the apex and the other in the lower cell on the side near the septum. They are borne singly on the tips of short, hyaline pedicels. They have no dormancy and germinate immediately.

Mode of survival, spread and epidemiology. The disease perpetuates only by means of urediniospores. However, the urediniospores are short-lived in the infected crop debris and are unlikely to survive in the soil for more than a month. The practice of continuous cultivation of groundnut without any marked interval between successive crops in different parts of India may provide initial inoculum. The urediniospores produced from groundnut crops raised in the southern parts of India may be carried northward by wind currents during the monsoon period and may initiate the disease. Aeroscope studies have shown that rust urediniospores are carried over very long distances by wind currents and may even cross natural barriers between countries. The pathogen may also survive during the off season on volunteer groundnut plants. Movement of pods or kernels contaminated with urediniospores may also be another way of perpetuation of the disease. The fungus is reported to attack other species of *Arachis*, such as, *Arachis marginata, A. nambiquarae* etc.

Temperatures between 20° - 30°C and free water on the leaf surface favor infection and spread of the disease. A temperature range of 20° - 25°C favor germination of urediniospores. Bright sunlight inhibits spore germination and infection.

Disease management

Agronomic practices (i) Infected crop debris from the previous crop should be removed and destroyed. (ii) Volunteer groundnut plants and ground keepers should be removed and destroyed (iii) Use of seed materials contaminated with the spores should be avoided (iv) There should be a break of atleast 30 days between successive groundnut crops.

Chemical control. Spraying the crop with a mixture of carbendazim - 250 gm. + mancozeb - 500 gm. or wettable sulfur - 1000 gm. or chlorothalonil - 300 gm. or tridemorph - 250 ml. in 250 lit. of water per acre gives good control of the disease. The first spraying should be given as soon as the disease symptoms appear, followed by another spraying after 10 - 15 days.

Biological control. Mycoparasites, such as, *Verticillium lacanii* and *Penicillium islandicum* have been found to parasitize the pathogen.

3. Dry root rot or root rot of groundnut

Rhizoctonia bataticola

The disease is found in most of the groundnut growing countries, including Argentina, Israel, Gambia, the United States of America, Venezuela and India. It was first reported in India in 1927 and it is prevalent in almost all the states, where groundnut is cultivated.

Symptoms. The disease appears mostly during the seedling stage. The leaves of affected plants turn yellow in the initial stages and within a few days the plants dry and die completely. In the beginning, small, water-soaked, reddish or brown spots appear at the collar region at the ground level. The spots enlarge rapidly and girdle the collar region, leading to rotting of the collar region. The disintegrated and rotten tissues obstruct the movement of water and nutrients to the aboveground parts of the plant and consequently the leaves, branches and eventually the whole plant withers and dies. The roots are also affected and they rot and die. The bark at the collar region and roots are torn and shredded. The affected plants come off very easily when pulled slightly. On the stem and root portions, white or grayish, cottony, weft of mycelial growth is seen. In the mycelial weft, minute, pinhead-like, black, spherical sclerotia are seen in large numbers. The fungus may also affect the pegs and pods. When the affected pod is split open, numerous, minute, black sclerotia are found sticking on the inside of the pod shell and on the surface of the kernels. The kernels are under-developed and shrunken. Sometimes, the fungus grows in-between the cotyledons and produce sclerotia in large numbers **(Fig.47)**.

The causal organism. *Rhizoctonia bataticola* belongs to the group of fungi classified under Mycelia Sterilia. The mycelium is septate, much branched, inter- and intracellular. It does not form any kind of spores, asexual or sexual. Only the sclerotia serve as spores.

Mode of survival, spread and epidemiology. The disease is mostly soil-borne. The pathogen can survive on organic matter present in the soil as a facultative parasite. In the presence of favorable environmental conditions and suitable host, the fungus becomes parasitic and attacks the host. The sclerotia can remain in a viable state in the soil, as well on plant parts for prolonged periods and can withstand adverse environmental conditions. The sclerotia are also carried on the seeds.

The optimum temperature for seedling infection is 29°-30°C. Generally high temperature and low soil moisture favor the disease occurrence. While young plants are more vulnerable to attack by the fungus, plants at all growth stages may be affected. The disease occurs more in the rainfed than in the irrigated crops. The bunch varieties are more susceptible than semi-spreading and spreading types. Besides groundnut, the disease attacks green gram, black gram, red gram, cowpea, soybean, sesame, sunflower and several other crops.

Disease management

Agronomic practices (i) Disease-free, well-dried and clean seeds should be used for sowing (ii) The testa of groundnut seed is very thin and so, all precautionary measures should be taken to avoid injuries to the seed coat at the time of harvesting, shelling, transporting and storing (iii) Deep sowing should be avoided (iv) The seedlings or grown-up plants showing symptoms of the disease should be removed and destroyed by burning (v) In well-irrigated and well-maintained crop, the incidence of the disease is much less. So, proper maintenance of the crop is necessary (vi) When the temperature is high and soil moisture is less, the fungus becomes more virulent. So, adequate irrigation should be given (vii) The harvested produce should be well-dried before storage (viii) Long-term crop rotation, with non-host crops may be followed, wherever possible.

Seed treatment. The seeds should be treated with captan or thiram - 4 gm. or carbendazim or carboxin - 2 gm./ kg. of seeds, at least 24 hours prior to sowing. This will protect the seedlings for a period of 15 - 20 days from seed-borne and soil-borne infection. After chemical seed treatment, the seeds should be treated with Rhizobial culture. This will enhance the seedling vigor and encourage rapid growth of the seedlings and thus ward off disease infection.

Biological control (i) Treating the seeds with biocontrol agent *Trichoderma viride* at 4 gm./ kg. of seeds helps to destroy the fungus (ii) To encourage the growth of actinomycetous fungi present in the soil naturally, which are capable of destroying the fungus, well decomposed compost or farm yard manure should be applied to the field.

4. Seed rot or crown rot of groundnut

Aspergillus niger and *A. pulverulentus*

The disease, which was first reported from Java in 1927, is found in many other countries, such as Australia, Africa, Israel, Nigeria, the United States of America and India. In India, it occurs in the entire groundnut growing states to a larger or lesser extent.

Symptoms. The disease causes pre- and post-emergence rotting of seeds and seedlings. Under moist conditions, the fungus attacks the seeds before they germinate and causes rotting of seeds. If such ungerminated, rotten seeds are examined, they are found to be covered with masses of black, powdery spores, which give the seeds a sooty appearance. When the seedling start emerging from the seed, the fungus attacks the cotyledons and the infection spreads to the hypocotyl region. The infected areas become water-soaked, rot and turn brownish and are soon covered with black, powdery spores. Post-emergence infection results in rapid decay of the collar region, followed by desiccation and death of the whole plant. Older plants are also attacked sometimes. Lesions develop at the collar region and then spread along the branches and as a result, the branches or the whole plant may wilt, dry and die. The dried branches are easily broken and get detached from the disintegrated collar region. Later, the fungus may infect the pods and produces black, powdery masses of spores on the pod surface and on the kernels also **(Fig.48)**.

The causal organism. The mycelium of the fungus is hyaline or sub-hyaline, septate, inter- and intracellular. The conidial heads are borne at the terminal ends of septate conidiophores. The long, erect conidiophore terminates in a bulbous vesicle. All around the vesicle, radiating conidiogenous cells arise. These cells or sterigmata are usually produced in two layers one above the other. The secondary layer or phialides cut off a series of conidia, which are found in long chains. The conidia are globose, spherical, dark-brown in color and measure 2.8-4.6μ in diameter. The dark color of the conidia gives the conidial head a black color. The conidia germinate by producing germ tubes. Sexual reproduction is very rarely found.

Mode of survival, spread and epidemiology. The pathogen is commonly found everywhere and can survive in a number of substrates, including dead organic matter. Under favorable conditions, the facultative parasite invades the cotyledons or the hypocotyl of the germinating seedlings. Injuries on the seeds or seedlings make them more prone to attack by the fungus. The disease is also carried through contaminated seeds. The fungus may also be present inside the embryos of infected seeds and cause primary infection. Seeds get infected during maturation in the soil when wet weather prevails and also during harvesting, handling and shelling.

The fungi can tolerate a wide range of temperature and moisture conditions. But temperatures between 30°-35°C and fairly high soil moisture levels favor disease occurrence and spread. Conditions that prolong germination and emergence, such as deep sowing, high temperatures and fairly high soil moisture levels increase the chances of infection and severity of the disease. Spreading varieties are more susceptible than bunch varieties.

Disease management

Agronomic practices (i) Infected crop debris in the field should be removed and destroyed (ii) Disease-free, sound seeds should be used for sowing (iii) Care should be taken to avoid mechanical injuries to the produce at every stage (iv) Deep sowing should be avoided (v) Plants showing symptoms of the disease should be removed and destroyed (vi) The harvested produce should be properly dried before storage.

Seed treatment. The seeds should be treated with captan or thiram at 4 gm. or carbendazim or carboxin at 2 gm./ kg. of seeds, 24 hours prior to sowing.

5. Bud blight, bud necrosis or ring mosaic of groundnut

Tomato spotted wilt virus

The disease has been reported from the United States of America, Australia and India. In India, it is found in Punjab, Telungu Desam, Karnataka, Maharashtra, Uttar Pradesh and Tamil Nadu. It is one of the most important virus diseases of groundnut in India and causes considerable damage to the crop and yield loss.

Symptoms. The disease usually appears, when the crop is about 30 days old. Symptoms first appear on young leaflets as chlorotic spots or mottling and may develop into chlorotic ring spots or necrotic rings. The necrotic spots enlarge rapidly and the leaves at the top of the plant dry and die. This is soon followed by necrosis of the terminal bud. The necrosis starts from the terminal bud and progresses downward leading to total necrosis of the entire plant. In case of late infections, necrosis may occur in the tips of only some of the branches.

Stunting of early-infected plants and proliferation of axillary buds may also occur. Leaflets produced on such axillary shoots are very much reduced in size, malformed with distorted lamina and show mosaic mottling along with chlorosis. Such plants appear chlorotic, stunted and bushy. Infected plants fail to produce flowers and pods. Even if a few pods are formed, the seeds in such pods are smaller, shrivelled and discolored.

The causal organism. The disease is caused by *Tomato spotted wilt virus (TSWV)*, which belongs to the Genus - *Tospovirus*. The virus consists of isometric particles, 85 nm. in diameter. The - ssRNA genome is composed of three molecules of RNA. The virus is transmitted by sap and by thrips.

Mode of spread. The virus can be transmitted mechanically by sap transmission and by thrips in a persistent manner, but not through seeds. *Frankliniella schultzi* and *Scirtothrips dorsalis* are the two most efficient vectors transmitting the virus. Both nymphs and adults are capable of transmitting the virus. The virus has a host range of more than 160 plant species. The hosts include tomato, eggplant, green gram, black gram, cowpea, cluster bean, bitter gourd, zinnia, chrysanthemum and many weed plants. Environmental conditions, which favor multiplication and spread of thrips, are conducive for the spread of the disease also. Moderate temperature and slow wind currents favor mass flights of the vectors that aids the transmission and rapid spread of the disease.

Disease management

Agronomic practices (i) Plants showing the symptoms of the disease should be removed and destroyed (ii) Early sowing, immediately after the onset of monsoon reduces the incidence of the disease to a large extent (iii) Close planting should be avoided (iv) Weeds, which serve as alternate hosts of the disease should be removed and the field kept clean.

Chemical control (i) Spraying with fenthion - 400 ml. or phosalone - 500 ml. or phosphamidon - 125 ml. or endosulfan - 400 ml. in 200 liters of water per acre at 10 days interval, commencing from 20 days after sowing is effective in controlling the vector (ii) Spraying the crop with 'Antiviral principle' (AVP), prepared from, either coconut or sorghum leaves, on the 10th and 20th day after sowing reduces the virus infection to a large extent. Cut and dried leaves of coconut or sorghum are powdered. In 2.0 liters of water, 1.0 kg. of the leaf powder is put and heated for one hour at 60°C. The extract is filtered through a muslin cloth and the volume made up to 10 liters. This solution is then sprayed on the foliage. A quantity of 200 liters of the solution is required to cover one acre.

6. Groundnut rosette

Groundnut rosette virus

The disease is worldwide in occurrence. In India, the disease occurs in all the states to a more or lesser extent. In Tamil Nadu, the disease is more prevalent in the districts of Coimbatore and Pollachi.

Symptoms. The symptoms first appear on the young leaves, which show mottling and chlorosis. Infected plants appear stunted. The leaflets become small, erect, cupped, crinkled, malformed and chlorotic. The internodes get shortened. There is proliferation and rosetting of secondary shoots and the plant presents a miniature, bushy appearance. The leaves are closely packed, as in a rose flower. Flower and pod formation is reduced drastically.

The causal organism. The disease is caused by *Groundnut rosette virus*.

Mode of spread. The virus can be transmitted mechanically by sap inoculation and by aphid vectors. *Aphis craccivora* is the most efficient vector and can transmit the virus in a persistent manner. The acquisition feeding time is 30 minutes. Only after an incubation period of 24 - 48 hours inside the body, the vector is capable of transmitting the virus. The virus is not transmitted through seeds. Several weed host species belonging to the Genus - *Chenopodium, Trifolium, Nicotiana, Stylosanthes* etc. are known to harbor the virus. Volunteer and self-sown groundnut plants may also carry the virus. The vector also has several hosts and is found all through the year.

Disease management

Agronomic practices (i) Plants showing symptoms of the disease should be removed and destroyed periodically (ii) Volunteer groundnut plants and weed hosts should be removed.

Chemical control. Spraying with methyl demeton - 200 ml. or phosphamidon - 110 ml. or dimethoate - 200 ml. or neem oil - 2 % in 200 lit. of water per acre, twice, at 10 days interval commencing from 20th day after sowing controls the vector.

Minor diseases. In addition to the diseases discussed above in detail, several other diseases attack groundnut. 'Yellow mould' or 'Aflafoot disease' caused by *Aspergillus flavus* attacks the seeds before emergence, as a result they rot; *Rhizoctonia solani* causes 'damping off of seedlings'; 'Stem and pod rot' is caused by *Corticium rolfsii (=Sclerotium rolfsii)*; *Macrophomina phaseolina* causes 'Dry root rot'; *Alternaria alternata* causes 'Leaf spot'; 'Bacterial wilt' is caused by *Pseudomonas solanacearum*

Sesame or Gingelly *(Sesamum indicum)*

1. Gingelly phyllody

'Gingelly phyllody' or **'green flower disease'** is prevalent in many countries of the world, such as Sudan Venezuela, Uganda, Tanganyika, Egypt, Pakistan and India. It is widespread in all the gingelly growing states of India and occurs year after year in an endemic form. Affected plants remain partially or completely sterile and in case of severe incidence, the yield loss may go up to 100 per cent.

Symptoms. The symptoms may vary according to the growth stage of the crop at which infection occurs. Infection at the early stages of crop growth results in severe stunting and phyllody symptoms are expressed in the entire plant. However, typical symptoms of the disease can be seen only at the time of flowering. The characteristic symptom is the transformation of all the floral parts into green leafy structures. The calyx becomes polysepalous and the corolla becomes polypetalous. The veins in the floral parts become thick and quite conspicuous. Even the stamens and the pistil become long and leafy. The whole inflorescence appears like a bunch of green, short, twisted, leafy structure. Capsule setting is completely impaired. In the case of late infection, some normal capsules are formed on the lower portion of the plant, while on the upper portion, many phylloid flowers are formed and capsule setting is very poor.

The causal organism. The disease is caused by a *'Mycoplasma-like organism' (MLO).*

Mode of spread and epidemiology. The pathogen is transmitted by the leafhopper, *Orosius albicinctus.* The pathogen is not transmitted by sap or through seed. The acquisition feeding time is about 8 hours. After acquisition, an incubation period of 20 - 30 days inside the body of the vector is necessary before it can transmit the pathogen. A feeding time of 30 minutes is required for transmission of the pathogen into a healthy plant. Both males and females are capable of transmitting the pathogen in a persistent manner. The nymphs, though they can acquire the pathogen, cannot transmit them. A single leafhopper is capable of transmitting the pathogen to a number of plants in a single day. The leafhoppers prefer diseased plants to healthy plants.

The weather conditions that favor multiplication of the leafhopper vector, such as low temperature, especially night temperature, high humidity and dampness favor the development of the disease also. The disease attacks all species of sesame. The disease is also found to occur on about 51 plant species including sunnhemp, rapeseed, chickpea, broad bean and mustard. These alternate hosts, which may be present all the year round serve as reservoirs of the pathogen and may infect sesame.

Disease management

Agronomic practices. (i) Disease affected plants should be removed periodically and destroyed (ii) Early sowing may help sesame crop to escape infection

Chemical control. Spraying with monocrotophos - 400 ml. or phosphamidon - 200 ml. or fenthion - 250 ml. in 200 lit. of water per acre is effective in controlling the leafhopper vectors. Because the vectors are very active and quick moving insects and can transmit the pathogen to a large number of plants within a short time, quick acting insecticides should be used to mobilize and kill them at the shortest possible time.

2. *Alternaria* leaf blight

Alternaria solani

The disease occurs in Russia, Africa, Japan, the United States of America and most of the Asian countries. In India, the disease was reported in 1958. It is widely prevalent in the states of Uttar Pradesh, Bihar, Karnataka, Maharashtra, Orissa and Tamil Nadu.

Symptoms. Initial symptoms appear as small, water-soaked spots, 1.0 - 8.0 mm. in diameter. The spots gradually enlarge in size and become almost circular spots up to 2.0 cm. in diameter. The spots are dark- brown to black in color, and develop concentric rings, which give the spots a target board-like appearance. Lesions may coalesce to form large necrotic areas and the affected leaves turn yellowish, dry and fall off. Dark-brown, sunken lesions appear on the petioles, branches, stems and capsules. The lesions may spread along the entire length of the branches and stems and girdle them, resulting in the death of the branches or the entire plant. Infected capsules split open prematurely and the seeds are under-developed and shrivelled. The fungus may also infect the seeds **(Fig.49)**.

The causal organism. The mycelium is dark-colored, septate, inter- and intracellular. In older diseased leaves, the mycelium produces short, erect, simple, light-colored conidiophores, which emerge through the stomata singly or in groups. The conidiophores bear conidia singly or in chains of two. The conidia are detached easily, leaving a scar on the conidiophore at the point of attachment. Conidia are large, dark-colored, ellipsoid or pear-shaped, multicellular with 6 - 11 transverse septa and few longitudinal cross walls. They measure 90 - 260 x 14 - 33µ in size and have simple or rarely branched beaks **(Fig.49)**.

Mode of survival, spread and epidemiology. The fungus is seed-borne, air-borne and soil-borne to some extent. Air currents carry the conidia, which get detached easily. The conidia are present in the air and dust, and can survive for a long time. In the infected seeds, the mycelium remains in a dormant state under the epidermal layer of the seed coat. The mycelium may even be present in the embryo of the seed. Both the mycelium and the conidia can survive in the soil for a long time. The mycelium present in the diseased plant debris may continue to produce conidia, which may cause primary infection.

Low temperatures of 20°-25°C, high relative humidity and cloudy weather conditions favor the development and spread of the disease.

Fig. 46: Rust disease of groundnut-*Puccinia arachidis*

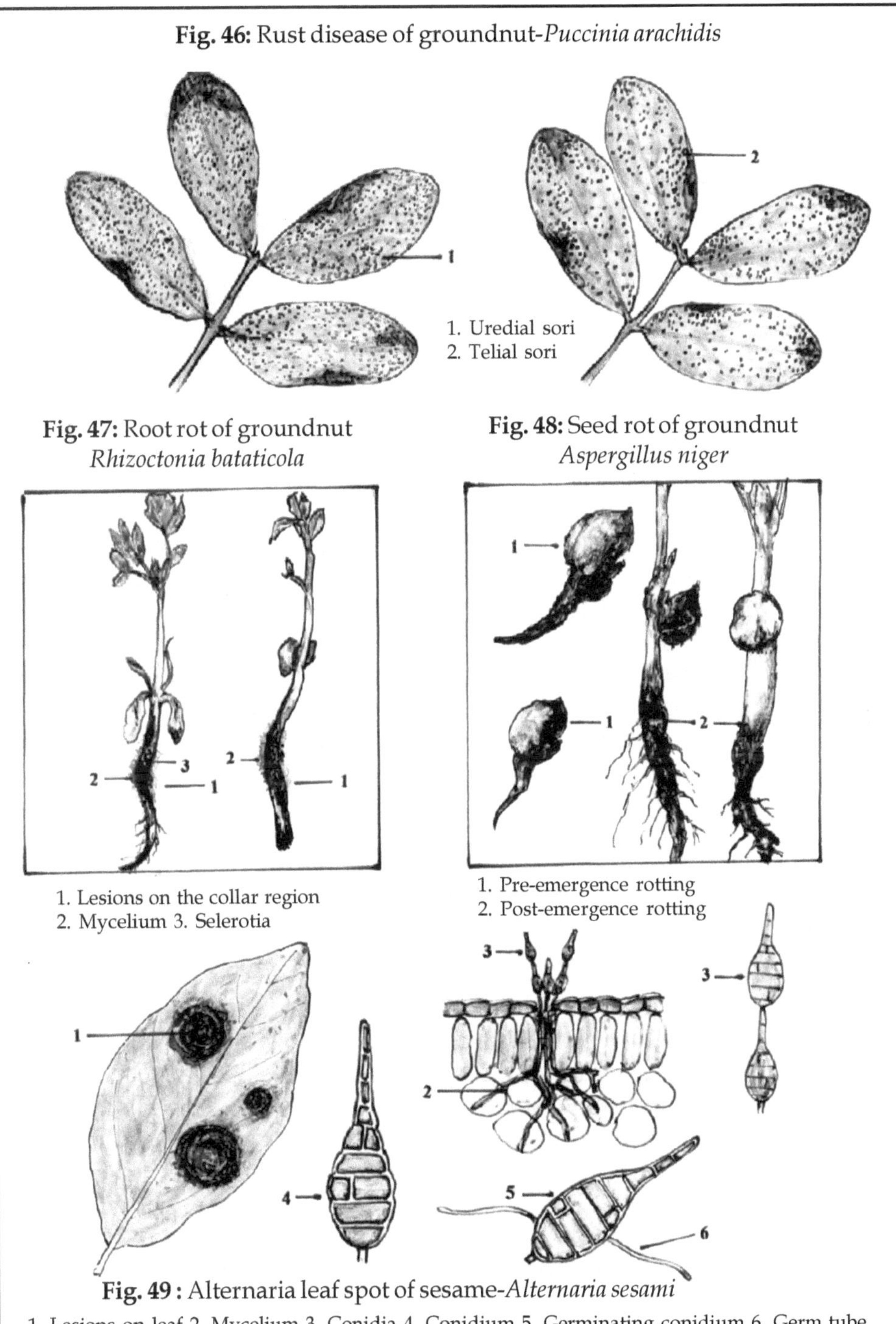

1. Uredial sori
2. Telial sori

Fig. 47: Root rot of groundnut *Rhizoctonia bataticola*

1. Lesions on the collar region
2. Mycelium 3. Selerotia

Fig. 48: Seed rot of groundnut *Aspergillus niger*

1. Pre-emergence rotting
2. Post-emergence rotting

Fig. 49 : Alternaria leaf spot of sesame-*Alternaria sesami*

1. Lesions on leaf 2. Mycelium 3. Conidia 4. Conidium 5. Germinating conidium 6. Germ tube

Disease management

Agronomic practices (i) Disease-free, good seeds should be used for sowing (ii) Diseased crop debris from the field should be removed and destroyed.

Seed treatment. Treating the seeds with captan or thiram at 4 gm./ kg. of seeds, 24 hours prior to sowing helps to prevent seed-borne infection.

Chemical control. Spraying the crop with copper oxychloride - 625 gm. or mancozeb - 500 gm. or Du-ter - 325 gm. or chlorothalonil - 325 gm. in 250 lit. of water per acre, 2 - 3 times, at 15 days interval from the time of appearance of symptoms is effective in controlling the disease.

Resistant varieties. TMV.1, TMV.2, Si.43, Si.247, Si.968, Si.972, Si.973, Si.994, Si.1001, Si.1683, T.12, T.12, TC.16 and TC.25 are found to be resistant to this disease.

Diseases of minor importance. The following diseases also attack sesame. *Macrophomina phaseolina* causes 'Charcoal rot' or 'Stem rot'; *Fusarium oxysporum* f.sp. *sesami* causes 'Wilt disease'; *Phytophthora parasitica* f.sp. *sesami* causes 'Leaf blight'; 'Powdery mildew' is caused by *Erysiphe cichoracearum*; 'Bacterial blight' is caused by *Xanthomonas campestris* pv. *sesami*; 'Bacterial leaf spot' is caused by *Pseudomonas syringae* pv. *sesami*; 'Leaf curl' is caused by *Tobacco leaf curl virus.*

Castor *(Ricinus communis)*

1. Seedling blight of castor

Phytophthora parasitica

The disease is of common occurrence in all castor-growing countries of the world. The disease was first reported from Bihar in 1931. The disease is prevalent in Uttar Pradesh, Telungu Desam and Tamil Nadu.

Symptoms. The disease occurs, when the seedlings are very young. Initial symptoms appear as small, round, yellowish, water-soaked spots on the cotyledons. The disease spreads to the point of attachment of the cotyledonary leaves and they rot completely. The infection then spreads to the true leaves and produce irregular, brown spots. The spots enlarge and finally turn black. When large areas of the leaves are thus attacked, they become necrotic and die. The infection further spreads to the stem and dark-reddish spots develop on the basal part of the stem region. Soon, the lesions spread and girdle the basal stem region of the seedlings, which rot, shrink and the seedlings topple down and die. The seedlings die in large numbers in patches within a few days of infection **(Fig.50).**

The causal organism. The mycelium of the pathogen is coenocytic, hyaline, inter- and intracellular. No haustoria are formed. From the infected areas, sporangiophores emerge through the stomata or directly by piercing the epidermis. They are long, slender, unbranched and are 100 - 300 μ long. Sporangia are borne singly, at the tips of the sporangiophores. Sporangia are hyaline, ovoid, papillate and measure 25 - 50 x 20 - 40μ in size. Zoospores are delimited within the sporangium and each sporangium may contain 5 - 45 zoospores. On maturity of the zoospores, they come out by breaking open the papilla. Zoospores are reniform in shape and possess two flagella. After swimming about for 20 minutes to two hours in the film of water on the host surface or in the soil, they encyst and become globular. After resting for a while, they germinate by producing a germ tube and attack the host. In the absence of sufficient moisture, free water and light, the sporangia germinate directly by producing a germ tube.

The fungus also forms chlamydospores. They are mostly spherical, thick-walled, smooth, yellow-colored, terminal or intercalary, 20 - 60μ in diameter and contain many oil globules.

Sexual reproduction is rarely found in nature. Oospores are spherical, hyaline, measuring 13-24μ in diameter and do not completely fill the oogonia **(Fig.50)**.

Mode of survival, spread and epidemiology. The disease is primarily soil-borne. The chlamydospores are resistant to adverse environmental conditions and can remain in a viable state for prolonged periods and initiate infection. The mycelium also can remain active in the soil for a long time. Secondary infection is by sporangia, which are spread by rain or by wind. Continuous rain, cool and moist conditions favor the spread of the disease. Free water is essential for the germination of sporangia by producing zoospores, thereby the disease can spread much faster.

High relative humidity, rainfall and low temperatures between 20°-25°C favor disease development. The disease occurs more in low-lying and ill-drained fields. Besides castor, the pathogen attacks tomato, eggplant, safflower, sesame, melon, citrus and betelvine.

Disease management

Agronomic practices (i) Disease affected plant debris in the field should be collected and destroyed (ii) Plants showing symptoms of the disease should be removed and destroyed (iii) Proper drainage facilities should be provided to the field (iv) Crop rotation may be followed with non-host crops.

Seed treatment. Treating the seeds with captan or thiram at 4.0 gm./ kg. of seeds, 24 hours prior to sowing protects the seedlings from infection for a period of about 15 - 20 days.

Chemical control. Spraying with copper oxychloride - 750 gm. or mancozeb - 600 gm. or captafol - 375 gm. in 300 liters of water per acre is found to be effective in controlling the disease.

2. Leaf spot of castor

Cercospora ricinella

This is one of the most common diseases occurring worldwide. It is prevalent in Brazil, Jamaica, Uganda, the United States of America, Venezuela and India. In India, it occurs in Telungu Desam, Bihar, Uttar Pradesh and Tamil Nadu.

Symptoms. The disease makes its appearance as minute, brownish, water-soaked spots on the leaves. The spots gradually enlarge in size and become brownish in color. Matured spots are irregular in shape, grayish-white in the center, with a dark-brown margin and are 2.0 - 5.0 mm. in diameter. Spots often occur in large numbers and they coalesce to form bigger spots or patches. The veins usually limit the spots. The tissues at the center of the older spots disintegrate and break off, leaving shot holes in the leaf. Under high disease intensity, older leaves turn yellow, wither and die. This disease does not affect young leaves **(Fig.51)**.

The causal organism. The mycelium of the pathogen is light-brown in color, septate, inter- and intracellular. The hyphae do not spread much and are found mostly around the points of infection. Before fructifications are formed, the hyphae collect beneath the epidermis in a stroma and from the stroma, conidiophores emerge in clusters of 10 - 20, either through the stomata or directly by penetrating the host epidermis. They are dark-brown at the base and light-colored at the tip and septate. Conidia are borne at the tips of the conidiophores singly. Conidia are sub-hyaline, thin-walled, long, obclavate, the basal portion slightly broader than the apex, 6 - 7 septate and measure 50.0 - 105.0 x 4.0 - 6.5μ in size. The spores detach easily, leaving a scar in the conidiophore at the point of attachment. Meanwhile, the conidiophore continues to grow further and produce another conidium at its tip. The conidia germinate by producing one or more germ tubes **(Fig.51)**.

Mode of survival, spread and epidemiology. The mycelium, as well as the spores can remain in a dormant state in the diseased plant debris for a long time and can cause primary infection. Secondary infection is caused by wind-borne conidia.

Fig. 50: Seedling blight of castor-*Phytophthora parasitica*

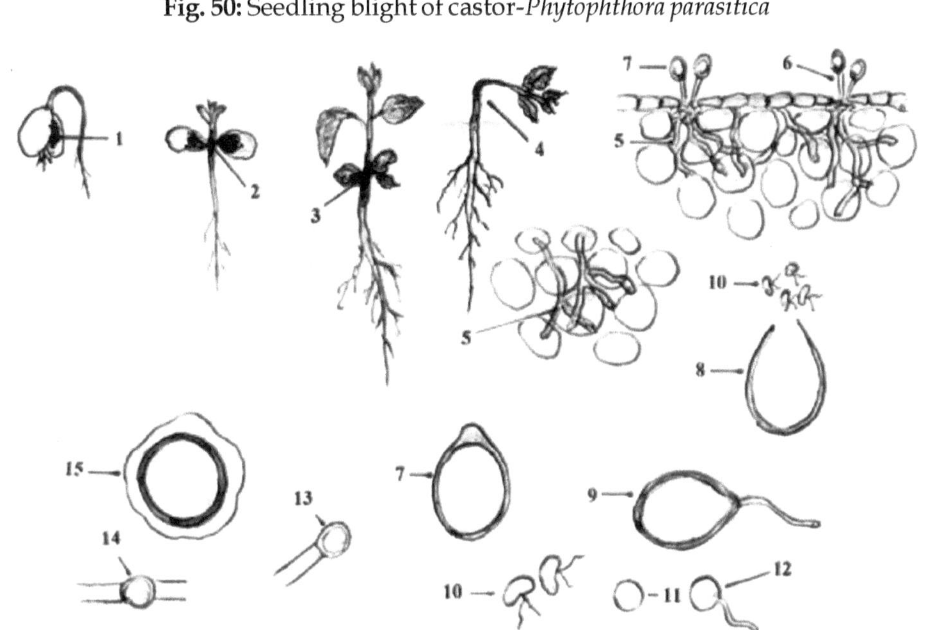

1. Lesion on cotyledon 2. Lesion extending to the stem from the cotyledon 3. Infection on the stem 4. Toppled diseased seedling 5. Inter and intracellular mycelium 6. Sporangiophore 7. Sporangium 8. Germination of sporangium by zoospores 9. Germination of sporangium by germ tube 10. Zoospones 11. Encysted zoospore 12. Germination of encysted zoospore 13. Terminal chlamydospore 14. Intercalary chlamydospore 15. Oospore.

Fig. 51: Leaf spot of castor *Cercospora ricinella*

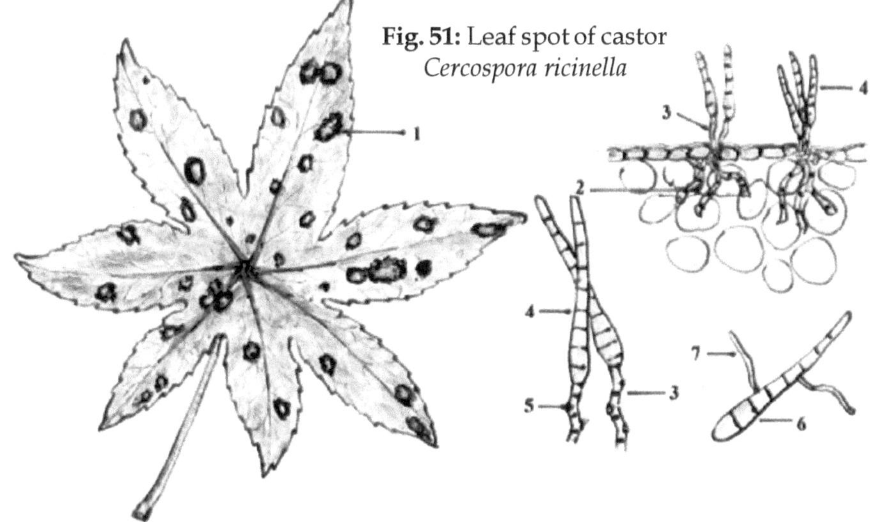

1. Spots on leaf 2. Inter and intracellular mycelium 3. Conidiophore 4. Conidium 5. Scar on the conidiophore 6. Germination of conidium 7. Germ tube.

Moderate temperature and relatively high humidity favor the development of the disease.

Disease management

Agronomic practices. Diseased plant debris in the field should be removed and destroyed to reduce the inoculum potential.

Chemical control. Spraying the crop with copper oxychloride at 2.5 gm. or wettable sulfur at 4.0 gm. per liter of water, twice, at fortnightly interval is effective in controlling the disease.

Diseases of minor importance. Diseases other than those mentioned above also attack castor and may cause considerable damage to the crop resulting in much yield loss. *Melampsora ricini* causes 'Rust disease'; *Alternaria ricini* causes 'Leaf blight'; 'Powdery mildew' is caused by *Leveillula taurica*; *Macrophomina phaseolina* causes 'Charcoal rot' or 'Stem rot'; *Xanthomonas campestris* pv. *ricini* causes 'Bacterial leaf blight'.

Rapeseed or Mustard *(Brassica campestris)*

1. *Alternaria* leaf blight of mustard

Alternaria brassisicola and *A. brassicae*

Symptoms. Cruciferous crops are attacked by more than two species of *Alternaria*, of which *Alternaria brassisicola* and *A. brassicae* are very important. These pathogens attack cabbage, cauliflower and knol kohl. Radish is not attacked by these pathogens, but is attacked by another species, *Alternaria raphani*.

The symptoms caused by *Alternaria brassisicola* appear as small, dark-colored spots on the leaves, which enlarge rapidly to form circular lesions up to 1.0 cm. in diameter. Under humid weather conditions, a bluish, velvety fungal growth is seen in the center of these spots. Sometimes, lesions are formed in concentric rings, like a target board. On the petioles, stems and seed pods, linear spots are formed.

The spots caused by *Alternaria brassicae* are almost similar to those produced by *A. brassisicola*, but are comparatively smaller and lighter in color **(Fig.52)**.

When many spots are formed on the leaves, the leaves turn yellow and dry prematurely. Formation of pods and seeds are adversely affected.

The radish leaf spot fungus, *Alternaria raphani* appears on older plants left for seed. The spots are yellow in color, slightly raised, round to elliptical and measure up to 1.0 cm. in diameter. Black sporulation may be seen on the spots. The tissues at the center of the spots soon dry, disintegrate and fall off.

The causal organism. The mycelium of the pathogens is inter- and intracellular and effused. Hyphae are olive in color, branched, septate and smooth. Conidiophores arise from hyphae, which gather beneath the host epidermis.

In ***Alternaria brassisicola***, the conidiophores arise singly or in groups of 2-12 or more through the stomata. They are olivaceous, septate, branched and measure 35.0-45.0 x 5.0-7.5µ in size. Conidia are produced in acropetal succession and are mostly formed in chains of 8 - 10. They are septate, muriform, with the basal cell rounded and the beak very short and almost non-existent. They are pale to dark-brown in color, with 1-11 transverse septa and 1-6 longitudinal septa. They measure 50-75 x 11-17µ in size **(Fig.52)**.

In ***Alternaria brassicae***, the conidiophores arise in fascicles through the stomata. They are simple, erect, septate, olivaceous and up to 170µ in length. Conidia are produced in acropetal succession, either singly or in short chains of up to 4. They are dark-colored, straight or slightly curved, smooth, obclavate to muriform, with 16 - 19 transverse septa and 0 - 8 longitudinal septa. They measure 125 - 225 x 16 - 28µ in size. The beak is rather long **(Fig.52)**.

Alternaria raphani is almost similar to *A. brassicae* in morphology, but the conidia possess a shorter beak than *A. brassicae*. The conidia measure 70 - 115 x 14 - 18µ in size. Further, *A. raphani* differs from the other two species in that, it produces chlamydospores. The chlamydospores are spherical, thick-walled and olive-brown in color.

Mode of survival, spread and epidemiology. The conidia and mycelium present in the diseased plant debris serve as the main means of perennation of the pathogen. Conidia are formed abundantly in moist atmosphere and are disseminated by wind. Seed-borne infection is of no significance. The weed hosts viz., *Anagallis arvensis* and *Convolvulus arvensis* and various other crop hosts provide enough inoculum all through the year.

Conidia are formed abundantly in moist atmosphere and are disseminated by wind.

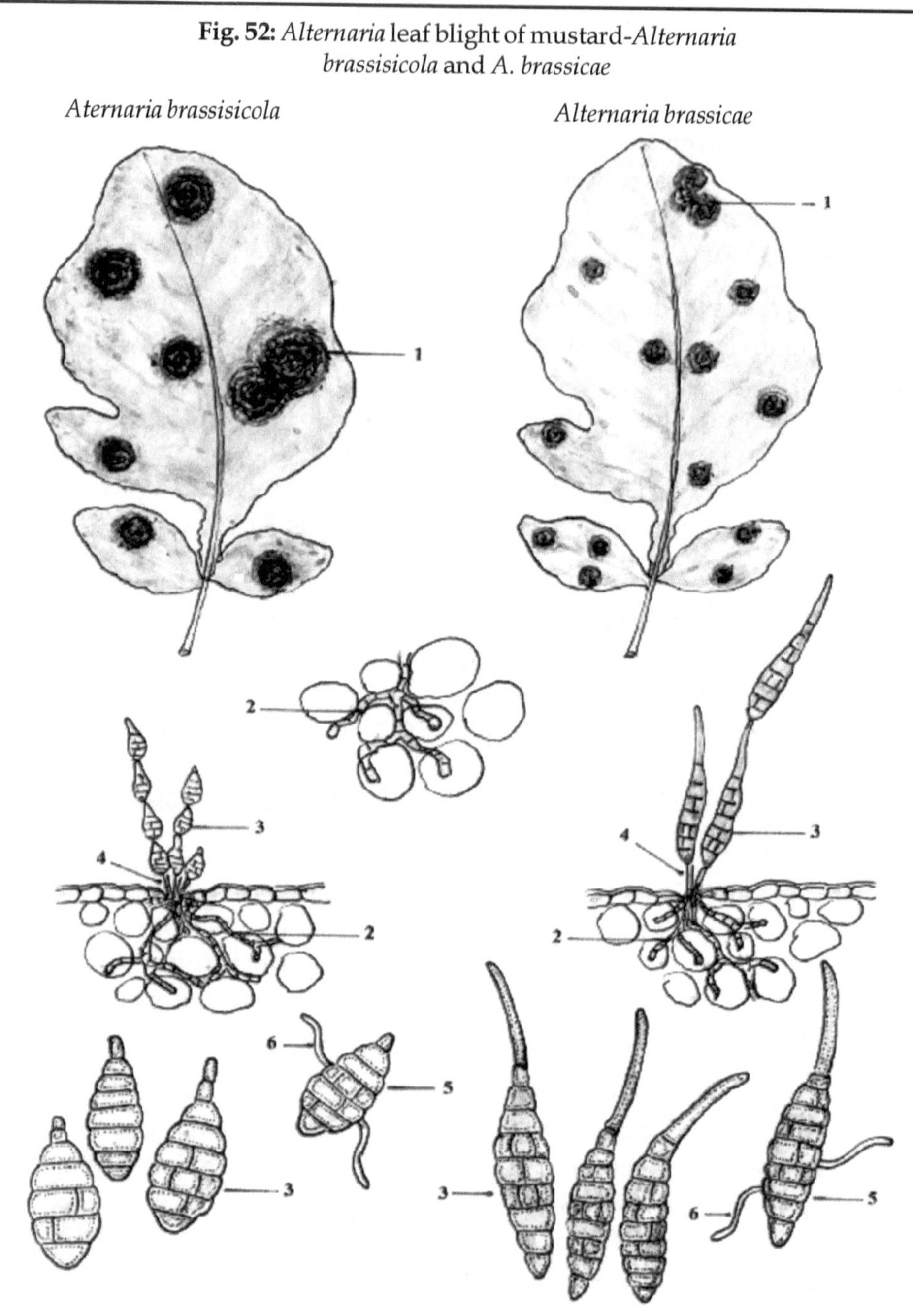

Fig. 52: *Alternaria* leaf blight of mustard-*Alternaria brassisicola* and *A. brassicae*

1. Spots on leaf 2. Inter and intracellular mycelium 3. Conidia 4. Conidiophore 5. Germinating conidium 6. Germ tube.

Disease management

Agronomic practices. Clean cultivation, field sanitation and crop rotation with non-host crops may help in minimizing the chances of disease occurrence.

Chemical control. Spraying the crop with copper oxychloride - 750 gm. or mancozeb - 600 gm. or captafol - 375 gm. in 300 lit. of water per acre gives adequate control of the disease.

2. White rust or white blisters of mustard

Albugo candida (Cystopus candidus)

'White rust' or **'white blisters'** is one of the most important diseases of mustard and occurs in all the mustard-growing states of the country.

Symptoms. Two types of infections viz., local and systemic may occur. All aerial parts of the plant may show symptoms. In the case of local infection, shiny-white or cream-yellow, isolated pustules of various sizes and shapes appear on the leaves and stems. These pustules or blisters are raised and may be 1.0 - 2.0 mm. in diameter. They may arise in close proximity and may ultimately merge to form larger patches. Often, the pustules develop in a circular arrangement around one or two central pustules. They usually develop on the under surface of leaves, but sometimes may appear on the upper surface also. On maturity of the spores, the host epidermis is ruptured and masses of white, powdery spores are exposed. When young stems and inflorescence are infected, the fungus becomes systemic in the tissues and stimulates the cells, which undergo hypertrophy and hyperplasia, resulting in various types of deformities. Blisters may be formed on the inflorescence and floral parts. Due to cell enlargement and proliferation, the floral parts become swollen and distorted. The peduncle, pedicel, sepals, petals and carpels become hypertrophied, thickened and leaf-like. The ovules and pollens are usually atrophied, resulting in sterility. The formation of chlorophyll in these parts, increases the photosynthetic activity and starch gets accumulated in these parts, which imparts a characteristic violet discoloration in these parts. In the hypertrophied parts, oospores are formed in large numbers in the intercellular spaces. When systemic infection occurs early, the entire plant becomes dwarfed and only small leaves are formed. Normally lateral buds may be stimulated and they grow into abnormal lateral shoots **(Fig.53)**.

The causal organism. The fungus causing the disease is an obligate parasite. The mycelium is intercellular and forms knob-shaped haustoria. Usually, many haustoria are found inside a single cell. Hyphae from the endophytic mycelium congregate beneath the epidermis to form the sporangial bed. From these hyphae, sporangiophores are formed by the vertical growth of broad, short, clavate, stalks under the epidermis. Because of the pressure exerted by these structures, the host epidermis is pushed up and gets separated from the underlying cells. The sporangiophores are free from each other laterally and are very thick, especially at the base. Sporangia are produced in basipetal succession in chains. Pads of a gelatinous material called **'isthmus'** are formed between every pair of sporangia, which function as disjunctor cells and aid in the discharge of the sporangia. When the host epidermis is finally ruptured due to pressure exerted by the continuous production of sporangia, they are liberated free. The sporangia are hyaline, nearly spherical, single-celled and measure 14 - 16 x 16 - 20μ in size. The sporangia usually germinate by production of zoospores. Before germination, a rounded papilla is formed at the apex of the sporangium. Zoospores are formed inside the sporangium and they emerge out by breaking open the papilla. Usually, 4 - 8 zoospores are formed in each sporangium. Sometimes, the papilla swells to form a bladder-like vesicle and the whole contents of the sporangium are emptied into the vesicle. The zoospores are delimited inside the vesicle and they come out after breaking the wall of the vesicle. The zoospores, after emergence encyst and after a period of rest germinate by producing a germ tube. However, under unfavorable environmental conditions, the sporangia germinate directly by means of a germ tube.

Oospores are formed in the intercellular spaces of the hypertrophied parts in large numbers. They are globose, thick-walled, tuberculate, dark-yellowish-brown in color and measure 40 - 45μ in diameter. They germinate after a long period of rest by the formation of a vesicle outside the oospore wall or a vesicle at the end of a short tube. The zoospores are formed inside the vesicle **(Fig. 53)**.

Mode of survival, spread and epidemiology. The pathogen perpetuates through oospores present in the diseased plant debris in the soil or in diseased plant fragments along with the seed. Secondary spread of the disease is by means of sporangia and zoospores, which are dispersed by rain or wind. The disease is found to occur on turnip, radish, cabbage and other *Brassica* species, besides a few weeds, such as *Cleome viscosa*.

The optimum temperature for germination of sporangia by zoospores is 10°C. Cool, moist weather conditions favor the development of the disease. Temperatures less than 15°C and relative humidity above 65 % are conducive for the occurrence of the disease.

Fig. 53: White rust of mustard-*Albugo condida*

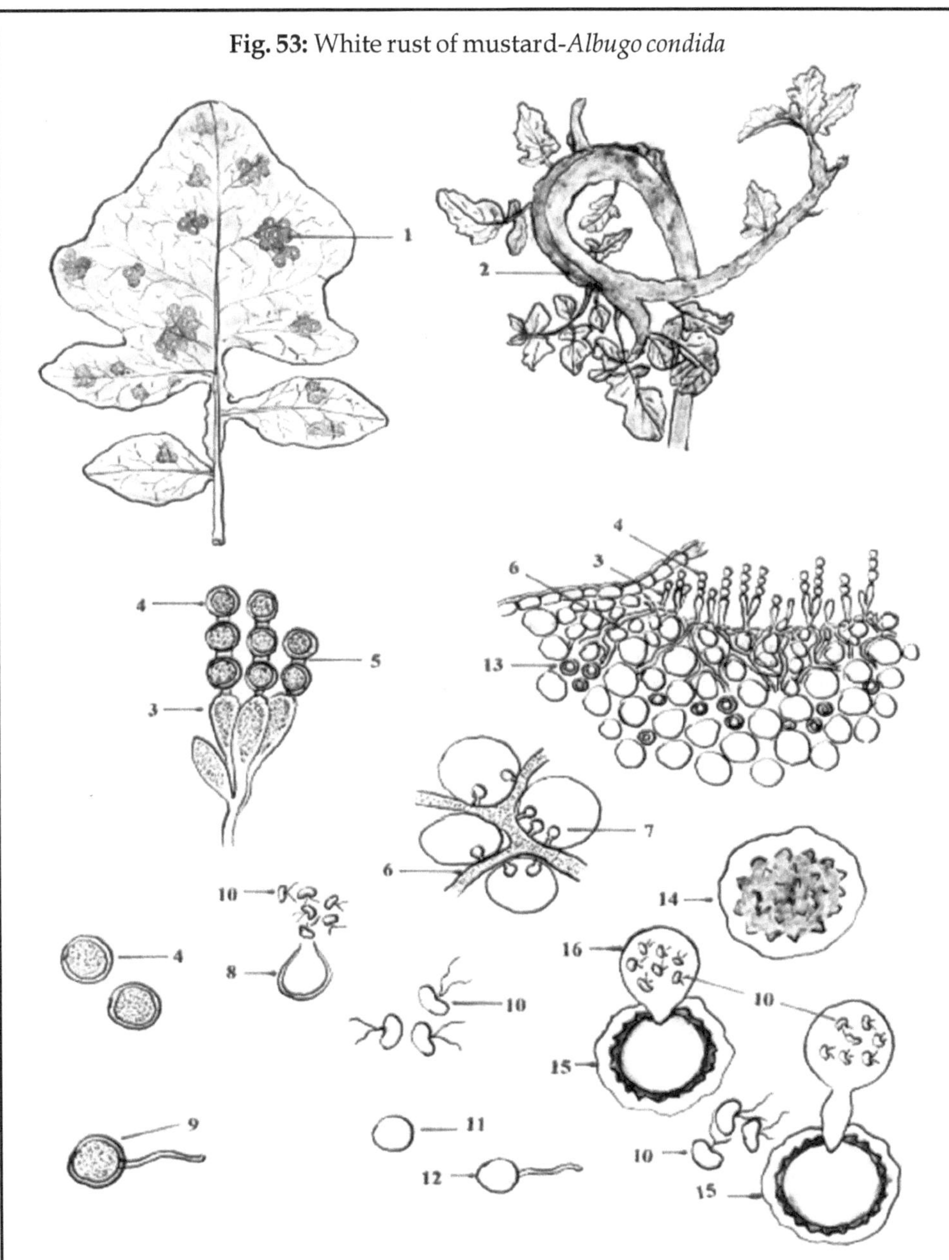

1. White rust pustules on leaf 2. Hypertrophied and deformed inflorescence 3. Sporangiophore 4. Sporangium 5. Disjunctor cell 5. Intercellular mycelium 7. Haustoria 8. Germination of sporangium by zoospores 9. Germination of sporangium by germ tube 10. Zoospores 11. Encysted zoospore 12. Germination of encysted zoospore 13. Oospores in the intercellular spaces 14. Oospore 15. Germination of oospore by zoospores 16. Vesicle.

Disease management

Agronomic practices (i) Seeds, free of contamination should be used for sowing (ii) Diseased plant debris in the field should be collected and destroyed (iii) Weeds in and around the field, which may harbor the pathogen should be removed and the field should be kept clean (iv) Disease affected plants in the field should be removed and destroyed (v) Crop rotation with non-host crops may be followed.

Chemical control. Spraying with Bordeaux mixture - 0.75 % or copper oxychloride - 750 gm. in 300 lit. of water per acre is found to give satisfactory control of the disease

3. Downy mildew of mustard

Peronospora parasitica (brassicae)

The disease, which was considered to be of minor importance, has assumed serious proportions in recent times and causes severe damage to the crop and yield loss. The disease occurs in most of the regions, where mustard is grown. In India, it is very prevalent in Uttar Pradesh and Bihar.

Symptoms. The disease usually attacks young seedlings and senescent leaves of grown-up plants. In mustard and radish the disease is found to attack the stem and inflorescence also. The symptoms appear as thin, grayish-white, scattered patches of fine, downy growth on the under surface of leaves. The upper surface of the leaves, corresponding to the downy growth on the under surface, is marked by whitish-yellow spots. In case of severe infection, the spots may be so crowded, that the leaf shrivels, tears easily and dries up. In seedlings, the whole under surface of leaves may be covered by the downy growth under favorable conditions of moisture, temperature and light intensity. Oospores may be formed in abundance inside the tissues of the cotyledons, while they are formed only sparsely in the affected leaf tissues. The stems may show deformities and small or several centimeters long swellings are formed and the stalks bent abruptly. The floral parts do not exhibit swellings, but the young ovary is much elongated into a twisted body. Very often, the floral buds are atrophied and seed setting is completely affected. Occasionally, the roots are attacked and they rot near the soil surface **(Fig.54)**.

Often *Peronospora parasitica* and *Albugo candida* occur together. *Albugo candida* produces a violet discoloration in the affected parts, but *Peronospora parasitica* never produces a violet discoloration. Further, *Albugo candida* produces the greatest deformities in the inflorescence, whereas *Peronospora parasitica* causes such deformities more in the stem region.

The causal organism. The mycelium of *Peronospora parasitica,* an obligate parasite, is strictly intercellular and the haustoria are large, elongated, club-shaped and often branched. Many haustoria may invade a single cell and fill the cell almost completely. At the fructification stage, numerous, erect, branched, sporangiophores emerge through the stomata on the under surface of the leaf, and stem and inflorescence directly from the endophytic mycelium. They are 200 - 300µ long and bear sporangia at the branched tips. The sporangiophores bifurcate 6 - 8 times at the top portion and the final branches, the sterigmata, which are at acute angle with each other are slender, pointed and bear a single sporangium at the tip. The sporangia are broadly oval, hyaline and 24 - 27 x 15 - 20µ in size. They germinate directly by a lateral germ tube and enter the host by penetrating the epidermis or through the stomata.

Later, oospores are formed in the intercellular spaces of the hypertrophied parts of the host. The oospores are furnished with a thick wall, with crest-like folds. The oospores are globose, yellowish-brown and 30 - 40µ in diameter. They germinate by a germ tube. No zoospores are produced either from the sporangia or from the oospores **(Fig. 54)**.

Mode of survival, spread and epidemiology. The fungus survives in the form of oospores released into the soil and in infected plant debris. Seeds contaminated with fragments of diseased plant parts may also carry the disease. Wild cruciferous host plants also serve as potential source of inoculum. Secondary spread is through sporangia dispersed by wind or rainsplash. The fungus is known to have different physiologic races. Besides mustard, the disease attacks cabbage, cauliflower, radish, turnip and several other cruciferous crops and weeds.

Low temperatures of about 14°C, high rainfall of about 15 cm. and good sunlight favor the development of the disease.

Disease management

Agronomic practices. Selection of quality seeds, field sanitation, destruction of wild hosts and crop rotation help to minimize the disease occurrence.

Chemical control. Spraying with captafol - 375 gm. or ziram - 600 gm. or metalaxyl - 300 gm. in 300 lit. of water per acre is effective in controlling the disease.

Diseases of minor importance. Many other diseases, besides the ones discussed above also attack mustard. *Erysiphe cruciferarum* causes 'Powdery mildew'; 'Stem and root rot' is caused by *Sclerotinia sclerotiorum*; *Xanthomonas campestris* pv. *campestris* causes 'Bacterial rot' that affects the basal stem region; *Macrophomina phaseolina* causes 'Charcoal rot'; several virus diseases are also found to attack mustard.

Sunflower *(Helianthus annuus)*

1. *Alternaria* blight of sunflower

Alternaria helianthi

The disease is widely distributed all over the world, wherever sunflower is grown. It is considered to be the most destructive disease of sunflower and causes considerable damage to the crop and yield loss.

Symptoms. The disease usually attacks the leaves and causes leaf spot. Initially, small, dark-brown to almost black, round or oval spots, ranging from 2.0 - 5.0 mm. in diameter appear scattered on the leaves. The spots gradually enlarge in size and become larger spots up to 2.0 - 3.0 cm. in diameter. These spots are dark-brown in color, with a light-brown margin, irregularly round and surrounded by a chlorotic or yellow halo. Later, the center of the spots turns grayish-white and becomes necrotic. The spots may show concentric rings. Under humid conditions, the spots coalesce, covering large areas of the leaves, resulting in blighting and defoliation. The pathogen may attack the leaf petioles and young stems producing linear, necrotic lesions. Spots may develop on the sepals and petals also **(Fig.55)**.

The causal organism. The mycelium of the fungus is much branched, septate, light-brown in color, inter- and intracellular. Conidiophores are produced from the hyphae, which gather below the epidermis of the affected area to form a stroma. Conidiophores emerge singly or in clusters through the stomata. They are erect, septate, olive-brown in color, simple or branched from the base, geniculate and are 25 - 80μ long. Conidia arise singly from the tip of the conidiophores. They get detached easily, leaving behind a scar on the conidiophore at the point of attachment. Conidia are not formed in chains. They are ellipsoid or barrel-shaped, light-brown in color, with broader, rounded base, narrowing towards the apex and have a short beak. They have 2 - 11 transverse septa and 0 - 2 longitudinal septa and show slight constrictions at the transverse septa. They measure 40 - 110 x 13 - 28μ in size. Germination takes place from any one or more cells by germ tubes **(Fig.55)**.

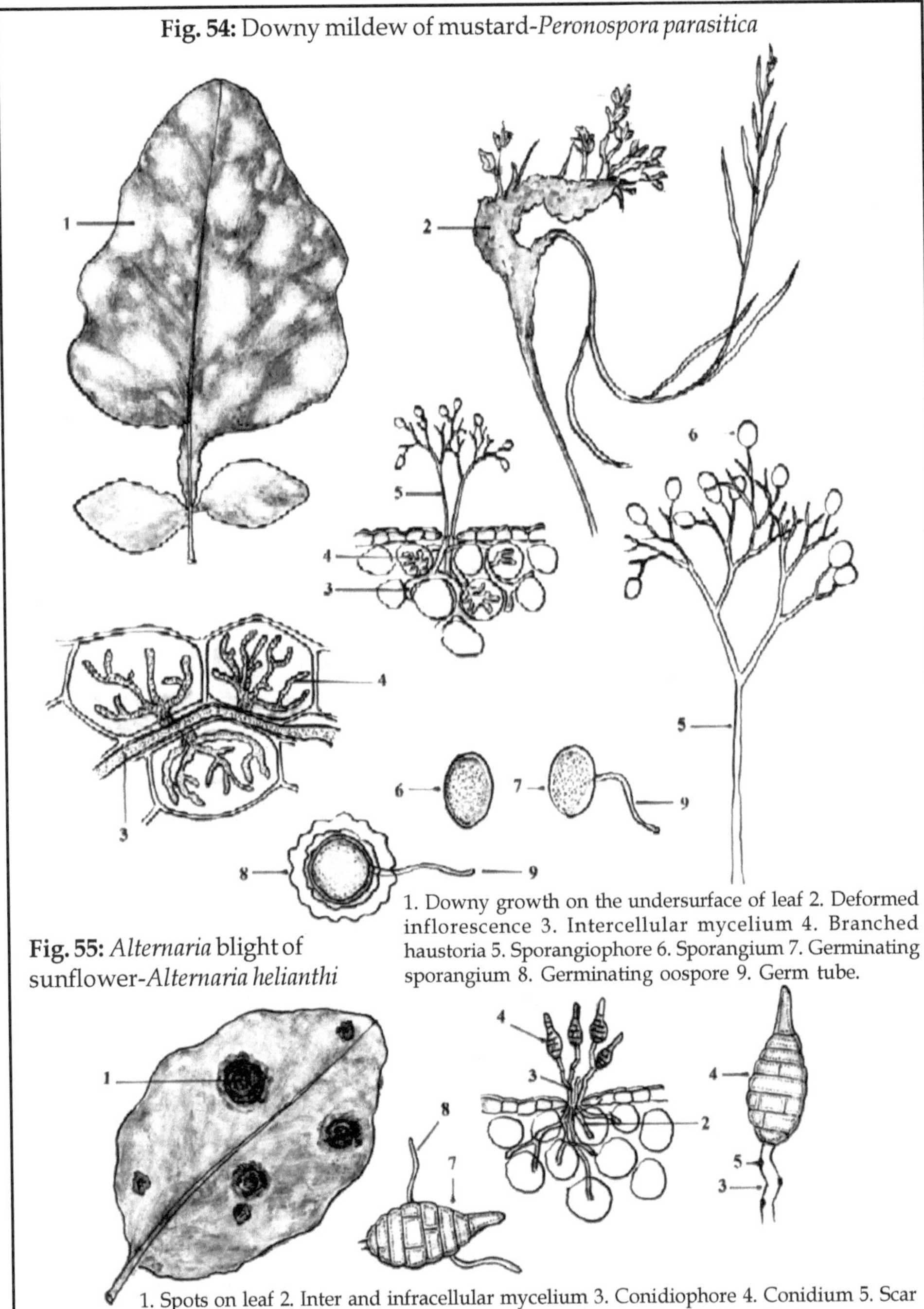

Fig. 54: Downy mildew of mustard-*Peronospora parasitica*

1. Downy growth on the undersurface of leaf 2. Deformed inflorescence 3. Intercellular mycelium 4. Branched haustoria 5. Sporangiophore 6. Sporangium 7. Germinating sporangium 8. Germinating oospore 9. Germ tube.

Fig. 55: *Alternaria* blight of sunflower-*Alternaria helianthi*

1. Spots on leaf 2. Inter and infracellular mycelium 3. Conidiophore 4. Conidium 5. Scar on the conidiophore at the point of attachment of the conidium 6. Germinating conidium 7. Germ tube.

Mode of survival, spread and epidemiology. The mycelium and conidia present in the diseased plant debris in the soil may remain in a viable state for several months. The mycelium may remain active in such diseased plant materials and continue to produce conidia. The disease spreads mainly through air-borne conidia, which are present in the atmosphere throughout the year. The disease may also perennate in alternate weed hosts. All cultivated varieties of sunflower are found to be susceptible.

The disease occurs more in the cooler winter seasons and spreads faster during the rainy periods.

Disease management

Agronomic practices. Field sanitation, clean cultivation and crop rotation may help to minimize the disease occurrence.

Chemical control. Spraying the crop with copper oxychloride - 750 gm. or mancozeb - 600 gm. or captafol - 375 gm. or ziram - 600 gm. in 300 lit. of water per acre, two or three times, at 10 - 15 days intervals is found to afford good control.

Seed treatment. Treating the seeds with captan or thiram at 4.0 gm./kg. of seeds, 24 hours prior to sowing protects the seedlings from initial infection.

2. Head rot of sunflower

Rhizopus nigricans

Symptoms. The disease occurs during the heading and grain formation stages. The affected heads show water-soaked, dark-brown, rotten patches on the under surface of the heads. The affected portions become soft and pulpy and saprophytic larvae may be found in the rotten portions. The infection gradually spreads from the head, down to the stalk to a length of 10 - 15 cm. Under humid conditions, typical cottony, coarse and stringy mycelium, with characteristic black sporangia cover the diseased tissues. After destroying the tissues of the head region, the fungus attacks the developing seeds also. In case of severe infection, all the seeds are destroyed and they become ill-filled, dark masses. The loss in yield in such cases may go up to 100 per cent. *Alternaria solani, Rhizoctonia bataticola, Fusarium* sp., *Aspergillus* sp. and *Penicillium* sp. may often be associated with head rot of sunflower **(Fig.56)**.

The causal organism. The causal organism is a facultative, weak parasite. The mycelium is coenocytic, olivaceous in color, both endophytic and ectophytic. The white, cottony, aerial, unbranched, arched hyphae or stolons are differentiated into long internodes and nodes. From each of the node, rhizoids are formed below, which enter into the substratum to absorb nutrition, while sporangiophores are formed above. The sporangiophores are restricted to the nodes only. They arise in fascicles and are short, stout, stiff and bear black sporangia, singly at their tips. Inside the sporangium, sporangiospores are formed in large numbers in-between the columella and the sporangial wall. The sporangial wall breaks up into fragments in dry air and the dry powdery, black, mass of spores is exposed. The columella collapses and the non-motile sporangiospores are blown away by wind to long distances and cause new infection. The sporangiospores are minute, oval, dark-pigmented and marked with striations.

As a result of sexual reproduction, the fungus produces zygospores. These resting spores are black in color, globose and have several layered, thick, warty wall. Under suitable environmental conditions, the zygospore germinates by producing a globular zygosporangium at the tip of a long stalk, in which the sporangiospores are formed. A columella is present inside the zygosporangium. On maturity of the sporangiospores, the outer wall breaks and the spores are liberated **(Fig.56)**.

Mode of spread and epidemiology. The disease is spread by air-borne sporangiospores, which are present in the atmosphere throughout the year. The pathogen can enter the host only through injuries or wounds caused by sucking insects or larvae, which attack the head. Once inside the head, the fungus attacks the tissues and causes rotting.

Moderate temperature, high relative humidity and intermittent rains at the time of heading favor the disease development.

Disease management

Agronomic practices. Diseased earheads should be removed and destroyed.

Chemical control. Spraying the head portion with an insecticide, fenthion - 450 ml. or endosulfan - 600 ml. + a fungicide, wettable sulfur - 1200 gm. or mancozeb - 600 gm. in 300 lit. of water per acre, twice, at 10 days interval after flowering is effective in controlling both the insect pests causing injury to the heads and also the disease.

3. Powdery mildew of sunflower

Erysiphe cichoracearum

'Powdery mildew' occurs very commonly on sunflower, besides plants of the cucumber family, potato, tobacco seedlings, lettuce, cineraria and a large number of hosts belonging to different families.

Symptoms. Under high moisture conditions, the disease appears as small, white patches on the leaves. Soon, the patches enlarge and cover the entire leaf area, on which the characteristic white, powdery coating consisting of the conidial fructifications of the fungus is seen. Often, the stem and other aerial parts of the plant may also be covered with the white fructifications. The parts overshadowed by the upper leaves, as well as leaves close to the soil and the under surface of the lowermost leaves are severely affected. From such spots, the mildew spreads rapidly to the upper surface of the leaves and to the younger leaves and stems. Severely affected leaves turn yellow, dry and fall off prematurely, leading to suppression in growth and yield loss.

The causal organism. The fungus forms a tangled web of ectophytic, septate, mycelium over the surface of leaves and stems. Spherical haustoria, produced from the mycelium penetrate the host epidermal cells, but do not penetrate deeper into the mesophyll tissues. The superficial mycelium gives rise to short, unbranched, conidiophores in profusion from the foot cells and they measure 33.6 - 60.0μ in length. From the conidiophores, conidia are abstricted in short chains. Conidia are unicellular, oval to elliptical and measure 25 - 45 x 14 - 26μ in size. They germinate directly by producing germ tube.

Cleistothecia are rarely formed in nature. If formed, they are found only on certain parts of diseased leaves as black, pinhead-like bodies embedded in the superficial mycelium on the host surface, especially during autumn. They are globose, becoming depressed or somewhat irregular in shape and 90 - 135μ in diameter. Appendages are numerous, mycelioid, hyaline to dark-brown and rarely branched. In each cleistothecium, 10 - 25 asci are found. They are ovate to broadly ovate, more or less stalked and are 60 - 90 x 25 - 50μ in size. Each ascus contains two, very rarely three ascospores. They are oval or sub-cylindrical and measure 20 - 28 x 12 - 20μ in size. They germinate by producing germ tube **(Fig.57)**.

Mode of survival, spread and epidemiology. In areas where cleistothecia are formed, they may serve to perennate the disease from one season to the next. During the off season of the host, the fungus may survive in its conidial state on wild hosts and the conidia produced from these sources may be carried by wind to cultivated areas. Secondary spread is by air-borne conidia.

Fig. 56: Head rot of sunflower-*Rhizopus nigricans*

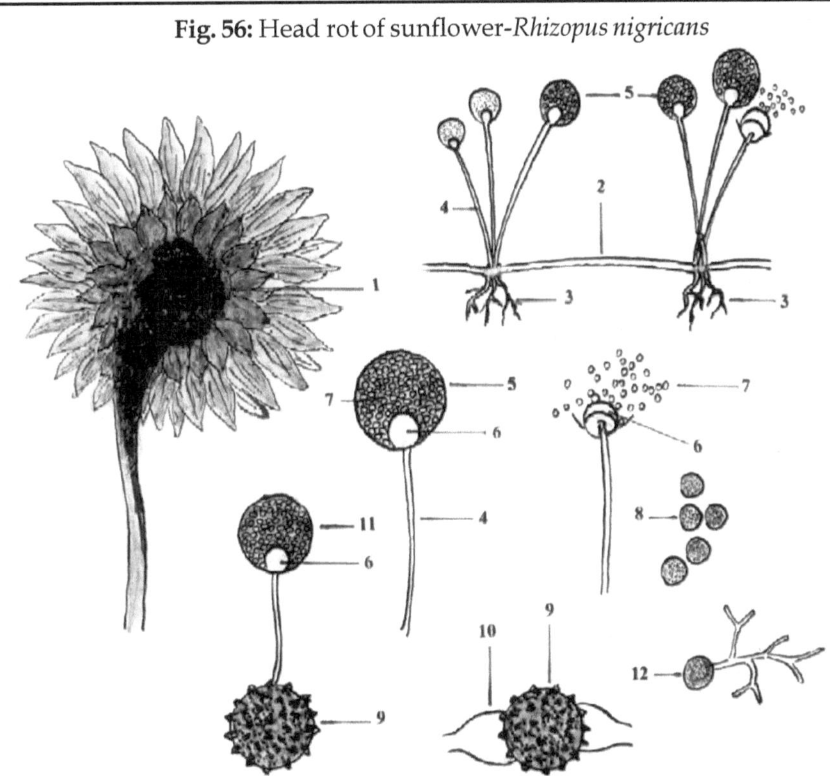

1. Diseased, rotten head and stem portion 2. Stolon 3. Rhizoid 4. Sporangiophore 5. Sporangium 6. Sporangiospores 7. Sporangiospore 8. Columella 9. Zygospore 10. Suspensor 11. Zygosporangium 12. Germinating sporangiospore.

Fig. 57: Powdery mildew of sunflower-*Erysiphe cichoracearum*

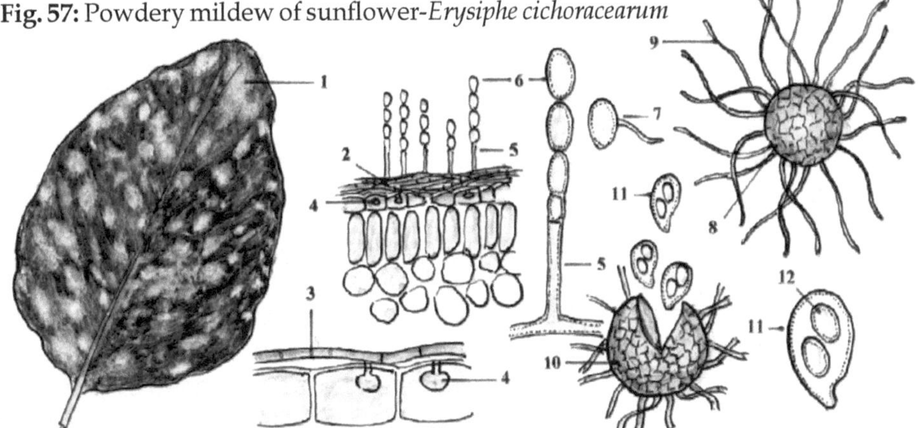

1. Powdery growth on the leaf surface 2. Superficial mycelium 3. Hypha 4. Haustorium 5. Conidiophore 6. Conidia 7. Germinating 8. Cleistothecium 9. Mycelioid appendages 10. Asci coming out of a disintegrated cleistothecium 11. Ascus 12. Ascospore.

Heavy dew, intermittent heavy rainfall, high atmospheric humidity of about 80 % and moderate temperatures of 24° - 25°C favor development and spread of the disease. Several physiologic races of the pathogen have been identified.

Disease management

Chemical control. Spraying with wettable sulfur - 1200 gm. or dinacap (Karathane) - 300 gm. or triadimefon (Bayleton) - 300 gm. or tridemorph (Calixin) - 300 ml. in 300 lit. of water per acre controls the disease effectively.

Minor diseases. Many other diseases also attack sunflower. *Puccinia helianthi,* an autoecious rust causes 'Rust disease'; 'Root rot' or 'Charcoal rot' is caused by *Macrophomina phaseolina; Plasmopara halstidii* causes 'Downy mildew' that affects seedlings, leaves and inflorescence; *Cercospora helianthicola* and *Septoria helianthi* cause 'Leaf spots'; *Sclerotinia sclerotiorum* and *Sclerotium rolfsii* cause 'Basal stem rot'; 'Bacterial leaf spot' is caused by *Pseudomonas syringae* pv. *helianthi;* several virus diseases, such as 'Mosaic', 'Leaf curl' etc. also attack sunflower.

Linseed *(Linum usitatissimum)*

1. Rust of linseed or flax

Melampsora lini

'Rust' is one of the most serious diseases of linseed and is a great handicap in most of the linseed growing countries of the world. The disease was recorded in Central India in 1918 and is prevalent in Madhya Pradesh, Maharashtra and Uttar Pradesh. The disease is of particular significance, since linseed is one of the major oilseed crops in India. Reduction in the photosynthetic area of leaves, rupture of the rust pustules and absorption of food reserves of the host by the pathogen result in debility of the infected plants, thereby the yield of seed is very much reduced. Stem infection leads to damage of the fiber. The yield of oil from seeds of infected plants is also very poor.

Symptoms. All the aerial parts of the plants are affected. The diseased plants appear bright-orange in color due to the formation of numerous uredia on the leaves and stems. The uredia occur on both the surfaces of leaves and on the surface of other aerial parts. On the leaves, they are small, nearly round, reddish-yellow in color and are surrounded by a chlorotic zone. On the stem, they are elongated and irregular in shape. Severely affected leaves become necrotic and fall off prematurely. Later on, telia may be formed on leaves, which have not fallen off, but more commonly on the stems. Often, the orange-yellow uredia are surrounded by reddish-brown telia. The telia do not break open the epidermis of the host, but remain covered and appear black and glossy **(Fig.58).**

The causal organism. *Melampsora lini* is an autoecious, macrocyclic rust and all the four fructification stages viz., pycnial, aecial, uredial and telial stages occur on linseed plants. The fungus is an obligate parasite. The pycnia are pale yellow, flask-shaped, sub-epidermal and are formed in groups of 5 - 10 closely. They are formed underneath the stoma and the stomatal opening serves as an ostiole. Sometimes, the pycnia are diffused and do not have any definite shape. The pycniospores are minute and oval to globose in shape. The aecia are orange-yellow and are scattered on the under surface of leaves or concentrically grouped to form a ring. They are formed on the stems also. They do not have a peridium and no paraphyses are formed. The aeciospores, which are formed in chains are elliptical or polygonal, 15 - 25µ in diameter and have thin, verrucose wall. Uredia are reddish-yellow in color and occur on both the surfaces of leaves, scattered or in groups. The uredia on the leaves are usually circular, but on the stems they are elongated and sometimes joined together. The uredospores are obovoid or ovate, 15 - 25 x 13 - 18µ in size and have a spiny surface. They germinate by producing germ tube. Numerous, capitate, hyaline, paraphyses, up to 80µ long are found intermingled with the uredospores. Telia, which are mostly formed on the stems are elongated, often fused with one another, black and sub-epidermal. The host epidermis is not broken to expose the teliospores. The teliospores are sessile, cylindrical, one-celled, and reddish-brown. They are arranged in a single layer and measure 46.0 - 80.0 x 8.0 - 20.0µ in size. They germinate by producing a basidium and basidiospores **(Fig.58).**

Mode of survival, spread and epidemiology. The uredospores and teliospores cannot survive the summer temperatures prevailing in the plains of India. However, in temperate countries, soil-borne teliospores found in the crop refuse are the source of primary infection. The teliospores germinate to form sporidia, which infect young host leaves. In India, the uredospores and teliospores survive throughout the year in the hilly regions of Chaubattia, Simla etc. The rust may survive on linseed and other alternate hosts in its uredial and telial stages at high altitudes. From these regions, the uredospores may be carried to the plains by wind. Once the disease appears, it spreads very rapidly and within a few days, most of the fields in the locality get affected.

Germination of uredospores is favored by the presence of free water on the host surface and a temperature of 15° - 16°C. The disease occurs on *Linum catharticum* and several other *Linum* species. The fungus is known to have many physiologic races and in India, 18 such races have been identified.

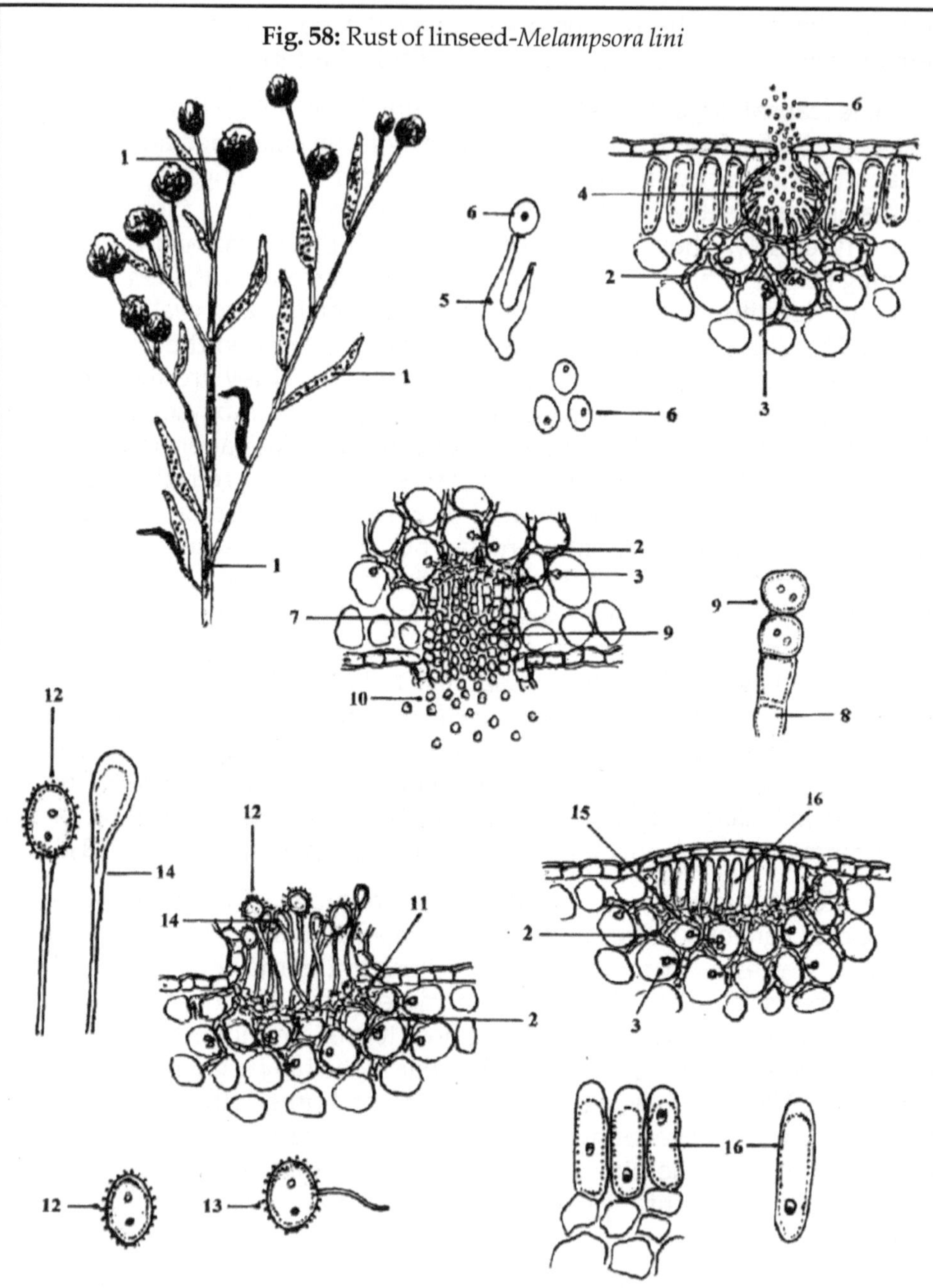

Fig. 58: Rust of linseed-*Melampsora lini*

1. Rust pustules on leaves, stem and flowers 2. Intercellular mycelium 3. Haustoria 4. Pycnium 5. Pycniophore 6. Pycniospores 7. Aecium 8. Basal cell 9. Chain of aeciospores 10. Aeciospores 11. Uredium 12. Uredospore 13. Germinating uredospore 14. Paraphysis 15. Telium 16. Teliospore

Disease management

Agronomic practices. Removal and destruction of diseased plant debris from the field by burning, adopting field sanitation methods and crop rotation help to avoid the occurrence of the disease.

Chemical control. Spraying the crop with wettable sulfur - 1200 gm. or mancozeb - 600 gm. or tridemorph (Calixin) - 300 ml. or triadimefon (Bayleton) - 300 gm. or ehterimol (Milstem) - 300 gm. or Daconil - 375 gm. in 300 lit. of water per acre, 2 or 3 times, at 10-15 days intervals controls the disease.

Diseases of minor importance. Most of the diseases that attack other oilseed crops attack linseed also. *Rhizopus solani* causes 'Seedling rot' that kills the seedlings; *Fusarium oxysporum* f.sp. *lini* causes 'Wilt disease'. This vascular disease affects the plants at all growth stages leading to wilting and death of the affected plants; 'Foot rot' and 'Root rot' is caused by *Sclerotium rolfsii*; *Alternaria lini* causes '*Alternaria* leaf spot'; *Oidium lini* causes 'Powdery mildew'; *Colletotrichum lini* causes 'Anthracnose'; a few virus diseases also attack linseed.

Safflower *(Carthamus tinctorius)*

Safflower, which is yet another important oilseed crop, is subjected to attack by most of the diseases that attack other oilseed crops. *Pythium aphanidermatum* and *P. de baryanum* cause 'Root rot' and 'Damping off of young plants'; *Phytophthora palmivora* causes 'Seedling and leaf blight'; 'Wilt disease' is caused by *Fusarium oxysporum* f.sp. *carthami;* 'Charcoal rot' or 'Collar rot' is caused by *Rhizoctonia bataticola*; *Sclerotium rolfsii* also causes 'Collar rot'; *Puccinia carthami*, an autoecious rust causes 'Rust disease'; *Erysiphe cichoracearum* causes 'Powdery mildew'; *Alternaria carthami, Ramularia carthami, Septoria carthamicola* and *Ascochyta carthami* cause 'Leaf spots'; 'Bacterial blight' is caused by *Pseudomonas syringae*; 'Mosaic' caused by virus and 'Phyllody' caused by Mycoplasma-like organism are also found to occur on safflower.

CASH CROPS

Sugarcane *(Saccharum officinarum)*

1. Red rot of sugarcane

Colletotrichum falcatum
(Glomerella tucumanensis)

Among the diseases that attack sugarcane, **'red rot'** is one of the most serious and important diseases that attacks planting setts, standing canes and leaves. The disease, which was first recorded in Java (Indonesia) in 1893, is prevalent in all countries, where sugarcane is cultivated on a large scale. In India, the disease occurs in an epiphytotic form during some of the years and causes considerable losses to the growers and sugar factories. The disease occurred in a serious, epiphytotic form in Uttar Pradesh and Bihar during 1939 - '40 and 1946 - '47 and caused severe damage that threatened the very existence of many sugar factories. In these epidemics, some of the very promising Coimbatore varieties, such as Co.210, Co.213 and Co.312 were found to be highly susceptible and had to be withdrawn from cultivation. Localized epidemics occur almost every year in some parts of the country. Only in 1974, the disease was first recorded in Tamil Nadu on a large scale in the South Arcot district and the variety, Co.658 was found to be highly susceptible. Many promising varieties, such as Co.318, Co.419, Co.453, Co.671, Co.785, Co.997, CoC.671, Bo.10, Bo.11, Bo.17, Bo.54 etc. had become vulnerable to attack by the pathogen. Early infection after planting causes damage to the buds, which fail to germinate. In a susceptible variety, if the disease establishes in the early stages of growth, it spreads very rapidly leading to large-scale mortality of young canes. Infection at a later stage results in the reduction of juice content. The quality of the juice and sucrose content in the juice are also adversely affected. Setts obtained from disease affected canes are the main source of primary infection in the planted fields.

Symptoms. It is difficult to recognize the disease in the early stages of growth. First outward symptoms are visible after the rainy season, when active plant growth ceases and sucrose formation begins in the stem region. Loss of color and drooping of third leaf from the top are the earliest symptoms. This is followed by withering of the whole crown in about 8 to 10 days. In later stages, characteristic symptoms are noticed in the stem region. When withering of leaves starts, if the diseased canes are split open longitudinally, the pith is found reddened. Characteristic bands of white areas are found running transversely across the reddened pith. In advanced stages of the disease, the red color may become dirty brown and the white bands may not be very conspicuous. Cavities filled with grayish or white mycelium may also be found in the pith region.

The juice emits a bad, pungent odor, as a result of conversion of sucrose into **'glucose'** and **'alcohol'** due to enzymatic action of the pathogen. In the final stages of disease development, the canes become shrivelled and the rind develops longitudinal wrinkles. Such canes are lighter in weight and are easily broken. Minute, velvety, dark, pinhead-like dots, which are the acervuli of the pathogen, are formed around the nodal region and also in the shrunken areas. Pink, spore masses are also seen extruded from the acervuli. Acervuli are also produced in large numbers in the mycelial growth in the cavities formed in the pith region. The pathogen also produces small, reddish lesions, about 2.0 - 3.0 x 0.5 mm. on the center of the mid rib of leaves, which gradually elongates. Later on, the spots become dark-red in color, with a dark-brown margin and straw-colored center. The infected leaves may break at the lesions and hang down. In the central portion of the lesion, minute, black, pinhead-like dots are produced, which represent the acervuli.

Causal organism. Once the fungus, the causal agent of the disease gains entry into the host, it grows rapidly, producing septate mycelium, which is both inter- and intracellular. Because of the presence of the fungus, the sugarcane tissues react vigorously, as a result some physiological changes occur in the host cells. The protoplasm in the cells changes color and a dark-red, gummy substance oozes out of the cells, filling the intercellular spaces. This soluble pigment is absorbed by the cell walls, which gives the characteristic red rot appearance. After growing for a period within the host tissues, the hyphae produce large number of chlamydospores in the pith region. They are round and formed by the segmentation of the hyphae. These chlamydospores can persist in the soil for a long time. As the disease advances, hyphae collect beneath the epidermis and form a stroma of densely packed cells. Long, dark-brown, bristle-like setae, 100 - 200 μ long, with four septa arise from the stroma of the acervulus and push their way through the epidermis. Sickle-shaped conidia, measuring 20.0 - 80.0 x 5.0 - 7.0 μ, with a large oil globule in the center, which are borne on short, small, conidiophores are exposed. Wind, rain, raindropsplashes and insects disseminate them effectively. The conidia are short-lived and germinate in the presence of moisture. On germination, they give rise to a germ tube and then an appressorium at the tip of the germ tube. The appressorium becomes thick-walled and function like a chlamydospore.

The perfect stage, which was earlier reported as *Physalospora tucumanensis* is now considered as a saprophyte and so, the perfect stage has been now confirmed as *Glomerella tucumanensis*. The perfect stage is rarely met with and that too in dry and decomposing sugarcane leaves. Perithecia are globose, ostiolate, superficial, with the bottom embedded in the host tissue and measure 150 - 300 μ in diameter. Asci are numerous, hyaline, clavate, unitunicate and measure 49.0 - 66.0 x 7.0 - 10.5 μ. Mixed with these asci, are numerous, hyaline paraphyses. Each ascus contains eight ascospores. They are single-celled, hyaline, elliptical and measure 17.5 - 21.0 x 5.3 - 7.0 μ **(Fig.59).**

Fig. 59: Red rot of sugarcane-*Glomerella tucumanensis (Colletotrichum falcatum)*

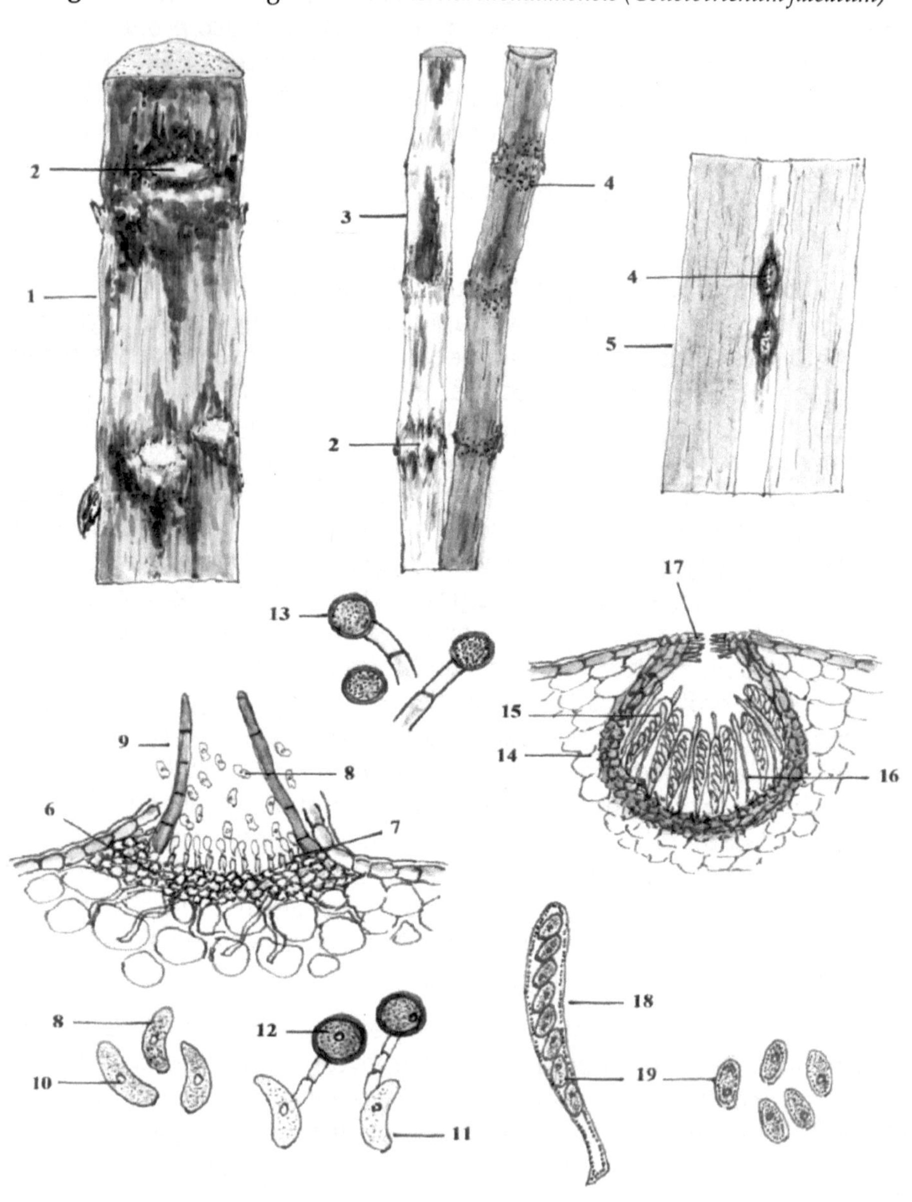

1. Split cane showing reddening of the pith with transverse white blotches 2. White blotches 3. Dry, shrivelled cane, showing acervuli 4. Acervuli 5. Leaf mid rib lesion showing acervuli 6. Acervulus (stroma) 7. Conidiophores 8. Conidia 9. Seta 10. Oil globule 11. Germinating conidia 12. Appressorium 13. Chlamydospores 14. Perithecium 15. Asci 16. Paraphyses 17. Periphyses 18. Ascus 19. Ascospores.

Mode of survival, spread and epidemiology. Seed-setts from diseased canes are the main source of primary inoculum and the means of survival of the pathogen. Incipient infection of the bud scales, leaf scars, root primordia or injuries caused by insect pests may also serve in initiating the disease. The chlamydospores, which can survive in the soil for 60 - 150 days may also serve as primary inoculum. Once the disease affected shoots appear in the field, secondary infection is caused by conidia produced in acervuli and are disseminated through various means. When the conidia come in contact with the nodal buds, infection is caused. The thick-walled appressoria can perpetuate the fungus in the nodal buds for long periods. Ratooning of susceptible sugarcane crops also serves in perennation and multiplication of inoculum.

The conidia produced on the mid-rib lesions also cause fresh infection on standing crops. The acervuli and chlamydospores in leaves pass through the digestive system of cattle, uninjured and may be disseminated through manure. The pathogen is known to survive in some of the collateral hosts, such as *Sorghum vulgare, S. halepense, Saccharum spontaneum, Leptochloa filiformis, Miscanthus* sp. etc.

High humidity, water-logging, poor crop growth, monocroppimg, presence of susceptible varieties and ratooning of susceptible varieties are some of the main factors leading to occurrence and build-up of inoculum, which may cause epidemics.

Disease management

Agronomic practices (i) As the accumulation of inoculum year after year results in epiphytotics, much importance has to be given to prevent such build-up of inoculum. As soon as isolated cases of the disease are noticed in a field, the entire clump should be dug out and burnt. After harvest, crop residues, such as stubble, dry leaves, dead and dried up canes etc. should be removed from the field and burnt (ii) Healthy cane-setts should be used for planting. Setts should be obtained from disease-free fields or areas (iii) Setts showing any reddening of the pith should be discarded. Setts with hollows or red patches at the cut ends should be rejected. Setts with damaged buds, insect injuries or other injuries should not be used (iv) For seed purposes, canes should be grown in selected, disease-free plots under proper supervision. This will help in avoiding chances of occurrence of other diseases, such as smut, wilt and virus diseases also, besides red rot (v) Proper drainage should be provided to avoid water-logging (vi) In disease-prone areas and in the case of susceptible varieties, ratooning should be avoided (vii) Growing of a particular variety, especially a susceptible variety year after year in the same field should not be encouraged (viii) Crop rotation will help in checking inoculum build up in the soil. A crop rotation of 2 to 3 years, incorporating rice and green manure crops in the crop sequence is ideal. Wet rice cultivation for 2 or 3 consecutive seasons helps in eliminating the pathogen due to anaerobic condition.

Disease resistant varieties. Use of resistant varieties is by far the most ideal means of controlling the disease. However, due to continuous evolution of new physiological races, the resistance of a particular resistant variety may be broken in the course of time. So, breeding for resistance to red rot should proceed continuously and new resistant varieties have to be evolved to replace the earlier resistant varieties, which have become susceptible. Co.6806, Co.7704, CoC.771, CoC.772, CoC.62198, CoC.90003 and CoC.91061 have been reported to be resistant. CoC.8001, CoC.8201 and CoC.85061 are found to be moderately resistant. Varieties, such as CoSi.767, CoSi.802, Co.975, Co.7219, Co.7314, Bo.91, Bo.99 and CoJ.58 have also been found to have resistance to red rot. CoC.86062 is tolerant to this disease. Co.449, Co.658, CoC.671, CoC.8001 and CoC.86062 are some of the very popular varieties, which have become highly susceptible.

Seed treatment. Soaking the setts in 0.05 % carbendazim + 1.0% urea solution or 0.25 % emisan solution for 20 minutes, prior to planting is effective in controlling the disease.

2. Smut or whip smut of sugarcane

Ustilago scitaminea

Sugarcane smut is widely distributed and occurs in most of the sugarcane growing countries of the world including Java, Egypt, East and South Africa, Madagascar, the Philippines, Australia, Mauritius, Italy, South America, British guano and many Asian countries. The disease was recorded in India in 1906 and it is widespread in all the states where sugarcane is cultivated. Wild canes and those, which are more akin to wild canes, are more susceptible than the improved thick canes. However, some of the improved cultivated varieties are found to be highly susceptible to the disease. The disease appears at all growth stages of the crop.

Symptoms. The most characteristic symptom of the disease is the production of long, whip-like, black, sooty shoot, which may be often several feet in length and very much curved. This is probably the floral shoot that gets transformed into the whip-like structure. A thin, silvery membrane covers the whip in its earlier stages, which breaks into flakes exposing dense, black, powdery masses of spores. This membrane is actually the epidermis of the host part. The exposed dusty spores are easily blown away by wind and cause secondary infection. Systemically infected canes produce lateral shoots from buds found on the lower part, which also produce the smut whips usually. If the whip at the main shoot is cut off or if the main shoot of an infected plant is damaged due to attack by insects or by any mechanical means, production of lateral shoots is stimulated and in such cases almost all the lateral shoots invariably produce smut whips. In the case of localized infection, the main shoot may not produce the whip, but the laterals, which arise from the lower part of the main stem may produce whips. Affected plants usually have slender, thin, taller canes than the healthy plants and can be easily identified before they produce smutted whips. The spore masses are confined to a few outer cortical layers of the whip, while the internal tissues consisting of parenchyma and vascular bundles remain normal. Primary infection results in the development of smutted whips early in the season, while in the case of secondary infection the smutted whips appear late in the season **(Fig.60).**

The causal organism. The mycelium of the fungus is intercellular and sends haustoria into the host cells to obtain nourishment. The mycelium aggregates in the tissues of the cane below the whip-like structure and in the meristem. The hyphae collect in dense masses towards the surface of the spore-bearing shoot, where the spores are formed in abundance, which appear as black, dusty mass. The spores are globose to sub-globose, reddish-brown in color, smooth or punctate-walled and measure 5.0 - 10.0µ in diameter. They are loose and very light and are easily carried by wind. Under moist conditions, they germinate forming short promycelium, with usually 4 cells. From each cell, a sporidium is formed at the end of a short stalk. The sporidia are elongate, single-celled and hyaline. They get detached easily and germinate to produce the infection hypha. Under favorable conditions, the sporidia may produce more sporidia in short chains by budding. Sometimes, instead of producing sporidia the promycelium grows into a branched hypha and functions as an infection hypha **(Fig.60).**

Mode of survival, spread and epidemiology. The disease is primarily seed-borne. The mycelium present in the setts of smutted canes causes systemic infection. The spores adhering to the buds of setts used for sowing also cause primary, systemic infection. The pathogen is also soil- and air-borne. Under dry conditions, the spores may remain in a viable state for more than 7 months, whereas under moist conditions, they loose their viability within 3 weeks. Spores coming in contact with young germinating shoots infect them. Any injuries on the scales of eyes facilitate the entry of the pathogen. The disease may also attack standing canes through the buds. Spores carried by wind may fall on the exposed buds. They germinate on these buds and cause infection. Some of the buds thus infected may produce smutted shoots in the same season, while others carry the dormant mycelium to the next season in the setts. Sugarcane crop is found all through the year in the sugarcane growing tracts and as such the spores are always present in the air and cause fresh infections. Ratooning of sugarcane crop, especially crops affected by smut, increases the severity of the disease. The disease attacks some of the grasses, such as 'Kans grass', which may serve as collateral hosts.

Optimum temperature for spore germination is 25° - 30°C. High relative humidity of 100 % is essential for spore germination. No germination occurs when the relative humidity is 90 % or less. Germination of spores stimulates secretion of certain **'diffusates'** from buds, which favor infection. The optimum temperature and relative humidity for the germination of smut spores, is highly favorable for infection and development of the disease also.

Disease management

Agronomic practices (i) In case of mild attack, the smutted whips can be cut before the membranous covering ruptures and destroyed by burning. To avoid shedding of spores, the whip may be completely covered with a paper bag before cutting. After this, the entire clump should be dug out and burnt (ii) Seed- setts should be selected from disease-free fields. It is advisable not to select setts from fields, if nearby fields are affected by the disease (iii) Ratooning of crops affected by the disease should be avoided (iv) In areas where the pathogen is known to be well established, growing of susceptible varieties in such areas should be avoided.

Sett treatment (i) Steeping the setts in a solution of mercuric chloride - 0.1 % or formalin - 1.0 % or Bordeaux mixture - 0.75% for 5 minutes, prior to planting is effective in eradicating the spores present on the surface of setts or buds (ii) Dipping the setts in a solution of systemic fungicides, such as carbendazim at 1.0 gm. or carboxin at 1.0 gm. or benomyl at 1.0 gm. per lit. of water, prior to planting controls both the external and internal inoculum (iii) Treating the setts in hot water at 50°C for 2 hours or hot air at 54°C for 8 hours or aerated steam at 52°C for 1 hour is also found to be effective in controlling both the external and internal inoculum. After heat treatment, the setts are cooled and planted immediately.

Resistant varieties. Varieties, such as Co.449, Co.527, Co.6806, Co.7704, Co.7807, Co.8139, Co.62175, Co.62198, Co.C.671, Co.C.771, Co.C.772, Co.C.773, Co.C.86062, Co.C.90063, Co.C.91061 and Co.S.86071 have been found to be resistant. Co.419, Co.740 and Co.975 are highly susceptible.

3. Wilt disease of sugarcane

Cephalosporium sacchari

The disease is prevalent in many countries of the world, including South Africa, Argentina, Columbia, Barbados, Mexico, the Philippines, West Indies, Uganda, the United States of America and India. The disease was first recorded in India in 1913 and now, it occurs as a common and destructive disease in the states of Uttar Pradesh, Bihar and Tamil Nadu. In Tamil Nadu, it is found in all sugarcane growing districts. Very often, the disease occurs in association with red rot and *Fusarium* stalk rot.

Symptoms. The earliest symptoms are stunting and retardation in growth of a few plants or clumps, scattered in the fields. The symptoms become manifested, when the crop is about 4 - 5 months old. However, in the early stages of the disease, no discoloration or damage is observed in the root system. When the crop is nearing maturity, typical symptoms, such as sudden yellowing and withering of the crown leaves appear, followed by rapid drying of the canes. The affected canes become light and hollow.

When canes in the early stages of disease development are split longitudinally, the internal tissues, especially of the lowest internodes show brick-red discoloration, while individual vascular strands appear dark-red in color. The reddening may be confined to a few lower internodes or extend to the full length of the cane. But, as in the case of red rot disease, no transverse white bands are seen. Brown patches may appear on the rind, where the underlying tissues are dead. In the reddened pith region, the mycelium grows in the cells in profusion in all directions. Gradually, the pith dries and becomes hollow like a boat. In the hollowed areas, dense growth of grayish-white mycelium and large number of conidia are seen. At the later stages, when the stem tissues start drying, the underground roots and the aerial roots, which may develop from aboveground nodes also become red and die. Inside the lumen of the vascular elements, fungal hyphae and a gummy substance are found plugging the vessels **(Fig.61).**

The causal organism. The mycelium of the pathogen is effuse, white, sparsely septate and much branched. The conidiophores, which arise from the hyphae are aseptate, tapering towards the apex, simple or verticillately branched with a few branches and measure 6.0 - 30.0 x 3.0 - 4.0μ. Conidia are produced in succession in large numbers from the tip of the branches. They collect at the apex of the branches and held together in a head with a mucilaginous substance, but are separated easily. They are hyaline, ovoid or oblong-ellipsoid, single-celled and measure 4.0 - 12.0 x 1.0 - 3.0μ in size **(Fig.61).**

Mode of survival, spread and epidemiology. The mycelium of the pathogen can survive in the vascular elements of apparently healthy canes for long periods and gets distributed through setts. Because of its poor saprophytic survival capacity, it cannot survive in the soil for prolonged periods. Infection may take place in the seed pieces after planting. But, no infection occurs through roots or through the rind. Injuries in the underground parts of stem or roots may facilitate entry of the pathogen into the host. Crops affected by red rot are more prone to attack by this pathogen.

High temperature, low relative humidity and soil alkalinity favor disease development.

Disease management

Agronomic practices. Use of disease-free planting setts, uprooting and destroying diseased plants, providing proper drainage facilities, applying balanced doses of fertilizers and protecting the crop from insect or mechanical injuries are methods that can be adopted to minimize disease occurrence.

Sett treatment. Steeping the setts in a solution of carbendazim + urea for 15 minutes prior to planting is effective in controlling the disease. In 100 liters of water, carbendazim - 50 gm. and urea - 1.0 kg. are mixed and in this solution, setts required to plant one acre of field are steeped.

Resistant varieties. Varieties, such as Co.356, Co.370, Co.393, Co.395, Co.617, Co.859, Co.1158 and BP.7 are found to be tolerant to the disease.

4. Sett rot or 'pineapple' disease of sugarcane

Thielaviopsis paradoxa

(Ceratocystis paradoxa/ Ceratostomella paradoxa/ Ophiostoma paradoxa)

The disease caused by *Ceratostomella paradoxa / Ceratocystis paradoxa / Ophiostoma paradoxa* was first reported from Indonesia in 1893. It occurs in all sugarcane growing countries of the world. In India, the disease is prevalent in Punjab, Maharashtra, Karnataka, Tamil Nadu, Kerala and Pondichery. The disease primarily attacks the setts and rarely the stem region. It causes rotting of setts and germination failure leading to gappiness and poor crop stand.

Fig. 60: Whip smut of sugarcane
Ustilago scitaminea

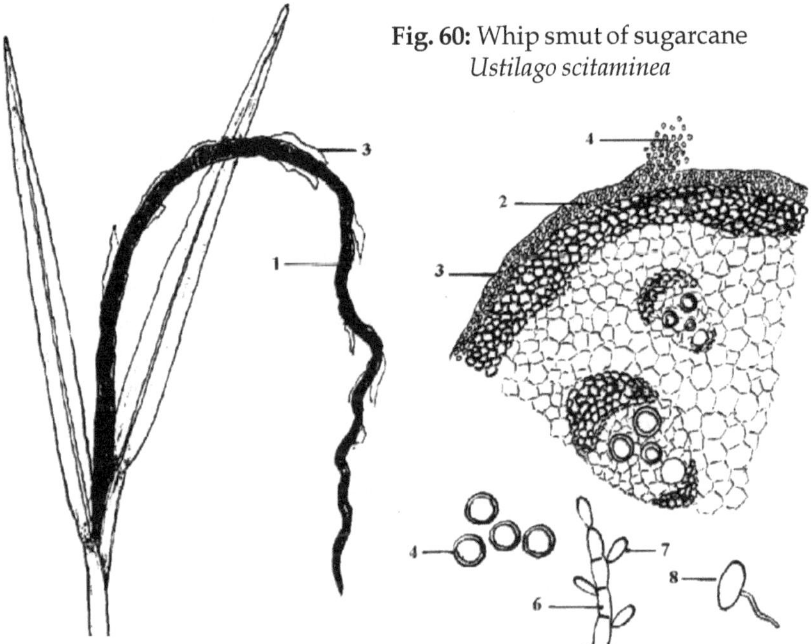

1. Smutted shoot 2. Transverse section of whip showing teliospores 3. Host epidermis 4. Teliospores 5. Germinating teliospore 6. Basidium 7. Sporidium 8. Germinating sporidium.

Fig. 61: Wilt disease of sugarcane
Cephalosporium sacchari

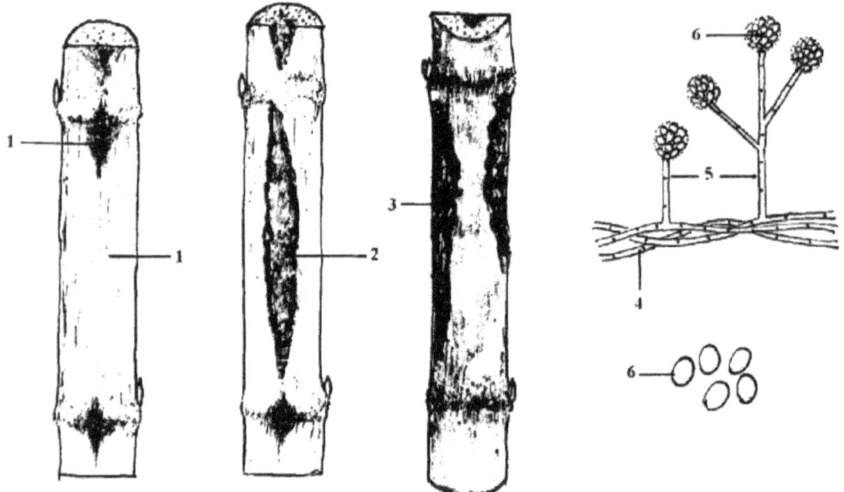

1. Early symptoms showing reddening of internodal tissues and vascular strands 2. Hollowed, boat-like depression of the pith region 3. Brown patches on the rind 4. Mycelium 5. Conidiophore 6. Conidia.

Symptoms. The disease primarily attacks the setts after planting. The fungus enters the setts through the parenchymatous tissues of the cut-ends and causes rotting of the setts. Affected setts fail to germinate. Such setts, when split open longitudinally in the early stages of infection emits an odor of fresh pineapple and hence the disease is also called **'pine apple disease'**. As a result of the fungal attack, when the tissues start rotting, the sugars present in the setts are converted into **'ethyl acetate'**, which emits the pine apple odor. The rotten areas turn red, with scattered black patches. However, the fibrovascular tissues are not affected. As the disease progresses, the core of the internodal regions becomes hollow and black. The black color is due to the formation of large number of dark-colored microconidia. Sometimes, the setts may germinate and in such cases, the shoots are also attacked. The tissues of the affected shoots turn red, the leaves become chlorotic and the plants die. The disease may occasionally affect standing canes, which have been injured by stem borers, rats, white ants or by mechanical means. Such affected plants become stunted and chlorotic.

The causal organism. The mycelium of the fungus is hyaline or light-greenish in color and septate. In its asexual stage, the fungus produces both micro- and macroconidia. Microconidia are produced endogenously in succession and come out from the conidiophores in long chains. Conidiophores are slender, aseptate, hyaline to pale-brown, about 100µ in length and usually occur in the form of synnemata. The microconidia are very small, oblong, thin-walled, hyaline at first and become black as they mature and measure 10.0 - 15.0 x 3.6 - 5.6µ in size. Macroconidia are produced on short lateral conidiophores, which arise almost perpendicular to the hyphae and are about 20 - 80µ long. The macroconidia are oval to obovate, thick-walled, light-green to black in color and measure 16 - 19 x 10 - 12µ in size. About 20 macroconidia may be formed in chains from each conidiophore. The presence of large number of conidia inside the setts gives the black coloration.

In the sexual phase *(Ceratocystis paradoxa)*, the fungus produces perithecia. They are flask-shaped, brown and measure 200 - 350µ in diameter. The perithecia, which have long and narrow neck, with a small, round ostiole, measure 800 - 1200 x 30 - 40µ. Asci are formed at the inside bottom of the perithecia. They are club-shaped and are about 25 x 10µ in size. In each ascus, there are 8 ascospores, which are oval, thin-walled, hyaline and measure 7.0 - 10.0 x 2.5 - 4.0µ in size. The asci break open, while still inside the perithecia, liberating the ascospores, which emerge through the ostiole and are held together in a drop of viscous fluid at the ostiole. The ascospore germinates by producing a germ tube **(Fig.110)**.

Mode of survival, spread and epidemiology. The macroconidia remain in a viable state for long periods in the soil and in the diseased plant residues and behave as chlamydospores. The microconidia cannot remain alive in the soil for a long time. The perithecia formed in the rotten seed pieces can survive in the soil for a long time and perpetuate the disease through ascospores. The fungus can also lead a saprophytic life in the soil in the absence of the host. The weak parasite, gains entry into the planted setts only through the cut-ends and grows rapidly in the internodal tissues. However, the nodes rarely stop its growth. Under favorable conditions, the entire seed piece is invaded and destroyed. Usually, the buds are destroyed before they sprout. The conidia are disseminated by wind, irrigation water or rainwater from field to field. The pathogen can also gain entry into the host through injuries caused by sugarcane early shoot borer *(Chilo infuscatellus)* and internode borer *(Chilo sacchariphagus-indicus)*. The pathogen has several collateral and alternate hosts, such as coconut, arecanut, date palm cocoa, coffee, mango etc.

The disease occurs more in heavy clay soils. The incidence of the disease is more in alkaline and acid soils, and in water inundated fields.

Disease management

Agronomic practices (i) Care should be taken to select good quality setts for planting (ii) Proper drainage facilities should be provided (iii) Setts infested with scale insects, mealy bugs, borer pests and white ants should be rejected (iv) Setts with damaged buds, cracks, rooted nodes etc. should not be used for planting (v) Irrigation through diseased field to other fields should be avoided (vi) It is preferable to use setts with 3 or 4 nodes in places where the disease occurs regularly.

Sett treatment. Steeping the setts in a solution of carbendazim + urea for 15 minutes is effective in controlling the disease. In 100 liters of water, carbendazim - 50 gm. and urea - 1.0 kg. are mixed and in this solution, setts required to plant one acre of field are steeped.

5. Ratoon stunting disease of sugarcane

Clavibacter xyli sub sp. *xyli*

The disease was first reported from Australia in 1950 and from India in 1956. Now, it is found in most of the sugarcane growing countries of the world. In India, it occurs in all the states. The disease causes considerable reduction in cane yield and the percentage of sugar in the diseased canes is also very much reduced.

Symptoms. The disease occurs mostly in the ratoon crop. The infected plants are stunted and weak, with much shortened internodes. The symptoms resemble those caused by wilt disease. Tillering is considerably reduced. Striking symptoms can be seen, when the stems of affected plants are split open longitudinally. In young plants, pinkish spots are seen in the nodal region, while in older plants the internodal regions below the nodes also turn pinkish to a length of about 3.0 mm. In matured canes, the vascular strands become reddish or brown. The xylem vessels are plugged with bacteria contained in a gummy substance. Setts from diseased canes have poor viability. The germination is very much delayed and the growth of the plants is very poor.

The causal organism. The disease, which was originally thought to be caused by a virus or by a mycoplasma-like organism, has now been definitely identified as *Clavibacter xyli* sub sp. *xyli*.

The fastidious, xylem-inhabiting, gram positive, coryneform bacteria, measure 1.0 - 4.0 x 0.3 - 0.5μ in size. They are non-motile and non-spore forming. Sometimes, they occur in chains and appear like filaments.

Mode of survival and spread. The pathogen overseasons in infected sugarcane plants and propagative materials, such as seed-canes. Plants growing from infected setts develop the disease invariably. The bacterium is sap transmissible and is spread by cutting knives, by implements used in cultural operations and also by harvesting equipments. Rodents, such as field rats, rabbits, wild pigs etc. also aid in the transmission of the bacteria through injuries they cause to the standing canes. The disease continues to spread to different countries through infected sugarcane germplasm.

Disease management

Agronomic practices. Use of disease-free setts for planting, sanitation of cutting equipments, control of rodents and use of resistant cultivars are some of the methods that can be adopted to control the disease.

Sett treatment. Treating the setts in hot water at 50°C for 2 hours or hot air at 54°C for 8 hours or aerated steam at 52°C for 1 hour is found to be very effective in controlling the disease. After heat treatment, the setts are cooled and planted immediately.

6. Grassy shoot disease of sugarcane

Mycoplasma-like organism

'Grassy shoot disease' (GSD) is also known as **'bunchy disease'** or **'albinism'**. It is prevalent in Burma, Thailand, Sudan, Sri Lanka, Taiwan and

India. In India, it is found in the states of Uttar Pradesh, Madhya Pradesh, Maharashtra, Telungu Desam, Orissa, Bihar, Punjab, Rajasthan, Haryana, Karnataka, West Bengal and Tamil Nadu. Ratoon crops are more susceptible to this disease.

Symptoms. The characteristic symptom of the disease is the production of large number of lanky tillers and the affected plant presents a bushy appearance. The leaves of such plants are small and narrow like the leaves of grass and the stem is also thin, with short internodes and the clump looks grass-like. Shoots growing from diseased setts remain stunted and weak. Infected plants show varying degrees of chlorosis. The leaves of some of the tillers in a clump totally lack chlorophyll and become completely white and hence the disease is also called **'albinism'**. The development of root system in infected plants is also very poor. When many tillers in a clump are affected badly, the entire clump may die. Typical symptoms are produced in the case of late infection. In the newly formed leaves, long, white or whitish-yellow streaks are formed along the veins and from the basal part of the stem, many grassy shoots may arise. In some cases, aerial roots may be produced from the basal nodes. Such plants become stunted and only a few millable canes are produced from such clumps.

The causal organism. The disease is caused by a *Mycoplasma-like organism*. The pathogen is found in the sieve tubes of phloem vessels and moves from cell to cell through pores in the sieve plates.

Mode of survival and spread. The disease is primarily sett-borne. It is also transmitted through sap and by a few insect vectors. Infected canes used as seed pieces for planting invariably carry the pathogen and cause systemic infection. Sap transmission to setts from healthy canes takes place while using same cutting knives to cut both infected and healthy canes. The aphid vectors, *Aphis sacchari, A. idiosacchari* and *Rhopalosiphum maydis* transmit the pathogen. The minimum acquisition feeding time is about 35 seconds and the optimum is about 15 minutes. The inoculation feeding time is about 30 seconds. Vector transmission occurs in a non-persistent manner. Winged forms are more effective in spreading the disease in the standing crop. The disease can also be transmitted by the plant parasite, *Cuscuta campestris* Ratooning of affected crop serves as a potential source of perennation of the pathogen. The disease is also found to attack sorghum.

Disease management

Agronomic practices. Removal and destruction of diseased clumps, use of disease-free, healthy seed canes, sanitation of cutting equipments by dipping them in phenol or lysol frequently, avoiding ratooning of disease affected crop and control of insect vectors in the early stages of crop growth by spraying insecticides, such as methyl demeton - 400 ml. or dimethoate - 400 ml. or phosphamidon - 220 ml. in 400 liters of water per acre, using a high volume sprayer are methods that can be adopted to control the disease.

Sett treatment. The disease can be effectively controlled by heat treatment of setts in hot water or hot air or aerated steam.

7. Iron chlorosis of sugarcane

Iron deficiency

'Iron' is a catalyst of chlorophyll synthesis and is mainly concerned with the process of photosynthesis. It also forms a part of many enzymes. Plants need only a very small quantity of iron, however non-availability of iron leads to acute chlorosis. Although iron may be present in the soil in sufficient quantities, under some conditions, it is not made available to the plants or it is immobilized within the plant. Presence of excess of lime in the soil, converts the available ferrous salts into unavailable ferric salts due to alkalinity and this condition results in **'lime-induced iron chlorosis'**. In acid soils, iron, which is usually available, is made unavailable, when excess of soluble phosphate is present in the soil. Excess of manganese in the soil also makes the iron unavailable to plants, as manganese interferes with the role of iron in chlorophyll synthesis.

The disease is more severe and the symptoms more pronounced in ratoon crops than in planted crops. Further, varieties with broad leaves are more susceptible to the disease than varieties with narrow leaves.

Symptoms. The symptoms are more pronounced in young leaves, which become severely chlorotic. In older leaves, while the veins remain green, the interveinal regions become chlorotic and whitish. Thus, the leaves appear with alternate green and chlorotic stripes extending to the full length of the leaves. Later, the entire leaf lamina becomes chlorotic. Chlorosis of leaves without any mottling symptoms is characteristic of the disease. Ultimately, the chlorotic leaves dry and the affected plants may die.

Disease management

Iron deficiency can be corrected by the application of iron through the foliage by spraying or through soil application. In the case of young plants, which are less than 120 days old, foliar spraying with ferrous sulfate - 1.25 kg. in 200 liters of water per acre corrects the deficiency. In the case of older plants, which are more than 120 days old, 2.0 kg. of ferrous sulfate may be sprayed. The spraying should be continued at fortnightly intervals, till newly emerging leaves become green. During the last spraying, 5.0 kg. of urea may be added along with ferrous sulfate. Instead of foliar application, ferrous sulfate may be applied to the soil at 4.0 kg. per acre. Application of chelated iron compounds to the soil is more effective than simple inorganic iron compounds, especially in acid soils

8. Phanerogamic parasite on sugarcane

Striga species

'*Striga* species' commonly known as **'witchweeds'** are found in Australia, Burma, India, Africa, America, Madagascar and Mauritius. It was first found in America in 1956. *Striga* species, such as *Striga densiflora, S. lutea* and *S. euphrasioides* parasitize sugarcane, tobacco, cowpea, rice, sorghum, maize, lesser millets and some grasses. These semi-root parasites cause serious damage to the crop during some seasons.

Symptoms. The parasites appear in clusters as green, erect herbs around sugarcane plants and absorb nourishment from the host by attaching themselves to the host roots. Attacked plants become stunted and weak due to continuous drainage of water and mineral nutrients by the parasite. The leaves turn yellow, wilt and wither. When young plants are attacked, they wither and die. When older plants are attacked, they may not die, but the cane yield and quality are badly affected. If drought conditions prevail, the host plants are destroyed completely, as witch weeds increase the water stress in the host plants.

The parasite. The parasites have chlorophyllus green leaves and are capable of manufacturing their own requirements of carbohydrates. The pretty, erect, herbaceous parasites grow to a height of 22 - 60 cm. It produces many branches from the base and from the top. The surface of the stem is covered with short, erect, stiff, hair-like structures. The leaves are scale-like, narrowly lanceolate, entire, green and are 20 - 60 x 3.0 - 8.0 mm.in size. Leaves on the lower portion of the stem are opposite, while those on the upper portion are alternate. Floral buds are formed between the underground stem and leaf axils. Flowers are sessile, longer than the leaves, red, yellowish or whitish in color and always with yellow center. Cylindrical capsules are formed from the flowers, which contain large number of tiny, brown seeds. When the capsules dehisce, the seeds are dispersed on the soil surface. A single plant may produce 50,000 to 5,00,000 seeds. The seeds usually require a dormancy period of 15 - 18 months, but they retain their viability and remain in the soil for a period of 12 - 40 years.

The exudates coming out of the host roots stimulate the germination process of seeds of the parasite. After germination, the parasite makes contact with the host. As soon as the rootlet of the witchweed comes in contact with the host root, its tip is transformed into a bulb-shaped haustorium, which penetrates the host root. The enzymes produced by the roots of the parasite assist in the penetration of the host root. The leading cells of the haustorium, usually tracheids, reach the vessels of the host roots and absorb water and nutrients. The xylem of the parasite is connected to the vessels of the host and the parasite gets established. The germinated *Striga* seedling grows below the soil level for 4 - 6 weeks and develops root system, underground stem and floral buds, at the same time many connections are made between the roots of the parasite and those of the host. Young *Striga* seedlings, as long as they are subterranean are completely parasitic and get their entire requirements of nutrients from the host tissues and vascular elements. It is at this stage, they cause maximum damage to the host. However, after they emerge aboveground and produce aerial leaves, they manufacture their own requirements of carbohydrates and exist as a semi-root parasite. It takes about 2 - 4 months for the parasite to complete its life cycle.

Mode of survival and spread. The seeds of the parasite that are shed in the soil are disseminated over short distances through rain and irrigation water, cattle, agricultural implements etc. and over long distances by wind, floods etc. Seeds, even at a depth of 35 cm. in the soil can germinate however, the seeds have to be close to the host, preferably within 3.0 - 4.0 mm. for quick establishment. The witchweed seedlings, which fail to make contact with the host die eventually.

Management

Agronomic practices (i) The parasites should be uprooted and destroyed before seed formation. Within 2 months of planting of sugarcane, 4 - 6 weeding may be needed (ii) Crop rotation with non-host trap crops, such as cotton, soybean, sunflower or groundnut helps to control the parasite. The root exudates of these crops stimulate *Striga* seeds to germinate. But, as these crops do not support the growth of the parasite, they are starved to death for want of suitable hosts.

Chemical control (i) Spraying the weedicide 2-4-D (Fernoxone) at 450 gm. in 500 liters of water on the soil surface immediately after planting of sugarcane destroys the parasite. But, this weedicide is highly toxic to crops, such as cotton, tobacco, soybean etc.(ii) Soil injection with the fumigant, ethylene at 400 ml. per acre prior to planting of sugarcane, when the soil is in a moist condition stimulates the germination of the *Striga* seeds within a few hours and the germinated seedlings die of starvation (iii) Spraying a solution of 20 % urea on the emerged *Striga* plants before seed formation kills them by scorching.

Diseases of minor importance. Sugarcane is subjected to attack by many other fungal, bacterial and virus diseases. *Gibberella moniliformis (=Fusarium moniliforme)* causes '*Fusarium* sett rot' and 'Stem rot'; *Puccinia melanocephala* and *P. kuehnii* cause 'Rust disease'; *Mycosphaerella koepkei (=Cercospora koepkei)* causes 'Yellow spot'; *Drechslera (Helminthosporium) sacchari* causes 'Eye spot'; 'Pokkah Boeng' or 'Twisted top' is caused by *Gibberella fujikuroi* sub sp. *subglutinans (=Fusarium moniliforme)*; 'Red stripe and Top rot' is caused by the bacterium *Pseudomonas rubrilineans;* 'Mosaic' is caused by *Sugarcane mosaic virus*; Root knot nematode, *Meloidogyne javanica* causes 'Root knot disease' that affects the roots.

Cotton *(Gossypium species)*

1. *Fusarium* wilt of cotton

Fusarium oxysporum f.sp. *vasinfectum*

'*Fusarium* wilt' is one of the most serious diseases of cotton and occurs in almost all the cotton growing countries of the world. This vascular disease was first reported from Mexico and in India, it was reported in 1908 from Madhya Pradesh. It is widely prevalent in Tamil Nadu, Telungu Desam, Karnataka, Maharashtra, Gujarat and Madhya Pradesh. It occurs more in the heavy, black cotton soils than in sandy loam soils. In case of severe incidence, the entire plant wilts and dies and in such cases, the yield loss may go up to 100 %. In Tamil Nadu, it appears sporadically, sometimes in a severe form. *Gossypium arboreum* and *G. herbaceum* types of cotton are more vulnerable to attack by the disease.

Symptoms. The disease attacks the crop at all growth stages. The first symptoms appear at the time of emergence of the seedlings, when the cotyledons are attacked and they become yellow and then turn brown. The base of the petioles also becomes brown and girdled. Wilting and drying of the affected seedlings follow this. In the field, the plants show wilting symptoms in patches. First of all, the older leaves become yellow, loose their turgidity, wilt, turn brown and droop. This is followed by wilting of the younger leaves. Abscission layers are formed at the base of the petioles and the leaves fall off. The color of the stem also becomes brown or black. Soon, the branches and eventually the entire plant dries and dies. Mostly, all the leaves drop off and only the stem remains standing in the field. Sometimes, the disease may attack one side of the plant alone and in such cases, partial wilting of the branches on the attacked side may occur. The taproot is also discolored, stunted and only fewer lateral roots are found. If the bark of an affected plant is peeled off, brown to black, continuous lines or streaks are seen extending upwards to the branches and downwards to the roots. Similarly, if the stem is split open longitudinally, black streaks are seen extending to the branches, petioles and even to the bolls, and downward up to the lateral roots. This is due to the production of a reddish-brown pigment in the vascular vessels by the pathogen. The browning of the vascular vessels is characteristic of the disease. Later infections lead to stunting of plants and formation of fewer and ill-developed bolls **(Fig.62).**

The causal organism. The wilt fungus is a facultative parasite. It enters the host through the root system, when the plants are 1 - 3 weeks old and the symptoms of wilt appear when the plants are 5 - 6 weeks old.

The mycelium is both inter- and intracellular. After entry into the host, the mycelium reaches the vascular region, grows and multiplies in the xylem vessels and plugs the vessels, thereby obstructing the movement of water and mineral nutrients to the aboveground parts. This is one of the reasons for the wilting of plants. Further, the pathogen excretes certain toxic substances in the plant tissues, in the rhizosphere and in the soil. The wilt toxin, **'fusaric acid'** produced by the pathogen in the vascular system is translocated throughout the plant and this toxin is primarily responsible for the wilting. The pathogen is known to produce certain **'pectinolytic'**, **'cellulolytic'** and **'proteolytic enzymes'**, which cause disintegration of cell walls. These enzymes produce a gummy substance in the vascular vessels. This substance is also partly responsible for plugging the vessels.

In advanced stages of infection, the pathogen forms whitish to grayish or bluish weft of mycelium on the collar region near the ground level. The mycelium forms stroma, which is brownish-white to violet and plectenchymatous. The pathogen produces both micro- and macroconidia. Macroconidia are formed on sporodochia and the pionnotes are light-buff to reddish-orange in mass. Sometimes, conidia are formed directly from the hyphae. Macroconidia are hyaline, sickle-shaped, curved inwards at both ends and have a pedicel at the base. Two types of macroconidia, 3-septate, measuring 27.0 - 40.0 x 2.5 - 4.0µ and 4 - 5-septate, measuring 32.0 - 48.0 x 3.5 - 4.5µ in size are formed. Microconidia are hyaline and oval in shape. They are either 0-septate, measuring 6.4 - 10.6 x 2.3 - 3.2µ or 1-septate measuring 13.0 - 20.0 x 2.5 - 3.4µ in size. Terminal or intercalary, thick-walled, spherical and single-celled chlamydospores are also formed. The perfect stage of this fungus has not been reported so far **(Fig.62).**

Mode of survival, spread and epidemiology. The fungus is mainly soil-borne. The mycelium can survive in the stems, roots and branches of diseased plants in the soil for a long time. The chlamydospores also remain in the soil in a viable state for a long time and initiate infection. The injuries caused by soil-inhabiting nematodes in the roots facilitate entry of the pathogen into the host roots. The fungus is also seed-borne to some extent.

The disease occurs more in heavy clay soils with a pH of 7.6 - 8.0. Soil temperatures of 24° - 28°C up to a depth of about 15 cm. and soil moisture levels of 80 - 90 % is favorable for the occurrence and development of the disease. Temperatures above 35°C inhibit the development of the disease. Besides cotton, the pathogen attacks pigeon pea, coffee, lady's finger, castor, alfalfa, Solanum sp. etc.

Disease management

Agronomic practices. In soils cultivated with cotton continuously for several years, the inoculum gets accumulated gradually and this leads to soil sickness. Such sick soils may become unfit for cotton cultivation. By adopting some agronomic practices, it is possible to reclaim the soil and make it suitable for cotton cultivation once again. (i) The fields should be ploughed during the hot summer months, so as to bring the mycelium and spores to the soil surface, so that they are exposed to direct sunlight and get killed (ii) Application of well-decomposed farm yard manure or compost at 5 tons per acre encourages the growth of antagonistic fungal and bacterial organisms, which may destroy the pathogen (iii) Application of sufficient quantities of potash helps to induce resistance in the plants against the wilt pathogen (iv) Whenever water is available in plenty, water should be impounded in the field continuously for a few days, when there is no crop. This will help to destroy the pathogen (v) Diseased plants should be removed and destroyed (vi) Diseased plant debris in the field should be removed and destroyed by burning (vii) Long term crop rotation with non-host crops may be followed (viii) Micronutrients, such as zinc, boron etc. should be applied as per recommendations.

Seed treatment (i) Since the disease is seed-borne to some extent, seed treatment will help in preventing the occurrence of the disease in the early stages of crop growth. First of all, the seeds should be acid delinted by using concentrated sulfuric acid at the rate of 1.0 kg. of acid per every 10 kg. of seeds. After delinting, the seeds should be treated with captan - 4 gm. or thiram - 4 gm. or carbendazim - 2 gm. or thiophanate methyl (Topsin M) - 2 gm. or carboxin (Vitavax) - 2 gm. or thiobendazole (Mertect)

- 2 gm. or chlorothalonil (Daconil) - 2.5 gm./ kg. of seeds, 24 hours prior to sowing (ii) Instead of dry seed treatment as mentioned above, slurry seed treatment may be done. In a small quantity of water, carbendazim - 2 gm. or thiophanate methyl - 2 gm. is added and made into slurry. In this slurry 1.0 kg. of seeds are mixed thoroughly, shade dried and then used for sowing.

Inducing resistance. To induce resistance in young plants, muriate of potash - 7.5 kg. and urea - 5.0 kg. are mixed in 400 lit. of water per acre and sprayed on the plants in the morning hours, when the plants are 2 - 3 months old.

Resistant varieties. While the Desi varieties of cotton are susceptible to this disease, the American cotton varieties belonging to *Gossypium hirsutum* and *G. barbadense* are resistant under Indian conditions. and are widely used in various breeding programs. Varieties, such as Varalakshmi, Vijay, Pratap, Jayadhar, Verum, C.C-1-35, JLA.101, AKH.590, SM.143, LD.254 and LD.327 are resistant to this disease.

2. *Verticillium* wilt of cotton

Verticillium dahliae

This is another typical vascular disease of cotton. Besides cotton, it occurs on eggplant, cowpea, black gram, tomato, potato, chillies, tobacco, bhendi, dahlia etc. The disease was first reported from the United States of America in 1914. It occurs in almost all the cotton growing countries of the world and is considered to be one of the major diseases of cotton in America and Russia. In India, the disease is prevalent in the states of Tamil Nadu, Telungu Desam, Karnataka and Maharashtra. The disease not only causes considerable loss in yield, but also affects the quality of the fiber.

Symptoms. The disease mostly attacks the plants, when they are about 3 months old and are in the flowering and boll-forming stages. Affected plants show chlorosis and crinkling of leaves. The veins become brown and the interveinal regions of the lamina and the margins of leaves turn yellow and gradually become necrotic, producing a **'tiger stripe'** effect. Often cupping of the leaves is also seen. Eventually, the leaves dry completely and drop off. The symptoms start from the lower leaves and progress gradually to the upper leaves. Most of the leaves are thus affected and they dry and fall off. Finally, only a few chlorotic, diminutive leaves alone remain at the tips of the branches. The squares and bolls are also shed prematurely. Even if a few bolls remain, they are smaller, ill-formed and have poor quality lint. However, the affected plants do not dry and die completely.

When the outer skin of the stem and roots of the affected plant is peeled off, pinkish or brownish, discontinuous streaks are seen underneath the skin. Such streaks may be seen even in the petioles. Similarly, if the stem and root portions are split open longitudinally, reddish-brown or brown streaks, which may be continuous or discontinuous are seen. This discoloration of the vascular elements is characteristic of the disease **(Fig.63).**

The causal organism. The mycelium is whitish to grayish, septate, inter- and intracellular. Fructifications appear on the surface of the stems at soil level and on the surface of diseased, dried leaves as small tufts of erect conidiophores, which break through the epidermis. The conidiophores are branched in characteristic fashion, forming 1 - 3 whorls of branches, with 1 - 7 whorled (verticillate) phialides in each whorl. From the tips of the main conidiophore and phialides, small, hyaline, one-celled conidia are produced singly or in dense aggregation, which are held together in mucilage. The conidia are elliptical and measure 4.2 - 5.5 x 1.6 - 2.3μ in size. They are short-lived. Besides conidia, the fungus produces microsclerotia. Microsclerotium is formed from the budding of a single hypha, becoming a dark, knob-like, thick-walled structure. They are not true sclerotia, since they do not possess an intertwining system of hyphae and a cortex. They are irregularly spherical or slightly elongated and measure 48 - 120 x 26 - 45μ in size **(Fig.63).**

Mode of survival, spread and epidemiology. The fungus mainly overwinters as microsclerotia in the soil. They can survive in the soil for up to 15 years without losing their viability. The fungus can also overwinter as mycelium in other alternate hosts or in diseased plant debris. It can also be carried through contaminated seeds. The microsclerotia may be disseminated through surface water and soil adhering to agricultural implements or feet of cattle and also by humans. The disease may spread from a diseased to a healthy plant by contact of their roots. Secondary spread of the disease is through conidia dispersed by wind or rain splash.

The fungus can enter the host by penetrating the roots directly or through wounds caused by insects or soil inhabiting nematodes. Once the roots are attacked, it interferes with the water supply to the growing plants, thus inducing wilting of the shoots. The fungus, which enters the wood vessels from the roots, stains the lignified walls brown and grows into the vascular bundles of the stem for long distances and this mycelium may also plug the xylem vessels. The organism also produces a gummy substance in the vascular elements, which may often plug the cells completely. However, the actual wilting is not due to the plugging of the water channels by the mycelium or by the gummy substance alone, but more due to the secretion of toxic substances, which is carried up to the leaves in the transpiration system, thereby causing their wilting and death.

Moderately cool temperatures of about 20°C are favorable for the occurrence of the disease. No infection occurs at temperatures above 30°C. The disease occurs more during the cold months of November and December, when the temperature is around 15° - 20°C. It is more common in clay loam and in soils rich in organic matter.

Disease management

Agronomic practices (i) Quality seeds should be selected from disease-free fields and used for sowing (ii) Diseased plant debris should be removed from the field and destroyed by burning (iii) Long term crop rotation, with non-host crops should be followed (iv) Summer ploughing should be encouraged (v) Diseased plants should be removed and destroyed (vi) Irrigation from diseased to other fields should be avoided (vii) Application of excessive doses of nitrogenous fertilizer should be avoided (viii) Proper drainage facilities should be provided (ix) Application of organic manure, such as farm yard manure or compost should be encouraged

Seed treatment. Seed treatment methods recommended for the control of *Fusarium* wilt may be followed.

Resistant varieties. Sujatha, CBS.156 and Suvin have been found to be fairly resistant.

3. Rust of cotton

Phakopsora gossypii (P. desmium)

The disease occurs in most of the tropical countries of the world, including India, the Philippines, Indonesia, Sri Lanka, Africa, West Indies and tropical parts of America. In India, it is prevalent in all the cotton growing areas.

Symptoms. The pathogen attacks the crop mostly at the later stages of growth and the disease is almost confined to the leaves, but may sometimes attack the bolls also. The symptoms appear as small, round, yellowish-brown spots, about 0.1 - 0.2 mm. in diameter on the upper surface of leaves. Larger spots up to 3.0 mm. may also be formed. On the corresponding under surface of these spots, orange-colored, erumpent pustules are formed. These pustules are the uredosori of the fungus. These sori rupture and release numerous, yellowish, uredospores. In case of severe infection, the pustules cover the entire leaf area, as a result, the leaves turn brown and drop off prematurely. Similar pustules may appear on the bolls also. Later on, telia may be formed on the leaves, which have not fallen down and sometimes on the bolls however, telia are rarely formed in nature. They are dark-brown in color, circular in shape, sub-epidermal and not erumpent **(Fig.64).**

The causal organism. The mycelium is septate and intercellular. Haustoria produced from the hyphae penetrate into the cells of the mesophyll tissues and absorb nourishment. The uredospores come out of the uredia as powdery masses, after breaking open the host epidermis. They are pedicellate, pale yellow in color, oval or broadly pear-shaped, finely echinulated and measure 19 - 27 x 16 - 19μ in size. Teliospores are sessile, single-celled, cylindrical, irregularly arranged inside the telial sorus, light-brown in color, not echinulated and measure 80 - 100 x 10 - 13μ in size. They do not come out by breaking open the host epidermis. The telial stage, which is rarely found, has not been reported in India. The pycnial and aecial stages of the fungus are not known **(Fig.64).**

Mode of survival, spread and epidemiology. The disease may occur only through uredospores, which may survive for a long time in a viable state in the diseased crop debris in the soil. Secondary spread is by wind-borne uredospores. Rainsplash and a few insects may also serve to disseminate the spores. The fungus attacks *Thespesia populnea* and *Azanza garkama* and the uredospores produced from these alternate hosts may also serve as primary inoculum.

High atmospheric humidity, heavy rainfall and moderate temperature favor the occurrence and spread of the disease.

Disease management

Agronomic practices. Diseased crop debris from the fields should be removed and destroyed by burning.

Chemical control. In case the disease appears early, the crop may be sprayed with wettable sulfur - 1200 gm. or carboxin - 300 gm. in 300 lit. of water per acre.

4. Gray mildew or areolate mildew

Septocylindrium (Ramularia) areola

(Mycosphaerella areola)

The disease, which was first reported from the United States of America in 1890, is now prevalent in almost all the cotton growing countries of the world. All Desi cotton varieties of *Gossypium arboreum* and *G. hirsutum*, as well as the American cotton varieties of *Gossypium barbadense* are found to be susceptible to the disease. It occurs widely in most of the cotton growing tracts of Tamil Nadu.

Symptoms. The disease usually appears, when the plants are nearing maturity and the symptoms are manifested on the older leaves initially. Small, irregular or angular, pale yellow spots, 1.0 - 10.0 mm. in diameter develop on the under surface of leaves. The smaller veins limit these lesions. The corresponding upper surface of the lesions appears yellowish-green. In a single leaf, more than 100 such spots may develop. Under favorable conditions, whitish to grayish mildew growth, which are the fructifications consisting of conidiophores and conidia develop, mostly on the under surface of the lesions and occasionally on the upper surface also. Eventually, the entire leaf surface is covered with the white to gray powdery growth. Badly affected leaves turn yellowish-brown, dry up from the margins and fall off prematurely. Lesions may occur on the bracts also. Severe defoliation results in premature boll opening and poor quality lint **(Fig.65)**.

The causal organism. The mycelium of the fungus is hyaline, septate, inter- and intracellular. The mycelium is not widely spread inside the tissues but is mostly restricted around the foci of infection. The endophytic mycelium produces conidiophores, which emerge through the stomata in clusters. They are hyaline, short and with a few branches arising from the base. Conidia are borne at the tips of the branches singly or in chains of a few conidia. The conidia are hyaline, irregularly oblong, with narrowing ends, sometimes with rounded or flattened ends, single-celled or with 1 - 3 septa and measure 10.0 - 35.0 x 4.0 - 5.0µ. in size.

The perfect stage of the fungus is *Mycosphaerella areola*. But the ascigerous stage has not been reported from India and is not important as a pathogen **(Fig.65)**.

Mode of survival, spread and epidemiology. The pathogen survives in the infected crop debris and may produce conidia. Perennial cotton plants and volunteer plants may also harbor the fungus and help in perpetuating the disease. Staggered sowing is another reason for perpetuation of the disease. The early sown crops get infected first and provide enough inoculum to infect the late sown crops. The perfect stage, which does not occur in India, has no part in the perennation of the disease.

Fig. 62: Fusarium wilt of cotton-*Fusarium oxysporum* f.sp. *vasinfectum*

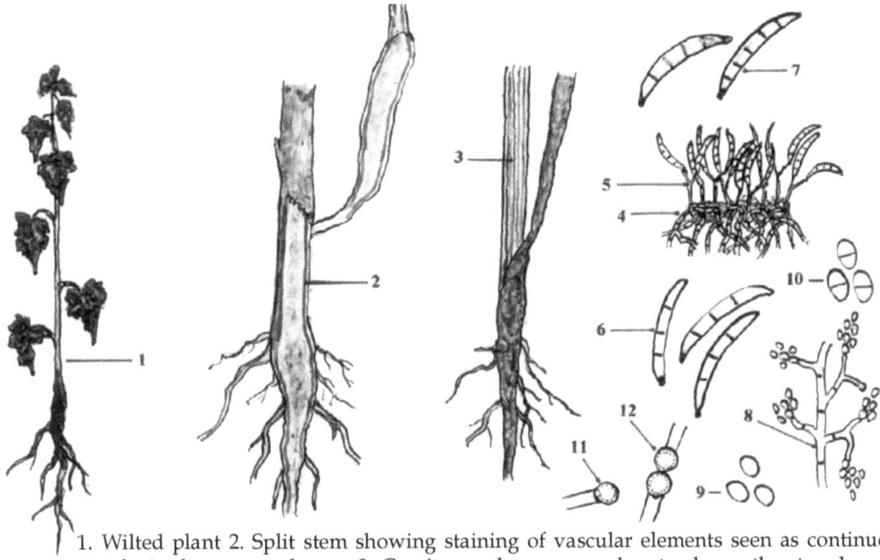

1. Wilted plant 2. Split stem showing staining of vascular elements seen as continuous streaks in the stem and roots 3. Continuous brown vascular streaks on the stem beneath the outer skin 4. Sporodochium 5. Conidiophore bearing macroconidia 6. Three-septate macroconida 7. Four to five-septate macroconidia 8. Microconidia from hyphal tips 9. Single-celled microconidia 10. Two-celled microconidia 11. Terminal chlamydospore 12. Intercalary chlamydospores.

Fig. 63: *Verticillium* wilt of cotton *Verticillium dahliae*

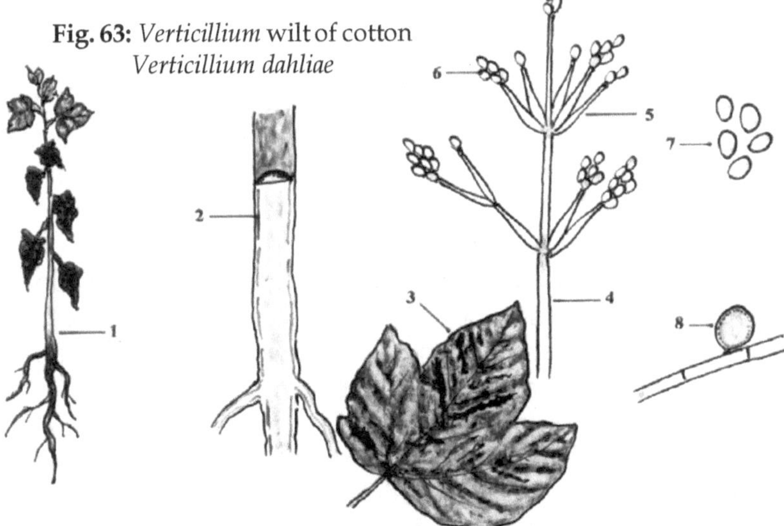

1. Wilted plant 2. Split stem showing staining of vascular elements as brown discontinuous streaks in the stem and roots 3. 'Tiger stripe' symptom on leaf 4. Conidiophore 5. Phialides 6. Conidia adhering together in mucilage 7. Conidia 8. Microsclerotium formed by budding from hypha.

Water inundation, wet and moist conditions favor the disease occurrence and development. Intermittent heavy rains, high humidity and a temperature range of 20° - 30°C are conducive for disease development. Application of excessive doses of nitrogenous fertilizer and close planting increase the chances of severity of the disease.

Disease management

Agronomic practices (i) Diseased plant debris in the fields should be removed and destroyed by burning or buried deep into the soil (ii) Volunteer plants should be removed and destroyed (iii) Application of excessive doses of nitrogenous fertilizer should be avoided (iv) Closer planting should be avoided (v) Proper drainage facilities should be provided.

Chemical control. Spraying the crop with Bordeaux mixture - 1 % or copper oxychloride - 825 gm. or carbendazim - 350 gm. or wettable sulfur - 1400 gm. or triadimefon (Bayleton) - 350 gm. in 350 lit. of water per acre controls the disease.

5. Alternaria blight of cotton

Alternaria macrospora (A. longipedicellata)

The disease is found in almost all the cotton growing countries of the world. It is prevalent all over India and is widespread in varieties of *Gossypium arboreum* and *G. hirsutum.*

Symptoms. The disease attacks the crop at all growth stages. However, 45 to 60 days old plants are more vulnerable to attack by this disease. The symptoms appear as small, dull-brown, circular or irregularly circular spots, 0.5 - 3.0 mm. in diameter on the upper surface of leaves. The spots gradually enlarge in size and become bigger spots, about 1.0 cm. in diameter. The spots show concentric rings and appear as a target board. The spots may coalesce, become irregular and occupy large areas of the leaves. Under humid conditions, fine, velvety growth of the fungus is seen on the surface of the spots. Severely affected leaves turn yellowish-brown, dry and fall off prematurely leading to acute defoliation. Older leaves are more commonly attacked than younger leaves. The petioles are also attacked and spots develop on them. On the stem region, cankers may be formed, resulting in cracking and breaking of the stem. At the later stages, when bolls are formed, they are also attacked and several spots may develop on the bolls. The affected bolls drop off prematurely. The infection may spread to the seeds also **(Fig.66)**.

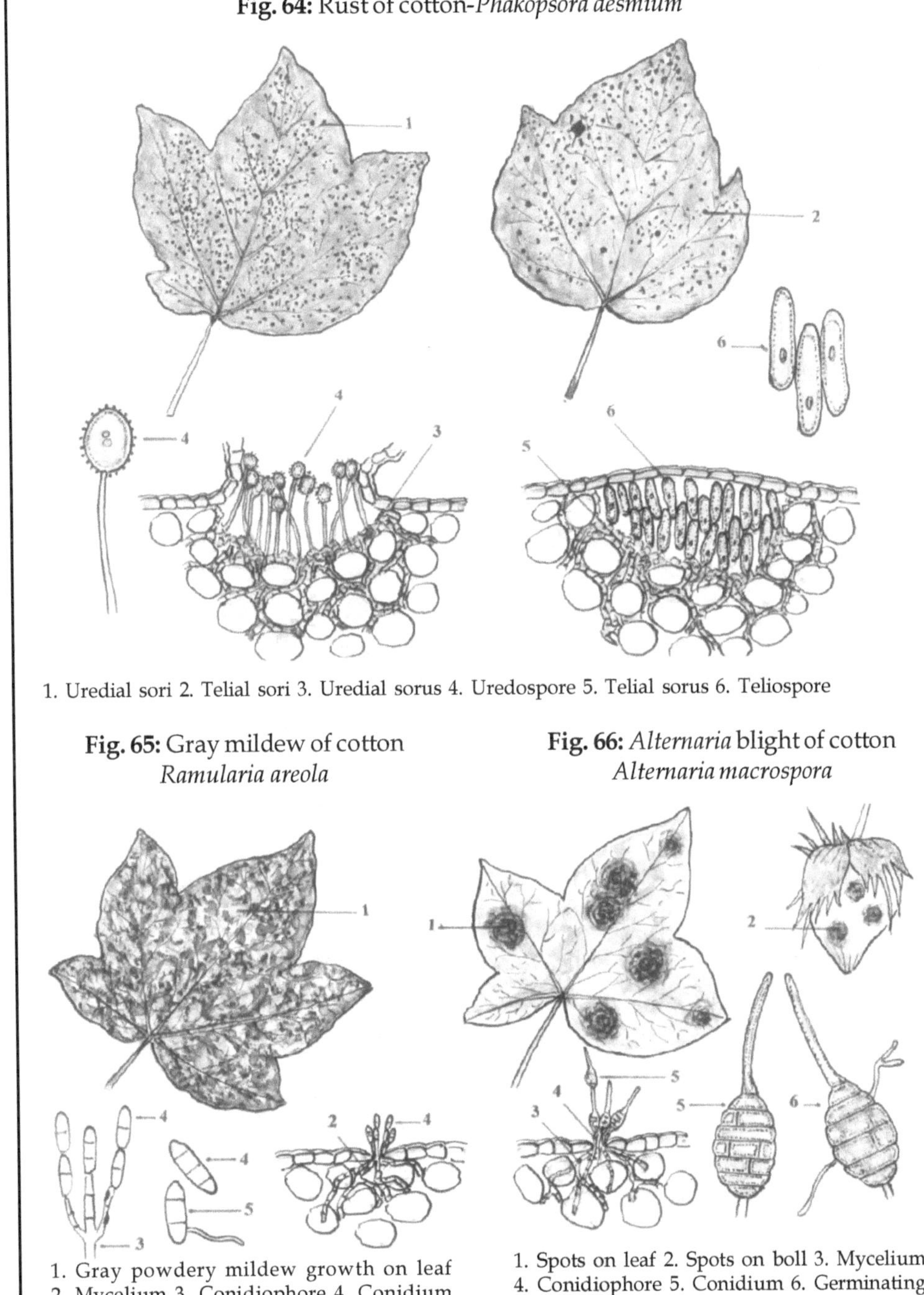

Fig. 64: Rust of cotton-*Phakopsora desmium*

1. Uredial sori 2. Telial sori 3. Uredial sorus 4. Uredospore 5. Telial sorus 6. Teliospore

Fig. 65: Gray mildew of cotton *Ramularia areola*

1. Gray powdery mildew growth on leaf 2. Mycelium 3. Conidiophore 4. Conidium 5. Germinating conidium.

Fig. 66: *Alternaria* blight of cotton *Alternaria macrospora*

1. Spots on leaf 2. Spots on boll 3. Mycelium 4. Conidiophore 5. Conidium 6. Germinating Conidium.

The causal organism. The mycelium of the fungus is light-brown in color, septate, much branched, inter- and intracellular. Before fructification, the hyphae aggregate underneath the stomata of the infected and dried leaves. From the stromata, conidiophores are produced, which emerge through the stomata singly or in groups. They are erect, simple, almost cylindrical or tapering slightly towards the apex, have one or more conidial scars and are up to 80μ long. Conidia are borne singly or rarely in chains of two at the tips of the conidiophores. They are straight or curved, dark-brown in color, ellipsoidal, with rounded bottom and tapering abruptly to a very narrow, long beak, equal or twice the length of the body of the conidia. They are septate, with 3 - 9 transverse septa and slightly constricted at the septa and have 1 - 4 longitudinal septa. They measure 90 - 180 x 15 - 22μ in size **(Fig.66).**

Mode of survival, spread and epidemiology. The fungus is a facultative parasite and can survive in the diseased plant debris in the soil as active mycelium and continues to produce conidia or remains as dormant mycelium for a long time. When the bolls are attacked, the mycelium and conidia stick on to the lint and become externally seed-borne. When the seeds are attacked, the mycelium may enter the seed and remain inside the seed in a dormant state and becomes internally seed-borne. Secondary spread is by means of wind and rainsplash.

High humidity, intermittent rainfall and moderate temperatures of 25° - 28°C favor disease occurrence and development.

Disease management

Agronomic practices. Removal and destruction of diseased plant debris, providing adequate drainage, proper crop maintenance and growing resistant varieties help in reducing the chances of disease occurrence.

Chemical control. Spraying the crop with copper oxychloride - 875 gm. or mancozeb - 700 gm. or chlorothalonil - 450 gm. in 350 lit. of water per acre, 2 or 3 times, at fortnightly intervals gives adequate control of the disease.

Seed treatment. Acid delinting, followed by seed dressing with captan at 4 gm. or thiram at 4 gm./ kg. of seeds, 24 hours prior to sowing prevents both internal and external seed-borne infections.

6. Seedling blight and root rot of cotton

Rhizoctonia bataticola (Macrophomina phaseolina) and *Rhizoctonia solani*

(Thanatephorus cucumeris)

The disease occurs in most of the cotton growing countries of the world, including Egypt, Greece, Israel, Pakistan, Sudan, the United States of America, Venezuela, India and many African countries. In India, it is commonly encountered in the states of Telungu Desam, Bihar, Gujarat, Punjab, Rajasthan and Tamil Nadu. American cotton varieties are more vulnerable to attack than the Indian cotton varieties.

Symptoms. The fungus attacks the crop at all growth stages. Germinating seedlings and one to two weeks old young plants are attacked at the hypocotyl region. The infection, which starts as small, black lesions, soon enlarge and girdle the stem, as a result the tissues rot and the seedlings collapse and die.

In grown-up plants nearing maturity, typical root rot symptoms are manifested. The most characteristic symptom is the sudden wilting and death of the affected plants. When one plant is affected, soon all the plants around that plant are attacked and appear as circular patches of wilted plants. The affected plants can be pulled out very easily. Such plants reveal the rotting of the entire root system, except the taproot and a few lateral roots. In advanced stages, all the lateral roots rot and disintegrate and only the taproot remains attached to the stem. The taproot appears water-soaked and sticky and the bark is shredded. The shredding of the bark may extend to the stem above the ground level. The woody portion underneath the bark appears black and becomes brittle. On the rotten, shredded bark and on the woody portion, white or grayish, loose, cottony weft of mycelial growth may be seen. In the mycelial weft, numerous, minute, pinhead-like, black sclerotial bodies are seen. Sometimes, sclerotia may be formed on the stem region also.

The causal organism. Both the fungi causing root rot of cotton are facultative parasites. The mycelium is colorless, septate, much branched and branches almost at right angles from the parent hyphae. The endophytic mycelium grows both inter- and intracellularly and are thin-walled, while the ectophytic mycelium is relatively thick-walled. The superficial hyphae aggregate and spin together to form minute, pinhead-like sclerotia in large numbers on the dying and dead tissues of the roots and stems. They are globose to irregularly globose in shape and about 100μ in diameter. The sclerotia of *Rhizoctonia bataticola* are jet black in color, while those of *R. solani* are dark-brown in color. The conidial stage of *R. bataticola* is not found usually. Whenever pycnidia are formed, they are produced singly or in groups on the stem region or on the leaves. They are sunken in the host tissues, nearly globose, dark-brown to black in color, ostiolate and measure 100 - 200µ in diameter. Conidia are produced in succession from the tips of short, hyaline, phialides, arranged like palisade cells layering on the inside of the pycnidia. The conidia are hyaline, single-celled, and ellipsoid to obovoid and measure 14.0 - 30.0 x 5.0 - 10.0µ in size. The basidial stage of *R. solani* is rarely formed and is not of any significance in the disease cycle (Fig. 43).

Mode of survival, spread and epidemiology. The disease is primarily soil-borne. The fungus can lead a saprophytic life in the diseased plant debris and in organic matter present in the soil for a number of years. The sclerotia, which are the resting bodies, can also remain in the soil in a viable state for a long period of time. The sclerotia are disseminated through irrigation water, soil sticking on to agricultural implements and feet of cattle, by various cultural operations and by strong winds.

Low soil moisture of 15 - 20 % and high soil temperature of 35° - 39°C favor disease occurrence and development. The disease occurs more in clay soils. Wounds caused by soil-inhabiting insects and nematodes facilitate the entry of the pathogens into the host. Once the pathogens gain entry into the host, they grow and spread rapidly in the tissues, reach the vascular elements and plug the xylem vessels, thereby obstructing the movement of water and mineral nutrients to the aboveground parts of the plants. This results in sudden wilting and death of the plants. The toxins secreted by the pathogens are also partly responsible for causing sudden wilting. Besides cotton, the pathogens attack several other vegetable crops, oilseed crops, pulses etc.

Disease management

Agronomic practices. The control measures suggested for the control of *Fusarium* wilt is applicable for controlling this disease also. Removal and destruction of diseased plant debris from the fields, application of farmyard manure or compost, seed treatment, crop rotation etc., reduce the chances of occurrence of the disease.

Biological control. Treating the seeds with *Trichoderma viride* at 4 gm./ kg. of seeds is effective in controlling the disease.

7. Bacterial blight of cotton

Xanthomonas campestris pv. *malvacearum*

The disease is also known as **'angular leaf spot'**, **'vein blight'**, **'black arm'** or **'boll rot'**. It was first reported from the United States of America in 1891. It is considered to be one of the most serious and destructive diseases of cotton and occurs all over the world. In India, it is prevalent in Telungu Desam, Maharashtra, Madhya Pradesh, Uttar Pradesh and Tamil Nadu. It was reported in 1918 in Tamil Nadu.

Symptoms. The disease attacks the crop at all growth stages, from emergence of the seedlings, till maturity. It attacks the germinating seedlings and causes **'seedling blight'**. The first symptoms are manifested on the under surface of cotyledons as small, water-soaked, circular or irregularly circular spots. The spots enlarge and become dark-brown to black, irregular spots. The cotyledons get deformed, distorted and dry. The infection then spreads to the stem region through the petioles and the terminal part is affected, as a result the seedlings die.

When grown-up plants are affected the symptoms develop as small, dark-green, water-soaked spots on the under surface of leaves. Soon, dark-brown spots appear on the upper surface of the leaves also. The spots enlarge, turn dark-brown or reddish-brown. The spots are limited by the veinlets and appear as angular spots. Hence it is called **'angular leaf spot disease'**. When numerous such spots are produced, they gradually enlarge and cover large areas of leaves, leading to withering and premature defoliation.

Gradually, the infection spreads to the veins and veinlets from the leaves along their edges, as a result they become thickened and blackened. So, the disease is called **'vein blight'** or **'vein necrosis'** or **'black vein'**. Vein blight causes crinkling and twisting of leaves inwards. From the veins the infection moves to the petioles causing blighting and defoliation. Bacterial ooze comes out from the leaves on the under surface, dries and forms crusts.

Elongated, sunken, dark-brown to black spots are formed on the stem and fruiting branches. These spots may enlarge and girdle the stems and branches, The stems may crack and a gummy exudate comes out through the cracks. Eventually, the branches break and hang down. All the leaves and bolls drop off and only a few dry, black twigs alone stand sticking out like a burnt arm. Hence this disease is also known as **'black arm disease'**.

The pathogen attacks the flowers and bolls. Basal infection of flower buds and young buds results in premature shedding. On the bolls, water-soaked, sunken, dark-brown or black, round spots develop. The infection may spread deeper into the bolls and causes **'boll rot'**. Older bolls, even if they are infected, may not drop off but remain attached to the plants. They fail to open normally and the lint is stained yellow and the quality is also very poor. The bacteria that multiply inside the bolls stick on to the lint in a gummy substance. They may also enter the seeds through the micropyle and remain dormant in the seeds.

The causal organism. The bacteria causing the disease are facultative saprophytes. The bacteria are rod-shaped, 1.3-2.7 x 0.3-0.6μ in size, gram-negative, occurring singly or in pairs and are motile with a single, polar flagellum. They are aerobic, non-spore forming and capsulated.

Mode of survival, spread and epidemiology. The disease is mostly seed-borne. The bacteria stick on to the lint in large numbers in a gummy substance. The bacteria that enter the seeds, remain in the embryo in a dormant state. When the seeds germinate, they attack the cotyledons and cause primary infection. The pathogen can also survive in the diseased leaves, branches and bolls in the soil for a long time. It has been established, that the bacteria remain alive in dried, diseased leaves for up to 17 years under dry conditions. However, under conditions of high soil moisture and temperatures between 31°-33°C they can live only for about 8 days. Secondary spread is by bacteria, which enter the host through stomata, hydathodes or wounds. Wind driven rains is mostly responsible for the dispersal of the bacteria

High soil moisture at the time of sowing and a few days after and a temperature of 28°C is highly favorable for the disease occurrence and development. Presence of moisture is very important for secondary infection. Dry and hot weather conditions are unfavorable for disease development. *Jatropha curcas, Thurbaria thespesioides* and *Eriodendron anfructuosum* are also attacked by this pathogen. In India, 16 physiologic races of the bacteria have been identified.

Disease management

Agronomic practices (i) Diseased plant debris should be removed and destroyed by burning or ploughed and buried deep into the soil (ii) Seeds should be selected from disease-free fields (iii) Alternate hosts that may harbor the pathogen during the off season should be destroyed (iv) Young plants showing symptoms of the disease should be removed and destroyed (v) Sufficient quantities of potash should be applied to induce resistance in the plants (vi) Inundating the fields with water continuously for a few days, when there is no crop, destroys the bacteria (vii) Long term crop rotation may be followed (viii) The bacteria in the seeds is known to remain inside the seeds in a viable state for up to one year. So, the seeds may be stored for about two years and then used for sowing. Storing the seeds for a long time may reduce the germination percentage. So, the seed rate should be increased, so as to maintain the plant population.

Seed treatment (i) Acid delinting with concentrated sulfuric acid, followed by seed dressing with captan or thiram eliminate seed-borne inoculum (ii) Hot water treatment at 56°C, continuously for 10 minutes destroys the seed-borne bacteria without affecting the viability of the seeds (iii) Treating the seeds with antibiotics, such as Streptomycin sulfate is also found to eliminate the seed-borne inoculum. Streptomycin sulfate is dissolved in water at 1.0 gm. per lit. of water and in this solution, seeds are kept immersed for a period of 4 - 8 hours, shade dried and then used for sowing.

Chemical control (i) Spraying with Agrimycin - 50 gm. + copper oxychloride - 875 gm. in 350 lit. of water per acre is very effective in controlling the disease (ii) Spraying with copper oxychloride - 975 gm. alone, also gives fairly good control of the disease (iii) Spraying with the antibiotic Bactrinol - 70 gm. in 350 lit. of water per acre is also effective. The sprayings should commence, when the plants are about 5 - 6 weeks old or when initial symptoms of the disease are noticed, followed by 3 - 4 more sprayings, at fortnightly intervals

Resistant varieties. The Desi cotton varieties of *Gossypium herbaceum* and *G. arboreum* are relatively more resistant than the American cotton varieties of *Gossypium hirsutum* and *G. barbadense*. Several resistant varieties have been evolved by hybridization and selection. HC.9, BJA.592, P.14-T.12, 101-102.B, Reba-B.50, 70-IH.480/2, 70-IH.480/3, 70-IH.480/9. K.4005, Badnawar-1, B.1007, Khandwa-2, DHY.286 etc. have been found to be resistant.

Diseases of minor importance. Several other diseases are also known to occur on cotton. However, most of them are of minor importance economically. *Mycosphaerella gossypii (=Cercospora gossypii)* causes '*Cercospora* leaf spot'; *Helminthosporium gossypii* and *Drechslera spicifera (=Cochliobolus spicifer)* cause 'Leaf spots'; *Colletotrichum capsici* and *Glomerella gossypii (=Colletotrichum gossypii)* cause 'Anthracnose', which affects the seedlings, as well as stems and bolls of grown-up plants; several fungi, such as *Fusarium moniliforme, Rhizopus nigricans, Aspergillus flavus, A. niger* etc. are responsible for causing 'Boll rot'; 'Leaf curl' is caused by a Virus.

Tobacco *(Nicotiana tabacum)*

1. Black shank of tobacco

Phytophthora parasitica var. *nicotianae*

The disease was first reported from Java in 1893. Now, the disease is found in many countries of the world, including America, Canada, East Indies, Uganda, Bulgaria, Rumania, Japan, Argentina, Venezuela, Indonesia, the Philippines, Mauritius and India. In India, the disease occurs in Telungu Desam, Gujarat, Karnataka and Tamil Nadu. It is more prevalent in regions of high rainfall over a continuous period of many days. Almost all cultivated varieties of tobacco are found to be susceptible to the disease. Besides tobacco, the pathogen attacks several other crops, such as tomato, eggplant, castor etc.

Symptoms. The disease appears both in the nursery and in the planted field. It attacks the plants at the collar region near the soil level. When the seedlings are attacked during the colder months, the stem region becomes water-soaked and rots completely, while in the hot season the seedlings turn brown, become blighted and die. Infected seedlings, when used for planting or when healthy seedlings are planted in sick soils, the disease spreads very rapidly. Typical symptoms of black shank are seen in slightly grown-up plants. In the infected plants, the lower leaves turn yellow, crinkle and hang down from the stem. Within a few days, the entire plant dries and dies. The stem portion up to a height of about 30 cm. and the roots become black and rotten. If the plants are infected in the early stages of growth, the entire stem region and the roots are completely rotten. During rainy seasons, the disease spreads to the leaves and large, round to irregular, brown spots appear and very soon, the leaves become necrotic and die. The stem of an old infected plant, when split open longitudinally, shows blackening of the pith region and formation of transverse, disc-like plates, one above the other at short intervals. This is a characteristic symptom of the disease. The stem above the collar region is also shrunk **(Fig.67).**

The causal organism. The mycelium of the fungus is coenocytic, hyaline, multinucleate, intercellular and sometimes intracellular. No haustoria are formed. The mycelium is seen on the surface, as well as inside the affected tissues, as fine, cottony growth. All the tissues including the fibrovascular bundles are. Invaded. From the hyphae, long, slender, unbranched sporangiophores are formed. From the tips of the sporangiophores, sporangia are formed singly. They are thin-walled, hyaline, oval or pyriform, with a prominent papilla and measure 25-50 x 20-40µ in size. Usually, the sporangia germinate by producing zoospores. The zoospores, on maturity, break open the papilla and emerge out. There may be 5 - 45 zoospores in a single sporangium. They are bean-shaped, with 2 flagella. After emergence, they swarm in free water for 20 minutes to 2 hours, lose their flagella, encyst and after a short period of rest germinate by issuing a germ tube. Under unfavorable conditions and in the absence of free water, the sporangia may germinate directly by producing a germ tube.

As a result of sexual reproduction, the fungus forms oospores. They are spherical, hyaline to straw-colored, with a hard inner wall and measure 15 - 20µ in diameter. Under favorable conditions and in the presence of suitable host, they germinate by means of a germ tube. Asexually, the fungus also produces terminal or intercalary chlamydospores. They are globose, thick-walled, and yellowish in color and measure 27 - 42µ in diameter **(Fig.67).**

Mode of survival, spread and epidemiology. The fungus is soil-borne and can live saprophytically in infected plant debris and decomposed organic matter in the soil for up to 5 years. Under favorable weather conditions and when suitable hosts are available, the fungus attacks them. The oospores and chlamydospores can withstand adverse conditions and remain in the soil in a viable state for long periods and cause primary infection. Secondary spread of the disease is through sporangia, which are disseminated by wind, rain and some insects.

High humidity, free water and good sunlight are necessary for the germination of sporangia by zoospores. High humidity, continuous wet weather and temperatures between 16° - 30°C are favorable for the development of the disease. The disease incidence is high in heavy soils. Injuries caused by the root knot nematodes, *Meloidogyn javanica* and *M. incognita* predispose the plants to black shank attack.

Disease management

Agronomic practices (i) Diseased plant debris and other organic matter should be removed from the fields and destroyed by burning (ii) Plants showing disease symptoms should be immediately removed and destroyed (iii) Nurseries should not be raised in fields infected by the disease or in sick soils (iv) Proper drainage facilities should be provided to the nursery beds. Raised beds are preferable for raising seedlings (v) Diseased seedlings should be rejected and should not be used for planting (vi) Resistant varieties may be grown as far as possible.

Chemical treatment (i) Seedbeds should be drenched with Bordeaux mixture - 1 % or copper oxychloride at 2 gm./ lit. of water, 2 days prior to sowing, so as to wet the soil thoroughly to a depth of 10 cm. (ii) To prevent seedling infection, soil drenching should be done, 10 - 15 days after sowing with Bordeaux mixture - 1 % or copper oxychloride at 2.5 gm. or captafol at 1.5 gm./ lit. of water, so as to wet the stems of seedlings and the soil around the stems (iii) Drenching the nursery bed with metalaxyl or thiobendazole at 1.0 gm./ lit. of water is very effective in controlling the disease (iv) In the main field, drenching with Bordeaux mixture - 1 % or copper oxychloride at 2.5 gm. or captafol at 1.5 gm./ lit. of water may be done, 30 days after planting, followed by 2 - 3 drenchings, at fortnightly intervals, so as to cover the stems and the soil around the stems. A quantity of 1,000 liters of fungicidal fluid may be required to cover one acre of field.

Seed treatment. Treating the seeds with thiram or captan at 4 gm./ kg. of seeds, 24 hours prior to sowing protects the seedlings from infection in the early stages.

2. Frog-eye leaf spot of tobacco

Cercospora nicotianae

The disease attacks the crop both in the nursery and transplanted fields. It causes reduction in yield and quality of the leaves, resulting in lower market value of flue-cured and wrapper tobacco. Almost all the cultivated varieties of tobacco are found to be susceptible to the disease.

Symptoms. The disease usually attacks the lower and matured leaves. Initial symptoms appear as small, reddish-brown, circular spots on the leaves. The spots gradually enlarge and the center of the spots becomes ashy-white, surrounded by a brown border. Later, the center becomes white and necrotic. The spots are round and are 5.0 - 10.0 mm. in diameter. A typical spot appears white in the center, surrounded by a gray or brown ring, which is surrounded by a dark-brown or black ring and appears like a frog's eye. More than 100 such spots may be produced in a single leaf. The spots at the tip and margin of the leaf coalesce, forming big patches and the leaf starts drying inwards. Eventually, the affected leaf dries completely. In the later stages of the crop, spots may be formed on the bracts, calyx and capsules **(Fig.68).**

The causal organism. The mycelium of the fungus is usually found only around the points of infection. It is septate and olivaceous in color. The mycelium forms stromata below the host epidermis at the points of infection prior to sporulation. From the stromata, conidiophores are produced, which come out in clusters through the stomata. They are simple, septate, brown in color, geniculate and measure 70.0 - 100.0 x 4.0 - 5.0µ in size. Conidia are formed singly from the tips of conidiophores. From each conidiophore,

2 - 3 conidia may be formed. A distinct scar is seen in the conidiophore, at the points from where the matured conidia had fallen down. Conidia are hyaline, slender, straight or slightly curved, thin-walled, 3 - 6 septate and measure 40.0 - 75.0 x 3.0 - 4.0µ in size **(Fig.68).**

Mode of survival, spread and epidemiology. The disease is mainly air-borne. The conidia produced from diseased plant debris in the soil infect the seedlings initially. The conidia produced from such infected seedlings are carried by wind and infect other plants. Planting of disease infected seedlings, carry the inoculum to the crop in the main field.

Heavy, intermittent rainfall, high humidity, frequent irrigation and moderate temperature favor the development and spread of the disease. Close planting increases the humidity level within the crop canopy, which favors initial infection of lower leaves. Under favorable conditions, even the younger leaves get infected. Though free water is necessary for the germination of conidia and penetration into the host, dew or mist is enough to cause heavy infection. Optimum temperature for disease development is 20° - 30°C. Mosaic affected plants are more prone to attack by this pathogen.

Disease management

Agronomic practices (i) Diseased plant debris should be removed and destroyed (ii) In heavily infected fields, cultivation of tobacco should be avoided (iii) Disease-free, healthy seedlings should be used for planting (iv) Close planting and frequent irrigation should not be done (v) Application of heavy doses of nitrogenous fertilizer should be restricted and sufficient quantities of potash should be applied.

Chemical control. Foliar spraying with Bordeaux mixture - 1 % or copper oxychloride - 2.5 gm. or mancozeb - 2.0 gm. or benomyl - 1.0 gm. or carbendazim - 1.0 gm. or thiophanate methyl - 1.0 gm./ lit. of water affords adequate control of the disease. The fungicide should be applied at fortnightly intervals, from the time initial symptoms are noticed. During rainy season, sprayings should be given at weekly intervals.

3. Damping off of seedlings of tobacco

Pythium species

The disease is worldwide in occurrence. It is serious in tobacco nurseries and in heavy clay, and water-logged soils. In India, it occurs in all tobacco growing states.

Fig. 67: Black shank of tobacco-*Phytophthora parasitica* var. *nicotianae*

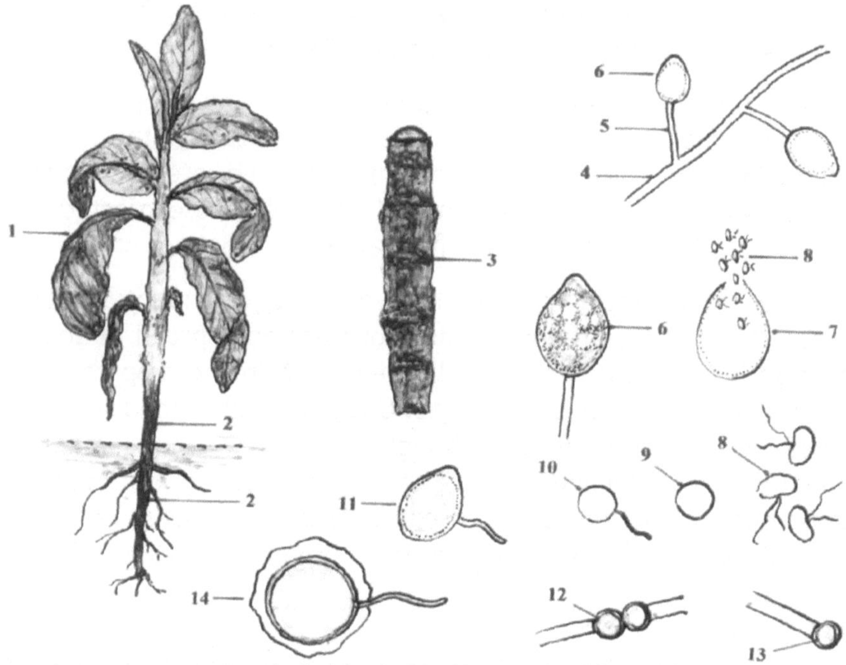

1. Infected plant 2. Rotten stem region and roots 3. Split stem showing transverse disc-like plates of the pith 4. Hypha 5. Sporangiophore 6. Sporangium 7. Sporangium germinating by zoosores 8. Zoospores 9. Encysted zoospore 10. Germinating encysted zoospore 11. Sporangium germinating by germ tube 12. Intercalary chlamydospore 13. Terminal chlamydospore 14. Oospore germinating by germ tube.

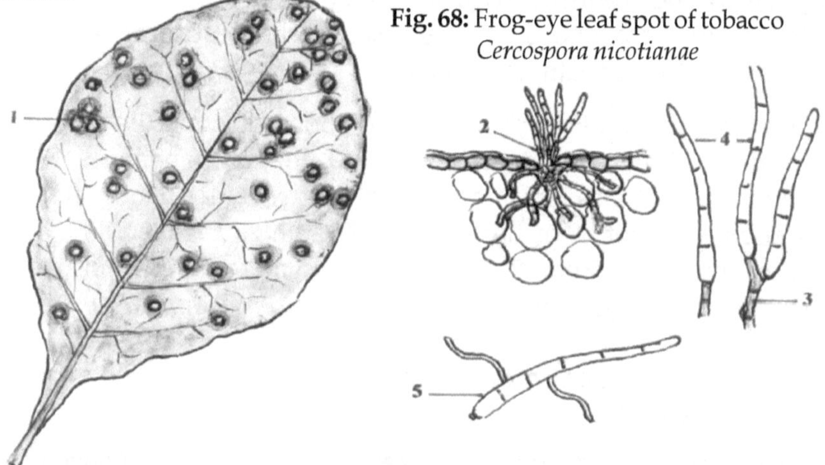

Fig. 68: Frog-eye leaf spot of tobacco *Cercospora nicotianae*

1. Spots on the leaf 2. Conidiophores emerging through the stoma 3. Conidiophore 4. Conidia 5. Conidium germinating by germ tubes.

Symptoms. The disease usually occurs in the nurseries. The pathogens cause pre-emergence and post-emergence damping off. Post-emergence damping off, when the seedlings are 5 - 6 weeks old, causes considerable damage. Diseased plants turn pale-green and show a girdle of water-soaked, brown color near the surface of the soil in the region of the hypocotyl. The stem rots and the seedling collapses and topples over due to disintegration of the tissues at the base of the stem. Death of large number of seedlings in patches causes gappiness in the nursery.

The causal organism. *Pythium aphanidermatum, P. de baryanum* and *P. myriotylum* cause damping off of tobacco seedlings. The morphology of the first two organisms have been described under damping off of seedlings of eggplant (Page 339), **(Fig.85).**

The mycelium of *Pythium myriotylum* is coenocytic, hyaline, multinucleate and much branched. The hyphae form numerous, knob-like appressoria. Sporangia are terminal or intercalary. Sometimes, they are formed on long filaments from which swollen, lobulate outgrowths are produced laterally. Zoospores are formed within the sporangium or in a vesicle formed apically from the sporangium. In each sporangium, up to 45 zoospores are formed. The zoospores are reniform, with 2 lateral flagella, inserted laterally on the concave portion and measure 10 - 12μ in diameter. They encyst and after a period of rest, germinate by a germ tube. The oospores are pale yellow in color, sub-spherical and 12 - 37μ in diameter.

Mode of survival, spread, epidemiology and disease management. Refer damping off of seedlings of eggplant (Page 339).

4. Powdery mildew of tobacco

Erysiphe cichoracearum

The disease is also known as **'white mould'** or **'ash disease'**. It is prevalent in all parts of the world, where tobacco is cultivated and attacks all types of tobacco. In India, it occurs in the states of Bihar, West Bengal, Karnataka, Telungu Desam and Tamil Nadu. In the diseased crop, the yield, quality and market value of the produce are adversely affected.

Symptoms. The initial symptoms appear as small, white to grayish patches of size 0.5 - 1.0 cm. in diameter. Under favorable conditions, the patches rapidly enlarge and cover large areas of the leaves. Such patches may also be formed on the upper surface of the leaves, when conditions become highly conducive for the pathogen. As the patches develop, they become powdery, as a result of formation of conidiophores and conidia abundantly. In advanced stages of infection, the leaves turn yellow, wither and dry. Usually, the infection starts from the lower leaves and spreads to the upper leaves. In the later stages of disease development, the ascigerous stage may be formed on the superficial mycelial mat as small, dark- brown to black, pinhead-like bodies, which are the cleistothecia.

The causal organism, mode of survival, spread, epidemiology and disease management. Refer powdery mildew of sunflower (Page 198), **(Fig.57)** and powdery mildew of cucurbits (Page 361), **(Fig. 88).**

5. Angular leaf spot and wild fire of tobacco

Pseudomonas syringae pv. *tabaci*

The disease is prevalent in all tobacco growing countries of the world. It is endemic in some areas and is very destructive during certain seasons. The disease is considered to be one of the worst diseases of tobacco in the United States of America. In India, it is reported from Karnataka and Tamil Nadu. Besides tobacco, the pathogen is known to attack some other crops, including bean, soybean etc.

Symptoms. The disease attacks the crop at all growth stages, including seedlings in the nursery. On seedlings in the nursery, two to three, small, dark, angular spots are formed on the leaves. In the main field, characteristic symptoms develop as numerous, small, angular, dark-brown to black spots. The spots gradually enlarge to become larger spots, which are limited by veins. The spots are surrounded by a chlorotic, yellow halo. The spots may coalesce covering large areas of the leaves. In wet weather the disease spreads in a rapid manner and hence this disease is called **'wild fire'** of tobacco.

The causal organism. *Pseudomonas syringe* pv. *tabaci* causing wild fire produces a phytotoxin called **'tabtoxin'** or **'wild fire toxin'**, which is non-host specific. This toxin is also produced by other strains of *P. syringae* pv. *tabaci* attacking other hosts, such as bean, soybean etc. Mutants of *P. syringae* pv. *tabaci* that have lost their ability to produce 'tabtoxin' show reduced virulence and cause necrotic spots, without a yellow halo. The pathogen causing angular leaf spot of tobacco was previously named *Pseudomonas angulata*. But, this pathogen cannot be distinguished from *P. syringae* pv. *tabaci*. *P. angulata* is now considered to be a non-toxogenic form of *P. syringae* pv. *tabaci*. The toxin produced by the toxogenic form of *P. syringae* pv. *tabaci* destroys the chloroplasts in the host tissues, thereby causes chlorosis, which is characterized by the formation of the yellow halo and eventually necrosis. The bacteria measure 2.0 - 2.5 x 0.5μ in size and have 1 - 6 polar flagella.

Mode of survival, spread and epidemiology. The diseased plant debris in the soil may harbor the bacteria, which may cause primary infection. The bacteria may be carried to the main field on the roots of seedlings at the time of transplanting and cause infection later on.

Wet weather, accompanied by lashing winds results in heavy splashing of water, which aids in the dissemination of the bacteria. Continuous rainy weather, high humidity and moderate temperature favor disease development. Application of excessive doses of nitrogenous fertilizer predisposes the plants to attack by the pathogen.

Disease management

Agronomic practices (i) Diseased plant refuse should be removed and destroyed (ii) Proper drainage facilities should be provided (iii) The nursery site should be changed every year (iv) Plants showing disease symptoms should be removed and destroyed. Such seedlings should not be used for planting (v) Seeds should not be collected from diseased fields.

Chemical control (i) Spraying the seedlings in the nursery with Bordeaux mixture - 1 % or copper oxychloride at 2.5 gm./ lit. of water checks the disease (ii) Spraying the crop with Streptocycline at 200 ppm (2.0 gm./10 liters of water) is also effective in controlling the disease.

6. Mosaic disease of tobacco

Tobacco virus -1 or *Nicotiana virus*

'Tobacco mosaic disease' was the first virus disease discovered and studied in much detail. It is a very common disease found in all tobacco growing countries of the world and causes severe damage to the crop, resulting in substantial yield loss. It occurs both in the nursery and in the main filed.

Symptoms. The disease occurs at all growth stages of the crop. In the early stages of infection, the interveinal regions become pale green in color. This is followed by formation of light- and dark-green, irregular, indistinct patches. Usually, the areas adjacent to the veins remain green. Dark-green blisters and enations (outgrowths) may appear on the lower surface of leaves due to hyperplasia and hypertrophy of cells, resulting in crinkling of the leaf surface. Sometimes, necrotic, dark-brown spots or scorched patches known as **'mosaic scorch'** or **'mosaic burn'** may appear during hot spells. Plants infected at the early growth stages, are very much stunted. When the disease intensity is high, the leaves become thin, puckered, distorted, malformed and narrow, with elongated central veins, giving the appearance of **'rat tail'** to the leaves. The disease also causes partial sterility.

The causal organism. The virus causing the disease is single, rigid, rod-shaped, helical particle, measuring 300 nm. in length and 15 - 18 nm. in diameter. It contains a single molecule of ssRNA. The protein sub-units of the virus are arranged on the helical RNA molecule, with a distinct central hole. The protein capsid functions as a protective covering around the RNA.

Mode of spread. The pathogen is sap transmissible and highly contagious. It enters the host through any kind of injuries caused by mechanical means or by insects. It is not transmitted by insect vectors or through seeds. The virus can be completely destroyed when subjected to a temperature of 90°C for a continuous period of 10 minutes, but is stable even at 65°C. They can survive for a period of up to 50 years when stored under dry conditions. They are capable of withstanding highly adverse environmental conditions and remain virulent for a long time and can cause infection. The viruses present in the sap from an infected plant, even at a dilution of 1 : 1,000,000 can cause fresh infection.

Natural forces, such as strong wind, lashing rain etc., which make the leaves to rub against one another cause invisible injuries, through which the virus gains entry into the host. Touching a diseased leaf and then touching a healthy leaf can transmit the virus. Cultural operations, such as topping, clipping etc., which are usually followed, result in wounds that serve as portholes of entry for the virus into the host. Air-dried tobacco is a common source of infection. The virus is also carried in various tobacco products, such as cigarettes, beedis, pipe tobacco, chewing tobacco, snuff etc. The virus present in the diseased crop debris buried in the soil also serves as potential inoculum. Over 116 plant species, belonging to 29 families, such as beans, tomato, eggplant, chillies, *Datura* etc. serve as alternate hosts of this pathogen.

Disease management

Agronomic practices (i) Diseased plant debris in the field should be collected and destroyed (ii) Diseased seedlings should be discarded (iii) During field operations like planting, weeding etc., the laborers should wash their hands thoroughly with soap and running water (iv) Diseased plants should be uprooted and destroyed (v) After handling diseased plants, healthy plants should not be handled (vi) The fields and surrounding areas should be kept clean and free from solanaceous weeds (vi) While working in the fields, use of tobacco products in any form should be prohibited (vii) To avoid leaf contact between plants, wider spacing should be given (viii) The knives used for topping and clipping operations should be disinfected by dipping them frequently in formalin or lysol solution (ix) In tobacco mosaic endemic areas, crop rotation should be followed.

Phytocontrol. Leaf extracts of some plant species, such as *Agave americana, Bougainvillea spectabilis, Clerodendron fragrans, Azadirachta indica, Thevetia neriifolia, Carica papaya* etc. inhibit *Tobacco mosaic virus.* Prophylactic spraying with 1 % leaf extract of any one of the above plants reduces the incidence of the disease.

Resistant varieties. Jayasri MR, Godavari Special, L.1158, TMVRR.2, TMVRR.2a and TMVRR.3 are found to be resistant to this disease.

7. Tobacco leaf curl virus disease

Tobacco leaf curl virus (TLCV) or *Nicotiana virus* -10

The disease is widespread and is found in all tobacco growing countries of the world, including India. In India, the disease is more prevalent in North India and Gujarat.

Symptoms. The disease usually appears in crops, 4 - 6 weeks after planting however, sometimes it appears late in the nursery. The symptoms first appear on young leaves, which show reduction of size, downward curling of whole or part of the leaf lamina, thickening and yellowing of veins. The affected plants become stunted. The leaf edges curl downwards and the leaf edges almost join together lengthwise. When the intensity is high, all the leaves curl downwards and the plants become very much stunted. Enations are seen along the veins on the under surface of leaves. The inflorescence is very much reduced in size and the floral parts become green in color.

The causal organism. The virus causing the disease belongs to the Genus - *Geminivirus*. The viruses have unique particular morphology in that, their isometric particle size of 18 - 20 nm. diameter occur mainly in pairs. They have a monopartite genome, consisting of ssDNA. They infect several dicotyledonous hosts and are transmitted by white fly vectors.

Mode of spread. The virus is not sap or seed transmissible, but is transmitted by white flies and by grafting. The white fly - *Bemisia tabaci* is responsible for transmitting the virus. The vectors need a minimum acquisition feeding time of 15 minutes to 2 hours on an infected plant and an incubation period of 12 - 33 days to become viruliferous. A minimum feeding time of 10 minutes is required for transmission of the virus on to a healthy plant.

The insects remain viruliferous for a period of 7 days after acquisition. A single vector can transmit the virus to several plants in a single day. The vectors are very active during dry periods, after the monsoon rains. The virus has a wide host range of over 63 plant species, distributed over 14 families. Tomato, *Datura*, *Petunia*, *Physalis*, *Rhynchosia*, *Sida*, *Solanum*, *Vernonia* and *Zinnia* are a few of the hosts.

Disease management

Agronomic practices (i) Solanaceous weed hosts in and around the fields should be removed and destroyed (ii) Healthy and disease-free seedlings should be used for planting (iii) Infected plants should be removed from the nursery and main field and destroyed in the early stages of infection (iv) Application of required quantities of phosphorus fertilizer increases resistance to the disease.

Chemical control. Because the white fly vector transmits the virus, its population should be monitored from the beginning by installing yellow sticky traps in the field at 5 traps per acre. If the number of vectors exceeds 100 per sticky trap, then control measures have to be taken. Foliar spraying with methyl demeton - 375 ml. or dimethoate - 500 ml. or phosphamidon - 125 ml. or monocrotophos - 375 ml. or acephate - 500 gm. in 250 liters of water per acre with a high volume sprayer, in the evening time, controls the vector. As the vector is capable of transmitting the virus to a large number of plants within a short time, chemicals having quick knockdown effect must be used. First spraying should be given before pulling out the seedlings and subsequently, 2 or 3 sprayings should be given in the main field at weekly intervals, from the time the seedlings have established. Using different insecticides during each of the spraying is better than using the same chemical repeatedly.

8. Phanerogamic parasite - 'Broomrape'

Orobanche cernua and *O. ramosa*

The parasites attack tobacco plants to a large extent and hence it is commonly known as **'pugaiyilaikkalan'** in Tamil. In Telungu it is called **'bodu'** or **'malli'**. Besides tobacco, the parasite attacks eggplant, tomato, cabbage, cauliflower, turnip, sunflower and several other solanaceous and cruciferous plants.

Symptoms. Plants attacked in the early stages of growth, are usually stunted and show typical wilting symptoms. However, only 5 - 6 weeks after transplanting tobacco seedlings, young *Orobanche* sprouts emerge in clusters from the soil around the base of tobacco plants. Plants attacked late in the season do not show marked visible symptoms, but the yield and quality of leaves are reduced.

The parasite. These flowering, holo root parasites do not produce green leaves and so, obtain their entire requirements of nutrients from the host. The stem of the parasite is stout, thickened at the base, succulent, and light-yellow or dark-brown in color and 10 - 40 cm. in height. Instead of leaves, the stem is covered with dense, boat-shaped, scale-like structures, with yellow base and dull brown tip. Tubular, white flowers, longer than the scales appear in the axils of the scales and they are 20.0 x 5.0 mm. in size. The single bract of the flower is boat-shaped. The calyx has two separate sepals. The corolla is long, tubular, with 5 united petals and white in color. The fruit is a capsule and contains numerous, minute, dark-brown seeds. When the capsules dehisce, the seeds are shed on to the soil. Maximum number of seeds is found in the top 5.0 cm. layer of the soil. The seeds of the parasite germinate, when the host plants are about 2 weeks old. The root exudates from the hosts stimulate germination of the seeds of the parasite. When the primary root of a germinating broomrape seed contacts the fibrous root of the host, it forms a nodule of tissue, which fuses with the tissues of the host root. New roots and a stem of the parasite develop at this point. While the stem of the parasite emerges above the ground, the roots of the parasite intertwine around the host roots and form new contacts with other host roots. Thus, a large number of parasitic plants appear at the base of the affected host plant. Up to 20 parasites may be seen around the base of a single host plant. The roots of the parasite drain all their requirements of food from the host roots. When the tobacco plants are 5 - 6 weeks old, the parasites start appearing above ground. The life cycle of the parasite is completed in 12 - 14 weeks.

Mode of survival and spread. The seeds of the parasite remain viable in the soil for 4 - 5 years. The seeds are minute and light and can be easily carried and disseminated by wind. The seeds, which are shed and remain in the soil are carried by rainwater, irrigation water or drainage water from one field to another. The seeds are also disseminated along with soil by cattle, agricultural implements and by man. Irrigated crop grown during November - February are more prone to attack by the parasite than the summer crop grown during March - June.

Parasite management

Agronomic practices (i) The best method to eradicate the parasite is to destroy it before seed formation. The parasites may be uprooted and destroyed by burning at weekly intervals before seed-setting (ii) Leaving the fields fallow during the summer months, summer ploughing and burying the seeds deeper into the soil help in reducing the emergence of the parasite (iii) Crop rotation may be followed for a few years with non-host crops, such as black gram, green gram, gingelly, sorghum and chillies (iv) Growing of crops, such as chillies, sorghum and gingelly in parasite infested fields, stimulates the parasite seeds to germinate by their root exudates. The parasites emerging from the soil may be removed and destroyed before seed setting. This practice continued for a few years may result in the complete elimination of the parasite (v) Application of 2 to 3 drops of neem oil or castor oil or linseed oil on young broomrape shoots before flowering kills the parasites.

Chemical control (i) Spraying the soil with 25 % copper sulphate has been found to destroy the parasite (ii) The parasite can be destroyed by the application of the weedicide, Glyphosate.

Diseases of minor importance. Besides the diseases detailed above, several other diseases are known to attack tobacco, some of which are of minor importance. *Colletotrichum tabacum* causes 'Anthracnose', which attacks young seedlings in the nursery and grown-up plants in the main field; *Alternaria longipes* causes 'Brown spot' that affects the leaves; *Curvularia verrucolosa, Phyllosticta tabaci* and *Ascochyta nicotianae* cause 'Leaf spots'.

Betelvine or Pan *(Piper betle)*

1. Leaf and foot rot disease of betelvine

Phytophthora parasitica var. *piperina*

'Betelvine' or **'pan'** is an important commercial crop of India, Bangladesh and Sri Lanka. It is also cultivated in Burma, Malaysia, Singapore, Thailand, the Philippines and New Guinea. In India, it is grown extensively in the states of West Bengal, Madhya Pradesh, Uttar Pradesh, Maharashtra, Assam and Tamil Nadu. The disease, which is also known as **'collar rot'** or **'wilt disease'**, poses a big threat to the successful cultivation of the crop.

Symptoms. The pathogen infects almost all the parts of the plant. Initial symptoms of the disease appear as loss of luster of leaves, followed by darkening of the stem, drooping of the upper leaves and tender shoots of the vine. Soon, the entire plant turns yellow, wilts and dries up. The underground parts are almost completely rotten at this stage. The smaller roots are infected first and gradually the rotting spreads to the bigger roots and from there to the collar region. The stem becomes weak and brittle and may break at any point near the ground level. As the rotting develops, the fungus attacks the soft tissues. Unlike *Fusarium* wilt disease, which attacks the vascular elements, here the vascular tissues are not affected, but due to rotting of the roots and the basal stem or collar region, wilting is caused. The fibrous remains in the collar region are exposed and appear partially shredded and slimy. Actually, the disease starts from the smaller roots much before the symptoms appear on the aerial parts.

In the leaf rot phase, the first symptoms appear as fairly large, circular, black or brownish, water-soaked spots or blotches. Under continuous wet and humid conditions, the black spots enlarge rapidly and the wet rot may cover large areas of the leaf. The under surface of the blotches may be covered with a delicate, downy growth of the fungus. More than one spot is often found on a leaf and the spots may coalesce to form bigger blotches. The infection spreads from the leaf to the petiole and then to the stem. The infection may spread more rapidly along the veins. If dry period follows the wet spell, the diseased areas show concentric puckering and wrinkles due to unequal growth of the leaf tissues. The lower leaves near the ground level are infected first and from there the infection spreads to other leaves. The top-most leaves are rarely infected **(Fig.69).**

The causal organism. *Phytophthora parasitica* var. *piperina* is the major cause of this disease. However, *Phytophthora capsici* is also associated with the disease. The mycelium of *P. parasitica* var. *piperina* is hyaline, coenocytic, mostly intercellular and sometimes intracellular. The fungus produces no haustoria. Sporangia are borne singly at the end of short stalks. They are pear-shaped or broadly ovate, with a prominent papilla and measure 30.0 - 63.3 x 20.4 - 40.8μ in size. Zoospores are formed within a vesicle formed by the swelling of the papilla or within the sporangium itself. The zoospores are liberated by dissolving the wall of the vesicle or the papilla as the case may be. Zoospores are bean-shaped and are motile by means of 2 flagella, arising laterally from the concave portion. They swarm about in free water for a short time, come to rest and encyst. The encysted zoospores are spherical and measure 2.2 - 5.7μ in diameter. They germinate by means of a germ tube. The fungus produces both terminal and intercalary chlamydospores. Oospores, which are produced within the host tissues are spherical, almost filling the oogonium, smooth, thick-walled and measure 17.8 x 53.1μ in diameter. They germinate by means of a germ tube **(Fig.69).**

Mode of survival, spread and epidemiology. The fungus is mainly soil-borne. It is a facultative parasite and can survive in the soil as a saprophyte in the absence of the host. Oospores, which are produced in the leaf tissues may remain in the soil in a viable state for long periods of time and initiate infection. The chlamydospores may also serve as a source of primary infection. The inoculum is generally carried through diseased vines used for planting. The disease is also spread through irrigation water and through contaminated soil carried by cattle, agricultural implements etc. Secondary spread is through sporangia carried by wind, rain or irrigation water. The pathogen attacks wild *Colocasia* plants, which serve as collateral host

Optimum temperature for sporangial development in nature is 20°-31°C, when the relative humidity is 100%. Free water is essential for the liberation and movement of zoospores, which is maximum at 21°-23°C. The intensity of the disease is high, when the night temperature is below 23°C for a continuous period of several days. Low-lying, ill-drained soils, with excessive free moisture around the root zones are highly favorable for disease development. Leaf rot is usually high during the monsoon seasons.

Disease management

Agronomic practices (i) Dead and dying plants and diseased plant refuse in the soil should be removed and destroyed by burning or buried deep in the soil outside the field premises along with lime (ii) Disease-free cuttings should be used for planting. Cuttings with discoloration or spots on the surface should be rejected. Top portion of vines, up to 1.5 meters should be preferred for getting cuttings for planting (iii) Water-logging around the vines should be avoided by providing adequate drainage facilities (iv) Planting should be avoided during periods of excessive rains (v) Collateral hosts, such as *Colocasia* species growing around the gardens and in irrigation channels should be removed and destroyed (vi) Judicious water management should be followed to prevent dissemination of the pathogen (vii) Application of excessive doses of nitrogenous fertilizer should be avoided. Application of phosphorus and potash in sufficient quantities increases resistance to the disease (viii) Crop rotation with rice and banana has been found to be very effective in reducing the inoculum in the soil.

Chemical control (i) Soil drenching with Bordeaux mixture - 1%, 4 times, at monthly intervals and foliar spraying with Bordeaux mixture - 0.5 %, 8 times, at fortnightly intervals, starting just prior to the onset of monsoon rains is found to be very effective in controlling the disease and increasing the leaf yield (ii) Soil drenching around the stems of vines with metalaxyl at 1.0 gm. or chlorothalonil at 1.25 gm./ lit. of water gives very good control of the disease.

Biological control. Dipping the cuttings in a spore suspension of *Trichoderma viride* before planting controls the disease to a large extent.

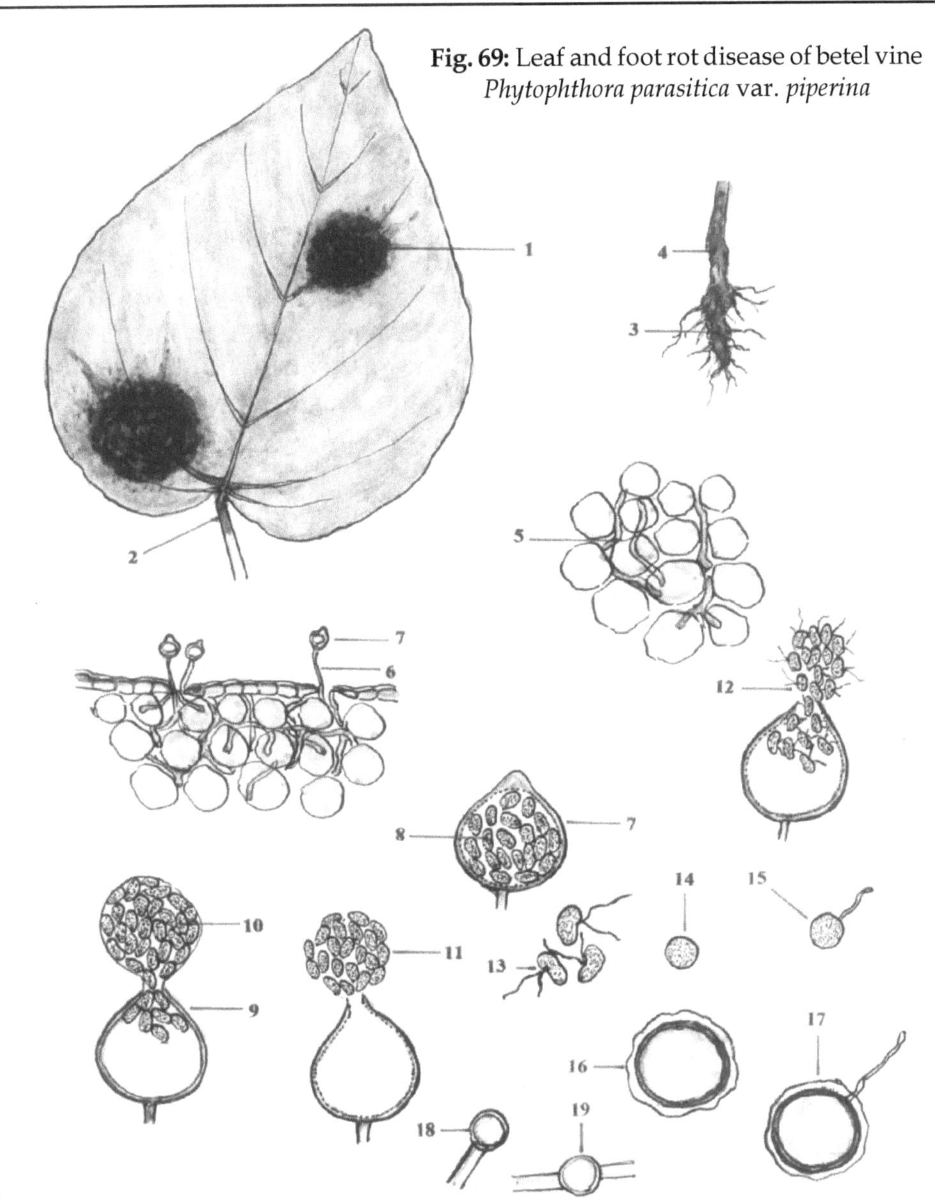

Fig. 69: Leaf and foot rot disease of betel vine
Phytophthora parasitica var. *piperina*

1. Leaf rot - spots on leaf 2. Infection on the petiole 3. Rotten roots 4. Rotten collar region 5. Inter and intracellular mycelium 6. Sporangiophore 7. Sporangium 8. Zoospores inside the sporangium 9. Germination of sporangium by formation of a vesicle 10. Zoospores moving into the vesicle 11. Zoospores held together after the vesicle wall is dissolved 12. Zoospores liberated directly through papilla 13. Zoospores 14. Encysted zoospore 15. Germination of encysted zoospore 16. Oospore 17. Germination of oospore by germ tube 18. Terminal chlamydospore 19. Intercalary chlamydospore.

2. *Sclerotium* wilt of betelvine

Sclerotium rolfsii (Corticium rolfsii)

The disease is prevalent in all the countries, where betelvine is cultivated. In India, it is found in the states of Assam, Madhya Pradesh, Maharashtra, West Bengal and Tamil Nadu. Besides betelvine, the pathogen attacks a wide variety of plants, including chillies, tomato, chickpea, groundnut, eggplant etc. and sometimes causes severe damage to the crop.

Symptoms. The disease attacks vines at all stages of growth. The infection starts on the succulent stem as a dark-brown lesion just below the soil level. As the disease progresses, the lower leaves and then the upper leaves turn yellow and wilt. Meanwhile, the fungus grows upward in the plant, covering the stem lesion with a fine, cottony, white, weft of mycelium. The infection rapidly spreads downward into the roots also and the entire root system is destroyed. The white mycelium is always present in and on the infected tissues. It also grows over the soil to adjacent plants, starting new infections. Invaded stem regions become dark and soft, but not watery, shrink and the barks peel off easily. The roots rot, disintegrate and are replaced by debris, interwoven with mycelium. On the infected tissues and on the nearby soil, numerous sclerotia are formed.

The causal organism. Individual hypha is hyaline, but in mass the hyphae are white to ash-colored. Sclerotia are small, spherical, mustard seed-like, smooth, almost of uniform size and measure 0.34 - 6.00 mm. in diameter. They are white, when immature, becoming dark-brown to black on maturity. The sclerotium is differentiated into a hard outer rind, a middle cortex and an innermost area of loosely arranged hyphae. The perfect stage of this fungus is *Aethalium rolfsii*. In the perfect stage, which is found occasionally, it produces basidiospores at the margins of the lesions under humid conditions.

Mode of survival, spread and epidemiology. The fungus is mainly soil-borne. It can survive in the soil and can lead a saprophytic life, when there is plenty of organic matter and moisture in the soil. The fungus may also survive in the infected plant tissues left over in the soil. But, the disease is perpetuated mainly through sclerotia present in the soil and infected plant refuse. The fungus is capable of entry into the host by direct penetration. However, the fungus secretes certain toxic substances, such as oxalic acid, pectinolytic, cellulolytic and other enzymes that kill and disintegrate the host tissues prior to penetration.

The sclerotia are spread by running water, through contaminated soil carried by cattle, man, agricultural implements and infected cuttings used for planting.

High soil moisture and high temperatures between 30°- 35°C are favorable for disease occurrence and development.

Disease management

Agronomic practices (i) Sterilization of the surface soil by burning trash prior to planting helps to destroy the inoculum present in the soil (ii) Affected plants should be uprooted along with the soil around the root zone and destroyed by burning (iii) Diseased plant debris should be collected and destroyed (iv) Application of organic manure or green manure helps to increase the antagonistic microbial population in the soil, which may destroy the sclerotia found in the soil.

Chemical control. Drenching the soil around the base of the stem region with pentachloro nitrobenzene at 1.0 gm. or chloroneb at 1.0 gm./ lit. of water is effective in controlling the disease.

Biological control. Soil application of antagonistic fungi, such as *Trichoderma harzianum, T. viride* etc. gives good control of the disease by parasitizing and killing the pathogen.

3. Bacterial leaf spot of betelvine

Xanthomonas campestris pv. *beticola*

The disease was first reported from Sri Lanka in 1926. In India, the disease occurs in all the states where pan is grown, especially in Kerala, West Bengal, Madhya Pradesh and Tamil Nadu.

Symptoms. The disease usually appears in the wet seasons. Initial symptoms appear as minute, water-soaked, angular spots on the lower surface of leaves. The spots enlarge and become roughly circular or angular, with a yellow halo. The spots are mostly limited by the veins. When the spots enlarge and cover large areas of the leaves, they turn yellow and drop off prematurely. During wet weather, a slimy bacterial ooze is seen on the under surface of the spots. The infection then spreads to the stem region and maximum damage is caused at this stage. The vines above the points of infection wilt and die. The infection may spread rapidly through injuries caused during the tying operation of the vines.

The causal organism. The bacterium causing the disease is rod-shaped, motile by a single polar flagellum, gram-negative and non-spore forming.

Mode of survival, spread and epidemiology. The bacteria enter the leaves through natural openings, such as stomata, hydathodes and through injuries. Stem infection occurs mainly through injuries. The bacteria may gain entry into the vines through wounds caused while plucking of leaves. They may be harbored in the diseased leaves and pieces of vines fallen to the ground. Secondary spread may occur through bacteria carried by lashing winds, rainsplash, irrigation water and by some insects.

Wet spells, damp and cloudy weather and moderate temperature favor the occurrence and development of the disease.

Disease management

Agronomic practices (i) Diseased leaves and fragments of vines should not be allowed to remain in the beds or in the trenches. They should be periodically collected and destroyed (ii) Disease affected and dead vines should be rouged out and destroyed.

Chemical control. Spraying the crop with copper oxychloride - 2.5 gm. + Agrimycin - 0.6 gm./ lit. of water or Bactrinol - 1.0 gm./ lit. of water is very effective in controlling the disease. The sprayings should be given at fortnightly intervals. During rainy season, sprayings should be given at weekly intervals.

Minor diseases. Betelvine is subjected to attack by many other diseases, besides the ones described above. *Rhizoctonia bataticola* and *R. solani* cause 'Root rot'; *Fusarium oxysporum, F. moniliforme* and *F. solani* cause '*Fusarium* wilt'; 'Powdery mildew' is caused by *Oidium piperis*; *Cercospora piperis-betle* causes '*Cercospora* leaf spot'; 'Anthracnose' is caused by *Colletotrichum capsici*; *Alternaria alternata* causes '*Alternaria* leaf spot'.

Mulberry *(Morus alba)*

1. Powdery mildew of mulberry

Phyllactinia guttata / P. corylea / P. suffulta

The disease is known to occur in more than 100 host species, including *Dalbergea sissoo* and mulberry. It is found in all the districts of Tamil Nadu, such as Kancheepuram, South Arcot, Salem etc., where mulberry is grown as a food for silk worms.

Symptoms. The disease appears as small, white patches on the under surface of leaves. Under highly moist conditions, the patches enlarge and form white, powdery patches, sometimes covering the entire leaf area. The corresponding upper surface of the leaves turns yellowish-brown. Severely affected leaves dry and drop off prematurely. The white, powdery coating consists of the mycelium, conidiophores and conidia, which are produced profusely. Dark-brown to black, pinhead-like structures, representing the cleistothecia are formed on the surface of the mycelial weft. The disease makes the leaves unsuitable for feeding silk worms **(Fig.70).**

The causal organism. The fungus causing the disease is an obligate parasite. The mycelium consists of branched, septate hyphae. Most of the hyphae of the pathogen are superficial however, they do not develop any haustoria like most Genera of Erysiphaceae. The fungus sends special hyphal branches into the host through the stomata. These hyphae come in contact with the mesophyll cells, send haustoria into them and obtain nourishment from the host cells. From the superficial hyphae, large number of long, hyaline, erect conidiophores are formed in profusion. From the tips of the conidiophores, conidia are formed singly. They are single-celled, hyaline, clavate and more or less diamond-shaped. The conidia fall off, as soon as they are formed and no conidial chains are formed.

At the later stages of the disease, when conidial production slows down, cleistothecia are formed on the white, mycelial weft. They are white at first, becoming dark-brown to black on maturity and are sub-globose. Matured cleistothecia are provided with characteristic appendages, which are rigid, long, spear-like, with a bulbous base and pointed tip and are arranged equatorially. The cleistothecia also bear a crown of short, apical, branched appendages. The long appendages bend and lift the cleistothecium above the surface, which may aid in the dissemination of asci and ascospores. Each cleistothecium contains many club-shaped asci. Each ascus contains, 8, hyaline, single-celled, oval ascospores **(Fig.70).**

Mode of survival, spread and epidemiology. In warm, tropical climate, cleistothecia are formed very rarely and as such, the pathogen perpetuates in its conidial stage. The hosts, such as mulberry are perennial plants and are grown in rotation throughout the year to provide leaves for feeding silk worms. So, the pathogen is present all through the year in its conidial stage and can cause fresh infection, when conditions are favorable. Secondary spread occurs through wind-borne conidia or ascospores.

The disease usually occurs in a severe form after a spell of rains for a few days or at the end of the monsoon seasons. High humidity and moderate temperature favor the disease occurrence and development.

Disease management

Agronomic practices. Severely affected leaves may be plucked and destroyed.

Chemical control. Spraying the plants with wettable sulfur - 4.0 gm. or dinacap 1.0 gm. or tridemorph - 1.0 ml./ lit. of water controls the disease.

The leaves should be harvested for feeding silk worms, only after a period of 15 days from the time of spraying.

Diseases of minor importance. A few other diseases also attack mulberry plants, but they are not of much significance. *Gibberella moricola* causes 'Twig blight'; *Alternaria alternata* causes 'Leaf blight'; *Cercospora moricola* causes 'Leaf spot'.

Jute *(Corchorus capsularis* and *C. olitorius)*

1. Seedling blight, stem and root rot of jute

Rhizoctonia bataticola

Macrophomina phaseolina

The disease occurs in all jute growing regions of India, including West Bengal, Assam, Bihar, Orissa, Tripura and Uttar Pradesh. Both the production of jute fiber and quality of the fiber are badly affected.

Symptoms. The disease attacks the crop at all growth stages. The initial symptoms appear on the cotyledons and hypocotyl as brownish-black lesions. Affected young seedlings rot, collapse and die. In grown-up plants, the leaves are shed, the stems rot and the plants die prematurely. In older plants, brownish-black lesions are formed on the collar and stem regions, followed by rotting of the affected regions, as well as the roots, shredding of the bark of roots and collar regions and eventual death of the plants. Later, when the pathogen infects the capsules, they become black and the seeds get discolored and shrivel. The sclerotia of the pathogen are often seen on the roots, collar regions and capsules.

The causal organism. The fungus produces both sclerotia and conidia. Sclerotia are formed on the shredded portions as minute, pinhead-like structures. They are jet black in color, spherical to irregularly spherical and measure about 100μ in diameter. The pycnidia are formed singly or in groups and are sunken in the stem tissues. They are usually globose, dark-brown to black in color, ostiolate and measure 100 - 200μ in diameter. Conidia, which are produced in succession from the tips of phialides inside the pycnidium are hyaline, single-celled, and ellipsoid to obovoid and measure 14.0 - 30.0 x 5.0 - 10.0μ in size.

Mode of survival, spread and epidemiology. The disease is both seed- and soil-borne. The mycelium can lead a saprophytic life in the infected plant debris and in the organic matter present in the soil. The sclerotia can also remain in a viable state in the soil for a considerably long period of time and cause primary infection. The sclerotia may also be carried along with the seeds as contaminant. Secondary infection is caused by conidia dispersed by wind or rainsplash.

Low soil pH, high temperature, water-logging, lack of adequate quantity of potash and excess of nitrogen predispose the plants to attack by the pathogen. The pathogen has a very wide host range, including cotton, tobacco, sesame, potato, eggplant, mulberry etc.

Disease management

Agronomic practices. Use of disease-free seeds, application of adequate quantities of potash, application of micronutrients, such as zinc, iron and boron, and crop rotation with rice as one of the crops in the crop sequence are useful in minimizing the occurrence of the disease.

Seed treatment. The disease is both externally and internally seed-borne. Seed treatment with carbendazim at 2.0 gm./ kg. of seeds is found to give protection from seed-borne infection.

Resistant varieties. Varieties of *Corchorus capsularis* jute viz., Halmhera, C.58-9433, Bangkpk, JRC.9826 and Patchy Albino have been found to be resistant. In *C. olitorius* jute, the variety JRO.3331 has field resistance.

Diseases of minor importance. Some other diseases also attack Jute. 'Wilt' caused by a microbial complex, including *Macrophomina phaseolina, Fusarium solani* and *Pseudomonas solanacearum* is mostly found in jute-potato fields; *Colletotrichum gloeosporioides* causes 'Anthracnose'; *Physoderma corchori* causes 'Stem galls' on the seedlings; *Sclerotium rolfsii* causes 'Soft rot' on the collar region and eventual death of affected plants; 'Leaf mosaic' is caused by *Pollen-transmitted seed-borne virus*, which is also transmitted by the white fly insect vector, *Bemisia tabaci*.

Fungal spoilage of grains during storage

Post-harvest fungal spoilage, deterioration and rot of grains and legumes occur, as a result of infection by the pathogens in the field and continue to develop after harvest. Such spoilage and rotting occur commonly in storage bins, ware-houses and godowns under conditions of high moisture and high temperature. Spoilage of agricultural products, such as bread, flour, hay, silage, feed and foodstuffs is also quite common in storage. Although several fungi, such as *Alternaria, Fusarium, Colletotrichum, Cladosporium, Diplodia, Curvularia* and *Cochliobolus* belonging to either Ascomycetes or Fungi Imperfecti attack grains and legumes in the field, they require a fairly high moisture content of 24-25 % in the grains for their survival and cause spoilage, deterioration and rotting while in storage. Grains and legumes are dried to bring down the moisture content to 12 - 14 % before storage. Under such low levels of moisture content, these organisms may not be able to thrive and may die after a few months in storage or become very weak to cause fresh infection. However, a few field pathogens may cause discoloration of kernels, kill the embryos, cause shrivelling of the grains and in some cases may produce toxic substances, known as mycotoxins before they die. Many of the field pathogens may not be able to attack grains in storage.

Most of the decay, spoilage and deterioration of grains and legumes in storage and in transit are caused by several species of *Aspergillus*. Sometimes a few species of *Penicillium* and *Fusarium* may also cause post-harvest spoilage of cereals, legumes and oilseeds in storage at low temperatures and slightly higher moisture levels. Certain species of *Aspergillus*, especially *Aspergillus flavus*, often infect corn grains and groundnuts in the field and the infection may persist in the produce, while in storage and cause spoilage and deterioration. Injuries caused by insect pests in the field or while in storage by storage pests, predispose the produces to infection by fungal pathogens. When the moisture content of the produces increases even slightly above the critical level and when the temperature also rises, the fungal invasion becomes severe. When moistened grains are infected, the temperature also rises to a considerable extent due to the heat generated from respiration of the vigorously growing microorganisms. This leads to rapid spread of the disease, resulting in more spoilage and deterioration of the produce.

Certain fungi, causing post-harvest diseases of grains cause diseases in animals, humans and birds when such feeds and foods are consumed. Certain toxic substances called **'mycotoxins'** produced by these fungi cause these diseases, which are known as **'Mycotoxicoses'**. The mycotoxin problems came to prominence during World War II, when numerous humans and animals died of necrosis of the skin, hemorrhage and liver failure after consuming mouldy grains. In 1960, large number of turkeys died in England after they were fed with contaminated peanut feed. These led to intensive research and the cause was established to be due to certain toxic substances called mycotoxins present in the feed. Mycotoxins pose a constant threat to humans, animals and birds throughout the world, especially in the developing countries. Common and widespread mould fungi, such as *Aspergillus*, *Penicillium* and *Fusarium* produce most mycotoxins. *Aspergillus* and *Penicillium* produce their toxins mostly in stored grains, hay and silage and also in commercially processed feeds and foods. *Fusarium* produces its toxins mainly in corn and other grains. Many other commonly occurring fungi are also known to produce mycotoxins in agricultural commodities.

One of the mycotoxins known as **'aflatoxin'** is produced by *Aspergillus flavus* and many other species of *Aspergillus*. Aflatoxins are produced in cereal seeds and legumes, but mostly at concentrations, which are non-toxic. Under certain conditions, the toxins produced may be very high and prove to be lethal, when consumed. In peanuts, cotton seed, fishmeal, as well as in some other grains and nuts, aflatoxins are produced at very high concentrations under humid and warm conditions and cause chronic Mycotoxicoses in humans, domestic animals and birds, which may even be fatal. Some of these toxins, when ingested with the feed by dairy cattle are secreted in their milk in toxic form.

Certain species of *Fusarium* produce toxins, such as **'zearalenones'**, **'trichthecenes'** and **'fumonisins'** mostly in moulded corn. These toxins are known to cause serious diseases in pigs, horses, cattle etc. Certain species of *Penicillium* produce toxins, such as **'ochratoxins'**, **'citreoviridin'**, **'citrinin'** etc. in rice, barley and corn and these toxins may cause nervous and circulatory disorders and degeneration of kidneys and liver. **'Patulin'** another mycotoxin produced by certain species of *Penicillium* and *Aspergillus* in moulded bread, bakery products etc. affects the lungs, kidneys and motor nerves.

Control of fungal spoilage of grains in storage

Post-harvest deterioration and spoilage of grains, legumes, fodder, commercial feeds and foodstuffs in storage by fungi can be avoided to a large extent by adopting certain preventive measures.

The harvested produces should be stored at levels of moisture content below the minimum required for the growth of the common storage fungi. A few of the hardy *Aspergillus* species are capable of growing and causing spoilage of starchy cereal grains at moisture content as low as 13.0 - 13.2 % and of soybean at a moisture content of about 11.5 - 11.8 %. Groundnuts, cottonseeds, fishmeal, oilcakes and flour are also attacked at such low moisture content. Other storage fungi require minimum moisture content of 14 % or more to cause deterioration and spoilage.

The temperature of stored grains should be kept as low as possible, since most storage fungi grow much rapidly at temperatures between 30° - 55°C. The storage fungi grow very slowly at temperatures between 12° - 15°C and their growth is almost arrested at 5° - 8°C. Low temperatures also slow down the rate of respiration of grains and prevent increase of moisture in the grains.

Infestation of stored grains and stored products by storage pests should be avoided or kept to a minimum using suitable insecticides or fumigants. This helps in preventing fungi from causing fresh infection and growth through insect injuries. Infected grains and pest infested grains and products should not be stored along with sound and disease-free seeds and other commodities. Unripe or too old grains should not be stored for long periods. Healthy, clean seeds, with good germinability may be stored for a comparatively longer period. Such seeds may resist infection by storage fungi. Mechanically damaged seeds, cracked and broken grains, immature and partially filled grains are more vulnerable to attack by storage fungi.

The storage bins, ware-houses, godowns, gunny bags etc. used for storing grains and other produces should be thoroughly cleaned and disinfected before storage. The ware-houses, godowns or store-houses should be well ventilated. Proper ventilation and airflow helps to remove excess moisture and heat, which are conducive for infection by the fungi.

Diseases due to abiotic factors or Non-infectious diseases

Adverse soil conditions, such as salinity and alkalinity, acidity, deficiency, toxicity, water inundation and water deficiency, as well as adverse environmental conditions, such as excess heat, frost, smoke and poisonous gases are some of the factors that may cause certain symptoms similar to those caused by biotic agents. These are usually not called diseases, but termed as disorders. Soil and environmental factors have a great impact on crop growth and disease occurrence. When environmental factors are favorable for the pathogens, disease occurrence, development and spread is considerably more, resulting in severe crop damage and yield loss. On the contrary, when such factors are favorable for the crop growth, disease incidence is usually less, because of the vigour of the plants. Abiotic factors produce certain changes in the physiological functioning of the plants, which are exemplified as symptoms as those caused by biotic agents.

Soil salinity and alkalinity. The hydrogen ion concentration (pH) of saline and alkaline soils is higher than the pH of normal soils, which is about 7.0. In **'saline soils'** there is accumulation of chlorides, sulfates and nitrates of sodium, potassium and magnesium. In **'alkaline soils'** there is accumulation of carbonates and bicarbonates of sodium and potassium. In saline or alkaline soils with a pH of 8.5 or more, crop growth is very much affected. In such soils, the root development is badly affected, as a result the capacity of the roots to absorb water and mineral nutrients from the soil is reduced. Under such conditions, plant growth is adversely affected, the leaves become small and discolored and deficiency symptoms appear.

Soil acidity. In **'acid soils'**, the hydrogen ion concentration is less than 7.0. In the mineral salts present in the soil or applied to the soil, the mineral ions, such as calcium, magnesium, potassium, sodium etc., dissolve easily in the presence of excess water and are leached out, leaving behind the insoluble acidic components. The accumulation of such acidic remnants over a long period of time results in soil acidity by combining with hydrogen to form acids. Continuous application of chemical fertilizers, such as ammonium sulfate, acid rains etc. may also contribute to increase in soil acidity. Most of the acid soils are found to be deficient in calcium. Soil acidity retards root development and growth of most of the cereals, vegetables etc. and the plants exhibit deficiency symptoms. Soil acidity interferes with the uptake of nutrients from the soil by the plants and also spoils the soil texture.

Deficiency and toxicity. For normal growth and production, plants require several mineral nutrients in a balanced proportion. Lack of one or more of these mineral elements may cause certain abnormalities in the plants known as **'disorders'** commonly referred to as **'hunger signs'** or **'deficiency diseases'**. Similarly, presence of one or more of these elements in excessive quantities in the soil may cause **'toxicity'** in plants. Nitrogen, phosphorus and potassium are the main elements that are required in larger quantities by plants and are termed as **'macronutrients'**. However, several other mineral elements are required by plants in very small quantities to promote ideal growth and are considered as **'micronutrients'** or **'trace elements'**. Lack of these micronutrients may also result in certain growth abnormalities. Sometimes, presence of one or more of these macro- or micronutrients in excessive quantities in the soil may indirectly affects the availability of other nutrients, either by locking them in the soil in an unavailable form or by immobilizing them in certain plant parts. The abnormalities are usually manifested in the form of specific symptoms.

Nitrogen. Nitrogen is vital for the overall growth of plants. It plays a major role in cell formation. Plants suffering from lack of nitrogen are weaker and stunted. The leaves become smaller and yellow. On the contrary, abundance of nitrogen results in more succulent vegetative growth and delayed maturity of plants and in some crops the yield is also reduced. Luxuriant, vegetative growth due to application of large quantities of nitrogen, increases the susceptibility of plants to several foliar diseases, while dearth of nitrogen results in poor growth and weaker plants, which may increase the susceptibility of many crop plants to wilt, collar rot and root rot diseases. Nitrogen deficiency can be corrected by the application of nitrogenous fertilizers, such as ammonium sulfate, urea, diammonium phosphate or ammonium chloride to the soil as top dressing. Foliar spraying with urea - 1% or ammonium sulfate - 2% gives quick response.

Phosphorus. Phosphorus is another important macronutrient required for plant growth and yield. Due to deficiency of phosphorus the color of the leaves turn bluish-green and eventually turns violet. Development of root system is badly affected, the plant growth becomes poor and the yield is reduced to a considerable extent. Phosphorus deficiency can be corrected by the application of phosphatic fertilizers, such as super phosphate or diammonium phosphate. In diammonium phosphate, the phosphorus is in a readily available form and can be absorbed by the plants quickly.

Potassium. Potassium is another macronutrient required by the plants and its deficiency leads to poor development of the stem region. When the deficiency is more, the leaves start drying from the tips. Older leaves show abnormal discoloration. The color of the leaves turns red from the margin inwards. Marginal reddening is very characteristic of potassium deficiency. In the central area of the leaf, reddish, rusty spots may develop. Eventually, the leaves scorch, dry and fall off prematurely. Potash deficiency has been found to predispose the plants to attack by various diseases. However, excess of potassium in the soil interferes with the uptake of phosphorus by the plants. Potash deficiency can be corrected by the application of potassic fertilizers, such as muriate of potash or potassium sulfate in appropriate quantities.

Iron. Iron is a micronutrient or a trace element, which is required in very small quantities by the plants. Although iron is present in sufficient quantities in most of the soils, due to some conditions, such as presence of excess amount of calcium or manganese, the availability of iron in a suitable form to the plants is reduced. Iron deficiency is characterized by chlorosis in several horticultural and agricultural crops, such as vine, pear, apple, cherry, orange, lemon, pineapple, sugarcane, rice, maize, cowpea, some vegetables and conifers. The non-availability of iron due to the presence of excessive amount of calcium in the soil is known as **'lime-induced chlorosis'** and that due to the presence of large amount of manganese in the soil as **'manganese-induced chlorosis'**. In calcareous soils, the available ferrous salts are converted into unavailable ferric salts due to alkalinity. In acid soils, when there is excess of soluble phosphate, the iron is made unavailable. Manganese interferes with the role of iron in the formation of chlorophyll. Iron is an important component required for the formation of chlorophyll and plays a vital role in photosynthesis. Due to iron deficiency, the leaves become yellowish-white without any mottling and the plants become very much stunted in growth. Acute deficiency leads to scorching of the leaves and eventual death of the plants. Soil application or foliar application of ferrous sulfate can correct the disorder. Foliar spraying with chelated form of ferrous sulfate - 1 % is very effective.

Boron. Boron deficiency affects mostly the storage organs of the plants, such as roots, tubers, fruits, nuts etc. Lack of boron causes hypertrophy, degeneration and disintegration of cambium cells in the meristematic tissues, leading to necrosis of the tissues. However, in some plants boron deficiency shows marked symptoms in the foliage also. 'Heart rot'and 'Dry rot' of beets, 'Corky core' of apple, 'Fruit drop' and 'Die-back' of terminal branches of apple trees, 'Cracking of fruits' and 'Die-back' in pears and cherries, 'Brown rot' of cauliflower, 'Vascular browning' of potato tubers, 'Top disease' of tobacco, 'Reduction of pod formation' in groundnut, 'Cracking and shedding of buttons', 'Formation of sterile nuts and spongy kernel' in coconut are mainly due to boron deficiency. Various types of yellowing, reddening and bronzing of the foliage in several herbaceous plants may be caused due to boron deficiency. The yellowing occurs in-between the veins of leaves as diffused patches or streaks, while the main veins remain green. The leaf margins may die and curl up. In sunflower, the leaves show marked abnormalities and the apical growth is arrested. Generally, high soil pH or alkalinity reduces the availability of boron. Boron deficiency can be corrected by soil application of borax at 8 - 10 kg./ acre in the case of field crops. In the case of trees, such as coconuts, apples etc. borax may be applied at 100 - 200 gm./ tree/ year, in two split doses.

Manganese. Manganese deficiency is widespread and is known to cause injuries to several cultivated crops, such as cereals, potatoes, sugarcane, sugarbeet etc. 'Gray speck disease' of oats, 'Pahala blight' of sugarcane, 'Marsh spot' of pea etc. are some of the diseases caused by manganese deficiency. Potatoes, tomatoes and onions are very sensitive to manganese deficiency. Fruit trees, such as oranges, lemons and peaches are also affected badly. Manganese deficiency results in chlorosis and stunting. The symptoms on the leaves appear as pale-green specks, which gradually enlarge in size and eventually turn buff or light- brown in color. In calcareous soils, the manganese present in the soil is converted into an insoluble form and is rendered unavailable to plants. Some of the soil bacteria are also capable of precipitating soluble manganese compounds into insoluble oxides. Manganese deficiency can be corrected by foliar spraying with manganese sulfate - 2 % solution.

Zinc. Zinc deficiency is quite common in many soils and affects several crops. Diseases, such as 'Rosette' or 'Little leaf' of apple, 'Mottle leaf' of citrus, 'Little leaf' of stone fruits, peach, apricot, plum and grapevine, 'Khaira disease' of rice etc. are caused by zinc deficiency. In the fruit tress, the characteristic symptoms are the development of prominent veins in young leaves, the color of leaves turning pale yellow, sometimes with dead patches, marked reduction in the size of leaves and crinkling of leaves. Bronzing of the leaf lamina may also occur. The internodes become shortened and the tree appears bushy with small, bunched, deformed leaves. In 'Khaira disease' of rice, the leaves become chlorotic at the base and numerous, small, brown or bronze, rusty spots develop on the leaf lamina. The spots may enlarge, coalesce and form bigger spots, covering almost the entire leaf area and eventually, the entire leaf gets bronzed. The root system is badly affected and most of the finer roots die. The plants become stunted in growth and weak. The yield is also considerably reduced. In open sandy soils, soluble zinc is readily leached, while in heavy clay soils and in soils rich in humus, zinc is fixed to the clay particles and is not available to the plants. In continuously water inundated soils also, zinc is not readily available to the plants. Too much of phosphates in the soil, increase the fixation of zinc. The deficiency can be corrected by foliar spraying with zinc sulfate - 1.0 kg. + slaked lime - 500 gm. in 200 lit. of water/ acre.

Copper. Copper is another element required for the formation chlorophyll in plants. Copper deficiency causes several disorders, such as yellowing of leaves near the tip region, die-back of branches and reduction in yield in cereals, crucifers, legumes and many fruit trees. In cereals, the leaves become pale- yellow, while the margins become deep-yellow in color and sometimes marginal drying occurs. Grain formation is very much reduced. Among the fruit trees, copper deficiency symptoms are more pronounced in citrus plants. Yellowing of leaves, die-back of branches, development of cracks in the bark of the stem region and exudation of a gummy liquid through the cracks are the common symptoms observed. Fruit production is very much reduced. In the yellow areas of the leaves, the chloroplasts are disintegrated. The disease is more in deep, acid, sandy, sandy loam and in poor soils. The deficiency can be corrected by foliar spraying with copper sulfate at very low concentrations. Higher concentrations of copper sulfate may produce phytotoxicity in many plants.

Magnesium. Magnesium is a constituent of chlorophyll and plays a major role in the photosynthetic activity of the plants, besides regulating the functioning of potassium in the tissues. Magnesium deficiency has been found to affect several crops, including cereals, tobacco, cotton, sugarcane, soybean, many fruit and forest trees and ornamentals. In cereals, the symptoms appear as a distinctive yellowish-green color of the foliage, with longitudinal, pale-green to white mottling between the veins of the leaves and the striations resemble the stripes on the skin of a tiger. The leaf tips are often reddened and in-rolled. Development of root hairs is very poor. The disorder is generally known as **'sand drown'**. In tobacco, chlorosis usually begins at the tip and margins of older leaves and spreads inwards and downwards. Later, the young leaves also show such symptoms. The interveinal spaces are whitish, while the main veins and adjacent areas remain green. In potatoes, the lower leaves become chlorotic and later turn brown and becomes thick and hard. In cotton, chlorosis of the leaves is followed by reddening of the leaves and is known as **'red leaf disease'**. In general magnesium deficiency causes reduced rate of growth, chlorosis of the older leaves, yellowing, followed by necrosis of the interveinal areas and puckering of leaves. Poor root growth and poor coloration of the flowers also occur. The disease occurs in soils deficient in magnesium and in acid soils. In sandy soils, magnesium is often leached out and becomes unavailable. The deficiency can be corrected by soil application of magnesium sulfate at 5.0 kg./acre or by foliar spraying with magnesium sulfate - 2 % solution. Application of organic manure or green manure helps to mitigate the disorder.

Adverse environmental conditions

High temperature. In general, most of the cultivated crops grow and yield better at temperatures ranging from 30° - 35°C. High temperatures beyond this limit is not tolerated by many plants. Due to high temperature, flowers, immature and matured fruits may become discolored and may fall off prematurely. **'Sun scald'** of apples and vegetables occur, when the temperature rises beyond the tolerance limit.

Frost. During periods of continuous frost, tips of branches, flowers and fruits of many plants may dry and die. Spots and cracks may develop on mature fruits. Frost may cause severe injury to potatoes and several winter crops. Severe frost may result in freezing of the sap in succulent plant cells, as a result the entire plant may die.

Smoke. Exposure to continuous smoke may cause certain disease-like symptoms on fruits. **'Black tip'** of mangoes is a typical example of smoke injury. The fruits of mango trees near brick kilns develop black, necrotic spot at the distal end of the fruit. The spot enlarges and the inner tissues rot and the fruits become unfit for consumption. The necrosis is said to be due to poisonous gases, such as sulfur dioxide present in the smoke. Further, the soot and other heavier carbon particles carried along with the smoke get deposited on the green surface of the plants and affect the photosynthetic activity and thus, the growth of the plants is also adversely affected.

Table 7 : Differences between diseases caused by abiotic / inanimate factors and viruses

Diseases due to abiotic / inanimate factors	Virus diseases
1. The symptoms caused by abiotic factors are usually termed as disorders.	The symptoms caused by biotic agents are referred to as diseases.
2. Disorders, which are sometimes called as diseases are caused by abiotic or inanimate factors, such as deficiency or toxicity due to lack of or excess of one or more micro- or macro-nutrients respectively; adverse soil conditions, such as soil salinity, alkalinity or acidity; or due to adverse environmental conditions, such as flooding, high temperature, frost, smoke, toxic gases etc.	Biotic / mesobiotic agents, such as viruses and mycoplasma-like organisms cause these diseases.
3. Besides affecting the crops in the fields, most of these diseases may continue to affect the plant produce also during storage, transit and marketing.	These diseases affect the crop only during the growth stages, as these agents causing the diseases can survive only in the living tissues of the plants (obligate parasites).
4. No pathogens are involved in causing such diseases. However, the affected plants may be more vulnerable to attack by several pathogenic organisms.	Specific viruses or mycoplasma-like organisms cause these diseases.
5. These diseases cannot be transmitted from diseased to healthy plants.	These diseases can be transmitted from diseased to healthy plants mechanically by sap, by grafting or by insect or non-insect vectors.
6. Diagnosis of these diseases are much more complicated and rather difficult, as more than one factor may be responsible for the disorder.	Diagnosis of these diseases is possible by serological tests, by Electron microscopy, or by vector transmission studies under controlled conditions in glass-houses.
7. These diseases affect almost all the plants in the field simultaneously	Usually these diseases occur in patches around several foci of infection in the field.

Contd...

Table 7. Contd...

8. These diseases can be corrected or cured by manipulating the nutrient status of the soils or by direct treatment of the plants or by adjusting or modifying the environmental conditions.	Affected plants cannot be cured.
9. These diseases cannot be controlled by the use of plant protection chemicals. Soil amelioration and protection of plants from exposure to extremes of environmental conditions are effective in controlling the diseases.	Occurrence and spread of these diseases can be controlled or checked by (i) using virus-free seed materials (ii) destruction of diseased plants (iii) controlling the vectors and by (iv) eradicating alternate and collateral hosts, which may harbor the viruses as well as the vectors.

Poisonous gases. Plants near about factories, industrial concerns etc. are exposed to a number of toxic gases, such as sulfur dioxide, hydrogen sulfide, chlorine, fluorine, ethylene, methane etc. emitted along with the smoke. These poisonous gases polluting the atmosphere, when taken by the plants during the process of respiration, interfere with the physiological functioning of the plants and thus, their growth is affected adversely.

Water inundation. Due to continuous inundation of water, there is lack of aeration in the soil. This affects the normal functioning of the roots and they rot. Further, in such soils poisonous gases, such as methane, carbon monoxide and hydrogen sulfide are produced in large quantities. These poisonous gases affect the growth of roots and their functioning, leading to rotting and death of the roots. This is followed by drying and rotting of the leaves and ultimately the entire plant may die. Further, continuous flooding makes the micronutrients present in the soil unavailable to the plants and the plants develop deficiency symptoms.

Lack of water. When there is lack of water in the soil, the plants are not able to get enough water and nutrients from the soil for their survival and may die of hunger.

Fig. 70: Powdery mildew of mulberry
Phyllactinia guttata

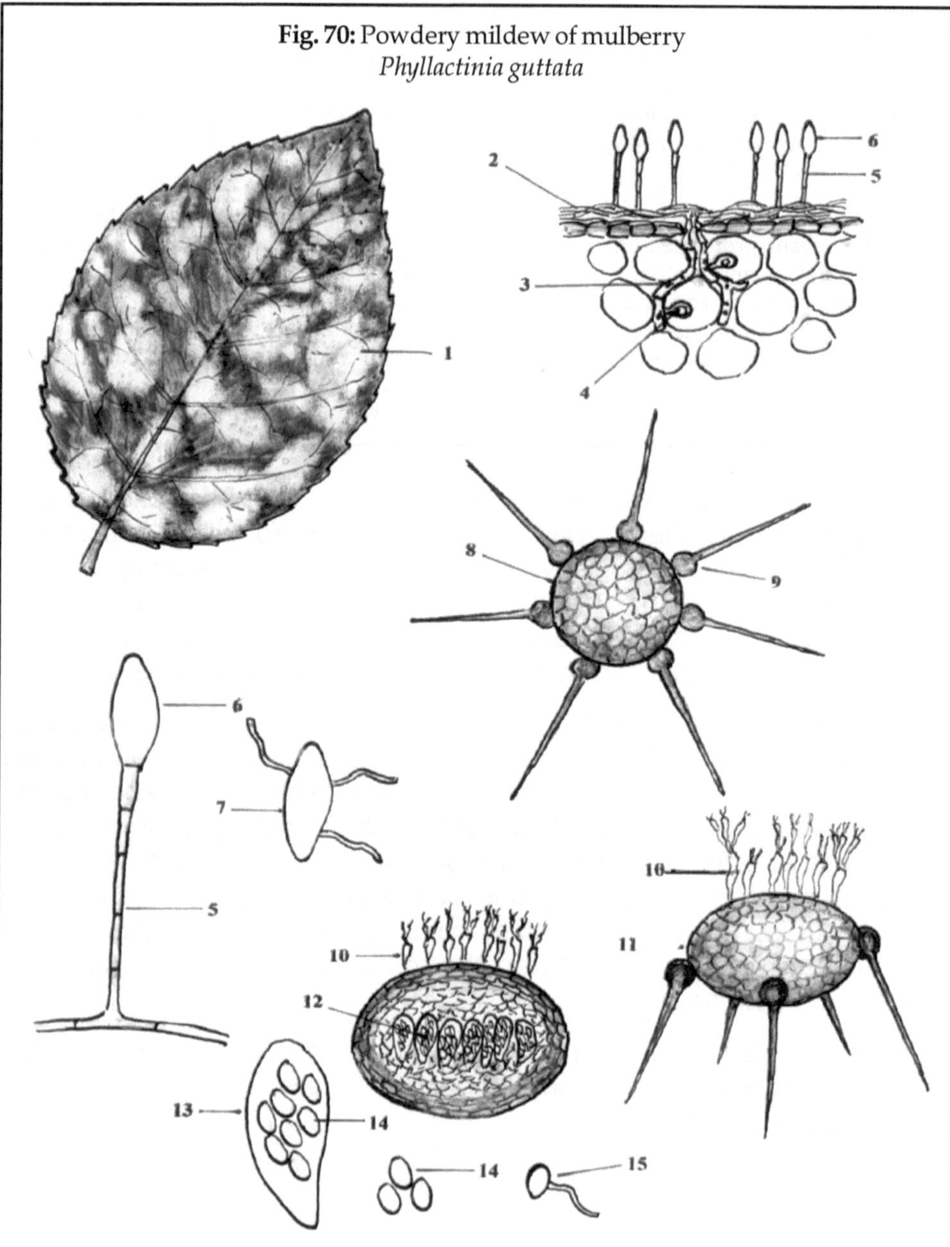

1. White powdery growth on the leaf 2. Super ficial mycclium 3. Intercellular mycelium 4. Haustorium 5. Conidiophore 6. Conidium 7. Germinating conidium 8. Cleistothecium 9. Appendages with bulbous base and pointed tip 10. Crown of apical short appendages 11. Cleistothecium lifted by the appendages 12. Asci and assospores inside the cleistothecium 13. Ascus with ascospores 14. Ascospores 15. Germinating ascospore.

Chapter 2

Diseases of Horticultural Crops

FRUIT CROPS

Banana (*Musa species*)

1. Panama disease of banana

Fusarium oxysporum f.sp. *cubense*

'Panama disease', **'*Fusarium* wilt'** or **'vascular wilt'** of banana is one of the most serious and destructive diseases of the crop. The disease was first reported from Australia in 1874. Now, it is widely distributed all over the world, including America, Africa, Sri Lanka, Burma, Thailand, Malaysia, Indonesia, Hawai, Fiji, the Philippines, India, Australia and New Zealand. In India, it is prevalent in the states of Assam, Bihar, Karnataka, Kerala, Tamil Nadu, Telungu Desam, Maharashtra and West Bengal.

Symptoms. Usually the symptoms are manifested, when the banana plants are about 5 months old. However, under highly favorable conditions for the pathogen, even 2 - 3 months old plants show wilt symptoms. Initial visible symptoms appear as pale yellow streaks, lengthwise in the petioles of the oldest, lowermost leaves. Two types of symptoms follow this. In the first type, the old leaves turn yellow progressively and finally they break from the petiole and hang down. In the second type, the leaves break from the petiole and hang down, without becoming chlorotic. Gradually, all the leaves collapse and hang down, except the crown, which alone stands upright. The newly formed leaves show yellow blotches and wrinkling of the lamina. Often, the leaf sheaths covering the pseudostem show longitudinal splitting above the ground level. But, sometimes this symptom may not appear. The affected plants may die completely in about 4-6 weeks after the appearance of yellow streaks on the petioles.

The vascular strands show discoloration varying from yellow to dark-brown, which is the characteristic internal symptom of the disease. The discoloration usually appears first in the outer leaf sheaths and extends to the vascular strands of the pseudostem. Vascular discoloration is more pronounced in the corms, but is not common in the roots. When the rhizome of an infected plant is cut transversely, the discoloration of the vascular bundles is distinctly seen as numerous, pin-point like dots all over the cut surface. The roots of affected plants become black and rotten. The suckers growing from the diseased corms are systemically affected and they wilt and die very soon **(Fig.71)**.

The causal organism. The mycelium of the fungus is hyaline, septate, much branched and mostly intracellular. Few hyphae may also be present in the intercellular spaces. The hyphae are largely confined to the vascular bundles, often filling the cavity of the vessels. The fungus produces three types of spores viz., macroconidia, microconidia and chlamydospores. The sporodochia bearing conidiophores and conidia emerge through the stomata on the petioles and leaves. The conidiophores are vertically branched. Macroconidia are borne at the apical ends of the main and lateral branches and are formed in succession one after another. They are thin-walled, hyaline, pedicellate, sickle-shaped, with both ends somewhat pointed, 2 - 5 septate, mostly 3-septate and measure 22.0 - 36.0 x 4.0 - 5.0μ in size. Microconidia are produced from the branches of conidiophores or from the tips of hyphae in very large numbers. They are single-celled or two-celled, thin-walled, hyaline, ovate or elongated and measure 5.0 - 7.0 x 2.5 - 3.0μ in size. Terminal or intercalary chlamydospores are formed from the hyphae, as well as from the conidial cells. They are oval or spherical, thick-walled and usually occur in pairs. They measure 7.0 - 13.0 x 7.0 - 8.0μ in size **(Fig.71)**.

The fungus is a facultative parasite. It can invade the host mainly through wounds in the corms or roots. After entry, the fungus develops extensively in the vascular tissue of the corm and then proceeds systemically to the leaf sheaths and pseudostem through the vascular system. The fungus colonizes the vascular bundles and produces masses of mycelium, which bear conidia and chlamydospores, filling the cavity of the vessels. The plugging of vascular elements by the fungus obstructs the translocation of water and nutrients to the aboveground parts of the plant leading to wilting and death of affected plants.

Mode of survival, spread and epidemiology. The fungus is soil-borne in nature. It can survive saprophytically in the diseased corms and other plant parts for prolonged periods and cause fresh infection. The chlamydospores can also remain in a viable state in the soil and diseased plant parts for a long time. The disease can be easily spread through spores present in the infected banana trash and soil, which may be carried in surface flooded water. Suckers used for planting from diseased areas to other places spread the disease easily. Contact of the roots of healthy plants with the roots of diseased plants carrying the spores also leads to fresh infection. The entry of the pathogen is mostly through wounds in the roots or corms, especially deep wounds up to the xylem vessels, caused incidentally during cultural operations or by soil-inhabiting pests, particularly nematodes.

Light textured, sandy loam soils, with acid reaction and low soil moisture of 25 % favor the occurrence and development of the disease. Ratooning of infected banana plants increases the disease incidence to a great extent and leads to continuous accumulation of soil inoculum.

Disease management

Agronomic practices (i) Diseased plants should be uprooted along with the corm and suckers and destroyed by burning. After uprooting the diseased plant, lime should be applied to the pit at 1.5 kg. and left as such for a few weeks. After that, healthy sucker may be planted in the pit (ii) Diseased and disintegrated plant debris found in the soil should be periodically removed and destroyed (iii) Healthy suckers obtained from disease-free fields should be used for planting (iv) Ratooning of diseased crop should be avoided (v) Flooding the fields followed by fallowing has been found to be very effective in reclaiming infected, sick soils. Whenever water is available in plenty, the field should be flooded and kept under inundated condition for a number of days and afterwards the field is left fallow for about six months prior to cultivation of banana in the field. Under inundated condition, toxic substances, such as **'acetic acid'** and other toxic substances are produced in the soil, which can destroy the fungus. Further, such conditions lower the oxygen availability in the soil, which also contributes to the destruction of the fungus (vi) Rotation with rice or sugarcane helps to eliminate the soil inoculum to a great extent (vii) Soil amendment with organic matter is known to encourage the growth of certain actinomycetes and soil bacteria capable of producing antibiotic substances, such as **'Musarin'** and **'Monamycin'**, as well as some **'lytic substances'**. These substances have antagonistic effect on the fungus (viii) During cultural operations care should be taken to avoid causing injuries to the roots and corms (ix) Soil-inhabiting nematodes, such as *Radopholus similis* burrow the roots and corms and thus predispose the plants to the fungal infection. Suitable nematicides should be applied to eliminate them (x) Knives, used for cutting off unwanted suckers, old leaves etc. should be disinfected by dipping them frequently in formaldehyde - 5 % or phenol - 5 % solution.

Chemical control (i) Carbendazim capsule application in the corm is found to be very effective in controlling the disease. Empty gelatin capsules are filled with carbendazim at 50 - 75 mg./capsule and covered with the lid. In plants showing symptoms of the disease, a hole is made with a big nail from the base of the plant into the corm diagonally at an angle of 45° to a depth of about 10 cm. One fungicide-filled capsule is introduced into the hole and the hole is sealed with clay mixed with copper oxychloride fungicide. Depending upon the disease intensity, one or two more applications may be given at monthly intervals. In case gelatin capsules are not available, the fungicidal solution may be introduced into the hole directly. For this carbendazim - 20 gm. is dissolved in 1.0 liter of water (2 %) and 3.0 ml. of this solution is introduced into the hole and the hole sealed (ii) To prevent fresh fungal infection and nematode injury, chemical treatment may be given at the time of planting. For this, carbendazim - 1.0 gm. is dissolved in 1.0 liter of water. A small quantity of clay is added to this solution to make it into a slurry. The corm is dipped into the slurry and then 40 gm. of carbofuran-3G is sprinkled over the corm. The granules stick on to the slurry. The suckers are then planted. Sucker treatment, followed by one or two bimonthly drenching with carbendazim - 0.1 %, commencing from the sixth month after planting is found to be very effective.

Resistant varieties. Cavendish, Rajavazhai, Moongil, Poovan and Vamankeli are resistant, while Gros Michel, Monthan, Karpooravalli and Rasthali are susceptible.

2. 'Sigatoka' disease of banana

Cercospora musae

(Mycosphaerella musicola)

The disease was first reported from Java in 1902. Now, it occurs in almost all banana growing countries of the world, including Columbia, Mexico,

Venezuela, Honduras, Tanzania, Uganda, Costa Rica, Guatemala, Haiti, Jamaica, Panama and India. In India, the disease occurs commonly in Assam, Bihar, Gujarat, Karnataka, Kerala, Telungu Desam, Maharashtra, West Bengal and Tamil Nadu.

Symptoms. The disease mostly attacks the older leaves. Initially, the symptoms appear as small, light yellow to light-brown streaks, about 5.0 - 10.0 mm. in length, in-between and parallel to the side veins of leaves. A few days later, the spots enlarge into long, spear-shaped spots, 1 - 2 cm. in length. The spots become brown in color, with a grayish, sunken center and are surrounded by a yellow halo. Adjacent spots coalesce to form large, irregular, necrotic patches and the leaves start drying from the leaf margin inwards. In case of severe infection, affected leaves dry and die within a few weeks. Destruction of most of the older leaves by the disease leaves only a few younger functioning leaves. Immature fruit bunches on such plants fail to develop and mature normally. If heavy infection occurs when the fruits are nearing maturity, they become undersized and angular in shape and their flesh develops a buff color **(Fig.72).**

The causal organism. 'Banana leaf spot' or **'Sigatoka disease'**, is caused by *Mycosphaerella musicola,* which has its conidial state in *Cercospora musae.* The pathogen produces spermatia (pycnospores) in spermagonia (pycnia), ascospores in perithecia and in the conidial stage *(Cercospora musae)* produces conidia in sporodochia. Pycnidia may be formed on both the surfaces of the leaf. They are sub-epidermal, dark, globose and 75 - 100µ in diameter. Pycnospores are single-celled, oval, hyaline and thin-walled. In the conidial state, successive crops of abundant conidia are produced on both the surfaces of leaves from sporodochia during the brown spot stage of the disease. Conidia are elongated, narrow, olive brown in color, multiseptate and measure 20.0 - 80.0 x 2.0 - 6.0µ in size. They dislodge easily and are disseminated by wind and rainsplash. Perithecia are produced during warm and humid weather. They are dark-brown to black in color, erumpent, ostiolate and measure 46.8 x 70.0µ in diameter. Asci are oblong, clavate and measure 28.8 - 36.8 x 8.0 - 10.0µ in size. Ascospores are shot out violently, when the perithecia are wetted. Ascospores are single-septate, hyaline, ellipsoid, the upper cell slightly broader and measure 14.4 - 19.0 x 3.0 - 4.0µ in size. Ascospores are spread by air currents and are responsible for long distance dissemination of the disease, while the conidia are the most important source of local spread. Infection occurs by, either conidia or ascospores **(Fig.72).**

Fig. 71: 'Panama disease' or *Fusarium wilt of banana-Fusarium oxysporum* f.sp. *cubense*

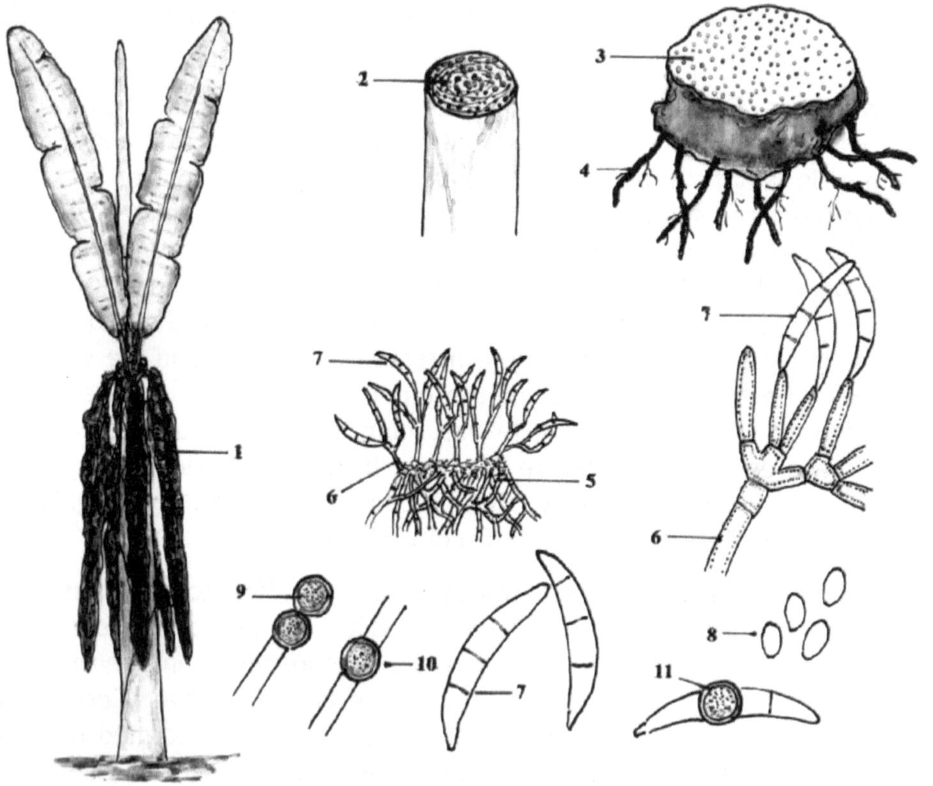

1. Infected plant showing collapsed peripheral leaves 2. Cut stem and 3. Cut corm showing vascular discoloration as dots 4. Blackened and rotten roots 5. Sporodochium 6. Conidiophore 7. Macroconidium 8. Microconidium 9. Terminal and 10. Intercalary chlamydospore 11. Chlamydospore from conidial cell.

Fig. 72: 'Sigatoka disease' of banana *Mycosphaerella musicola* (= *Cercospora musae*)

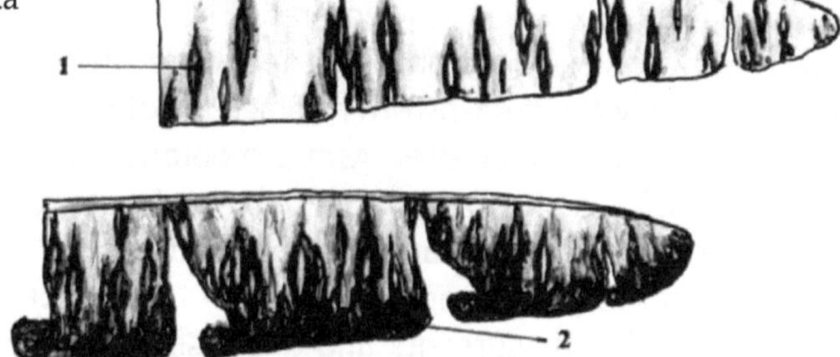

1. Leafspots on leaf in the early stages 2. Spots coalesce forming larger patches and drying of leaf from the margin toward.

Mode of survival, spread and epidemiology. The ascospores produced in perithecia formed on infected, old, dried leaves, which are left in the field, are the primary source of inoculum. Secondary spread occurs through conidia carried by wind and rainsplash. Banana is usually cultivated all through the year in some of the regions, as such the spores are always present in the air and may cause fresh infection.

Warm temperatures between 23° - 25°C and rainy, moist weather favor occurrence and development of the disease. The disease is more common in poor, ill-drained soils and in badly maintained gardens.

Disease management

Agronomic practices (i) Severely affected, old and dried leaves should be cut off and destroyed by burning (ii) Dried fragments of infected leaves and other plant debris fallen on the ground should be collected periodically and burnt.

Chemical control (i) Spraying with Bordeaux mixture - 1 % or copper oxychloride - 2.5 gm. + linseed oil - 0.2 % is recommended for the control of the disease (ii) Spraying with carbendazim - 1.0 gm. or thiophanate methyl (Topsin M) - 1.0 gm. or propiconazole (Tilt) - 1.0 gm. or chlorothalonil - 1.5 gm./ lit. of water has also been found to be effective. Adding wetting and spreading agents, such as Sandovit or Teepol or Triton at 1.0 ml./ lit. of spray fluid (0.1 %) prevents rolling down of the fungicidal solution from the leaf surface. Sprayings should be given at monthly intervals to keep the disease in check.

3. Anthracnose or fruit rot of banana

Gloeosporium musarum (Colletotrichum musae)

The disease occurs in most of the states of India, where banana is grown extensively, especially in Bihar, Karnataka, Kerala and Tamil Nadu. Most of the cultivated varieties of banana are susceptible to the disease to a larger or lesser extent.

Symptoms. Infection starts from the time of flowering and continues till the fruits mature. Initially, the flowers, skin and distal end of young fruits are attacked. The conidium germinates by producing a germ tube that forms an appressorium at its tip. The appressorium fastens the conidium to the host surface and an infection hypha is produced. The infection hypha penetrates the cuticle, develops into a mycelium and causes primary infection, leading to development of small, black, circular specks. As a result of numerous infection points, the distal end of small fingers becomes black first. This is followed by blackening of the entire fruit and the fruit stalk. Finally, the whole bunch is affected and all the fruits and fruit stalks become black. The fruits start ripening prematurely, shrink and dry. Under moist conditions, the fruits and sometimes the fruit stalks and the main stalk are covered with pinkish masses of spores. When conditions are not favorable, the infection hypha, which penetrates the cuticle develops a mycelium and remains dormant subcutaneously for about 5 months, till the fruits mature and cause latent infection. Under favorable conditions, the fungus becomes active and produces typical lesions on the fruit peel of matured fruits. Fresh infection may occur during or after harvest of the bunches and produce peel lesions without undergoing a dormant phase. The lesions appear as dark-brown to black, circular or irregularly circular, slightly sunken spots and the fruits get blemished. As the fruits ripen, the spots enlarge, coalesce and form larger spots on the peel. On these spots, minute, pinpoint-like dots are formed, which are the fructifications (acervuli) of the fungus. Soon, the surface of the spots is covered with pinkish, slimy masses of spores. In case of severe infection, the fruits are completely covered with dark blemishes and the flesh of the fruits softens and rots. The market value of the fruits is very much reduced and leads to significant loss.

The causal organism. The mycelium of the fungus is slender, hyaline, sparsely septate, inter- and intracellular. On the infected parts of the young and matured fruits, on the stalks of fingers and on the main stalk of the bunch, the fungus produces large number of acervuli, which appear as dark-brown to black pinpoint-like dots. The acervuli are sub-epidermal asexual fruiting bodies consisting of masses of entangled hyphae forming stromata. From the stroma, numerous, closely packed, erect, hyaline, unbranched, septate conidiophores arise. They taper towards the apex and are about 30µ in length. Conidia are produced at the tips of conidiophores in succession. They are released in slimy masses, after the epidermis is ruptured. The conidia, which appear pinkish in mass, are hyaline, single-celled, elliptical or cylindrical, with flattened or roundish ends, guttulate and are 11.0 - 17.0 x 3.0 - 6.0µ in size. Very rarely, setae are formed in the acervuli.

Mode of spread and epidemiology. The disease is mostly spread by air-borne conidia. They may also be disseminated by rainsplash. Numerous insects, that visit banana flowers for collecting nectar also spread the disease mechanically.

High temperatures of 30° - 35°C and almost saturated atmospheric humidity favor the occurrence and development of the disease. The disease is more severe during the rainy seasons. Wounds and bruises on the fruits, predispose them to fresh infection. Sweet varieties, such as Lacatan, Bongolan and Cavendish are more susceptible, while Gros Michel is less susceptible.

Disease management

Agronomic practices (i) Severely affected bunches, especially young ones should be removed and destroyed by burning to avoid build up of inoculum (ii) Infected plant materials and trash should be collected from the fields and destroyed (iii) Soon after all the hands have opened, the flower should be cut off to prevent insects visiting the flower (iv) Care should be taken to avoid causing injuries or bruises to the fruits at the time of harvesting, transporting and during storage (v) Cold storage of fruits at temperatures between 7°-10°C prevents disease occurrence.

Chemical control (i) Pre-harvest spraying with Bordeaux mixture - 1% or copper oxychloride - 2.5 gm. or carbendazim - 1.0 gm. or chlorothalonil - 1.5 gm./ lit. of water, 4 times, at fortnightly intervals, commencing from the time of flowering is very effective. Care should be taken to cover the bunches thoroughly while spraying (ii) Post-harvest dipping of fruits in Mycostatin - 450 ppm. or Aureofungin Sol - 100 ppm. or carbendazim - 400 ppm. or benomyl - 400 ppm. protects the fruits from fruit rot infection.

4. Bacterial wilt of banana

Pseudomonas solanacearum

'Bacterial wilt' or **'Moko disease'** of banana was first reported from Guyana in 1840. Now, it is prevalent in the tropics and in the warmer climates throughout the world and causes severe losses. In India, it occurs in West Bengal, Tamil Nadu and Kerala. In Tamil Nadu, the disease is found in most of the districts, where banana is cultivated, especially in Tirunelveli and North Arcot. Varieties, such as Robusta and Poovan are highly susceptible to this disease.

Symptoms. In infected young plants, the middle leaves break from the base, while still green and the plants wilt rapidly and die. In older plants, first the inner leaves turn yellow near the petiole. Then, the petioles break, the leaves hang down, wilt and die. In the meantime, the surrounding outer leaves also start breaking from the petiole, until all the leaves break and hang down around the pseudostem within a week's time and dry. The unfurled heartleaf alone may stand erect for sometime and eventually, that too breaks and topples over. The inner leaves of sword suckers from the infected rhizome, become yellow and necrotic. When such suckers are topped, the new leaves emerging from the suckers are shortened, turn black and become twisted and wrinkled. If diseased suckers are planted, the emerging leaves become necrotic and the plants die soon. When the pseudostem of an infected plant is cut transversely, the vascular strands appear yellow or dark-brown in color and grayish-brown bacterial ooze comes out of the vascular bundles as droplets. Such discoloration is more evident in the inner leaf sheaths and in the fruit stalks. The flower buds and fruit stalks turn brown and are shrivelled. From the fruit stalk and at the point of attachment of the fruits to the fruit stalk, bacterial ooze may come out. Pockets of bacteria and decayed plant tissue may be present in the pseudostem, in the rhizome and in individual banana fruit as a dark, gummy substance. The vascular discoloration is more pronounced along the central region in 'Moko wilt', while in *Fusarium* wilt, it tends to be more in the peripheral region. In the variety Gros Michel the symptoms are more pronounced in the fruit stalk and fruits. The fruits turn yellow to black before they mature and dry.

The causal organism. 'Bacterial wilt' of banana is caused by a specific race of *Pseudomonas solanacearum*. The bacterium causing the disease is a short rod, 1.5-0.5μ in size and motile by a single polar flagellum. The bacteria are gram negative, non-spore forming, not capsulated and are aerobic in nature.

Mode of survival, spread and epidemiology. The pathogen is primarily soil-borne. Diseased plant parts, such as rhizomes, roots, leaves etc. found in the soil may harbor the bacteria for almost indefinite periods. The disease spreads through infected suckers used for planting and is carried over long distance by this method. The sap containing the bacteria, which sticks on to the knives used for cutting the shoots of diseased suckers may spread the disease to healthy suckers. The bacteria may also be spread through irrigation water and mechanically by some insects. The bacteria enter the plants through injuries or wounds caused during cultural operations or by soil inhabiting insects and nematodes.

High temperatures ranging from 35°-37°C and high soil moisture favor the disease occurrence and development. The optimum temperature range for the growth and multiplication of the bacteria is 35°-37°C.

The bacteria multiply in the vascular vessels enormously, disintegrate the cell walls and destroy the tissues, besides plugging the vascular bundles, thereby obstructing the movement of water and nutrients to the aboveground parts of the plants. This leads to wilting and death of infected plants within a very short time. The bacteria also produce certain toxic substances, such as **'ethylene'**, which also cause yellowing and wilting of leaves.

Disease management

Agronomic practices (i) Diseased plants should be uprooted completely along with the corms and suckers and destroyed by burning (ii) Adequate drainage facilities should be provided and flowing of water from the diseased field to other fields should be avoided (iii) Suckers obtained from diseased fields should not be used for planting (iv) The bacteria can be easily destroyed by exposure to high temperatures. So, the fields should be ploughed deeply during the summer months, so as to bring the bacteria to the soil surface and get killed by direct exposure to the sun (v) After the last fruit bunch is formed, the flower should be cut and removed. This prevents insects that are attracted to the nectar in the flowers, thus preventing mechanical transmission of the bacteria to other plants (vi) Ratooning of affected crop should be avoided (vii) The knives used for topping suckers, cutting old leaves and harvesting the bunches should be disinfected by dipping them in 5 % lysol or 5 % formalin frequently.

Resistant varieties. Varieties, such as Giant Cavendish, Lacatan, Velery and Poyo are resistant to this disease to some extent.

5. Bunchy top of banana

Musa virus - 1

'Bunchy top', **'curly top'** or **'cabbage top'** disease was first reported from Fiji in 1889. It is widespread in many countries of the world, such as Australia, Borneo, Egypt, Fiji, Pacific Islands, Sri Lanka, Taiwan, the Philippines etc. In India, the disease was first reported from Kerala in 1940 and is now prevalent in the states of Bihar, West Bengal, Orissa, Maharashtra, Telungu Desam, Karnataka, Kerala and Tamil Nadu. In Tamil Nadu, the disease occurs commonly in the districts of Kanyakumari, Palani, Kodaikanal and a few of the most popular varieties of bananas became highly susceptible to the disease and their cultivation had to be restricted. The disease might have been introduced to India from Sri Lanka.

Symptoms. The symptoms may become evident in any growth stage of the plants, from very young plants to the fruit bearing ones. The disease is characterized by the bunching of leaves at the crown, forming a rosette. Younger plants, when infected become stunted in growth, their leaves narrower and shorter than the normal leaves, rigid, more erect and bunched together at the apex. The crown leaf starts unfurling before emerging out completely and gets torn. Appearance of green streaks on the under surface of secondary veins of leaf, on the mid rib and petiole, rolling of the leaf margin upwards and wrinkling of the lamina are other symptoms. The bunches become choked and may come out by splitting open the pseudostem. The fruits are much reduced in size and are not marketable.

The symptoms of systemic infection from diseased suckers are more typical. Diseased suckers produce short, narrow, erect chlorotic leaves with mosaic-like markings and they roll upward. The leaves are hard and brittle and are easily breakable. Many, dark-green spots or patches are seen on the leaves. The plants do not grow taller than 2 - 3 feet and they do not produce bunches, but the infected plants do not die soon.

The causal organism. The disease is caused by *Banana bunchy top virus (BBTV)* or *Banana virus* - 1 or *Musa virus* - 1. The virus consists of a geminate, single, double stranded isometric particle of size 18 - 20 nm. in diameter.

Mode of spread and epidemiology. All suckers produced by a diseased plant carry the virus and the disease is spread over long distances through such suckers. The virus is not sap-transmissible. The banana aphid vector, *Pentalonia nigronervosa*, transmits it. The aphids cause secondary spread to standing crops. They also transmit the virus causing **'marble mosaic disease'** in cardamom. The acquisition feeding time is about 17 hours. After acquisition, an incubation period of 1½ - 45 hours inside the body of the insects is required before they become infective. Afterwards, they remain viruliferous for a period of 13 days. The symptoms appear 35 - 45 days after inoculation of the virus. The aphids are usually found around the basal part of the plant and also on the apex of the crown around the heart leaf and at the base of petioles. Although a single aphid is capable of transmitting the virus, for maximum transmission, an optimum number of about 20 aphids are required. Once infected, the virus invades the phloem vessels mostly and moves through the vessels to all parts of the plant and to the suckers. Most of the cultivated varieties of banana are susceptible to the disease however, Viruppakshi, Hari Sal and Lab Velchi are highly susceptible. *Colocasia esculenta* has been found to be a symptomless carrier of the virus. The virus also infects *Musa cavendishii* and a few other *Musa* species.

In cool and wet season, which is conducive for the rapid multiplication of the vector, the occurrence of the disease is also more.

Disease management

Agronomic practices (i) Suckers for planting should be obtained from disease-free areas (ii) Plants showing symptoms of the disease should be uprooted completely along with the suckers and destroyed by burning periodically.

Chemical control (i) Before planting 40 gm. of carbofuran 3G is applied to each pit and mixed with the soil (ii) To eliminate the aphid vector, spraying with methyl demeton - 2.0 ml. or phosphamidon - 1.0 ml. or monocrotophos - 1.0 ml./ lit. of water may be taken up, so as to cover the crown and the pseudostem thoroughly. Three sprayings may be given at intervals of 21 days (iii) Stem injection at the base of the pseudostem with monocrotophos at 1.0 ml./ 4.0 ml. of water is also very effective in controlling the aphids.

This may be done at 45 days interval, from the third month of planting, till flowering. Banana injector, devised by the Tamil Nadu Agricultural University may be used for this purpose.

Quarantine measures. Movement of bunchy top infected planting materials from place to place or from one state to another should be prohibited by law enforcement.

Diseases of minor importance. Some other diseases are also known to attack banana plants. *Verticillium theobromae* causes 'Cigar-end rot' that affects the fruits at all stages of development; *Ceratocystis (Ophiostoma) paradoxa* causes 'Stem end rot' or 'Main stalk rot'; *Cucumber mosaic virus* causes 'Banana mosaic' or 'Virus sheath rot'; *Phyllostictina musarum* causes 'Freckle disease' that produces numerous, minute, dark-brown, erumpent spots on leaves; *Fusarium moniliforme* causes *Fusarium* fruit rot; *Xanthomonas musicola* causes bacterial leaf spot.

Mango *(Mangifera indica)*

1. Anthracnose of mango

Colletotrichum gloeosporioides

(Glomerella cingulata)

The disease is also known as **'blossom blight'**, **'leaf spot'**, **'twig blight'** or **'wither tip'**. It is a very common and widespread disease of mango and is prevalent in several mango growing countries, such as Brazil, France, the Philippines, Indonesia, Trinidad, Peru, the United States of America, Portugal, Hawaii and India. In India, it occurs in the states of Punjab, Uttar Pradesh, Bihar, Kerala, Maharashtra, Telungu Desam and Tamil Nadu. The varieties, Neelam and Bangalora are highly susceptible to this disease.

Symptoms. The disease attacks the leaves, tender shoots and the floral parts. On the leaves, numerous, brown to dark-brown, oval or irregular spots appear. Under humid conditions, the spots enlarge and form irregular, necrotic areas. Young leaves are more prone to attack by the pathogen. The affected leaves become crinkled. Severely affected leaves fall off, leaving the twigs bare. Wither tip or die-back symptoms appear at the tip of very young branches. Black, necrotic lesions appear at the tip of the branches, which dry from the tip downwards, accompanied by defoliation of the branches. During the flowering season, small, dark spots appear on the main stalk and lateral branches of the inflorescence. Individual flower stalks are also infected and minute, black spots are formed on the flowers, which dry and are shed prematurely. The inflorescence affected by blossom blight does not form any fruits. The disease attacks both young and matured fruits. Infection starts from the blossoming period, until the fruits are half-grown. The fungus enters the fruits through the pores, when they are still green and develops in the flesh during ripening. In case of early infection of the fruits, dark lesions develop on the fruits and they shrivel, turn black and drop off. In the matured fruits, dark-brown to black, round or irregular, slightly sunken spots are formed and the fruits get blemished. As the fruits ripen, the spots enlarge, leading to softening and rotting of the fruits. On the lesions and dead parts of the plant, minute, brown to black, pinhead-like dots, which are the acervuli of the fungus, are formed in large numbers **(Fig.73).**

The causal organism. The mycelium of the fungus is slender, sparsely septate, hyaline at first becoming slightly darker with age, inter- and intracellular. The fruiting bodies or acervuli are sub-epidermal, consisting of masses of entangled hyphae forming stromata. From the stroma, numerous, closely packed, erect, hyaline conidiophores arise. Conidia are produced at the tips of the conidiophores and are embedded in a viscous fluid that swells under moist conditions, rupturing the host epidermis and exposing the conidial mass. The conidia are held together in the mucilaginous secretion and form pink, slimy masses on the surface of the lesions. The conidia, which appear pinkish in mass, are hyaline individually, broadly oval to oblong, with rounded ends, single-celled and contain two prominent oil globules. They measure 12.0 - 16.0 x 4.0 - 6.0μ in size. The conidium germinates by means of a germ tube, which produces an appressorium. Setae are commonly found in the acervuli formed on the twigs and not on the fruits.

The perfect stage of the pathogen *(Glomerella cingulata)* is not commonly found. The perithecia, if produced are found on stromatic cushions in the infected twigs and dry leaves. They are sub-spherical and have prominent, ostiolar hairs. Asci are sub-clavate, slightly pedicellate, slightly curved and are 55 - 70μ long. The ascospores, which are eight in number in each of the ascus are hyaline, single-celled, 12.0 - 22.0 x 3.0 - 5.0μ in size and look very much like the conidia **(Fig.73).**

Mode of survival, spread and epidemiology. Diseased twigs, leaves and fruits fallen on the ground are a source of perennation and fresh infection. The fungus can also survive saprophytically on dead twigs, leaves and fruits in the soil. The conidia produced from the blighted peduncles, which remain attached to the branches also cause fresh infection. Under tropical conditions, fresh crops of conidia are continuously produced throughout the year. Secondary spread occurs through conidia disseminated by wind, rain and mechanically by some insects.

High relative humidity from 95 - 97%, misty conditions and moderate temperatures of about 25°C are favorable for the occurrence and development of the disease.

Disease management

Agronomic practices (i) Diseased and fallen leaves, twigs, floral parts and fruits should be removed from the orchards and burnt (ii) Tree sanitation is a very important practice. Diseased and dried twigs should be pruned, after the harvest of the fruits and burnt (iii) Plant vigor should be maintained by application of recommended doses of fertilizers and timely irrigation during the summer months. Spraying with Bordeaux mixture - 0.75 % or copper oxychloride at 2.5 gm. per lit of water should follow after pruning.

Chemical control. Spraying with Bordeaux mixture - 0.75 % or zineb at 2.0 gm. or carbendazim at 1.0 gm. or thiophanate methyl at 1.0 gm. or chlorothalonil at 1.25 gm. per lit. of water gives good control of the disease. Two sprayings should be given at the time of blossoming. Subsequently, 2 - 3 sprayings should be given at fortnightly intervals till the time of harvest.

Post-harvest treatment (i) After harvest, the fruits should be treated in hot water at 50°C for 15 minutes or in a solution of thiobendazole (Mertect) at 1.0 gm./ lit. of water for 5 minutes (ii) Treating the fruits in ammonia or sulfur dioxide gas also gives protection against the disease.

2. Sooty mould of mango

Capnodium ramosum

Symptoms. The symptoms appear on the leaves, tender stem portions, young fruits etc. The mycelium, which is dark, covers the affected plant surfaces and forms a dense sooty, black, papery film, which sometimes peels off. In case of severe infection, the sooty film or crust may cover the entire surface of leaves, branches and immature fruits. The mycelial layer harms the plant indirectly by reducing the light reaching the plant surfaces, resulting in the reduction of photosynthetic activity **(Fig.75).**

Causal organism. The fungus responsible for this melady is an obligate saprophyte and ectophytic. It thrives on the honey dew secretions of certain sucking insects such as, leafhoppers, scale insects, aphids etc. The dark mycelium consists of tubular, septate hyphae, with a mucilaginous coating on the surface, which helps the mycelium to stick on to the surface of the infected parts. No haustoria are produced and the fungus does not penetrate the host tissue to absorb nourishment from the host.

Asexual reproduction takes place by more than one means. The fungus produces large number of conidia from hyphal tips. These conidia possess cross walls on both the axes. Besides this, the organism produces superficial pycnidia in large numbers. The pycnidia are small, globose, ostiolate and have a dark wall. Pycnidiospores escape from the pycnidium through the small ostiole in large numbers. Pycnidiospores are dark colored, globose or oblong in shape and single-celled. These spores are embedded in the honey dew exudate of the insects and are transmitted through rain, wind or insects, which are attracted to the sweet honey dew secretion. The spores germinate by producing germ tube.

As a result of sexual reproduction, the fungus produces superficial, perithecioidal pseudothecia. A mucilaginous substance also covers the pseudothecia. The asci, which are produced in a fascicle at the bottom of the pseudothecia, are ovoid and bitunicate. Each ascus contains eight oblong, hyaline ascospores having one or more septa **(Fig.75).**

Mode of spread and epidemiology. In seasons, when sucking insects, especially leafhoppers are found in large numbers, sooty mould also appears on a large scale. When mango trees are in the blooming stage, leafhoppers appear abundantly. This seems to be highly favorable for the appearance and rapid spread of sooty mould also. Besides mango, the fungus affects many other fruit plants, such as sapota, cashew etc. and many plantation crops.

Rain accompanied by strong winds and presence of large number of insects are conducive for the spread of the menace.

Disease management

Chemical control (i) Sucking insects, such as leafhoppers, scale insects, aphids etc. should be controlled by using suitable insecticides, especially during the flowering season (ii) In 10 liters of water, 25 gm. of copper oxychloride fungicide and 50 gm. of soluble starch powder are mixed and sprayed on the foliage. When the starch - fungicide mixture dries in the sun, it peels off and falls down along with the fungal mass.

3. Powdery mildew of mango

Oidium mangiferae

The disease has been reported from all the states of India, including Uttar Pradesh, Punjab, Maharashtra, Telungu Desam, Karnataka and Tamil Nadu. It is found to attack all the cultivated varieties of mangoes and poses a great threat to mango production in some seasons, when conditions are favorable for the pathogen.

Symptoms. The pathogen attacks the leaves, flower scales, buds of tender flower heads, axils, panicles of the rachis, stalks and fruits. The affected parts are initially covered with wefts of white mycelium in patches. Soon, the whole surface is covered with a white, powdery coating, consisting of the ectophytic mycelium, conidiophores and conidia. The affected flowers are shed before fruit setting. The affected fruits do not develop normally and drop before attaining pea-size. Sometimes, the growing tips of the shoots alone are affected, while other parts of the tree remain free from infection **(Fig.74)**.

The causal organism. The fungus is an obligate parasite. The mycelium is hyaline, septate, much branched and spreads over the affected parts, forming a dense, white coating. From the ectophytic mycelium, saccate or lobate haustoria are produced, which penetrate into the host epidermal cells. The superficial mycelium produces numerous, simple, erect conidiophores. They are 63 - 163µ in length and have two or more basal cells. Conidia are cut off from the tips of the conidiophores and are found singly or rarely in chains of two. They are hyaline, thin-walled, single-celled, elliptical or barrel-shaped and measure 35 - 43 x 18 - 21µ in size. The perfect stage of this fungus has not been found yet **(Fig.74)**.

Mode of survival, spread and epidemiology. Because the host is a perennial one, the pathogen can survive in the diseased tissues throughout the year. The mycelium remains in a dormant state on the diseased shoots and leaves, when environmental conditions are not favorable. The fungus may also remain active on malformed inflorescence. When favorable conditions return, which usually coincide with the blossoming period, the dormant mycelium in the necrotic shoots and leaves gets activated and fresh crops of conidia are produced, causing fresh infection. Secondary spread is through conidia, which are blown over to new flushes and inflorescence.

Warm weather, with heavy morning dew and cloudy conditions predispose the trees to infection.

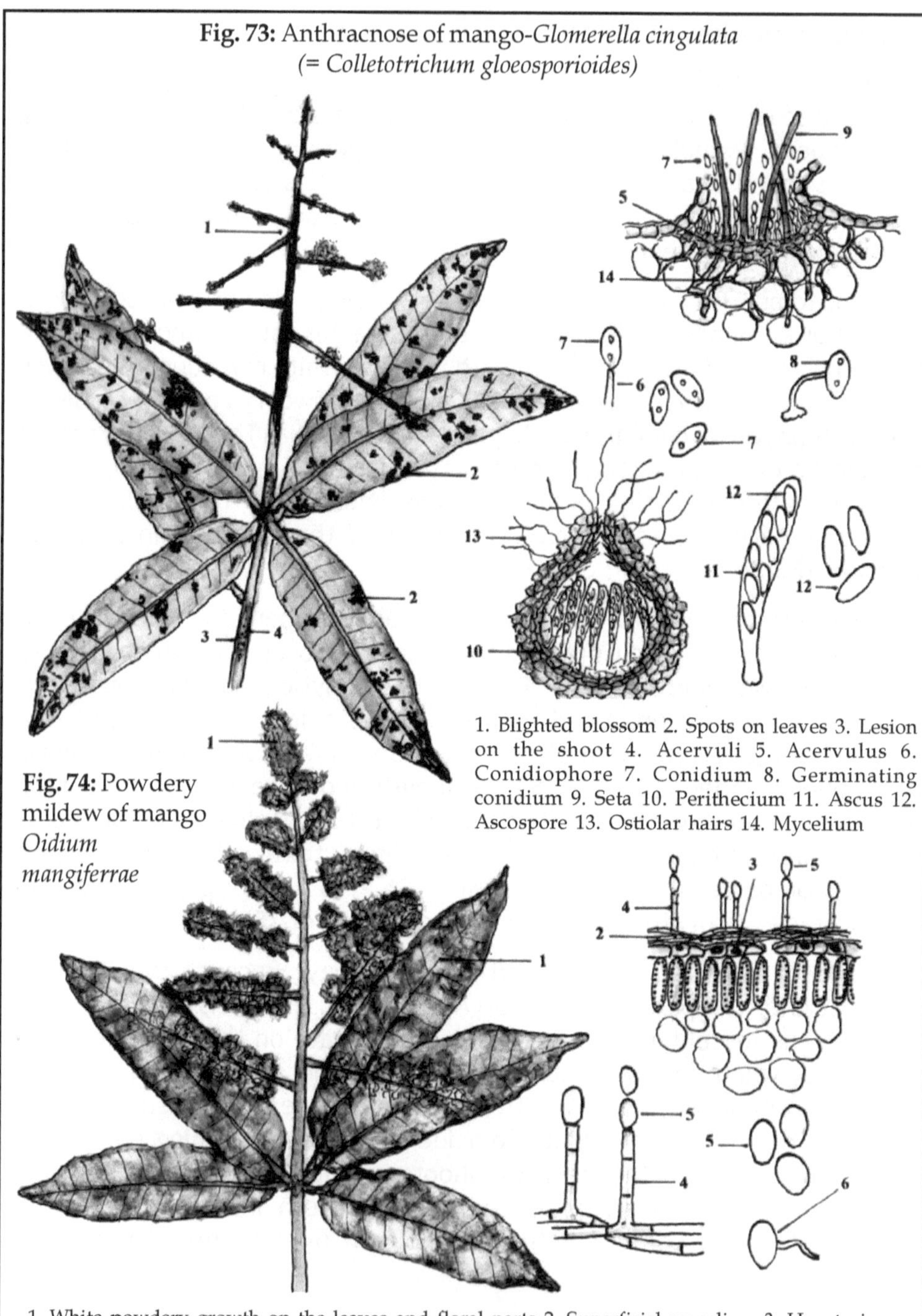

Fig. 73: Anthracnose of mango-*Glomerella cingulata* (= *Colletotrichum gloeosporioides*)

1. Blighted blossom 2. Spots on leaves 3. Lesion on the shoot 4. Acervuli 5. Acervulus 6. Conidiophore 7. Conidium 8. Germinating conidium 9. Seta 10. Perithecium 11. Ascus 12. Ascospore 13. Ostiolar hairs 14. Mycelium

Fig. 74: Powdery mildew of mango *Oidium mangiferrae*

1. White powdery growth on the leaves and floral parts 2. Superficial mycelium 3. Haustorium 4. Conidiophore 5. Conidium 6. Germinating conidium.

Disease management

Agronomic practices (i) Dried branches and panicles should be pruned and destroyed by burning (ii) Diseased, dried up leaves and other plant parts, that have fallen down to the ground should be collected periodically and burnt.

Chemical control. Spraying the canopy with wettable sulfur at 4.0 gm. or dinacap (Karathane) at 1.0 gm. or tridemorph (Calixin) at 1.0 ml. or carbendazim at 1.0 gm. per lit. of water effectively controls the disease. The fungicide should be sprayed at pre-blossom, full blossom and post-blossom stages, at fortnightly intervals. About 10 - 20 liters of spray fluid may be required to spray one tree.

4. Bacterial leaf spot of mango

Pseudomonas mangiferae pv. *indicae*

The disease occurs in almost all the states of India, where mango is grown and causes considerable loss in yield during some seasons.

Symptoms. The symptoms appear as minute, dark-brown or black, water-soaked spots in large numbers in groups, mostly near the leaf tips. The spots enlarge under favorable conditions and may reach 1.0 - 4.0 mm. in size. The spots become black and are surrounded by a yellow halo. As the spots enlarge, they are limited by the veins and become angular in shape. Many such spots may coalesce to form large, black patches. On the surface of the spots, bacterial ooze comes out, which on drying forms a rough coating. When large areas of the leaves are affected, they turn yellow and fall off prematurely. The bacteria attack the petioles, twigs of branches and fruits. On the surface of fruits in the early stages of development, small, dark-brown or black, water-soaked spots develop. Small cracks may be formed on the skin of affected fruits. Severely affected fruits drop off prematurely.

The causal organism. The bacteria causing the disease are rod-shaped, occurring singly or in short chains, motile by a single polar flagellum and measure 0.45 - 1.4 x 0.36 - 0.54μ in size. They are non-spore forming and are aerobic in nature.

Mode of survival, spread and epidemiology. The host being a perennial tree, the bacteria survive in the affected leaves and twigs of branches throughout the year. The pathogen is disseminated by strong winds, lashing rains and by insects mechanically. The bacteria enter the leaves through the stomata, through injuries caused by insect pests, especially the sucking insects and through lenticels in the twigs and fruits.

Fig. 75: Sooty mold of mango-*Copnodium ramosum*

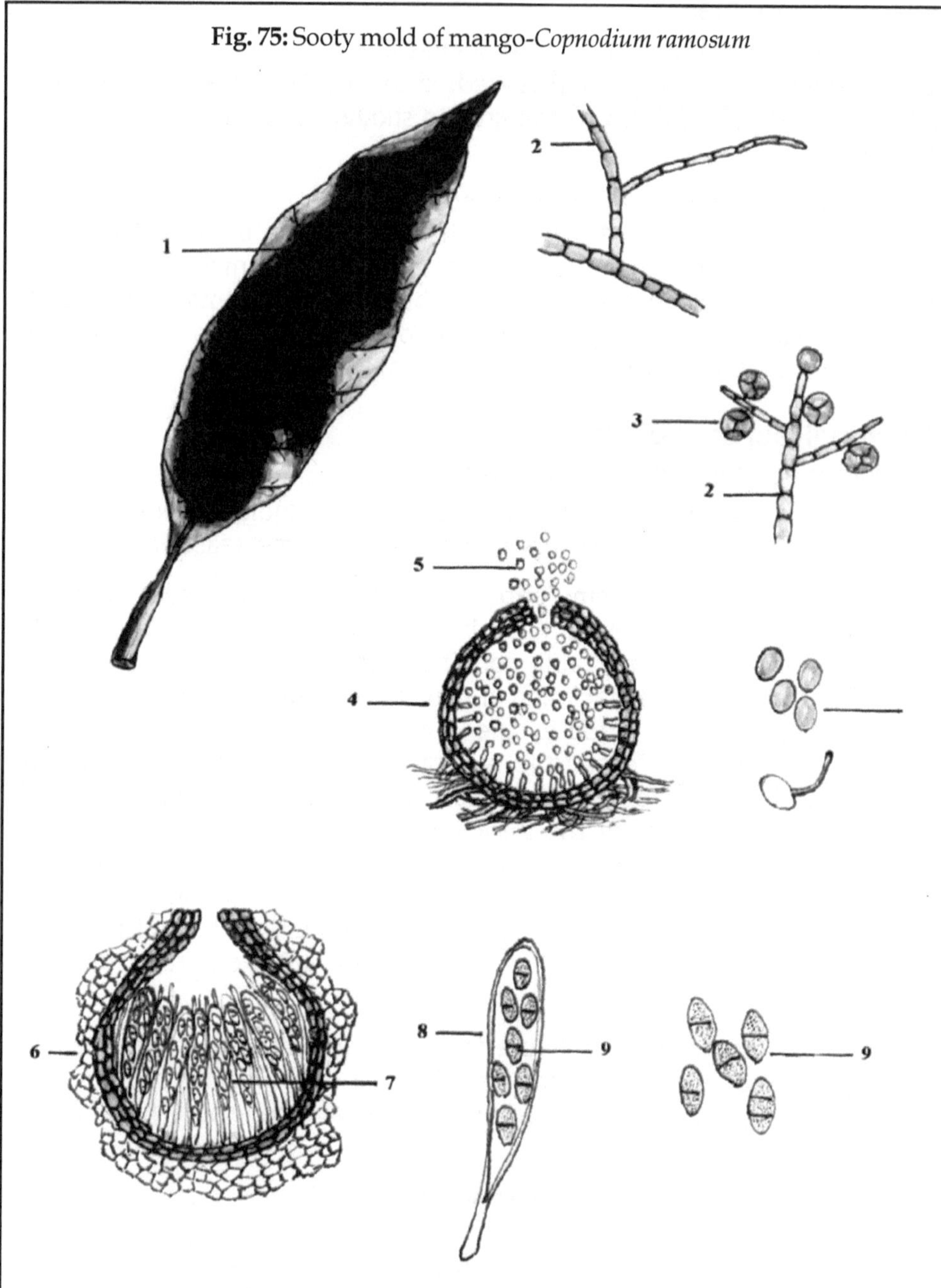

1. Sooty mold growth on the surface of leaf 2. Hypha 3. Conidia produced from hyphal tips 4. Pycnidium 5. Pycnidiospores 6. Pseudothecium 7. Hymenium (Asci and paraphyses) 8. Ascus with ascosporres 9. Ascospores.

Humid and moist conditions, continuous rains and moderate temperature favor occurrence and development of the disease.

Disease management

Agronomic practices (i) Severely affected branches may be pruned during the off season (ii) Healthy and disease-free seedlings should be used for planting.

Chemical control (i) Spraying the crop with Streptocycline at 100 - 200 ppm. (1.0 - 2.0 gm./ 10 lit. of water), twice, at 20 days interval, when the fruits are young reduces the fruit infection (ii) Suitable insecticides may be applied to control insect pests causing injuries to the leaves and young fruits, which helps to minimize infection by the bacteria.

5. Giant mistletoe

Dendrophthoe falcata

This phanerogamic parasite, commonly known as **'Loranthus'** is widespread and is found to attack mango trees of all ages and cause severe damage, especially trees in poorly maintained and neglected orchards. It is found commonly in almost all the mango growing states of India. This semi-stem parasite has well-developed, efficient leaves, with chlorophyll and are capable of manufacturing their own requirements of sugars and starch by photosynthesis. However, they depend completely on the host for their requirements of water and mineral nutrients, which they absorb from the host by sending in haustoria into the host xylem elements.

The parasite. The parasite attacks the stem, as well as the branches of trees. It is a bushy succulent plant, with fairly large, green leaves and has many branches. At the point of attack, the parasite produces haustoria, which penetrate the stem or branch and anchor the parasite firmly on to the host. The haustoria eventually reach the xylem vessels and absorb water and mineral nutrients from the host, at the same time it grows into a bushy plant from that point. The stems and branches of the host around the points of attack bulge out into irregularly shaped cankers. Very often, the branches of the parasite grow along the host stems or branches and parasitize them at several points and at each point, produces haustoria and aerial bushy growth. Due to continuous uptake of water and mineral nutrients from the host, the parts above the points of parasitization suffer from want of water and nutrients, as a result the branches become weak and their growth is very much retarded. The leaves become small and discolored. Newly formed branches are badly affected. Fruit formation is lowered considerably. Even if fruits are formed, they are smaller and are of poor quality. Severely attacked branches gradually wither and die. Sometimes, even big trees are killed.

The parasite produces long, tubular, greenish-white or red flowers in bunches. The fruits are barrel-shaped or slightly elongated with rounded ends, fleshy, sweet, reddish in color and single seeded. The flesh of the fruits is sticky in nature.

Mode of spread. Many kinds of birds and animals, such as squirrels are attracted to the fruits because of their attractive color and sweet taste and they relish the fruits. The seeds stick on to the beaks and feathers of birds and the fur coating of animals eating the fruits and are carried to other trees. The seeds, which happen to fall on the branching points of the hosts, germinate, produce haustoria and make contact with the branches. The seeds from fruits swallowed by birds come out along with the faecal matter, fall on the stem region of the host tree and parasitize the host. The seeds sticking on to the fur coating of animals are also disseminated similarly. They may be carried over long distances by birds. Besides mango, the fungus attacks guava, sapota, cashew, jackfruit, pomegranate, citrus plants and several other fruit trees, trees in wastelands, roadside shade trees etc.

Management

Agronomic practices. The branches attacked by the parasite should be cut off at about 30 cm. below the point of attachment, so that the haustoria inside the affected portion may be completely removed. The cut ends should be treated with Bordeaux paste immediately after cutting the branches.

Chemical treatment. Spraying a mixture of diesel oil and liquid soap can kill the parasite. In 1.0 lit. of water, 300 - 400 ml. of diesel oil and 5.0 ml. of liquid soap are added and mixed thoroughly. Then this solution is sprayed on the parasite.

6. Black tip of mango

Physiological disease or disorder

'Black tip', **'mango necrosis'** or **'tip rot'** is a physiological disease found only in India. It is a very common and serious disease of mango fruits in Uttar Pradesh, Bihar, Bengal, Punjab and in certain parts of Tamil Nadu and Telungu Desam. Some of the best mango varieties have been found to be highly susceptible to the disease. The disease is commonly found on fruits of trees, which are close to brick-kilns.

Symptoms. The disease is characterized by necrosis of fruits at the distal end. The initial symptom is the development of small, chlorotic area at the distal end of fruits. The area gradually enlarges in size, turns almost black in color and covers the tip completely. The tip gets flattened and the outer skin becomes hard and sunken. The inner tissues become soft and start rotting due to the attack of saprophytic bacteria. A dark-brown liquid oozes out from the necrotic area. The disease usually occurs, when the fruits are nearing maturity.

The cause. The disease is most common in orchards in the vicinity of brick-kilns. The smoke emanating from the kilns causes it. The smoke contains toxic gases like sulfur dioxide, which is responsible for causing necrosis of the tissues of fruits. It has been found that boron is deficient in the fruits exposed to brick-kiln fumes. The toxic gases in the smoke may interfere with the process of assimilation of boron and thereby indirectly cause necrosis of the fruits.

Disease management. Spraying the crop with borax at 60 - 80 gm./ 100 lit. of water at the early stages of fruit development reduces the incidence of the disease to a considerable extent. One or two sprayings may be given at fortnightly intervals.

Diseases of minor importance. In addition to the diseases described above, several other diseases also attack mango. *Pestalotiopsis mangiferae* causes 'Gray blight', which is characterized by the formation of brown spots on leaves and green fruits; *Diplodia natalensis* causes 'Stem end rot'; *Cephaleuros virescens*, an algal parasite causes 'Red rust', which is characterized by the formation of reddish, slightly raised, rusty spots on the leaves, young twigs and green fruits; 'Bacterial canker' caused by *Xanthomonas campestris* pv. *mangiferae-indicae* produces cankerous spots on leaves and fruits; *Botryodiplodia theobromae* causes 'Die-back' of branches; 'Black mould rot' is caused by *Aspergillus niger*.

Citrus plants *(Citrus species)*

Citrus plants include **'Lime'** *(Citrus aurantifolia)*, **'Lemon'** *(Citrus limon)*, **'Sweet orange'** *(Citrus sinensis)*, **'Mandarin orange'** *(Citrus reticulata)*, **'Kamala orange'** *(Citrus aurantium)*, **'Grape fruit'** *(Citrus paradisi)*, **'Malta'** *(Citrus medica)* and **'Pomela'** *(Citrus grandis)*. The diseases that attack one species of citrus mostly attack the other species also.

1. Gummosis, leaf fall and fruit rot of citrus

Phytophthora palmivora, P. parasitica and *P. citrophthora*

The disease is found in almost all citrus-growing parts of the world. In India, it is prevalent in many parts including Gujarat, Maharashtra, Telungu Desam, Karnataka, Assam and Tamil Nadu. Sweet varieties are more susceptible than the sour ones.

Symptoms. Gummosis occurs mainly at the basal part of the trunk and later spreads to the thicker branches also. Usually, infection starts from the ground level. The earliest symptom is the appearance of water-soaked patches in the basal part of the stem, followed by oozing out of a brownish, gummy exudate. The disease spreads up the trunk and down to the roots. When the stem portion above the ground level is attacked, droplets of gum trickle down the stem to the base. Exudation of gum is not conspicuous always, as it is washed down during the rains and the exudates near the ground get mixed up with the soil. However, during the summer months, the bark shows brown staining and hardened masses of gum on the surface. In course of time, longitudinal cracks develop in the bark, which peels off leaving the wood exposed. Cracks may develop on the branches also, through which gum is exuded. A major portion of the root system is also affected and it rots. As a result of severe gumming, the bark becomes completely rotten and the tree dies, as a result of girdling. Before death, the tree flowers profusely but the fruits are under-developed and drop off prematurely. The leaves show symptoms of nutritional deficiency, with yellow veins and premature leaf fall is common. Fruits at various stages of development are also attacked. Water-soaked spots develop on the skin. The fruits become soft and whitish, fungal growth appears on the skin. The fruits drop off prematurely and the fungus continues to grow on the fallen fruits.

The causal organism. The main causal agents of the disease have been found to be *Phytophthora palmivora* and *P. parasitica*. However, *P. citrophthora* has also been found to cause the disease to some extent.

For description of *Phytophthora palmivora*, refer 'Bud rot of Coconut' (Page 457) and for *P. parasitica*, refer 'Seedling blight of Castor' (Page 182).

The mycelium of *Phytophthora citrophthora* is coenocytic, slender, hyaline and intercellular. Sporangiophores are delicate, irregularly branched and with a swelling at the point of branching. Sporangia are pedicellate and vary much in size and shape, often with two widely divergent apices. They measure 40 - 45 x 27µ in size. They germinate by producing biflagellate zoospores. Chlamydospores are produced rarely. Sexual reproduction has not been found.

Mode of survival, spread and epidemiology. The fungi are soil-borne and may survive on fallen fruits, leaves and fragments of branches fallen on the ground. They may also survive in the bark and cracks of infected trees. The disease spreads through spores carried by irrigation water, rainsplash, wind and by some insects mechanically. Infection takes place through the tissues, where the scion has been budded, when the bark is wet. Infection may be initiated by the resting spores viz., oospores and chlamydospores and also by sporangia and zoospores.

Heavy clay soils, high soil moisture and temperatures ranging from 24°-28°C are conducive for disease occurrence.

Disease management

Agronomic practices (i) Proper drainage facilities should be provided and excessive irrigation should be avoided (ii) Resistant root stocks, such as sour orange *(Citrus aurantium)* and trifoliate orange *(Poncirus trifoliata)* may be used (iii) Diseased leaves, fruits and other plant parts fallen on the ground should be collected and destroyed by burning (iv) Irrigation water should not be allowed to flow from one tree to another (v) Deep planting should be avoided. While planting, the bud joint should be well above the ground level (vi) Thin branches showing symptoms of gummosis should be cut and destroyed.

Chemical control (i) Healthy trees should be protected from disease attack by painting the trunk region with Bordeaux paste up to a height of 60 - 75 cm. from the ground level every year (ii) During the summer and rainy season, the orchard should be sprayed with Bordeaux mixture - 1% or copper oxychloride - 2.5 gm./ lit. of water (iii) Planting pits may be dusted with a mixture of zinc sulfate, copper sulfate and lime in the ratio 5 : 1 : 4 parts prior to planting in gummosis prone areas (iv) On trunks and thick branches, the infected portions may be scraped thoroughly with a sharp knife to expose the healthy wood. Then the wound is cleaned with a solution of mercuric chloride - 0.1% or potassium permanganate - 1.0%, followed by application of Bordeaux paste to cover the wound (v) Spraying with a mixture of Aureofungin Sol - 3.0 gm.+ liquid soap - 30 ml. in 130 lit. of water, two or three times, at 15-30 days interval has been found to give complete control of the disease (vi) Spraying with Bordeaux mixture - 1% or copper oxychloride - 2.5 gm. or captafol (Difolatan) - 1.5 gm. or metalaxyl (Ridomil) - 1.0 gm. + mancozeb - 2.0 gm. or focetyl-Al (Aliette) - 1.0 gm./ lit. of water also provides effective control of the disease.

2. Citrus canker

Xanthomonas citri (X. campestris pv. *citri / X. axonopodis)*

'Citrus canker' is one of the most dreaded diseases of citrus and attacks all types of important citrus crops. The disease was found to be endemic in Japan and South East Asia. Diseased specimens of *Citrus medica* collected from India between 1827 - 1833 and *Citrus aurantifolia* collected from Java between 1842 - 1844 are found in the **'Kew herbarium'**. It is a disease of International importance and is quite serious in India, China, Japan, Java, Uruguay, Argentina and Brazil. The disease was introduced into Florida in 1912 with infected nursery plants from Japan. A massive eradication program was initiated in 1915 in the southern states of America, including Florida, Alabama, Georgia, Louisiana, Mississippi, Carolina and Texas and by 1947, it was totally eradicated from the United States of America by destroying hundreds of thousand bearing trees and nursery stock. However, the disease appeared again in Florida in 1986. Citrus canker has been eradicated from South Africa, Australia and New Zealand with concerted efforts. Still the disease is found in many of the citrus growing countries of the world. All citrus growing countries round the world enforce strict prohibition of import of citrus plants and fruits from countries, where the disease is found. Limes, grapefruit, sweet oranges and some lemon varieties are very susceptible.

Symptoms. The disease affects all aboveground parts of the plants and causes necrotic lesions on leaves, twigs, branches and even the thorns. On leaves, the symptoms appear as small, water-soaked, translucent spots. The spots are 1.0 - 9.0 mm. in diameter and are usually darker green in color than the surrounding areas, with a raised surface. The spots appear on the lower surface of leaves in the beginning, but appear on the upper surface also later on. As the disease advances, the spots become white or grayish and finally rupture exposing a light-brown, spongy mass in a crater-like formation and are surrounded by a yellow halo. Old lesions become corky and brown. Similar spots appear on young twigs, but in older twigs they are more irregular in shape, up to 1.0 cm. in length and cankerous. The canker lesions on the fruits appear similar to those on the leaves, but the yellow halo is usually absent and the corky, crater-like appearance is more pronounced. The canker spots may be scattered on the surface or adjacent spots may coalesce to form irregular, scabby patches. Cankers may be formed on the thorns also.

The causal organism. The bacterium causing the disease is short, rod-shaped, motile by a single polar flagellum and measure 1.5 - 3.0 x 0.5 - 1.5μ in size. It is gram negative and aerobic in nature.

Mode of survival, spread and epidemiology. The bacteria overseason in leaf, twig and fruit canker lesions. They can survive in the infected leaves for more than 6 months. During warm, rainy weather, they come out of the lesions as bacterial exudates. The bacteria enter the host through natural openings, such as stomata, lenticels etc. or through wounds caused by insects or aberrations caused by the thorns etc. After entry into the host, the bacteria multiply rapidly in the intercellular spaces, dissolve the middle lamella and get established in the cortex. Cankerous outgrowths develop from the points of infection, within which the bacteria multiply and come out along with the exudates. The disease is disseminated by lashing rains, strong winds and mechanically by some insects. Leaf miners are one of the chief agents of transmission of the disease. Movement of infected nursery stock by man is the major cause for dissemination of the disease over long distances.

The disease is favored by warm temperature and wet weather. Temperatures between 20° - 35°C, with an optimum of 30°C and free moisture for 20 minutes or more on the host surface are essential for infection.

Disease management

Agronomic practices (i) Movement of infected nursery stock from place to place should be banned by law enforcement (ii) Disease-free nursery stock should be used for new plantings (iii) Affected plant parts should be cut off and destroyed by burning (iv) Infected leaves, fragments of twigs and other plant debris should be collected and destroyed (v) Plant vigor should be maintained by timely irrigation and fertilization (vi) Insect pests causing injuries to the leaves, stems etc. should be eradicated by the application of suitable insecticides (vii) Resistant varieties should be grown in new gardens.

Chemical control (i) Spraying the crop with the antibiotics, Streptomycin sulfate at 5 - 10 gm./ 10 lit. of water (500 - 1000 ppm) or Phytomycin at 2.5 gm. in 10 lit. of water (2500 ppm), at 15 days interval has been reported to be effective in controlling the disease (ii) Spraying with a mixture of Streptomycin sulfate - 0.5 gm. + copper oxychloride - 2.5 gm./ lit. of water has given excellent control of the disease (iii) Spraying with neem seed cake broth is very effective in controlling the disease, as well as the leaf miner pest. Neem seed cake - 1.0 kg. is added to 20 liters of water and allowed to ferment for 7 days. After this period, the broth is filtered and sprayed. In a year, 12 sprayings have to be given, at monthly intervals (iv) Spraying with Bordeaux mixture - 1 % or copper oxychloride - 2.5 gm./ lit. of water also controls the disease to some extent.

3. 'Tristeza' or Quick decline of citrus

Citrus tristeza virus

The disease was first reported from Florida in 1950. Now, it is found in almost all citrus growing countries of the world, including Argentina, Australia, Brazil, India, Indonesia, Java, Malaysia, Pakistan, Panama, Sri Lanka, Thailand, Uruguay and Venezuela. It affects almost all kinds of citrus plants, but primarily orange, grapefruit and lime. Severe strains of the virus cause enormous loss in yield and result in a quick decline and eventual death of infected trees. The disease is very common and severe on trees propagated on sour orange rootstock.

Symptoms. The symptoms produced by the virus on various *Citrus* species vary with the strain of the virus and with the rootstock on which the scion is propagated. Severe strains cause yellowing of seedlings due to acute chlorosis and dwarfing of seedlings of sour orange, lemon and grapefruit. Young sweet orange, grapefruit and other citrus trees growing on sour orange rootstock, develop a quick decline within a few weeks, when infected by a severe strain of the virus. In such cases, the leaves turn yellow or brown, wilt and fall off, while the fruits continue to hang on the tree even after its death. Decline inducing *Tristeza virus* causes phloem necrosis at the graft union and interferes with the translocation of food to the roots. This results in poor root growth and the roots may die leading to decline of the aboveground parts of the tree. Some strains produce stem-pitting symptoms. Infected trees show deep, longitudinal pits in the wood under the bark of trunks, branches and twigs. Stem-pitting may occur even on the rootstock. Trees with stem pitting are stunted and produce less number of smaller and poor quality fruits. The twigs become brittle and break easily. Some other mild strains on the other hand develop slow decline or chronic decline. In such cases, the infected trees remain severely stunted, become less productive, decline slowly over several years and eventually die.

The causal organism. Citrus decline and stem pitting is caused by *Citrus tristeza virus*, which belongs to the Genus - *Closterovirus*. The viruses consist of very long, flexuous, rod-shaped particles measuring 600 - 2000 nm. in length and 10 nm. in diameter. The genome consists of a single ssRNA molecule.

Mode of spread. *Citrus tristeza virus* is transmitted by budding, grafting and by several species of aphids in a semipersistant manner. The aphids require an acquisition feeding time of 30 - 60 minutes and subsequently, they remain viruliferous for about 24 hours. Among the different aphid vectors, the citrus brown aphid, *Toxoptera citricida* is most efficient.

Compared to the other aphid species, it is 10 - 25 times more efficient in transmitting the disease. *Toxoptera aurantia, Aphis gossypii, Aphis craccivora, Dactynotus jaceae* and *Myzus persicae* are the other aphid vectors. The virus is also transmitted by the parasitic 'Dodder', *Cuscuta reflex.*

Disease management

Agronomic practices (i) Strict quarantine regulations should be enforced to prevent transport of disease infected seedlings to other places where the disease is not present (ii) Only tested, virus-free budwood should be used (iii) Trees showing symptoms of decline should be removed and destroyed (iv) Grafting or budding on sour orange stocks should be avoided. Scion varieties tolerant to *Tristeza virus* should be grafted on virus tolerant rootstocks (v) The plants should be maintained properly and should be kept healthy and strong by providing adequate irrigation and proper drainage facilities, application of recommended doses of fertilizers, adopting timely pruning and other field sanitation methods, which may induce resistance in the plants and ward off the disease incidence.

Chemical control. Periodical sprayings with systemic insecticides, such as monocrotophos - 1.5 ml./ lit. of water reduce the secondary spread of the disease by controlling the aphid vectors.

Cross protection. Seedlings can be cross-protected from severe strains of *Tristeza virus* for fairly long periods by inoculating them with certain mild strains of the virus or inactivated virus.

4. Citrus greening disease

Phloem-inhabiting fastidious bacteria

The disease is widespread in many parts of India, especially in the states of Jammu and Kashmir, Punjab, Haryana, Rajasthan, Uttar Pradesh, Madhya Pradesh and Karnataka. Citrus plants from the Coorge region has been completely wiped out due to this disease. The greening disease has been reported to be more devastating than 'Tristeza' in many parts of India. Often citrus greening disease and 'Tristeza' virus disease occur together and result in die-back and quick decline.

Symptoms. The symptoms caused by this disease are highly variable. Yellowing of the midrib and lateral veins of matured leaves is very characteristic. Young leaves of infected plants are mottled and chlorotic and resemble the symptoms caused by zinc deficiency. The yellow chlorotic leaf lamina is scattered with green, island-like patches. The midrib and the lateral veins mostly bound the yellow areas in leaves and the yellowing expands towards the leaf margin. The size of the leaves is considerably reduced and the leaves are thicker and remain erect. The internodes of branches become short giving the branches a bushy appearance. The diseased plants are stunted, flower earlier than the healthy plants and produce small fruits. The root system is also affected and the number of roots is reduced.

The causal organism. The disease, which was considered to be caused by a virus earlier, is reported to be caused by a *Phloem-inhabiting fastidious bacterium* with a double cell wall membrane and is distinctly different from mycoplsma-like organisms. The cells of the *Fastidious bacteria* are gram-negative, non-motile, non-pleomorphic, rigid rods and measure 1.0 - 2.0 x 0.25 - 0.5 µm. They are distributed in the sieve elements and vascular bundles.

Mode of spread. The pathogen is transmitted through vegetative propagation and by two species of citrus psyllids, *Diaphorina citri* and *Trioza erytreae*. These insect vectors are quite common in North and Central India. After acquiring the bacteria, the vector remains infections throughout its lifetime. An incubation period of 8 - 12 days inside the body of the vector is necessary before it becomes infective. Even though the nymphs can acquire the pathogen, they become infective only after they reach adulthood.

Disease management

Agronomic practices (i) Diseased plants should be eradicated from the orchards (ii) Disease-free seedlings raised from indexed stock should be used for planting.

Chemical control (i) Spraying with tobacco decoction is effective in controlling the vectors (ii) Spraying with nicotine sulfate 40 S - 1.5 ml. or monocrotophos - 1.5 ml. or methyl demeton - 1.5 ml. or diazinon - 2.0 ml./ liter of water controls the vectors.

Diseases of minor importance. Many other diseases also attack citrus plants and some of them cause serious damage to the plants. *Ganoderma* lucidum causes '*Ganoderma* root rot'; *Macrophomina phaseolina, Diplodia natalensis* and *Fusarium* species cause 'Dry root rot'; 'Anthracnose' or 'Die-back' caused by *Colletotrichum gloeosporioides* affects leaves, branches and fruits; 'Powdery mildew' caused by *Oidium tingitaninum* is a serious disease of citrus plants, especially mandarin and sweet oranges and affects all the aerial parts, such as leaves, twigs and fruits; 'Scab' is caused by *Elsinoe fawcetti* (=*Sphaceloma fawcetti*); *Pellicularia* (*Corticium*) *salmonicolor* causes 'Pink disease' that affects the stems, branches and twigs; *Capnodium citri* causes 'Sooty mould' mostly on lime and sweet oranges; *Penicillium italicum* and *P. digitatum* cause 'Blue and Green moulds' that affect citrus fruits during transit and storage; 'Red rust' caused by *Cephaleuros virescens* produces circular to irregularly circular, reddish-brown, slightly raised spots on leaves, twigs and fruits; Giant mistletoe, *Dendrophthoe ampullaceus, D. parasiticus* and a few other species attack citrus plants, especially in badly maintained orchards.

Grapevine *(Vitis vinifera)*

1. Powdery mildew of grape

Uncinula necator

This is the most destructive disease of grapevine and is distributed all over the world. It causes extensive damage to grapevines in Europe and the United States of America. The disease is also prevalent in France, Australia, Africa, Hungary, Syria, Afghanistan and India. In India, the disease appears in epidemic proportions in some years. It occurs commonly in the states of Maharashtra, Gujarat, Karnataka, Tamil Nadu and Telungu Desam.

Symptoms. The disease attacks the crop at all growth stages and all aboveground parts, such as leaves, stems, flowers and fruits are affected. Initially, small, whitish, powdery patches appear on the upper surface of young leaves. Sometimes, the whitish, powdery growth is seen on the under surface of leaves also. Under favorable conditions, these patches enlarge rapidly and cover large areas, sometimes the entire surface of the lamina. Distortion, malformation and discoloration of leaves, especially young leaves also occur. Such powdery patches are produced on the stems, tendrils, flowers and young berries. The powdery patches gradually turn gray and finally dark. Affected stem regions turn brown. Infected flowers fail to develop fruits. Young berries when infected, turn black, become irregular in shape and develop cracks. If fruits nearing maturity are infected, only few of them ripen normally, while the others rot. Diseased vines present a wilted appearance **(Fig.76).**

The causal organism. The fungus causing the disease is an obligate parasite. The mycelium consists of slender, whitish, much branched, septate hyphae, which are completely superficial and spread over the host surface. They are attached to the host surface by means of appressoria, from which haustoria are sent into the epidermal cells of the host to absorb nourishment. Sometimes, the haustoria may penetrate deeper into the mesophyll cells. Conidiophores arise from the superficial mycelium in large numbers. They are erect, simple and bear conidia in chains of 3 - 4 in basipetal succession. The conidia are oval to elliptical and measure 25- 30 x 15 - 17µ in size. They can resist desiccation to a large extent. They germinate by producing a germ tube.

The ascigerous stage is not found under conditions prevailing in India. However, cleistothecia have been found in other countries. The cleistothecia are embedded in the superficial mycelium on the leaves, on the nodal region and on buds. They are black, almost spherical, with a flattened top and measure 75 - 100µ in diameter. The peridium has 8 - 25, septate appendages inserted equatorially. They are coiled at the distal end. Each cleistothecium contains 2 - 8, broadly ovate asci, measuring 48 - 60 x 37 - 45µ in size. Each ascus contains 4 - 6, single-celled, thin-walled, hyaline, ovate ascospores, measuring 15 - 35 x 10 - 14µ in size **(Fig.76).**

Mode of survival, spread and epidemiology. Conidia and mycelium overwinter in diseased buds, fallen berries or on overwintered stem. Cleistothecia, when formed are also capable of overwintering and cause fresh infection.

Mycelial growth and conidial production is very rapid between temperatures ranging from 20° - 30°C, the optimum being 25°C. Temperatures above 35°C and bright sunshine are not conducive for the development of the disease. Warm, dry weather, low humidity and cloudy weather favor occurrence and development of the disease. The conidia being very light are easily carried by air currents and disseminated.

Disease management

Agronomic practices (i) Dense growth of vines should be avoided by proper and timely pruning (ii) Collection and destruction of fallen leaves and berries, cutting back of laterals and removal of diseased parts and destroying them help to minimize disease occurrence.

Chemical control (i) Spraying with wettable sulfur - 4.0 gm. or dinacap (Karathane) - 1.0 gm. or tridemorph (Calixin) - 1.0 ml. or thiophanate methyl (Topsin M) - 1.0 gm. or triadimefon (Bayleton) - 1.0 gm. or triadimenol (Bayton) - 1.0 gm. or etaconazole (Vangard) - 1.0 gm./ lit. of water has been found to be effective in controlling the disease. Three sprayings are to be given, the first one when the new shoots are 10 - 15 cm. long, the second at the time of flowering, followed by the third 40 - 45 days later.

2. Downy mildew of grape

Plasmopara viticola

This is another devastating disease of grapevine, but is comparatively less destructive than the powdery mildew. The disease is native to North America and has been found to occur in endemic form even before 1870. It was subsequently introduced into Europe in 1875 and it spread very rapidly throughout France and other European countries and caused widespread destruction of the crop and heavy losses to the wine industry. The disease also attacks other species of *Vitis*. It is prevalent in other humid parts of the world. Dry areas are almost free from this disease. In India, the disease occurs in Maharashtra, Karnataka, Telungu Desam and Tamil Nadu.

Symptoms. The disease attacks all succulent, tender parts of vines. However, the symptoms are more pronounced on leaves, young shoots and immature berries. Initial symptoms appear as, small, irregular, light yellow spots on the upper surface of the leaves and profuse white, downy growth on the corresponding under surface of the spots. The spots turn dark-brown due to necrosis of the tissues. Under favorable conditions, the mildew growth may cover the entire leaf lamina. Later on the white, downy growth turns dirty gray. Eventually, the leaves turn brown, wither and drop off. Infected shoots, tendrils and berries are also covered with the whitish growth of the fungus. Diseased shoots remain stunted. Due to hypertrophy of the cells of the affected parts, they become distorted or thickened. The shoots may show distortions in case of late or localized infection. When the floral parts are affected, the flowers die and are shed. In case of early infection of berries, the entire bunch of berries is destroyed. If infection occurs after the berries are half grown, the fungus grows mostly internally. Such berries become leathery, somewhat wrinkled, discolored and are often mummified, while still attached to the bunch. Sometimes, only a part of the bunch is affected, which may produce a few normal fruits **(Fig.77)**.

The causal organism. The fungus causing the disease is an obligate parasite. The mycelium is intercellular. The hyphae are coenocytic, thin-walled, hyaline and granulated. They produce globose haustoria, which grow into the mesophyll cells. Sporangiophores arise from hyphae congregated in the sub-stomatal cavities and emerge through the stomata in groups of 1 - 20 from the undersurface of leaves. Sporangiophores may also arise through stomata on the tendrils and stems and through lenticels in young fruits. Sometimes, they emerge directly by penetrating the epidermal cells. They are slender and 300 - 500μ long. Each sporangiophore produces 4 - 6 irregularly spaced branches, almost at right angles to the main stalk, the final branches arising from the apex. The lower branches produce 2 or 3 secondary branches in a similar manner. From the terminal end of each branch, 2 - 3 sterigmata arise, which bear sporangia singly. Sporangia are thin-walled, oval or lemon-shaped and measure 15 - 30 x 11 - 18μ in size. Sporangia germinate either by production of zoospores formed inside or outside in a vesicle or directly by germ tubes. The zoospores are pear-shaped, 7.0 - 9.0μ in size and have two flagella arising from the apex. They swim in a film of water for a short time, come to rest, encyst and then germinate to enter the host to cause fresh infection. At the end of the season, the fungus produces numerous oospores in the infected old leaves, mostly in tissues adjacent to the mid-rib and sometimes in the shoots and berries. They are spherical, thick-walled, with a rough epispore and measure 25 - 36μ in diameter. The oospore germinates after a period of rest by producing a short germ tube, which bears a terminal zoosporangium. Zoosporangia germinate in the normal manner by releasing zoospores **(Fig.77).**

Mode of survival, spread and epidemiology. In areas where grapevines are green all through the year, the fungus continues to produce conidia-like sporangia. Otherwise the disease is carried over from year to year through oospores, which overwinter in the old diseased leaves and shoots. During rainy periods, the oospores germinate and produce zoospores, which are carried to wet leaves near the ground by wind, rainsplash or water and cause infection. Soon, sporangiophores and sporangia are produced on the leaves. The zoospores produced from the sporangia cause secondary infection and then the disease spreads very rapidly. A disease cycle may be completed in 5 - 18 days depending upon the temperature, humidity and varietal susceptibility.

The most favorable temperature for the germination of sporangia is 10° - 23°C. Humid, cloudy and damp weather conditions and free water on the host surface is essential for the zoospores to cause infection.

Disease management

Agronomic practices (i) Fallen leaves, fragments of twigs and other plant debris should be collected and destroyed by burning (ii) While planting, proper spacing should be given between plants (iii) The vines should be trained and pruned properly, so that leaves do not remain near the ground.

Chemical control (i) Spraying with Bordeaux mixture - 0.75 % or copper oxychloride - 2.5 gm. or dithiocarbamates (zineb, maneb or mancozeb) - 2.0 gm. or chlorothalonil - 1.5 gm./ lit. of water has been found to be effective in controlling the disease. The spray applications should commence

Fig. 76: Powdery mildew of grapes-*Uncinula necator*

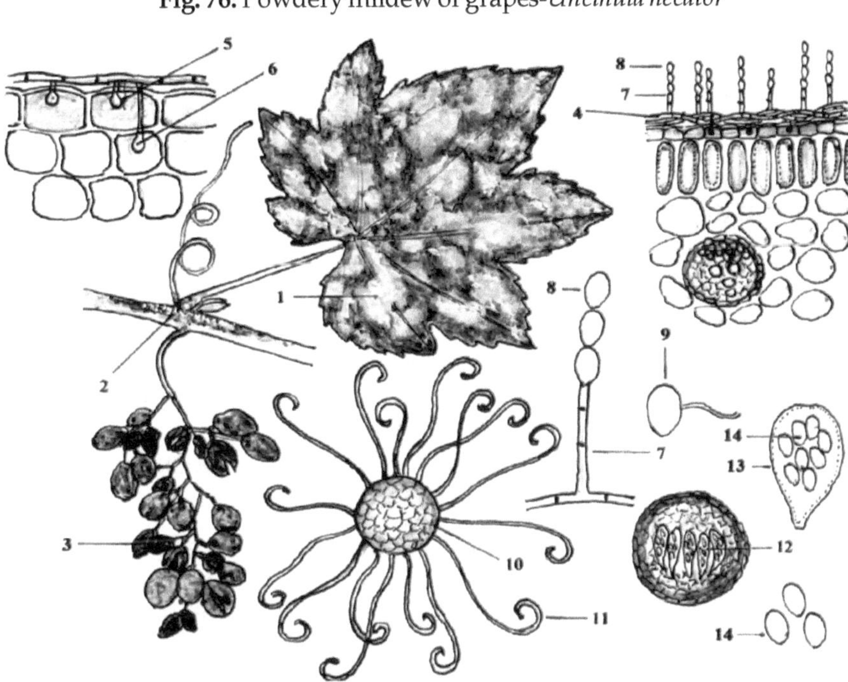

1. White powdery growth on the upper surface of leaf 2. Infection on the stem region 3. Infected berries 4. Superficial mycelium 5. Epidermal and 6. Sub-epidermal haustorium 7. Conidiophore 8. Conidium 9. Germinating conidium 10. Cleistothecium 11. Colled appendages 12. Asci with ascospores, inside the cleistothecium 13. Ascus with ascospores 14. Ascospores.

Fig. 77: Downy mildew of grapes *Plasmopara viticola*

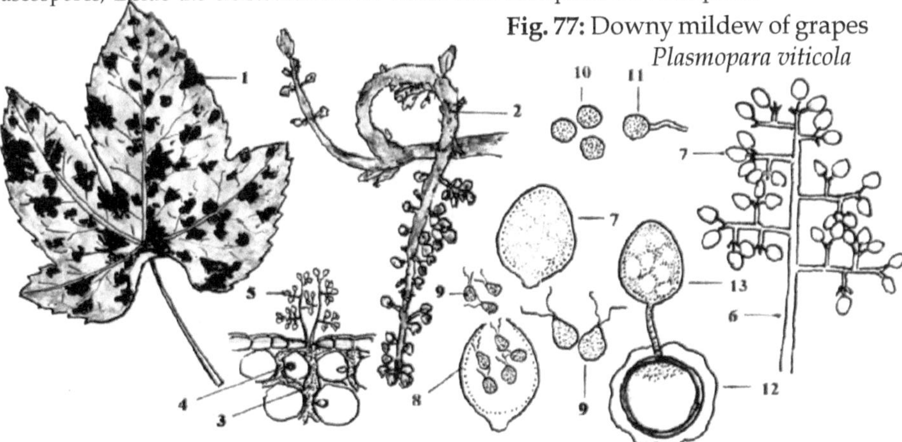

1. Brown, irregular spots on the upper surface of leaf 2. Infected young grape cluster showing distorted stalk and stem region 3. Intercellular mycelium 4. Haustorium 5. Sporangiophore with sporangia on the under surface of leaf 6. Sporangiophore 7. Sporangium 8. Zoospores emerging from sporangium 9. Zoospores 10. Encysted zoospores 11. Germinating encysted zoospore 12. Germinating oospore 13. Zoosporangium from oospore.

before bloom and continued at 7 - 10 days interval during the growing season. A quantity of 500 - 600 liters of spray fluid may be required to cover one acre of crop (ii) Combined spraying with a systemic fungicide, metalaxyl (Ridomil) - 1.0 gm. + a non-systemic fungicide, mancozeb - 2.0 gm./ lit. of water has been found to be highly effective.

3. Anthracnose or 'Bird's eye spot' of grape

Gloeosporium ampelophagum / Sphaceloma ampelinum (Elsinoe ampelina)

The disease is supposed to be native to Europe and from France or Italy, it had spread to America and Australia. It is one of the most serious of fruit diseases. The disease was first reported in India in 1903 and is now prevalent in the states of Telungu Desam, Haryana, Karnataka, Punjab, Rajasthan, Uttar Pradesh and Tamil Nadu. In Tamil Nadu, it occurs in all grape-growing districts, such as Coimbatore, Salem, Madurai, Dharmapuri, Dindigul and Kodaikanal.

Symptoms. All parts of the vines including leaves, tendrils, stems and berries are affected. Small, more or less circular spots, with a grayish center and brown margin, surrounded by a yellow halo appear on the leaves in large numbers. When too many spots appear crowded together, they remain small and do not enlarge much. However, isolated spots enlarge and the center turns light-gray, while the outer edge becomes dark-brown or purplish. The central part of the spot becomes sunken, dries and may drop out leaving a shot hole. The spots on the berry are similar to the spots produced on the leaves and they resemble the eye of a bird and hence it is called bird's eye spot disease. Small, elongated spots develop on the veins, which may join up, as a result the lamina becomes curled and torn. Similar elongated spots may develop on the stem region also. Young shoots are more susceptible than the leaves. Small, irregular, almost black spots develop on the shoots in large numbers and they become black, hard and dry. Such affected shoots may be killed within a short time. On older shoots, the spots enlarge, become light-brown with a dark-brown margin. The central part later becomes cracked and sunken, which forms crater-like cankers. Sometimes, the stem cankers cover long stretches of the vine. When tendrils are affected, they curl and dry. Lesions occur on the stalk of inflorescence and if the stalk is girdled, all the fruits may be lost. If individual flower stalks are girdled, the berries may drop at any stage of development. Infection of berries causes maximum yield loss. Small berries, when attacked dry and drop off or remain attached to the bunch, dry, shrivel and are mummified. Later attack causes the berries to crack and they become deformed **(Fig.78).**

The causal organism. The mycelium of the fungus is slender, septate, branched, inter- and intracellular and is mostly confined around the points of infection. The hyphae form stromata underneath the host epidermis and produce fruiting bodies (acervuli) in large numbers. The acervuli appear as black, pinpoint-like dots on the spots on leaves, berries and on the cankers. From the inside bottom of the acervuli, numerous, short, erect, unbranched conidiophores arise. From the tips of the conidiophores, single-celled, oblong-ellipsoid, bi-guttulate and hyaline conidia, which are slightly constricted at the middle, are produced in succession. They measure 5.0 - 6.0 x 2.5 - 3.5µ in size. Due to the pressure exerted, the host epidermis is ruptured and the conidia are exposed. The ascigerous stage, which is found in the United States of America, has not been reported in India. The perithecia, when produced are embedded in old cankers. They are ill-defined, pseudoparaenchymatous and open by an apical pore. Asci are ovoid to cylindrical and are found in fascicles. Ascospores are hyaline, 3-septate and measure 15.0 - 16.0 x 4.0 - 4.5µ in size **(Fig.78).**

Mode of spread and epidemiology. Young leaves are more susceptible to the disease. As the leaves get older, resistance to the disease increases. Primary infection occurs from conidia produced in cankers formed during the previous season. The spores are exuded from the acervuli in slimy masses and are splashed by rain or blown by wind. They fall on the tender new flush and germinate by a germ tube. From the germ tube an appressorium is formed, which fastens the conidium to the host surface. From the appessorium, an infection hypha develops, penetrates the host epidermis, enters the host and causes infection. Secondary spread takes place through conidia produced from acervuli on leaf spots, young shoots, tendrils and fruits.

Warm and wet weather favor occurrence of the disease. Heavy rains after pruning leads to severe infection on the newly emerging flush of foliage. Low-lying, ill-drained gardens are usually badly affected.

Disease management

Agronomic practices (i) Diseased shoots, stems, berries etc. should be removed and destroyed (ii) Proper drainage facilities should be provided to the orchard (iii) Disease-free planting materials should be used in new plantings.

Seed treatment. The cuttings used for propagation should be immersed in ferrous sulfate - 25 % solution prior to planting.

Chemical control (i) After cutting off diseased parts, the cut ends should be swabbed with a solution of ferrous sulfate - 2.0 kg. + concentrated sulfuric acid - 500 ml. in 4.0 lit. of water (ii) Spraying with Bordeaux mixture - 1 % or copper oxychloride - 2.5 gm. or mancozeb - 2.0 gm. or captafol - 1.5 gm. or chlorothalonil - 1.5 gm. or carbendazim - 1.0 gm. or bitertanol (Baycor) - 1.0 gm./ lit. of water, at 10 - 15 days interval effectively controls the disease.

Resistant varieties. Bangalore Blue, White Muscat, Angur Kalan, Beauty Seedless, Bharat Early, Delight, Golden Muscat, Golden Queen, Gulabi, Hussaini and Isabella are resistant to the disease.

4. Black rot of grape

Guignardia bidwellii

The disease, native to North America was introduced to France and then to most of the other European countries. It is now found in most of the grape growing countries of the world, including many parts of India. In Tamil Nadu, the disease is found in Madurai and other districts, where grape is cultivated.

Symptoms. The disease initially causes numerous, circular, reddish, necrotic spots on leaves. The spots enlarge and become brown to grayish-tan, while the margins of the spots form a black line. Minute, black, pinpoint-like dots representing *Phyllosticta*-type pycnidia are formed on the upper surface of leaves. The pycnidia are formed in a characteristic circular zone. Somewhat linear lesions appear on the tendrils, flower stalks and leaf veins and on these spots also pycnidia are formed in large numbers. Spots begin to appear on the berries, when they are about half grown. These are whitish at first, but are soon surrounded by a rapidly widening brown ring, with an outer black margin. The central area of the spots remains flat or sunken and dark pycnidia are formed near the center in a circular zone. The whole berry soon becomes rotten, shrinks and turns black and the entire surface of the mummified fruit is covered with numerous, black pycnidia **(Fig.79).**

The causal organism. The mycelium of the fungus is septate, hyaline when young and turns brown later on. The mycelium grows in the host tissue and kills a number of cells, causing necrotic spots on the leaves. The globose pycnidia formed on the leaves produce large number of pycnidiospores, which are exuded out through the pycnidial pore. The pycnidiospores are hyaline, sub-ovoid to elliptical and are 8.0 - 11.0 x 6.0 - 8.0μ in size. They germinate easily by producing germ tubes. The fungus also produces globose pseudothecia in the mummified, rotten berries and stem lesions. No paraphyses are present around the ostiole of the pseudothecia. The pseudothecia contain many asci, with 8 ascospores in each ascus. The ascospores are very unequally 2-celled, the basal cell appearing to be very short, stalk-like and sterile. They measure 12.0 - 25.0 x 5.0 - 7.0µ in size **(Fig.79).**

Mode of survival, spread and epidemiology. The fungus overwinters mostly as ascospores in pseudothecia in the mummified fruits and stem lesions, but conidia also survive the winter in most of the locations. So, both ascospores and conidia can cause primary infection. Ascospores and conidia are released only when the pseudothecia and pycnidia get thoroughly wet. The ascospores are shot out forcibly and are carried by air currents. The conidia, which are exuded in a viscid mass are washed down or splashed by rain. Primary infection takes place on young, rapidly growing leaves and on fruit pedicels. The conidia produced from these lesions spread the disease to other leaves, stems, berries and tendrils.

Frequent heavy rains, humid weather conditions and warm temperature favor disease occurrence and spread.

Disease management

Agronomic practices. Diseased berries, leaves, stems etc. should be removed and destroyed.

Chemical control. Spraying with Bordeaux mixture - 1 % or copper oxychloride - 2.5 gm. or mancozeb - 2.0 gm. or captafol - 1.5 gm. or chlorothalonil - 1.5 gm. or carbendazim - 1.0 gm. or bitertanol (Baycor) - 1.0 gm./ lit. of water, at 10 - 15 days interval commencing from the time the buds break and fresh shoots are formed gives effective control of the disease.

5. Fan leaf disease of grape

Grapevine fan leaf virus

The disease occurs worldwide. It was first reported in India in 1965.

Symptoms. The disease is caused by several strains of the virus and depending upon the strains, infected leaves show green or yellow mosaic rings, line patterns or flecks of green and yellow. The leaves become smaller and slightly asymmetrical. In most of the cases, the veins are spread abnormally and reduction of areas in-between the veins presents a half-

Fig. 78: Anthracnose or Bird's eye-spot of grapes-*Elsinoe ampelina* (= *Gloeosporium ampelophagum*)

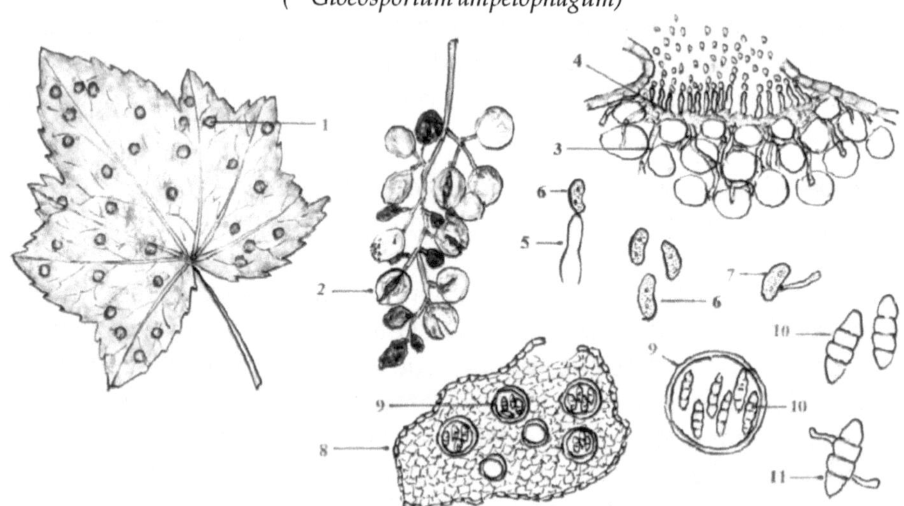

1. Spots on leaf 2. Infected bunch with mummified and cracked berries 3. Inter and intracellular mycelium 4. Acervulus with conidiophores and conidia 5. Conidiophore 6. Conidium 7. Germinating conidium 8. III-defined perithecium with asci and ascospores 9. Ascus with ascospores 10. Ascospore 11. Germinating ascospore.

Fig. 79: Black rot of grapes-*Guignardia bidwellii*

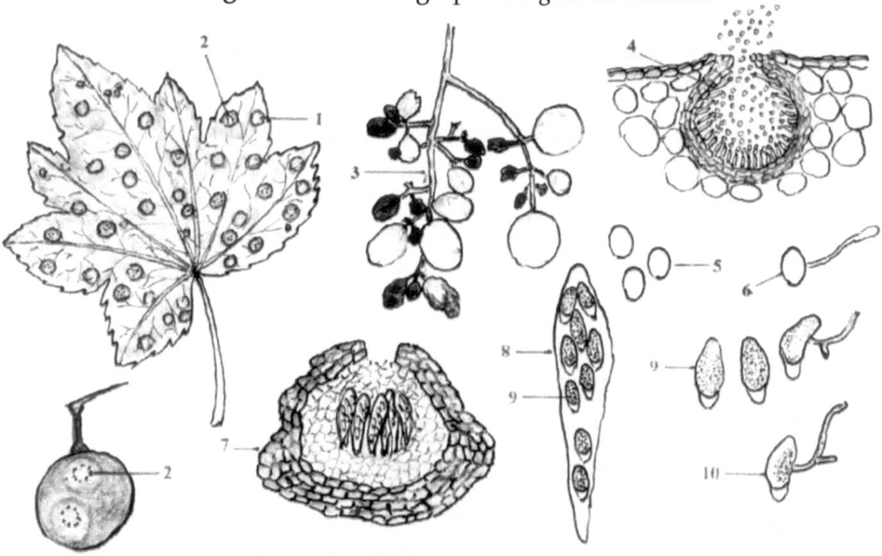

1. Spots on leaf 2. *Phyllosticta*-type pycnidia on the spots on leaf and fruit in a circular zone 3. Infected bunch with shruaken and rotten berries 4. Pycnidia with pycnidiospores 5. Pycnidiospore 6. Germinating pycnidiospore 7. Pseudothecium with asci and ascospores 8. Ascus with ascospores 9. Ascospore 10. Germinating ascospore.

closed, fan-like appearance. Affected leaves stand erect along the axis of young shoots. They may show a chrome-yellow mottling and the mottled areas later turn paler, necrotic and drop off. The stem region is often deformed with uneven internodal lengths, double nodes and pitting of the bark and wood. Many flowers are shed from the clusters and only some small, seedless berries are formed along with a few normal berries. The vigor and yield of grapevines are reduced gradually, the vines degenerate and die eventually.

The causal organism. The virus causing the disease infects only grapevines. It belongs to the Genus - *Nepovirus*. The angular, isometric particles are about 28 nm. in diameter. The genome is bipartite, consisting of two ssRNA molecules. The virus infects the parenchyma and phloem cells and move from cell to cell along the plasmodesmata.

Mode of spread. The virus is transmitted by budding and grafting, by cuttings and by the nematodes, *Xyphinema index* and *X. italiae*.

Disease management

Agronomic practices. Disease-free cuttings should be used for planting.

Chemical control. Controlling the nematode vectors transmitting the virus by soil application of carbofuran 3G - 13 kg. or phorate 10G - 5 kg. or aldicarb 10G - 5 kg./ acre reduces the spread of the disease.

Diseases of minor importance. Grapevine is subjected to attack by several other diseases, however most of them occur only sporadically and are of minor importance. *Mycosphaerella personata (=Cercospora viticola)* causes 'Brown leaf spot'; *Alternaria vitis* causes '*Alternaria* blight'; *Penicillium digitatum* causes 'Blue mould rot', which affects the fruits before and after harvest; *Rhizopus nigricans* causes '*Rhizopus* fruit rot'.

Papaya *(Carica papaya)*

1. Foot rot of papaya

Pythium aphanidermatun

'Foot rot', **'root rot'** or **'collar rot'** is a very serious fungal disease of papaya. The disease is widespread in India, Sri Lanka, Hawaii and South Africa. All cultivated varieties of papaya are susceptible to the disease and in case of severe incidence, almost all the plants in the orchard may be completely destroyed within a single season.

Symptoms. The disease affects young seedlings in the nursery, as well as grown-up plants. Plants that are two to three years old are more vulnerable to attack by the pathogen. The symptoms first appear as small, water-soaked patches on the stem at the ground level. These patches enlarge and girdle the entire collar region of the stem and the inner tissues start rotting and turn dark-brown or black. Simultaneously, the terminal leaves droop and wilt, turn yellow and drop off prematurely. If fruits are formed, they also drop off prematurely. Gradually, the rotting may spread up in the stem above the ground level and down to the roots. The root system is also badly affected and is destroyed. Due to rotting and disintegration of the parenchymatous tissues at the collar region, the entire plant topples over from the base and dies. If the bark is peeled off, the internal tissues appear dry and present a **'honey comb-like'** appearance. In the nursery, the pathogen may cause damping off of seedlings.

The causal organism, mode of survival and spread. In India *Pythium aphanidermatum* is reported to be the main causal organism of the disease, while in Hawaii *Pythium aphanidermatum* and *Phytophthora parasitica* are involved in causing the disease. For the description of *Pythium aphanidermatum* refer 'Damping off of seedlings of eggplant' (Page 339), **(Fig.85).**

Epidemiology. Usually the disease occurs during the monsoon seasons, especially during June - July, when there is plenty of rainfall and the temperature is also high. Optimum temperature for the development of the disease is 36°C. Presence of abundant moisture around the base of the stem predisposes the plant to infection and rapid development of the disease.

Disease management

Agronomic practices (i) Papaya plants should be grown on well-drained, loam soils and water-logging should be avoided (ii) Plants showing symptoms of the disease should be uprooted and burnt and the same pit should not be used for replanting papaya seedling immediately.

Chemical control (i) Seeds should be treated with captan - 4.0 gm. or thiram - 4.0 gm. or chlorothalonil - 2.0 gm./ kg. of seeds, at least 24 hours prior to sowing (ii) The affected tissues in the stem of affected plants may be scrapped off and Bordeaux paste may be applied on the exposed surface (iii) The soil around the stem, as well as the basal stem portion may be drenched with Bordeaux mixture - 1.0 % or copper oxychloride - 2.5 gm. or captan - 2.0 gm. or metalaxyl - 1.0 gm. or carboxin - 1.0 ml. or chlorothalonil - 2.0 gm./ liter of water.

2. Leaf curl of papaya

Tobacco leaf curl virus or *Nicotiana virus* 10

The disease is commonly found in kitchen gardens and commercial orchards in many regions of India, including Uttar Pradesh, Bihar and Tamil Nadu. The disease attacks all cultivated varieties of papaya.

Besides papaya, the virus is known to attack several other crops, such as tomato, tobacco, chillies, sunnhemp, zinnia, petunia, hollyhock, *Datura stramonium*, many ornamental plants and weeds.

Symptoms. The disease is characterized by severe crinkling and curling of leaves, accompanied by vein- clearing and reduction in leaf size. Leaves become leathery, hard and brittle. The interveinal areas of the leaf lamina are raised on the upper surface due to hypertrophy of the cells resulting in rugosity. The leaves roll downward and inward and the veins become thick and dark green in color. Almost all the leaves, including those at the top show such symptoms. The petioles are also twisted in an irregular and zigzag manner. Severely affected plants do not produce flowers and fruits.

The causal organism. The virus causing this disease belongs to the Genus - Geminivirus. The morphology of the virus particles is unique in that their isometric particles of size 18 - 20 nm. diameter occur mainly in pairs. The genome of the virus particle consists of ssDNA.

Mode of spread. The disease is not mechanically transmitted. In nature, the white fly vector, *Bemisia tabaci* transmits the virus.

Disease management

Agronomic practices (i) Diseased plants do not recover and do not produce fruits. So, such plants should be uprooted and destroyed (ii) Alternate hosts should not be allowed to grow near about papaya orchards.

Chemical control. The white fly vectors should be kept under check by periodical sprayings with methyl demeton - 1.0 ml. or dimethoate - 1.0 ml. or acephate - 2.0 ml. or phosalone - 2.0 ml or monocrotophos - 1.0 ml./ lit. of water.

3. Mosaic of papaya

Papaya mosaic virus

The disease is commonly found in many countries of the world, including Russia, the United States of America, Peru, India and Venezuela. In India, it was first reported from Maharashtra in 1948 and now it occurs in almost all the states, including Uttar Pradesh, West Bengal, Rajasthan, Orissa, Punjab, Bihar, Telungu Desam, Madhya Pradesh, Karnataka and Tamil Nadu.

Symptoms. The disease attacks the crop at all stages of growth, however about one year old plants are seriously affected by the disease. Almost all the cultivated varieties of papaya are susceptible to the disease. Typical mosaic symptoms, with yellow and green irregular patches, accompanied by mottling and puckering appear on the young leaves. The lower, matured leaves show no abnormalities. Newly formed leaves are severely affected and symptoms, such as chlorosis, reduction in size and malformation appear and the plants become stunted. Defoliation occurs in older plants and only a tuft of small, shoe string-like leaves remain at the top. Fruits are deformed and smaller in size. Sometimes, chlorotic spots or patches may develop on the fruits and stems also.

The causal organism. *Papaya mosaic virus* or *Carica virus* 1 causes the disease. The virus belongs to the Genus - *Potexvirus*. The virus particles are flexuous rods with helical symmetry, 470 - 580 nm. in length and 13 nm. in width. The genome consists of a single ssRNA.

Mode of spread. The virus can be transmitted by sap inoculation. In nature, the aphid vectors, *Myzus persicae, Aphis malvae, A. medicaginis* and *A. gossypii* transmit the virus in a non-persistent manner. It is not transmitted through seed. Besides cultivated papaya *(Carica papaya)*, other wild species, such as *Carica microcarpa, C. goudotiana, C. candamarcensis* and several cucurbits, such as *Trichosanthes anguina, Lagenaria siceraria, Cucurbita maxima, C. pepo, Cucumis sativus* and *Citrullus vulgaris* are also attacked by the virus.

Disease management

Agronomic practices (i) Disease-free and healthy seedlings should be used for planting (ii) Periodical rouging and destruction of diseased plants help to prevent the spread of the disease (iii) Growing alternate hosts near about papaya orchards should be avoided.

Chemical control. Periodical sprayings should be taken up with suitable systemic insecticides to keep the vectors in check.

Diseases of minor importance. Papaya is also subjected to attack by a few other fungal and virus diseases. *Pythium aphanidermatum* and *Rhizoctonia solani* cause 'Stem rot' that affects the plants at the ground level, leading to rotting of the stem and the root system and eventual death of the plants; *Oidium caricae* causes 'Powdery mildew' that affects the leaves, flower stalks and fruits; *Colletotrichum papayae (Colletotrichum gloeosporioides)* causes 'Anthracnose' of fruits; *Rhizopus stolonifer* causes 'Fruit rot'; *Papaya ring spot virus* causes 'Ring spot'.

Guava *(Psidium guajava)*

1. Fruit canker or scab of guava

Pestalotiopsis psidii

The disease is found in almost all the regions of India where guava is grown.

Symptoms. The symptoms appear on green fruits as minute, brown, circular and rough lesions. The lesions enlarge and become roughly circular, brown to rust-colored, scabby, cork-like spots, and 2.0 - 4.0 mm. in diameter. The center of the lesions becomes hard and develops cracks and the margin of the lesions becomes slightly raised. Lesions also appear on the leaves. The market value of the disfigured fruits is greatly reduced.

The causal organism. The mycelium of the fungus is septate, inter- and intracellular. Before fructification, the hyphae aggregate beneath the epidermis of the spots and form stromata. From the stromata, acervuli are formed subepidermally, which appear as minute, black, pinpoint-like dots on the spots. From the base of the acervuli, numerous, short conidiophores are formed, which bear conidia at their tips singly. Conidia are pedicellate, oblong, clavate, 5-celled and measure 13.0 - 31.0 x 5.0 - 10.0µ in size. The central three cells are light-brown in color, while the two terminal cells are hyaline. The apical cell bears three, long, slender, hyaline appendages.

Mode of spread. The conidia produced from lesions on the leaves and fruits are dispersed by wind and cause fresh infections, when conditions are favorable.

Disease management

Chemical control (i) Periodical spraying with Bordeaux mixture - 1 % or copper oxychloride - 2.5 gm./ lit. of water from the time of fruiting is effective in controlling the disease (ii) Post-harvest dipping of fruits in Aureofungin Sol - 200 ppm. (0.2 gm./ lit. of water) protects the fruits from infection.

2. Red rust of guava

Cephaleuros virescens

Symptoms. The disease attacks the plants at all growth stages and all the parts of the plants, such as leaves, stems, twigs and fruits are affected.

On the leaves, the symptoms appear as small, scattered, reddish, slightly raised, rough spots. The spots may enlarge to become roughly circular or irregular patches, about 1.0 cm. in diameter. Sometimes the spots are sunken, with slightly raised margins. Such spots or patches may appear on the stems and twigs also. Older patches on stems and twigs turn purplish-red and develop longitudinal cracks. Sometimes, the fruits are severely affected. The lesions on the fruits are usually smaller and reddish-brown to dark-brown in color, rough and corky in appearance.

The causal organism. The algal filaments penetrate the epidermal cells and grow in-between the palisade cells, but never extend deeper into the mesophyll tissues. The affected cells are killed and they turn brown. The pathogen attacks the bark of the stems and twigs and then penetrates deeper into the cortex. Under favorable conditions, fructifications appear as fine, cottony growth from the red patches. The fructifications consist of sporangiophores and sporangia. From the sporangia, which are formed at the tips of sporangiophores, biflagellate, spherical or oval, orange colored zoospores are liberated.

Mode of spread and epidemiology. When conditions are unfavorable, the algal filaments remain dormant in the leaf and stem lesions. When favorable conditions return, the algal filaments become active and produce fructifications

Rain accompanied by wind helps in the dissemination of the spores. Moist conditions and low temperatures between 20° - 25°C favor occurrence of the disease. Free water on the host surface is necessary for the zoospores to cause fresh infection.

Disease management

Agronomic practices. Field sanitation and proper maintenance of the crop to keep up the vigor and vitality of the plants are necessary to ward off the algal attack, as weak and unthrifty plants are more vulnerable to attack.

Chemical control. Spraying with Bordeaux mixture - 1 % or copper oxychloride - 2.5 gm./ lit. of water controls the disease.

Diseases of minor importance. A few other diseases are also known to occur on guava. *Fusarium oxysporum* f. sp. *psidii*, *Fusarium solani* and *Macrophomina phaseolina* cause 'Wilt'; *Gloeosporium psidii* causes 'Anthracnose'. 'Die-back', 'Fruit rot' and 'Twig blight'.

Sapota *(Achras sapota)*

Only a very few diseases are known to attack sapota plants and fruits and cause appreciable damage. *Phavophloeospora indica* causes 'Pink leaf spot'. The disease is common in the Karnataka state. The symptoms appear as small, pinkish to dark-brown, circular spots, scattered on both the surfaces of leaves. The spots may coalesce and cover large areas of the leaves, leading to premature leaf fall. The disease is more severe during October - December, when the humidity is high. The pathogen grows best at a temperature 25°C and 90 % relative humidity. Spraying with zineb - 2.0 gm. or ziram - 2.0 gm. or copper oxychloride - 2.5 gm./ liter of water, at monthly intervals controls the disease effectively; *Pestalotiopsis versicolor (Pestalotia sapotae)* causes 'Leaf spot' and 'Fruit rot'; *Glomerell cingulata (=Colletotrichum gloeosporioides)* causes 'Anthracnose'.

Pomegranate *(Punica granatum)*

1. Cercospora leaf spot of pomegranate

Cercospora punicae

Symptoms. The disease is characterized by the formation of light-brown, zonate spots on the leaves and fruits. Dark-brown to black, elliptical or elongated spots appear on the twigs. The affected areas become flattened or slightly depressed, with somewhat raised margins. The lesions may girdle the twigs and the twigs may dry and die.

The causal organism. The fungus produces conidiophores, which arise from the host surface in clusters through the stomata. The conidiophores are short, olive-brown in color and sparsely septate and produce conidia successively on the growing tips. The conidia are hyaline to pale olive-brown in color, long, slender, cylindrical, with slightly tapering apex, slightly curved, multicellular and measure 40.0 - 50.0 x 3.0µ in size. They are easily detached and often blown long distances by wind.

Mode of survival, spread and epidemiology. The disease is spread mostly by wind-borne conidia. The pathogen overseasons as stromata in old affected leaves or twigs.

The disease is most destructive during the summer months and in warmer climates. The spores require free water to germinate, penetrate the host and cause infection. Heavy dew is sufficient to cause severe infection. The fungus produces a toxin, **'cercosporin'**, which causes disruption of cell membranes and death of cells leading to necrosis.

Disease management

Agronomic practices. Diseased twigs should be pruned and destroyed.

Chemical control. Spraying with Bordeaux mixture - 1 % or copper oxychloride - 2.5 gm. or mancozeb - 2.0 gm. or chlorothalonil - 1.5 gm. or thiophanate methyl - 1.0 gm./ lit. of water controls the disease effectively.

2. Bacterial blight of pomegranate

Xanthomonas campestris pv. *punicae (X. axonopodis* pv. *punicae)*

Symptoms. The disease affects all aerial parts of the plants. On the leaves, the symptoms appear as small, irregularly circular, translucent, water-soaked spots, 2.0 - 5.0 mm. in diameter. The spots, which appear on the lower surface of leaves, may appear on the upper surface also later on. The spots turn brown to dark-brown and are surrounded by a water-soaked, yellow margin. The center of the spots becomes gray and necrotic. The spots may coalesce to form larger patches. Severely affected leaves drop off prematurely. On the branches and twigs, irregular, brown to black lesions appear on the nodal regions. The lesions enlarge and girdle the nodes, as a result the branches break and die. Brown to black, irregularly circular, water-soaked, scabby spots develop on the surface of the fruits also. The center of the spots becomes necrotic and develops cracks. Severely affected fruits dry.

The causal organism. The bacteria causing the disease are short, rod-shaped, motile by a single, polar flagellum and measure 1.0 - 2.5 x 0.5 - 1.0µ in size. They are gram-negative and aerobic in nature.

Mode of survival, spread and epidemiology. The bacteria survive in the lesions on the infected leaves, branches and twigs. They can also survive in the fallen leaves for about 4 months. During warm, rainy weather, the bacteria ooze out from the lesions and are disseminated by lashing rains. The bacteria enter the host only through natural openings or injuries. The disease is also carried through infected seedlings.

Warm and wet weather conditions favor the occurrence of the disease.

Disease management

Agronomic practices. Field sanitation and pruning and destruction of affected twigs and branches help to minimize the occurrence of the disease.

Chemical control. Spraying the plants with Bordeaux mixture - 1 % or copper oxychloride - 2.5 gm. or a mixture of Streptomycin sulfate - 0.5 gm. (500 ppm.) + copper oxychloride - 2.5 gm./ lit. of water is effective in controlling the disease.

Diseases of minor importance. A few other diseases are also known to attack pomegranate. *Colletotrichum gloeosporioides* causes 'Anthracnose', that affects the leaves and fruits; *Sphaceloma punicae* causes 'Leaf spot'; *Pestalotiopsis versicolor* causes 'Leaf and fruit spot'.

Anona *(Anona squamosa)*

Very few diseases occur on anona or custard apple. *Pellicularia salmonicolor* causes 'Pink disease', that affects the young woody branches leading to die-back; *Trichothecium roseum* and *Glomerella cingulata* cause 'Fruit rot'.

Jack *(Artocarpus heterophyllus / A. integrifolia)*

1. Inflorescence rot of jack

Rhizopus artocarpi

Symptoms. The fungus affects young female inflorescence and young fruits. Soft rot is caused on the affected parts and they drop off prematurely. Matured fruits are not affected. Dense, fleecy, whitish or grayish, cottony mycelium appears on the inflorescence and young fruits. The fungal growth soon spreads over the entire affected parts. The whitish mycelial growth turns black when sporangia are produced in large numbers from the aerial mycelium.

The causal organism. The fungus causing the disease is a weak, facultative parasite. The mycelium is coenocytic, hyaline, endophytic and epiphytic. The white, cottony aerial hyphae or stolons have nodes at several points, where they touch the host surface. From each of the nodes, rhizoids are formed below and sporangiophores above. The rhizoids grow down into the host tissues and absorb nourishment. The short, stout and stiff sporangiophores, which arise in fascicles from the nodes bear black sporangia singly. The sporangial wall breaks up into fragments in dry air and the dark-colored sporangiospores are blown away by wind. The fungus also produces black colored zygospores, as a result of sexual reproduction, which on germination produce zygosporangium at the end of a short stalk and non-motile sporangiospores are formed inside the zygosporangium.

Mode of spread and epidemiology. Being a facultative parasite, the fungus can survive in the soil for long periods and attack the inflorescence at the time of flowering. The fungus can also survive on the fallen floral parts and fruits. The fungus enters the host usually through natural openings or through injuries caused by insects. After gaining entry into the host, it attacks the inner tissues and causes rotting.

Moderate temperature, high humidity and intermittent rains at the time of flowering favor the occurrence of the disease.

Disease management

Agronomic practices. Infected floral parts and fruits fallen on the ground should be removed and destroyed periodically.

Chemical control. Spraying with Bordeaux mixture - 1% or copper oxychloride - 2.5 gm. or captan - 1.5 gm./ lit. of water, at the time of flowering, followed by one or two sprayings at 3 weeks interval controls the disease. Care should be taken to cover the inflorescence thoroughly with the fungicidal fluid.

2. Fruit rot of jack

Phytophthora palmivora

Symptoms. The disease usually appears two to three weeks after the beginning of monsoon rains. Water-soaked spots or patches develop on the outside of the fruits, mostly at the basal portion and the affected portions start rotting. As the disease advances, the fungus develops on the outside of the fruits. Attacked fruits begin to shed prematurely. On the fallen fruits, whitish, felt-like mass of mycelium is seen, which soon covers the entire surface of the fruits.

The causal organism. Refer 'Bud rot of coconut' (Page 457), **(Fig.111).**

Mode of spread and epidemiology. The fungus is soil-borne and can live saprophytically in the infected fruits, plant debris and organic matter in the soil. The oospores, if formed and the chlamydospores can withstand adverse weather conditions and remain in a viable state for prolonged periods and may cause primary infection. Secondary spread is by sporangia, which are disseminated by wind and rainsplash and mechanically by some insects.

High humidity, wet weather and low temperature favor the occurrence of the disease. Injuries caused by sucking insects and other insect pests predispose the fruits to infection.

Disease management

Agronomic practices. Field sanitation helps in reducing the occurrence of the disease to a large extent.

Chemical control. Spraying with Bordeaux mixture - 1% or copper oxychloride - 2.5 gm. or captan - 1.5 gm./ lit. of water, at the time of fruit formation, followed by two or three sprayings at 3 weeks interval controls the disease effectively. Care should be taken to cover the young fruits thoroughly with the fungicidal fluid.

Diseases of minor importance. Besides the above-mentioned diseases, a few other diseases also affect jack. *Pellicularia salmonicolor* causes 'Pink disease' that affects the young woody branches, leading to die-back; *Phyllosticta artocarpina, Pestalotiopsis elastica* and *Colletotrichum gloeosporioides* cause 'Leaf spots'.

Pineapple *(Ananas sativus / A. comosus)*

1. Basal rot, leaf spot and fruit rot of pineapple

Thielaviopsis paradoxa
(Ceratocystis paradoxa)

Symptoms. Basal rot affects the base of the plants soon after planting. The disease is characterized by rotting of the butt, followed by wilting of the leaves. The softer tissues of the butt are destroyed and only the stringy fibers remain without rotting. The diseased plants break off easily at the ground level. On the leaves, grayish spots with dark margin develop. As the disease advances, the spots become necrotic and the leaves are distorted. On the fruits, blisters may appear, followed by soft, water-soaked, rotten patches. The decaying flesh emits a sweetish odor. On the rotten fruits and leaves, the fungal growth and masses of black spores are seen.

The causal organism. Refer 'Pineapple disease of sugarcane' (Page 212), **(Fig.110).**

Mode of survival, spread and epidemiology. The macroconidia remain in the soil and in the diseased plant parts for a long time and cause fresh infection. The perithecia formed on the rotten plant parts and fruits also help in the perpetuation of the disease. Further, in the absence of the host the fungus can continue its life as a saprophyte in the dead plant tissues and organic matter present in the soil. The fungus is a weak, wound parasite and can gain entry into the host only through cut ends, cracks, injuries, bruises etc. Infection occurs during picking, packing and in transit. The spores are carried through wind, irrigation water and rain water from field to field and cause secondary infection. The pathogen has several alternate hosts, such as coconut, arecanut, date palm, cocoa, coffee, mango etc. and spores produced from these hosts may cause primary infection.

The incidence of the disease is more in heavy clay soils and in water inundated fields. The disease is more in alkaline or saline soils. Warm and wet conditions favor the occurrence of the disease.

Disease management

Care should be bestowed at every stage of the crop growth, from selection of planting materials, till the time of marketing of the fruits. (i) Pineapple is commonly propagated from suckers or slips. Suckers arise from the underground parts of the plants and slips arise from the fruiting stem and from the crown on top of the fruit. After the fruit is harvested, the stalks are cut into discs and are also used for propagation. Suckers or slips are first cured by stripping off the lower leaves, followed by drying in the sun or in partial shade for 3 to 4 days prior to planting. During all these stages, there is every possibility of the fungus infecting the planting materials through wounds. So, the places where planting materials are stored should be kept clean and storing the planting materials in heaps should be avoided (ii) Dipping the planting materials in a fungicidal solution of captafol - 1.5 gm./ lit. of water prior to planting helps to eradicate the spores adhering to the planting materials (iii) Proper drainage facilities should be provided in the fields and planting during wet periods should be avoided (iv) Disease affected plants in the field should be removed and destroyed (v) Plant debris in the field and in the vicinity of storing and packing sheds should be collected and destroyed (vi) Care should be taken, while handling the fruits at the time of harvest, storing and packing to avoid injuries (vii) Fruits should not be packed, when they are wet (viii) Dipping the fruits in a solution of thiobendazole - 1.0 gm. or carbendazim - 1.0 gm./ lit. of water for 3 minutes prior to packing helps to avoid infection of the fruits (ix) The store-houses and packing sheds should be disinfected periodically using formalin solution.

Diseases of minor importance. Besides the above-mentioned disease, a few more diseases affect pineapple. 'Heart rot' is caused by *Phytophthora cinnamomi* and *P. parasitica*; 'Wilt' is caused by *Pineapple wilt virus*.

Cashew *(Anacardium occidentale)*

1. Anthracnose of cashew

Colletotrichum gloeosporioides
(Glomerella cingulata)

The disease is known as **'brown leaf spot'**, **'blossom blight'**, **'die-back'** and **'twig blight'**. It occurs in all the regions, where cashew is grown. The disease is more severe in badly maintained and neglected gardens.

Symptoms. The disease affects leaves, twigs, shoots, inflorescence, fruits and nuts. On the leaves, many small, brown, circular to irregular spots, with dark colored margin appear. Under humid conditions, the spots enlarge and form irregular necrotic areas. When young, tender leaves are attacked, they become crumpled and dry. Severely affected leaves drop off prematurely. Twig blight or die-back symptoms appear at the tip of young branches. Reddish-brown, longitudinal, water-soaked lesions appear at the tip of the branches, followed by exudation of resin. Soon, the branches dry from the tip downwards, accompanied by defoliation. During the flowering season, small, dark, sunken spots appear on the main stalk of the inflorescence, lateral branches and individual flower stalks. Soon, the flowers dry and drop off, resulting in blossom blight. The fungus also affects the fruits and nuts. Affected young fruits and nuts are shrivelled and are shed. In the matured fruits, dark-brown to black, circular, sunken spots develop and the fruits rot. On the lesions and dead parts, minute, brown, pinpoint-like dots, which are the acervuli of the fungus, are formed in large numbers.

The causal organism, survival, mode of spread, epidemiology and disease management. Refer 'Anthracnose of mango' (Page 283), **(Fig.73).**

Diseases of minor importance. A few other diseases are also known to attack cashew. *Pythium* species, *Phytophthora palmivora* and *Cylindrocladium scoparium* cause 'Seedling blight'; *Corticium salmonicolor* causes 'Pink disease'; *Oidium anacardii* causes 'Powdery mildew'.

Ber *(Zizyphus mauritiana / Z. jujuba)*

A few diseases attack ber. *Oidium erysiphoides* f.sp. *zizyphi* causes 'Powdery mildew'; *Alternaria chartarum, Cercospora zizyphi, Septoria capensis, Pestalotiopsis subinae* etc. cause 'Leaf spots'; *Phakopsora zizyphi-vulgaris* causes 'Rust'.

Apple *(Pyrus malus)*

1. Apple scab

Spilocaea pomi
(Venturia inaequalis)

'Scab' is the most important disease of apples. Its primary effect is reduction of the quality of fruits. Scab also results in the reduction of fruit size and causes premature fruit drop, defoliation and poor fruit-bud development for the next year. It also reduces the duration for which infected fruits can be kept in storage. Losses due to scab of apple may be up to 70 per cent or more of the total fruit value. In most apple-producing areas, no marketable fruits can be harvested, if scab control measures are not taken.

Symptoms. The disease attacks the leaves, shoots, buds and blossoms, as well as fruits. Symptoms on the leaves consist of a number of scattered, roughly circular, brown or olive-green spots around which a dendritic margin (radiating fine branches) is seen due to the proliferation of mycelium radiating and branching all round the spot below the cuticle. Later, the spots turn gray and necrotic in parts, leading to malformation, thickening and puckering of the lamina around the infected areas Due to inequality of growth below the spots, the leaf may be covered in parts with irregular, diffused spots, which ultimately assume a brown, velvety texture or the blade may appear as if scorched. Spots may arise on both the surfaces of the lamina, but on the youngest leaves, as they emerge from the bud, the exposed lower surface is affected first. Infected young leaves remain small and curled and may later fall off. Lesions may remain distinct or they may coalesce. Small spots may appear on the bud scales and on adjacent parts of the stem. When flower buds are formed, similar spots may appear on the protecting sepals as well. Thus, bud-scale and shoot lesions play a vital role in carrying over the disease from year to year. When the trees are in bloom, lesions may be seen on the sepals and on the flower stalk. In case of severe infection, the stalks may be girdled, as a result, flowers and young fruits may drop.

The fruit is the part to suffer most from the ravages of scab. Infected fruits develop circular scab lesions similar to the leaf lesions and are velvety, olive-green at first, but later becoming darker, scabby and sometimes cracked. The cuticle of the fruit is ruptured at the margin of the lesions. Fruits infected early become misshapen, cracked and often drop prematurely. Fruits infected when approaching maturity form only small lesions, which may develop into dark scab spots during storage **(Fig.80).**

Causal organism. The mycelium of the pathogen is located only between the cuticle and epidermal cells. There are two stages in the life history of the pathogen, the more important conidial stage *(Spilocaea pomi / Fusicladium dendriticum)*, which appears on all affected living parts of the host and the less significant perithecial *Venturia* stage, occurring on the overwintered leaves on the ground. The conidial stromata are sub-cuticular. The brown, wavy conidiophores abstrict conidia from the apex. The conidia, which are produced one at a time, are ovoid, olive-brown and measure 28.0 - 40.0 x 7.0 - 10.0 µ in size. Conidial pustules on the fruit show the dark olive conidia, surrounded by a fringe of white, torn cuticle.

Before leaf-fall, the mycelium of the sub-cuticular stromatic tissue on the leaves extends down into the host mesophyll to form primordia of the future pseudothecia, but the pseudothecia attain full development only after the leaves have overwintered on the ground. These fructifications appear as small, dark-brown or black pimples, embedded in the leaf, opening by a short beak. They are spherical, 90-170μ in diameter and have a dark wall. Each pseudothecium may contain 50-100 asci. The eight-spored asci, elongate considerably on maturity and discharge the uniseriate ascospores through the ostiole. The ascospores are pale-green, two-celled, with the upper cell smaller than the lower cell and measure 12.0-15.0 x 6.0-7.0μ. in size **(Fig.80)**.

Mode of survival, spread and epidemiology. Conidia and ascospores are capable of causing primary infection in the spring. The conidial pustules, which remain protected, overwinter on the young shoots or bud scales, provide the earliest inoculum for spring infection. However, under favorable weather conditions, the ascospores serve as the primary source of infection.

The conidia are easily dislodged from the pustules on twigs and bud scales and are disseminated in splashes of rain or the leaves may become infected by direct contact with the pustules. Secondary infection of foliage and fruits takes place during wet periods on a wide scale. Young leaves are more susceptible and the leaves must be thoroughly wet prior to infection and penetration, which is direct. The germ tube forms a flange-like appressorium, which is held fast to the leaf by a mucilaginous sheath, while a narrow penetration hypha is pushed through the cuticle.

The optimum temperature for germination of conidia is about 22°C and for the ascospores, it ranges from 10-18°C. Humid, cool weather in spring and early summer is highly favorable for the incidence of scab disease. Densely foliaged trees are often heavily infected because of the cool microclimate within the canopy. Fruit infection takes a long period of time for incubation and longer periods are necessary for the development of the disease as the fruit develops, while leaf infections take a shorter period.

Disease management

Agronomic practices (i) Since primary attacks are often traceable to scabbed twigs and buds, these sources of infections should be removed before the pustules break open in the spring and release conidia (ii) Fallen leaves and other debris should be collected in the autumn and destroyed by burning or ploughed deeply into the soil to prevent spread by ascospores.

Chemical control (i) For an effective apple scab control program, apple trees should be sprayed or dusted before, during or immediately after a rain, from the time of bud-break, until all the ascospores are discharged from the pseudothecia. Several fungicides, such as captan, mancozeb, benomyl, thiophanate-methyl (Topsin M), fenarimol (Rubigan), myclobutanil (Rally) and dodine (Cyprex) give excellent control of apple scab. (ii) Apple plants should be completely dipped in Bordeaux mixture before export or distribution to the grower.

Some of the new strains of *Venturia* are found to have developed resistance to some of the fungicides, such as dodine, benomyl and a few other systemics. So, they must be applied in combination with one of the broad-spectrum fungicides, such as captan or mancozeb

Resistant varieties. Several apple varieties resistant to scab are available. However, all the popular varieties are moderately to highly susceptible to this disease.

2. Powdery mildew of apple

Podosphaera leucotricha

The disease is of common occurrence in all apple-growing countries of the world. In the United States of America, Britain and Czechoslovakia, it causes severe damage in some seasons. The disease is widely prevalent in the apple orchards of Northern India. It damages nursery plants more severely than older ones. The disease also attacks pear and quince.

Symptoms. The disease usually appears, as soon as the buds burst into new leaves and shoots. Buds severely infected in the previous season are killed. Infected buds, even if they sprout and grow, produce only weakened and mildewed shoots. Initial symptoms on the leaves from such buds appear as small patches of white or gray mildew on the under surface of leaves, but later both the surfaces are covered with a whitish, powdery coating consisting of mycelium, conidiophores and conidia of the fungus. Affected leaves become longer and narrower than healthy ones and their margins curl and appear to be somewhat blistered or crinkled. Subsequently, the leaves turn brown from the tip downwards. Soon, the discoloration spreads over the entire leaf lamina and the leaves present a scorched and bronzed appearance. Eventually, the leaves die. Such foliage attack leads to partial defoliation of the branches. Severe foliage attack may result in complete defoliation. The powdery lesions appear on young shoots also. In case of early infection of young nursery plants, growth is very much retarded and wood formation may be completely prevented. Fruit buds are more vulnerable to attack than vegetative buds. Affected fruit buds fail to blossom. The flowers of affected blossoms are usually malformed, with distorted sepals and pedicels and narrowed petals. The pistils and stamens become sterile, as a result fruits are not formed. When young fruits are attacked, they remain diminutive, deformed and develop a roughened surface **(Fig.81).**

Fig. 80: Apple scab-*Venturia inaequalis (=Spilocaea pomi)*

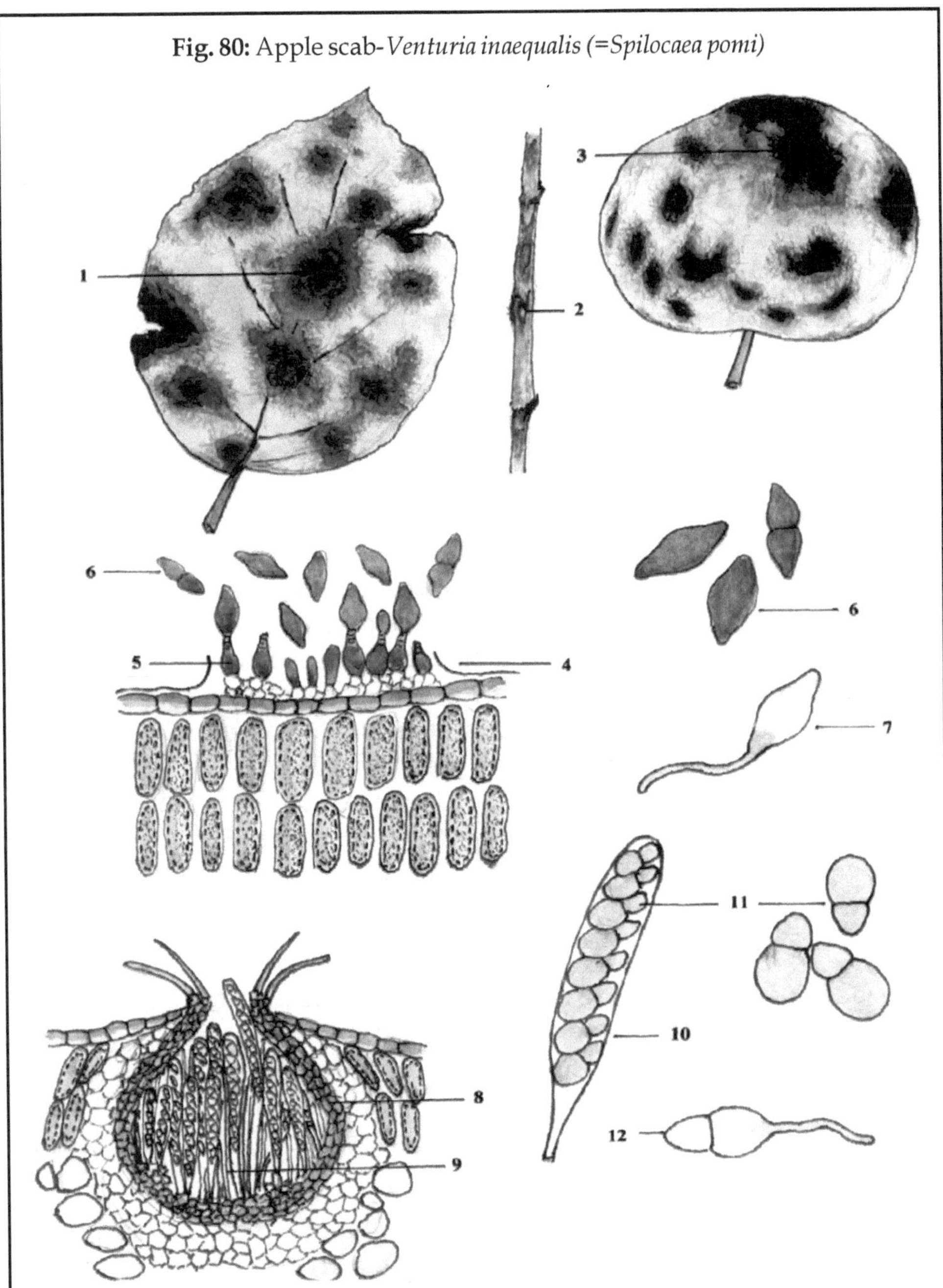

1. Spots on leaf with dendritic margin 2. Spots on the twig 3. Spots on the fruit 4. Conidial stage 5. Conidiophores 6. Conidia 7. Germinating conidium 8. Pseudothecium 9. Hymenium 10. Ascus with ascospores 11. Ascospores 12. Germinating ascospore.

The causal organism. The superficial mycelium of the fungus sends saccate haustoria into the epidermal cells of the host. Erect, aerial conidiophores arise in profusion from the ectophytic mycelium on the leaves and young shoots all through the season. Each conidiophore produces a chain of conidia, which are set free and dispersed by wind. The conidia are hyaline, oval and are 28 - 30 x 10 - 12μ in size. The germ tube of the germinating conidium penetrates the epidermis directly without forming an appressorium and a simple, spherical haustorium gets established in the penetrated cell. In the leaf, the infection is mostly restricted to the epidermal cells and the developing hyphae spread out over the leaf to form the superficial mycelium. In severe attacks, the fungus may sometimes penetrate deeper into the mesophyll.

The ascigerous stage is rarely found in India, but occurs in some other countries. They occur mostly on the wood of the current season, as well as on suckers, on the mid rib and large veins of leaves or on the petioles. These fructifications (cleistothecia) are sub-globose, black, partially embedded in the mycelial mat and measure 75 - 96μ in diameter. Two kinds of appendages are present on their surface. Those on the apical, depressed part are long, stiff and bristle-like, while the others emerging near the base are short and tortuous and help to secure the cleistothecium to the mycelial mat. Each cleistothecium contains a single ascus measuring 55 - 70 x 44 - 50μ. Usually, the ascus is ejected forcibly from the cleistothecium and when it comes in contact with water, explodes to set free the ascospores. Each ascus contains eight, hyaline, single-celled ascospores, measuring 20 - 26 x 12 - 14μ in size. Cleistothecia play a little part in the perennation of the fungus in some countries **(Fig.81).**

Mode of survival, spread and epidemiology. Usually, the fungus survives in the form of dormant mycelium or saccate haustoria in the buds and cause primary infection in the ensuing season. Once the mycelium comes out after the buds burst and infects the new leaves and shoots, conidiophores and conidia are produced in abundance and are dispersed by wind to cause secondary spread of the disease.

High atmospheric humidity, rather than actual moisture on the leaf surface is essential for germination of conidia and penetration into the host cells. The optimum temperature range for germination of the conidia is between 19° - 25°C and that for the superficial growth is 20°C. The incidence of the disease is more in poor and heavy clay soils.

Disease management

Agronomic practices. Proper pruning to stimulate vigorous growth, removal and destruction of all shoots mildewed during the previous season, cutting off all shoots showing signs of mildew, adequate manuring etc. help in the control of the disease.

Chemical control. Spraying the plants with wettable sulfur - 4.0 gm. or dinacap - 1.25 gm. or carbendazim - 1.0 gm./ lit. of water controls the disease. Three sprayings should be given, commencing from the time the buds start swelling.

3. Fire blight of apple

Erwinia amylovora

The disease was first reported from the United States of America in 1780. It causes serious damage to apple and pear orchards in many parts of the world. Several other species of the pome fruit group including stone fruits and many ornamentals are also affected. In India, it was reported in 1943 and is prevalent in Himachal Pradesh, Jammu and Kashmir and Uttar Pradesh.

Symptoms. The pathogen attacks flowers and twigs and kills them. Large branches and trunks may be girdled and eventually killed. Young trees may be killed in a single season.

Infected flowers become water-soaked, shrivel, turn brownish-black and fall down or remain clinging to the tree. The blighted flowers appear scorched, as if burnt by fire and hence the disease is called **'fire blight'**. Soon, leaves on the same branch or nearby twigs develop brown to black blotches along the mid rib and main veins or along the margins. As the blackening progresses, the leaves curl, shrivel, hang down and cling to the curled, blighted twigs. Terminal twigs wilt from the tip downward, their bark becoming brownish - black and soft at first, but shrinks and hardens later. From the fruit spurs and twigs the symptoms progress down to the branches, where cankers are formed. The bark of cankers is water-soaked in the beginning and becomes darker, sunken and dry later on. The cankers may enlarge and encircle the branch. In such cases, the branches above the point of infection die. Infected, small, immature fruits become water-soaked, turn brown to black, shrink and may cling to the tree for several months.

Under humid conditions, droplets of milky white, sweet, sticky bacterial ooze may appear on the surface of recently infected parts and may run down the surface of the infected twigs. The ooze usually turns brown soon after exposure to air **(Fig.82).**

Fig. 81: Powdery mildew of apple-*Podosphaera leucotricha*

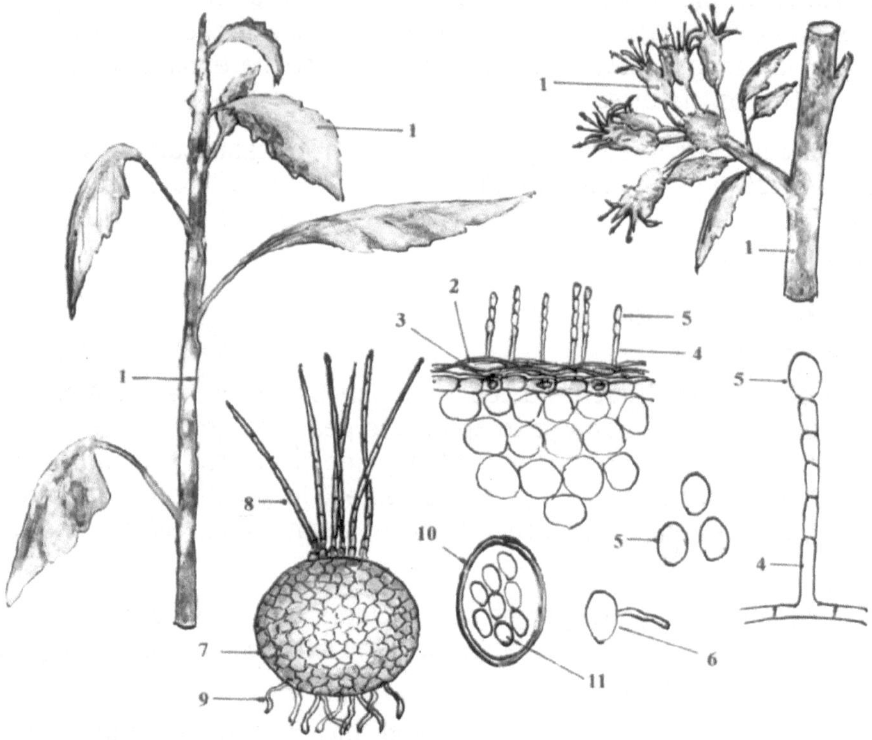

1. White powdery mildew growth on leaves, stem region and flowers 2. Superficial mycelium 3. Haustoria 4. Conidiophores 5. Conidium 6. Germinating conidium 7. Cleistothecium 8. Apical appendages 9. Basal appendages 10. Ascus 11. Ascospores.

Fig. 82: 'Fire blight' of apple-*Erwinia amylovora*

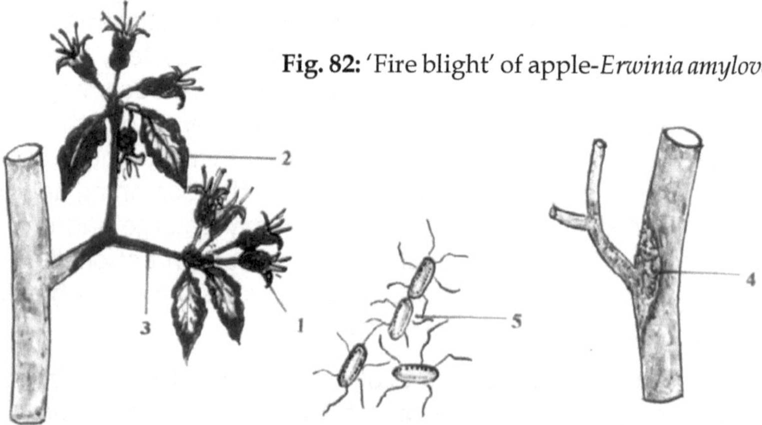

1. Blighted flowers 2. Blotches on leaves 3. Blighted twigs 4. Canker on the stem region 5. Bacteria causing the disease.

The causal organism. The bacteria causing the disease are rod-shaped, occurring singly or in pairs and measure 1.0 - 2.0 x 0.8 - 1.2μ in size. They are anaerobic, non-spore forming, encapsulated and motile by 2-8 peritrichous flagella **(Fig.82).**

Mode of survival, spread and Epidemiology. The bacteria overwinter at the margins of cankers, in buds and in the wood tissue. In the spring, they become active again, multiply and spread to the adjoining healthy bark. During humid or wet weather, the bacteria exude through lenticels and cracks usually at the time, when the flowers are opening. Various insects, such as bees, flies, ants etc. are attracted to the sweet, sticky, bacterial exudate and carry the bacteria sticking on to their bodies to flowers they visit afterwards. The bacteria may also be disseminated on to other flowers by lashing rains. The dry bacterial ooze may also be carried by wind to flowers. The bacteria multiply rapidly in the nectar and through nectarthodes enter the floral tissues. The bacteria multiplying inside the floral tissues kill the flowers. Then, the bacteria move through the intercellular spaces or through the macerated middle lamella and flower cells down the pedicel into the fruit spur. Infection of the fruit spur, results in the death of all flowers, leaves and fruits on it. The bacteria may also enter the leaves through stomata, hydathodes or through wounds. From the leaf, the bacteria pass into the petiole and the stem. After entering the tissues, they move through the vessels, attack and kill other cells, causing blight and canker symptoms in the process.

Warm, humid conditions favor infection and development of the disease.

Disease management

Agronomic practices (i) During the winter, all blighted twigs, branches and cankers should be cut about 10.0 cm. below the point of infection and destroyed by burning. The cut ends should be smeared with Bordeaux paste or copper oxychloride paste (ii) The cutting tools should be disinfected after each cut with mercuric chloride (1 : 1000) solution or commercial sodium hypochlorite - 10 % solution (iii) The crop should be maintained properly by judicious application of fertilizers (iv) Suitable insecticides should be applied to control insects, so as to prevent the spread of bacteria to succulent twigs and flowers by such insects (v) Resistant varieties should be grown in new areas.

Chemical control. Spraying with Streptomycin or Oxytetracycline at 100 ppm is effective in controlling the disease to some extent. Spraying with Bordeaux mixture - 0.75 % or copper oxychloride at 2.5 gm./ lit. of water affords fairly good control of the disease.

4. Blue mould or soft rot of apple

Penicillium expansum

The disease occurs all over the world. It is often responsible for serious losses of all kinds of apples, while in storage and transit.

Symptoms. The disease is characterized by a rapid form of soft rot, which may start at any point on the surface of the fruit. The affected part is soon covered with tufts of bluish-green fructifications of the fungus. The disease occurs at all times on harvested apples and many other fruits. The rotten area becomes soft and watery. A peculiar, characteristic odor emanates from such rotten fruits. The outer skin of the rotten areas becomes brown and wrinkled, sometimes in concentric circles.

The causal organism. The conidiophores arise from the mycelium as blue-green tufts or coremia around the points of infection. The conidiophores are long and intertwined into fascicles. The ultimate branches of the conidiophore are arranged close together and finally terminate in narrow phialides, which bear long, loose, chains of conidia. The conidia are broadly elliptical, green or bluish-green and measure on an average 3.5 x 3.2μ in size. The ascigerous stage of this fungus has not been found.

Mode of spread. Infection of the fruit usually takes place through wounds in the skin, which may be caused by insects, careless picking or rough treatment in the process of washing or storing. Infection also occurs through lenticels in the epidermis of the fruits. On the epidermis of young fruits, hairs are present and if these hairs are broken near the base, infection may take place through the hair pits. After picking, infection may occur through the fruit stalk also. The rotting tends to increase with the length of the storage period. Once the organism gains entry into the fruit, the **'pectic enzymes'** produced by it dissolve the middle lamella, resulting in a mass of soft, disorganized, water-soaked tissue and consequent rotting of the fruit in that area.

Disease management

Precautionary measures (i) The disease may be checked to a large extent, if reasonable care is taken to avoid injuring fruits at the time of harvesting and handling during storage, grading and packing (ii) The materials used for packing should be disinfected with sodium hypochlorite prior to packing

Chemical control (i) Treating the fruits with sulfur dioxide fumes protects the fruits from deterioration (ii) Steeping the fruits in diphenylamine - 1000 ppm solution for 5 minutes or Aureofungin-sol - 500 ppm for 20 minutes affords best control of soft rot.

Diseases of minor importance. Apple is also susceptible to several other diseases. But some of them are not of much importance in India. *Phytophthora cactorum* causes 'Collar rot' and 'Fruit rot'; *Nectria galligena* causes 'Apple canker'; *Glomerella cingulata (=Gloeosporium album)* causes 'Bitter rot'; *Rosellinia (Dermatophora) necatrix* causes 'White root rot'; *Sclerotinia laxa* f. sp. *mali* causes 'Blossom wilt' and 'Spur canker'; *Sclerotinia fructigena* causes 'Brown rot' and 'Spur canker'; *Leucostoma (Valsa)* sp. causes 'Canker'; *Agrobacterium tumefaciens* causes 'Crown gall'; *Apple mosaic virus* causes 'Mosaic'.

Pear *(Pyrus communis)*

1. Pear scab

Fusicladium pirinum

(Venturia pirina)

'Pear scab' is widely distributed in Europe and in the United States of America. In India, it occurs in the hilly regions of Northern and Southern India.

Symptoms. The symptoms of pear scab are quite similar to those of the apple scab, but are more distinctly defined. Leaves, twigs, bud scales, flower stalks, sepals, petals, main axis of the inflorescence and the fruits are all attacked in the same way as that of the apple scab, but here twig infection is far more extensive.

The causal organism. The pathogen causing the disease is almost identical with that of the apple scab fungus, but differs in certain morphological characteristics. The fungus of pear scab is more aggressive, penetrating the host more quickly and more extensively. It has more sporulating capacity and sporulation lasts for a much longer period. The conidial stage *(Fusicladium pirinum)* is the most important, parasitic stage of the pathogen. The conidiophores are brown, septate and are more zigzag. In addition to a terminal conidium, a number of lateral conidia arise in a sympodial manner, as against terminal conidial formation in succession in the apple scab fungus. The conidia are dark-brown or olivaceous, ovoid, unicellular or single-septate and measure 15.0-40.0 x 6.0-10.0μ in size. The perithecia are smaller in size than those of the apple scab fungus and are scattered or grouped together on old, fallen leaves. The ascospores, which are eight in number in each ascus, are transversely septated into two unequal cells. Unlike the apple scab fungus, in *Venturia pirina*, the larger cell is at the top, while the smaller cell is at the base. The germ tube is formed only from the larger cell.

Mode of spread, epidemiology and disease management. Refer 'Apple scab' (Page 323).

Minor diseases. Many of the diseases that attack apple are found to attack pear also. *Phytophthora cactorum* causes 'Collar rot' and 'Fruit rot'; *Podosphaera leucotricha* causes 'Powdery mildew'; *Leucostoma (Valsa)* sp. causes 'Canker'; *Erwinia amylovora* causes 'Fire blight'; *Penicillium expansum* causes 'Blue mould' or 'Soft rot'; *Glomerella cingulata (=Gloeosporium album)* causes 'Fruit decay'; few virus diseases also attack pear.

Peach *(Prunus persica)*

1. Peach leaf curl

Taphrina (Exoascus) deformans

The disease was first recorded in Britain in 1862. It is now widely distributed in Europe, America, parts of Asia, especially in China and Japan and also in Africa, Australia and New Zealand. In India, it is prevalent in the peach orchards of Kumaon, Kulu, and other Sub-Himalayan regions. Besides peaches, apricots and other *Prunus* species are also attacked by the pathogen.

Symptoms. The first symptoms appear in the early spring, soon after the leaves come out of the buds. Some of the leaves become distorted and fold over, so that the tips are directed backwards, while others curl and the whole lamina except the tip becomes puckered and blistered. In some leaves, only part of the lamina may show such distortions. The blistered portions are thicker and softer than the normal portions of the leaf blade. The affected areas may remain green for a while, but gradually become yellow and finally turn reddish-purple, which makes the affected leaves very conspicuous against the green color of healthy foliage. The reddish-purple surface of the lamina soon becomes covered with a whitish-gray bloom consisting of the fructifications of the fungus on the upper surface, but sometimes on the lower surface also. Sometimes, the young shoots may be attacked and they become swollen and distorted. No lesions are usually formed on the fruits.

Affected leaves fall off prematurely and in case of severe infections, the trees may suffer from acute defoliation in the late spring. Recurrent attacks of leaf curl from season to season, results in weakening of the trees and consequent loss in yield and quality of the fruits **(Fig.83).**

The causal organism. The fungus does not form true conidia. The whitish-gray bloom formed on the leaf surface consists of a large number of asci, which are not developed and protected within a fruiting body, but merely break through the leaf cuticle under which they arise from the mycelium within the leaf. The hyphae enter the leaf mostly by cuticular penetration, but stomatal entry also occurs. Since early infections take place as soon as the buds break into leaves, infections occur largely through the lower leaf surface, but later on infection may take place through the upper surface also. After entry into the leaf, the mycelium travels across the mesophyll intercellularly and spreads out chiefly between the palisade layer and the upper epidermis, where it develops widely. Infection of young leaves stimulates cell multiplication in the mesophyll tissue, resulting in distortion of the leaf surface. All the starch contents in the leaf tissue are used up by the fungus and with the gradual degeneration of chloroplasts, the reddish-purple pigmentation appears in the infected cells.

Prior to the reproductive stage, the hyphae below the epidermis move up to form a net work of short cells beneath the cuticle. When the cuticle is pushed up and finally ruptured due to the pressure exerted, these cells form a hymenium on the blistered area. From the hymenium a large number of ascogenous cells are formed, which ultimately develop into asci. They are naked and appear more like terminals of hyphae than the usual club-shaped asci. The asci measure 25.0 - 40.0 x 8.0 - 11.0µ in size. Each ascus contains eight or fewer globose ascospores, 3.0 - 4.0µ in diameter. The ascospores bud frequently like yeasts, while still within the ascus to form minute secondary spores in large numbers and fill the entire ascus. These 'sprout cells', called 'conidia' are slightly oval in shape and measure 2.5 - 6.0 x 4.8µ in size. This phase actually represents the asexual or conidial stage of the fungus. The ascospores are ejected forcibly from the asci in early summer. They come out as tiny spore balls and are carried by wind and rainsplash. Some of these sprout cells possess thicker walls than the normal sprout cells and contain dense protoplasm. These serve as resting spores. They germinate by producing very short germ tubes of a swollen, vesicular nature and from these, thin-walled conidia are budded again **(Fig.83).**

Mode of survival, spread and epidemiology. Primary infection may occur through resting conidia, which have been sheltered on some part of the host during the winter. The conidia, which are produced in masses and have yeast-like consistency, can adhere to the buds and cause primary infection during the ensuing spring. It is also probable that the fungus can survive during winter in the form of agglutinated masses of conidia on bud scales and twig surfaces, as they are capable of withstanding desiccation and low temperatures.

The optimum temperature for growth of the fungus is 20°C. Cold and wet weather conditions at the time of bud burst to form new leaves favor severe infection.

Disease management

Agronomic practices. Orchard and tree sanitation, removal and destruction of fallen leaves and other plant refuse, clean cultivation etc. help to minimize the occurrence of the disease.

Chemical control. Thorough spraying with Bordeaux mixture - 1 % or copper oxychloride - 2.5 gm. or mancozeb - 2.0 gm. or captan - 1.25 gm. or tridemorph - 1.0 ml. or chlorothalonil - 1.25 gm./ lit. of water before bud-break in early spring, followed by a second spraying after a fortnight controls the disease.

Minor diseases. Peach is subjected to attack by a few other diseases. *Clasterosporium carpophilum* causes 'Shot hole disease'; *Venturia carpophila* causes 'Scab'; *Puccinia pruni-spinosae* causes 'Rust'; *Sphaerotheca pannosa* var. *persicae* causes 'Powdery mildew'; *Leucostoma (Valsa)* sp. causes 'Canker'; *Pseudomonas mors-prunorum* causes 'Bacterial leaf spot'.

Plum *(Prunus domestica* sub sp. *insititia)*

1. Plum rust

Puccinia (Tranzschelia) pruni-spinosae

Plum rust is widely distributed in Europe, America, Australia, New Zealand, Tasmania, Egypt, South Africa and other plum growing countries of the world. In India, it occurs mostly in the hilly regions of Northern India. Besides plum, it also occurs on apricot, almond, peach, prune etc., but is not common on cherry.

Symptoms. The disease occurs mainly on the leaves, but sometimes the twigs are also attacked. The initial symptoms appear as small, yellow spots on the upper surface of leaves and brown pustules, which are the uredosori on the corresponding under surface of the spots. The lesions on the twigs form small cankered areas causing the bark to split longitudinally. Uredosori are produced in such cracks and crevices also. The blemishes on the fruits consist of small, circular, sunken spots or cracks. The teliosori, which are also formed on the under surface of leaves, on the twigs and on the fruits appear as black pustules. The disease usually causes partial defoliation and in case of severe infection complete defoliation, resulting in loss of vigor in the trees and reduction in yield.

The causal organism. The fungus causing the disease is heteroecious and macrocyclic. The uredial and telial stages are found in *Prunus* hosts, while the pycnial and aecial stages are found on various species of *Anemone*, chiefly on *Anemone coronaria*. The uredosori are hypophyllous, cinnamon - brown and are pulverulent. The uredospores are oblong-clavate or oblong-fusiform, 24 - 42 x 15 - 23μ in size, with a brownish-yellow wall, smooth above, but echinulated below and have 3 - 5 equatorial germ pores. The uredospores are interspersed with capitate paraphyses. The teliosori on leaves are pulvurulent, dark chocolate-brown masses, forming compact almost black clusters. The teliospores are two-celled, the upper cell globoid and the lower cell globoid to irregularly globoid, narrower than the upper cell and contracted at the base. The wall of the upper cell is usually thickened at the apex and much darker than the wall of the lower cell and usually smooth. They measure 26 - 39 x 15 - 23μ in size. The spermagonia, which appear as minute dots on the upper surface of leaves are usually epiphyllous and scattered.

Mode of survival. The rust is capable of surviving from season to season through perennial mycelium in the twigs of the host or in the alternate hosts. It can also survive the winter as uredospores in the lesions on twigs, bark, buds or leaves.

Disease management

Agronomic practices (i) Growing of *Anemone* plants in the vicinity of plum orchards should be avoided (ii) Orchard and tree sanitation should be followed.

Chemical control. Spraying the crop with wettable sulfur - 4.0 gm. or chlorothalonil - 1.25 gm. or carboxin - 1.0 gm./ lit. of water is effective in the control of the disease.

Diseases of minor importance. A few other diseases are also known to attack plum and may become serious under favorable conditions. *Sterium purpureum* causes 'Silver leaf disease'; *Fomes pomaceus* causes '*Fomes* disease' or 'Heart rot'; *Verticillium albo-atrum* causes '*Verticillium* wilt'; *Taphrina deformans* causes 'Leaf curl'; *Pseudomonas syringae* pv. *syringae* causes 'Bacterial canker' and 'Gummosis'; *Leucostoma (Valsa)* sp. causes 'Canker'; few virus diseases also attack plum trees.

Gooseberry *(Emblica officinalis / Phyllanthus emblica)*

1. Powdery mildew of gooseberry

Sphaerotheca mors-uvae

This disease is considered to be the most serious and destructive disease of gooseberry. Currants are also susceptible to the disease. It is found in almost all the countries of the world, where gooseberry is grown.

Symptoms. The symptoms appear as white patches of mycelial growth on the tips of young shoots and both the surfaces of leaves. Under favorable conditions, the patches enlarge in size covering almost the entire leaf area and spread to the growing shoots, leaves and berries at all growth stages. The white surface becomes powdery due to the formation of conidiophores and conidia in profusion. The disease persists throughout the growing season on twigs, leaves and fruits. Severely affected leaves and fruits drop off prematurely **(Fig.84)**.

The causal organism. The white, superficial mycelium of the fungus sends haustoria into the epidermal cells. Erect conidiophores arise from the mycelium in large numbers producing long, chains of hyaline conidia, which are easily dispersed by wind and cause secondary infection. Later in the summer, the white fungal growth gradually changes to brown. The discoloration is more evident on the twigs and berries rather than on the leaves. Numerous, black cleistothecia are found embedded in the brown weft of mycelium. The cleistothecia are sub-globose, 76 - 100μ in diameter and have few simple, pale brown, contorted and somewhat rudimentary appendages on the surface. Each cleistothecium contains a single ascus, which almost fills the interior of the cleistothecium. The asci are usually sub-globose, 70 - 92 x 50 - 62μ in size. Each ascus contains eight ascospores, which are ellipsoid and measure 20 - 25 x 12 - 15μ in size. The matured cleistothecium swells and splits at the top, when moistened. The apex of the ascus projects out through the slit, dehisces and the spores are discharged simultaneously into the air **(Fig.84)**.

Mode of survival and spread. The cleistothecia are capable of producing and discharging ascospores in the same season. However, they are mostly formed late in the season and remain dormant during the winter on the leafless bushes, cast-off twigs and on diseased berries fallen on the ground. The ascospores released from such sources first infect the lowermost leaves and twigs. Secondary infection takes place on a wide scale through conidia produced on these parts. The conidia produced from leaves, which have not been shed, may also cause primary infection. The presence of the pathogen on the host surface affects the physiological functioning of the leaves, retards photosynthesis and frequently causes extensive defoliation. Repeated attacks from season to season results in overall weakening of the trees and consequent loss in yield.

Disease management

Agronomic practices. Adopting correct pruning techniques to encourage open growth, timely pruning to avoid formation of cleistothecia, collecting and burning of infected plant debris and pruned parts, avoiding application of excessive doses of nitrogenous fertilizer etc. help in minimizing the occurrence of the disease.

Chemical control. Spraying the crop with wettable sulfur - 4.0 gm. or dinacap - 1.0 gm. or tridemorph - 1.0 ml. or benomyl - 1.0 gm. or carbendazim - 1.0 gm. or thiophanate methyl - 1.0 gm./ lit. of water affords control of the disease. First spraying should be given before the flowers open, the second one after the fruits have set and the third spraying 3 weeks later.

Diseases of minor importance. A few more diseases are also known to attack gooseberry. *Puccinia pringsheimiana* causes 'Cluster cup rust'; *Botrytis cinerea* causes 'Die-back'; *Glomerella cingulata* causes 'Anthracnose'; *Phoma putaminum, Alternaria alternata* and *Pestalotiopsis cruenta* cause 'Fruit rot'; *Penicillium islandicum, P. citrinum* and *Phomopsis phyllanthi* cause 'Soft rot of fruits'.

VEGETABLE CROPS

Eggplant or Brinjal *(Solanum melongena)*

1. Damping off disease of seedlings of eggplant

Pythium species

'Damping off of seedlings' is very common all over the world. The disease is found to attack several vegetable crops, field crops, ornamental plants and forest trees in the tropical and temperate climates. It occurs in all kinds of soils and all through the year in every nursery and green-house.

Symptoms. Two types of damping off of seedlings are known to occur viz., **'pre-emergence damping off'** and **'post-emergence damping off'**. In the case of pre-emergence damping off, the seed and the radicle rot before the seedling emerges from the soil, while in the case of post-emergence damping off, the newly emerged seedling is killed after its emergence from the soil. Stems of the affected seedlings rot at the ground level and they collapse, topple over and die. Post-emergence damping off is more conspicuous than pre-emergence damping off. Infection usually occurs at or below the ground level before the stem has hardened sufficiently to resist attack by the pathogens. The tissue in the infected area becomes soft and water-soaked. As the disease advances, the stem becomes constricted at the base, turns dark and the seedling topples over, rots and dies. The disease is very much evident in nursery beds and green-houses, where the symptoms develop suddenly, killing large number of seedlings in patches around the foci of infection **(Fig.85).**

The causal organism. The most important pathogens causing damping off are *Pythium debaryanum, P. aphanidermatum and P. ultimum.* Besides the above pathogens, many species of *Phytophthora, Rhizoctonia, Corticium* and *Fusarium* also cause damping off of seedlings of various crops.

The mycelium of *Pythium de baryanum* is coenocytic, multinucleate, hyaline and much branched. The ends of the terminal hyphae penetrate the cell walls of the hypocotyl and ramify within and between the tissues of the cortical parenchyma. The sporangia, which are produced from the hyphae, are terminal or intercalary. They are spherical, oval or barrel-shaped depending upon the place of origin and measure 15 - 26µ in diameter. Before germination, a prominent papilla is formed on the sporangium, followed by a thin-walled, globular vesicle, which is called the **'secondary sporangium'**. The sporangial contents pass into the vesicle. Inside the vesicle, zoospores are differentiated. On maturity of the zoospores, the wall of the secondary sporangium is broken and the zoospores emerge out. The zoospores are reniform, with two lateral flagella and measure up to 8.0µ in diameter. After swimming about in water present in the soil for a short while, the zoospores lose their flagella, encyst, become globular and after a short period of rest germinate by producing infection hyphae and infect a new host. When weather conditions are not favorable, the sporangia germinate directly by producing a germ tube instead of producing secondary sporangia and zoospores.

Sexual reproduction is oogamous. Oospores are globular, smooth, thick-walled, aplerotic and measure 12 - 20µ in diameter. They are the resting spores and can remain in the soil in a viable state for a prolonged period. On germination of the oospore under favorable conditions, a globular vesicle is produced at the end of a short stalk and zoospores are differentiated inside the vesicle. These zoospores are similar to those produced by the sporangium. When weather conditions are not favorable, the oospore also germinates directly by producing a germ tube and in that case no zoospores are produced.

Fig. 83: Leaf curl of peach-*Taphrina deformans*

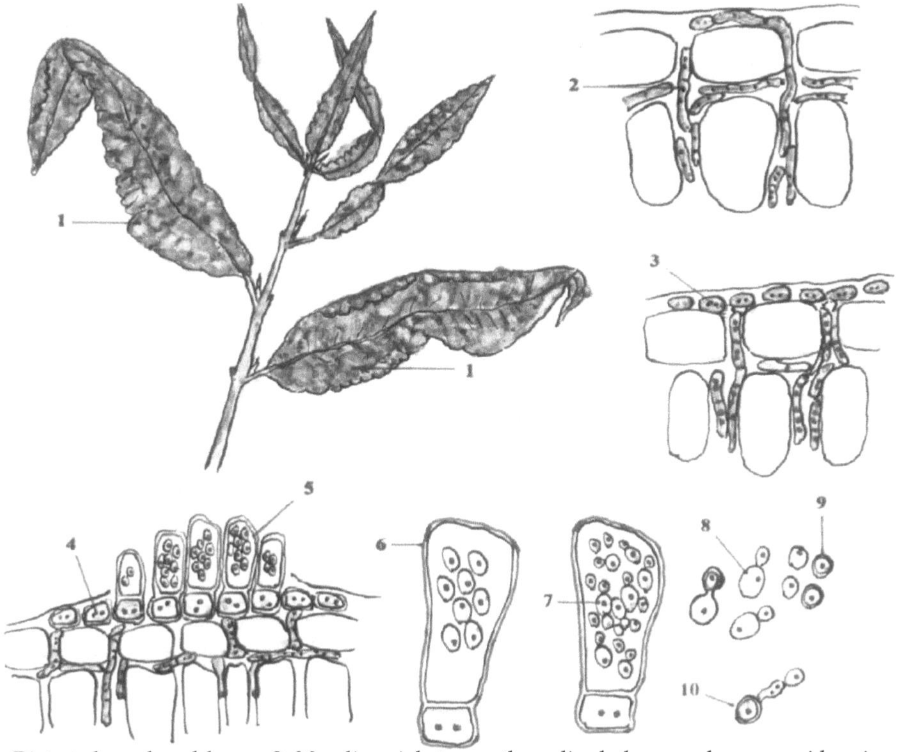

1. Distorted, puckered leaves 2. Mycelium inbetween the palisade layer and upper epidermis 3. Net work of short cells beneath the cuticle 4. Ascognenous cells 5. Asci 6. Ascus with ascospores 7. Budding of ascospores inside the ascus to form conidia 8. Budding conidia 9. Resting conidia 10. Germinating resting conidium budding off conidium from short germ tube.

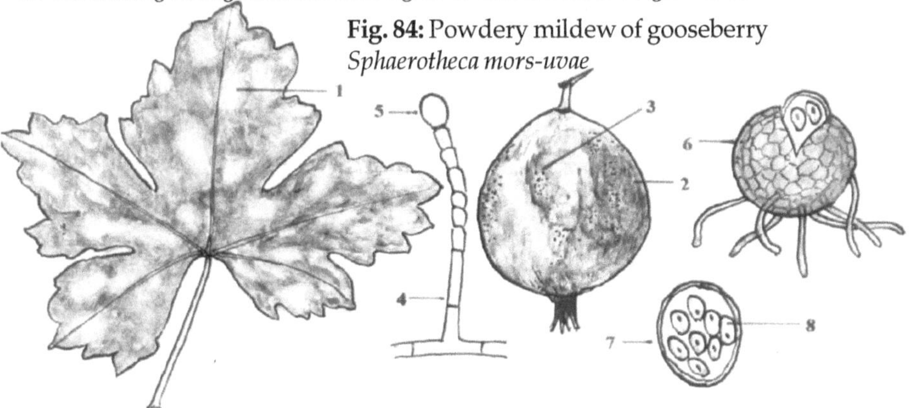

Fig. 84: Powdery mildew of gooseberry *Sphaerotheca mors-uvae*

1. White powdery mildew growth on the leaf 2. Brownish mycelial growth on the fruit 3. Cleistothecia embedded in the mycelial weft on the fruit 4. Conidiophore 5. Conidium 6. Cleistothecium with a single ascus 7. Ascus 8. Ascospores.

Pythium aphanidermatum closely resembles *P. de baryanum* in most of the characters except for the sporangia. The sporangia, which are either terminal or intercalary, are inflated, lobulate, branched or unbranched. The lobulate sporangium produces several globular secondary sporangia at the ends of short stalks. From each of the vesicle, 30 - 45 zoospores are produced. They measure 7.0 - 12.0μ in diameter. The sporangia do not germinate directly by producing a germ tube. The oospores are globular, smooth, thick-walled, aplerotic and measure 17 - 19μ in diameter. They germinate, either by producing zoospores or by germ tube depending upon weather conditions.

In *Pythium ultimum*, the sporangia produced are mostly terminal, spherical in shape and measure 12 - 28μ in diameter. Intercalary sporangia are occasionally formed, which are barrel-shaped and are 22.9 - 27.8 x 14.0 - 17.0μ in size. Oospores are spherical in shape, smooth, and aplerotic and are 14 - 18μ in diameter. The sporangia, as well as the oospores germinate directly by producing germ tubes and they do not form vesicles with zoospores **(Fig.85).**

The damping off fungi, besides causing damping off disease in several crops, such as tobacco, tomato, chillies, sugarbeet, cotton etc., cause 'root rot', 'foot rot' and 'soft rot' of vegetables and fruits and 'rhizome rot' of many plant species.

Mode of survival, spread and epidemiology. The damping off fungi are natural inhabitants of soil. They survive on dead plant and animal wastes and refuses as saprophytes or as parasites on fibrous roots of plants. When spore-germ tubes or mycelium of *Pythium* comes in contact with a germinating seed or seedling tissues of host plants, they enter the host tissue by direct penetration. **'Pectinolytic'**, **'proteolytic'** and **'cellulolytic enzymes'** secreted by the pathogens dissolve the cell walls, grow between and through the cells and cause disintegration of the cell walls. As a result of maceration of the tissues, the germinating seeds and young seedlings collapse, rot and die.

The oospores produced by these fungi are resistant to adverse soil temperature and moisture conditions and remain dormant in the soil for prolonged periods and are the most important source of primary infection. The type of germination of the sporangia and oospores is governed to a large extent by the prevailing soil temperature and soil moisture.

Temperatures between 10° - 18°C induce germination by means of zoospores, while temperatures above 18°C favor germination by germ tubes.

Fig. 85: Damping off of seedlings of eggplant-*Pythium de baryanum, P. aphanidermatum* and *P. ultimum*

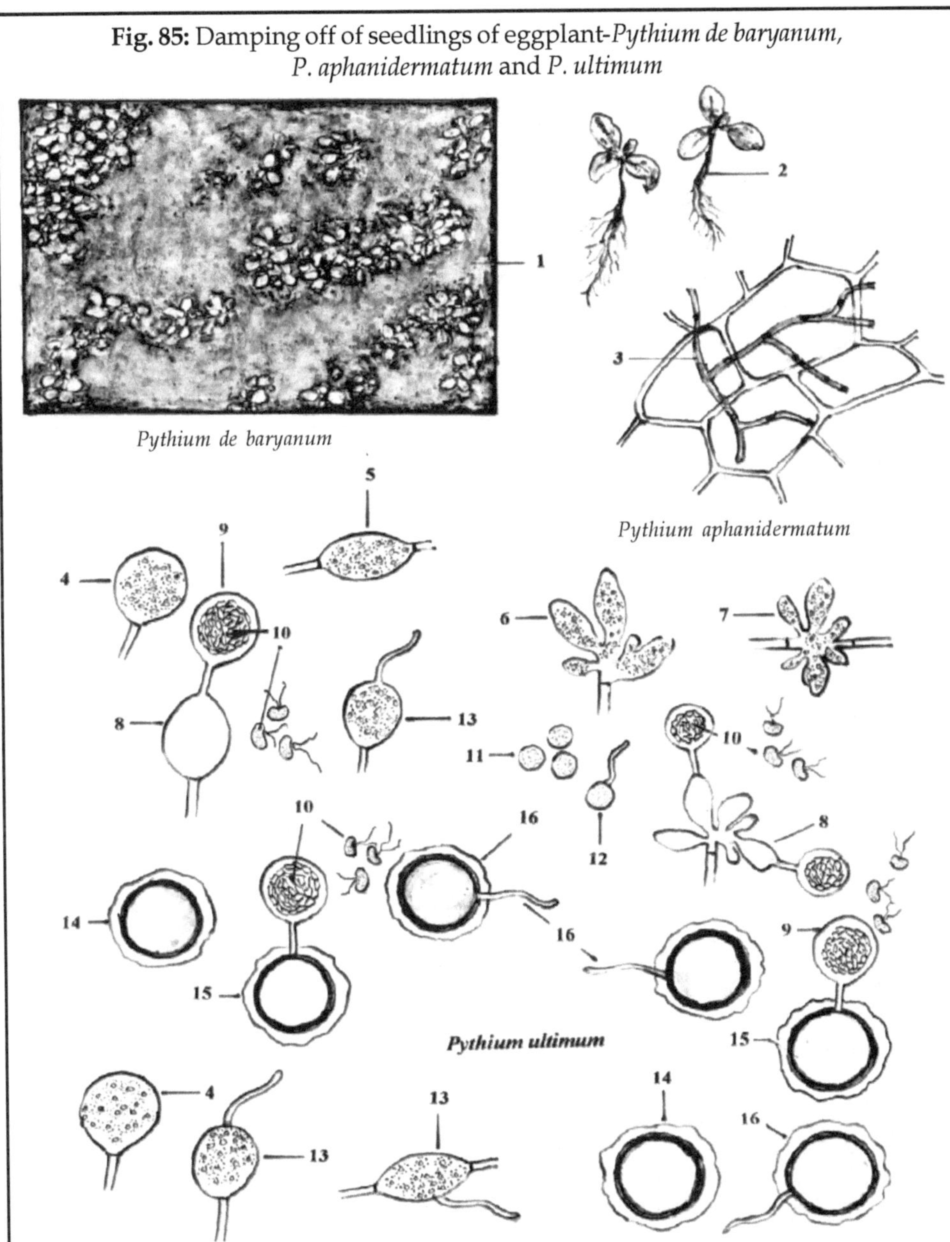

1. Damping off infection in seed bed 2. Damping off disease affected seedlings (Post-emergence damping off) 3. Intercellular mycelium 4. Terminal sporangium 5. Intercalary sporangium 6. Terminal sporangium (Lobulate) 7. Intercalary sporangium (Lobulate) 8. Germination of sporangium by zoospores 9. Vesicle (Secondary sporangium) 10. Zoospores 11. Encysted zoospores 12. Germinating encysted zoospore 13. Germination of sporangium by germ tube 14. Oospore 15. Grermination of oospore by zoospores 16. Germination of oospore by germ tube.

The disease is very destructive at soil temperatures of 24°- 30°C and high soil moisture content. In ill-aerated, ill-drained and poor, hard clay soils the incidence of the disease is most severe.

Disease management

Agronomic practices (i) The nurseries should be situated in elevated places, with good drainage facilities (ii) The soil in the nursery should be sandy-loam and heavy clay soils should be avoided (iii) Raised nursery beds are preferable (iv) Thick sowing should be avoided (v) Only well-decomposed farm yard manure or compost should be used in the nursery beds (vi) Light and frequent irrigation should be given (vii) Deep summer ploughing may be followed, so that the spores are exposed to direct sunlight and are destroyed.

Seed treatment. Seeds should be treated with seed protectants, such as thiram - 4 gm. or captan - 4 gm. or benomyl - 2 gm. or metalaxyl - 2 gm./ kg. of seeds, at least 24 hours prior to sowing to provide protection to the seedlings from pre-emergence damping off.

Nursery treatment (i) The soil in the nursery may be drenched with 2 % formalin solution (2 parts of formalin in 100 parts of water) at the rate of about 2 liters/ sq. ft. of soil, so that the soil is soaked to a depth of about 10 cm. The treatment should be given 15 days prior to sowing. A few days before sowing the soil is stirred well, so that the soil is completely free from any residual formalin. If traces of formalin is present in the soil at the time of sowing, seed germination will be adversely affected (ii) After sowing, the soil may be drenched with 0.2% solution of captan or thiram or 0.3% copper oxychloride. The treatment may be repeated again after emergence of the seedlings, so as to cover the stems and root portions of the seedlings.

Biological control. Seed treatment with spores of *Trichoderma harzianum* or *Penicillium oxalicum* or cells of *Pseudomonas fluorescens* may serve as effective biocontrol agents.

2. Leaf spot and fruit rot of eggplant

Alternaria melangenae and *Alternaria solani*

The disease occurs all over the world, wherever eggplant is cultivated. In India, it is prevalent in Karnataka, Tamil Nadu, Maharashtra and other states of the country.

Symptoms. The disease attacks the leaves and fruits. Initial symptoms appear as small, scattered, circular, pale brown, necrotic spots, mostly on the older leaves and gradually spreads to the younger leaves. The spots enlarge forming concentric rings and appear as target boards. The center of the spots becomes necrotic and sometimes the central dried portion disintegrates and falls off leaving a shot hole. As the spots enlarge further, they become irregularly circular in shape and attain a diameter of up to 4.0 - 8.0 mm. Adjacent spots may coalesce and cover large areas of the leaves, as a result the leaves turn yellow and drop off prematurely. Similar spots may develop on the fruits also. The spots on the fruits are fewer and are usually larger in size. They are sunken, scab-like, the skin becoming hard and leathery. Sometimes, a single spot may enlarge and cover a major portion of the fruit or the entire fruit. Affected fruits turn yellow and drop off prematurely.

The causal organism. Two species viz., *Alternaria melangenae* and *A. solani* are involved in causing the disease. The mycelium of the fungi consists of septate, branched, light-brown hyphae, which become darker with age. The hyphae are intercellular in the beginning, becoming intracellular later on and penetrate into the cells of the invaded tissues. Conidiophores emerge through the stomata from the dead, center of the spots in groups.

In *Alternaria melangenae,* conidia are produced in chains of 2 - 5. They are dark-brown in color, 28.0 - 84.0 x 7.0 - 17.5μ in size, with a rather short beak of length 10.5 - 52.5μ in length and have 7 - 14 cross septa and 0 - 5 vertical septa. Under moist conditions they germinate by producing 5 - 10 germ tubes.

In *Alternaria solani*, the conidiophores are dark-colored, 50 - 90μ in length and bear conidia singly in nature and in short chain of 2 to 3 in culture. The conidia are beaked, muriform, dark colored, with 5 - 10 transverse septa and a few longitudinal septa. They measure 120 - 296 x 12 - 20μ in size, including the beak and germinate by putting forth 5 - 10 germ tubes **(Fig.94)**.

Mode of survival, spread and epidemiology. The pathogens are mainly soil-borne. The mycelium of the fungus remains dormant in dry, infected leaves and fruits for more than a year. Conidia can also retain their viability for a long time under dry conditions. The mycelium and conidia, which remain in the soil in the diseased plant debris cause primary infection. Collateral hosts, such as tomato, chillies etc. play an important role in the perpetuation of the disease.

Under warm and moist conditions, the conidia germinate in about 35 - 45 minutes. Secondary spread of the disease is through conidia dispersed by wind, rain and by some insects mechanically. Moderate temperatures between 28° - 30°C and high atmospheric humidity favor the occurrence and spread of the disease.

Disease management

Agronomic practices (i) Diseased plant debris fallen on to the ground should be collected and destroyed by burning (ii) Growing of alternate hosts, such as tomato, chillies etc. in the vicinity of eggplant fields should be avoided.

Seed treatment. The seeds should be treated with captan or thiram at 4.0 gm./ kg. of seed, at least 24 hours prior to sowing.

Chemical control. Spraying with copper oxychloride - 750 gm. or dithiocarbamates - 600 gm. or organo- tin compounds - 450 gm. in 300 lit. of water per acre, covering the leaves and fruits thoroughly affords satisfactory control of the disease.

Resistant varieties. The variety 'Pusa Purple' is moderately resistant to the disease.

3. *Verticillium* wilt of eggplant

Verticillium dahliae

The disease is very common in many Western countries and in India. It was first reported from Poona in 1938 and now it is prevalent in several states of India, including Karnataka, Maharashtra, Tamil Nadu etc.

Symptoms, causal organism, mode of spread, epidemiology and disease management. Refer '*Verticillium* wilt of Cotton' (Page 224).

4. Little leaf of eggplant

Mycoplasma-like organism

The disease is prevalent in India, Sri Lanka and other neighboring countries. In India, it occurs in all the regions, wherever brinjal (eggplant) is cultivated and the disease poses a big threat to brinjal cultivation during some seasons.

Symptoms. The characteristic symptom of the disease is the marked reduction of the size of the leaves. The newly formed leaves become progressively smaller. The petioles are very much shortened and the leaves appear oppressed to the stem. The leaves become narrow, soft, glabrous and yellowish in color. The internodes get shortened and the axillary buds are stimulated to sprout and they grow into short branches, with very small leaves. The affected plant presents a miniature, bushy appearance. Usually, the affected plants fail to flower. Even if flowers are formed, they are modified into green structures. Fruit formation is very rare and the fruits formed are tough, leathery and may fail to mature normally.

The causal organism. The agent causing the disease is a '*Mycoplasma-like organism*', also called as '***Phytoplasma***'. The organisms have no rigid cell wall. They are spherical or oval in shape, measuring 40 - 300 nm in diameter and are sensitive to antibiotics, such as Tetracycline.

Mode of spread. The disease is transmitted by the leafhoppers, *Cestius phycitis* and *Empoasca devastans*, but the former is the most effective vector. The vectors acquire the pathogen after feeding on the infected plant for several hours or days. The nymphs can acquire the pathogen better than adult hoppers and the pathogen can remain in the body of the vectors through subsequent moults. Only after an incubation period of 10 - 45 days, the vector is capable of transmitting the pathogen to new plants and this can continue for the rest of its life. However, the pathogen is not passed on to the next generation through eggs. The disease also occurs on *Datura fastuosa, D. metal, Vinca rosea* and many other plants.

Disease management

Agronomic practices (i) Weed hosts in and around the fields should be eradicated (ii) Infected plants should be rogued out and destroyed.

Chemical control (i) The leafhopper vectors transmitting the pathogen can be controlled by spraying with endosulfan - 600 ml. or fenitrothion - 300 ml or malathion - 600 ml. in 300 lit. of water per acre (ii) Spraying with a solution of Tetracycline - 50 ppm controls the disease to some extent (iii) Dipping the seedlings prior to planting in a solution of Tetracycline - 50 ppm protects the seedlings from infection in the early stages of growth.

Resistant varieties. The varieties Arka Sheel, Aushy, Banaras Giant, Manjari Gota and Pusa Purple Cluster are moderately resistant to the disease, while Black Beauty, Brinjal Round and Surati are tolerant. Pusa Purple Long and Selection T are highly susceptible.

Diseases of minor importance. Besides the diseases discussed above in detail, eggplant is subjected to attack by a few more diseases. *Fusarium solani* causes '*Fusarium* wilt', 'Rust' is caused by *Puccinia penniseti*; *Erysiphe polyphaga* causes 'Powdery mildew'; *Colletotrichum melangenae* causes 'Anthracnose'; *Macrophomina phaseolina* causes 'Dry root rot'; *Pseudomonas solanacearum* causes 'Bacterial wilt'; 'Mosaic disease' is caused by *Potato virus* Y and *Tobacco mosaic virus; Orobanche cernua*, the phanerogamic holo root parasite also attacks the crop.

Tomato *(Lycopersicon esculentum)*

1. Early blight of tomato

Alternaria solani

The disease is of common occurrence in all the regions of India, where tomato is cultivated. All commercial varieties of tomatoes are susceptible to the disease. Besides tomato, the disease is found to occur in potato, chillies, *Atropha belladona, Nicotiana alata* etc.

Symptoms. On the leaves, brown spots, marked with concentric rings, which present a target-board appearance are formed. The spots are oval to angular in shape and up to 3.0 - 4.0 mm. in diameter. Usually a narrow chlorotic zone is found around the spots, which fades into the normal green color of the leaf. The lower leaves are affected first and gradually the disease spreads to the upper leaves. As a result of infection, the leaves droop exposing the fruits to the sun, resulting in sunburn of the fruits. Dark colored, target board-like spots develop on the stem region also. If such lesions appear at the junction of the stem and side branch, the branch may break down. The fungus also causes stem canker or collar rot of seedlings, as well as sunken spots on older stems, leading to flower and fruit drop. Dark, leathery spots, usually near the point of attachment of the fruit with the branch may also appear. Most of the spots, which appear as concentric rings, become much darkened with the production of conidiophores and conidia in large numbers under moist conditions. The black or brown, sunken lesions on the fruits may enlarge to such an extent to cover major areas of the fruits.

The causal organism, mode of spread, epidemiology and disease management. Refer 'Early blight of potato' (Page 384), **(Fig.94)**.

2. Late blight of tomato

Phytophthora infestans

The disease occurs in many parts of Northern India, especially in the hilly tracts, where rainfall is abundant. In South India, the disease is prevalent in the hilly tracts of Tamil Nadu and Karnataka. It is less common in the plains. The disease also attacks potato.

Symptoms. Leaves, stems and fruits are attacked by the disease. Late blight may kill the foliage and stems of plants at any time during the growing season. Light-brown, water-soaked patches develop on green fruits in the field and the fruits rot, either in the field or while in storage. The disease may cause total destruction of all the plants in a field within a week or two, when the weather is cool and wet. The symptoms on the foliage and stems resemble those of **'Potato late blight'**.

The causal organism, mode of spread, epidemiology and disease management. Refer 'Late blight of potato' (Page 379), **(Fig.93).**

3. Damping off of seedlings of tomato

Pythium aphanidermatum and *P. myriotylum*

The disease is quite common in tomato nurseries in many parts of India and elsewhere. Both pre-emergence and post-emergence damping off occurs. *Pythium aphanidermatum (P. indicum), P. myriotylum, Phytophthora parasitica* f.sp. *nicotianae* and *Rhizoctonia solani* are associated with the disease.

Symptoms, causal organism, mode of spread, epidemiology and disease management. Refer 'Damping off of seedlings of tobacco' (Page 241), **(Fig.85).**

4. *Fusarium* wilt of tomato

Fusarium oxysporum f.sp. *lycopersici*

This is one of the most common and destructive diseases of tomato and is found to occur all over the world. In India, the disease occurs in all the states, wherever tomatoes are cultivated.

Symptoms. The disease affects the plants at all growth stages. Initially the symptoms appear as mild vein clearing on the outer, younger leaflets. Subsequently, the petioles of older leaves droop and the leaflets curl downward abnormally. Plants infected at the seedling stage wilt and die soon after the appearance of the symptoms. Older plants may wilt and die within 5 - 7 days, when the infection is severe and weather conditions are favorable for the pathogen. Usually, older plants develop symptoms, such as vein clearing, curling of leaflets downward, stunting, yellowing of lower leaves, wilting of leaves and young stems, defoliation, marginal necrosis of the upper leaves, followed by death of the plants. Occasionally, adventitious roots are produced from the basal stem region. Sometimes, such symptoms appear on one side of the stem and progress upward and the branches and foliage on that side alone are killed. The fruits, when affected rot and drop off. The root system is also affected and the roots rot. If the stem portion is split open longitudinally, continuous brown to black streaks are seen extending up to the branches, petioles and even the flower stalks and down the main and lateral roots. This discoloration is due to the production of a reddish-brown pigment in the vascular vessels by the fungus.

The causal organism. The mycelium of the fungus is hyaline at first, but becomes creamy white, pale yellow or pale pink with age. The fungus produces three kinds of asexual spores viz., microconidia, macroconidia and chlamydospores. The microconidia, which are produced in abundance from hyphal tips are single-celled, hyaline, ovoid to ellipsoid and are 6.0 - 15.0 x 2.5 - 4.0µ in size. They are produced even inside the vessels of infected host plants. Macroconidia are typical ***'Fusarium'*** type of spores. They are produced in sporodochia-like structures on the surface of plants killed by the pathogen. They are 3 - 5 celled, sickle-shaped, with gradually curved and pointed ends, hyaline and measure 25.0 - 33.0 x 3.5 - 5.5µ in size. Chlamydospores are single-celled, thick-walled and round. They are produced on older mycelium and are either terminal or intercalary. Chlamydospores may be formed in the macroconidia also.

Mode of survival, spread and epidemiology. The fungus is a soil-inhabitant. It survives in the infected plant debris in the soil as mycelium and spores, but most commonly as chlamydospores. The pathogen enters the host through root tips directly or through wounds or at the point of formation of lateral roots. The mycelium advances through the root cortex intercellularly and penetrates into the xylem vessels. Afterwards, the mycelium remains exclusively in the vessels and travels through them mostly upward. While inside the vessels, the mycelium ramifies and produces microconidia, which are carried upward in the sap stream and produce new mycelium. The mycelium also advances laterally into the adjacent vessels. Clogging of vessels by mycelium, spores, as well as gummy substances **(tyloses)** produced by the host affects the translocation of water and mineral nutrients to the aboveground parts of the plant leading to wilting and eventual death. The fungus also produces toxins, such as **'lycomarasmin'** and **'fusaric acid'**, which also induce wilting. Further, some enzymes excreted by the fungus disintegrate the walls of the vessels and adjacent parenchyma cells and thus hasten the death of the infected plant. After the death of the plant, the fungus reaches the surface and sporulates profusely. Secondary spread of the disease may occur through spores carried by wind, water, soil, agricultural implements etc. Usually once an area becomes infected with *Fusarium*, it remains in the soil indefinitely.

Low soil moisture and temperatures around 28°C are favorable for the disease occurrence. It is more damaging in warm climates and is more serious in sandy loam soils.

Disease management

Agronomic practices and chemical control. Refer '*Fusarium* wilt of red gram' (Page 119).

Resistant varieties. Marglobe, Rutgers, Pritchard etc. are resistant to this disease.

5. *Verticillium* wilt of tomato

Verticillium albo-atrum and *V. dahliae*

The disease occurs in several countries of the world, including Britain, the United States of America, Canada, Holland, New Zealand, India etc. In India, it is prevalent in many regions and is severe during some seasons. The disease has several other hosts, such as cotton, egg plant, cowpea, black gram, potato, chillies, tobacco, bhendi, dahlia etc.

Symptoms. The disease may attack the crop at any growth stage. It attacks the roots and by interfering with the water supply to the growing plants, induces wilting of the shoots. Affected plants may recover during the nighttime, when transpiration is reduced considerably, but lose turgidity and wilt again during the daytime, when transpiration increases. Wilting of older plants starts from the lowermost leaves and proceeds upwards. Affected plants show chlorosis and crinkling of leaves. The veins become brown and the interveinal areas of the lamina, as well as the leaf margins turn yellow and gradually become necrotic producing a tiger-stripe effect. Ultimately, such leaves dry completely and fall off. In advanced stages of the disease, only a few, chlorotic, diminutive leaves alone are left at the top. The flowers and fruits are also shed. However, the affected plants do not die soon, as in the case of *Fusarium* wilt. Diseased plants show considerable browning and rotting of the roots at the tips. If a diseased plant is split longitudinally from the root to the stem, brown discoloration can be traced from the root to the base of the stem and sometimes all the way up to the tip of the stem and even into the petioles. This discoloration of the vascular elements is characteristic of the disease. Besides the brown discoloration of the wood vessels, a brown gummy substance is frequently seen to line the inside of the vascular bundles and may block some of the vessels more or less completely, thereby interfering with the translocation of water and mineral nutrients to the aboveground parts of the plant. However, the actual wilting is due to the secretion of toxic substances by the fungi, which are carried up to the leaves through the transpiration system.

The causal organism. For the description of the fungus *Verticillium dahliae* refer '*Verticillium* wilt of Cotton' (Page 224). *Verticillium albo-*

atrum and *V. dahliae* are identical in almost all respects, except the type of resting bodies produced by the two fungi differ. *Verticillium albo-atrum* develops a resting mycelium composed of dark, thick-walled, septate hyphae, while *V. dahliae* produces better organized bodies of the nature of microsclerotia, which are not actually sclerotia but **'pseudosclerotia'**. Further, *Verticillium albo-atrum* is more pathogenic than *V. dahliae*. The fungi are also known to produce thick-walled chlamydospores. Eleven strains of *Verticillium albo-atrum* and four strains of *V. dahliae* have been identified.

Mode of survival, spread and epidemiology. The fungi overseasons in the soil or on the diseased host debris in the form of the dark, thick-walled resting mycelium or microsclerotia (Pseudosclerotia) as the case may be and also as chlamydospores. The two fungi are also capable of saprophytic existence on organic matter and host remains in the soil. The conidia formed at the base of dead plants are short-lived. However, they germinate and thrive upon the decayed debris and produce resistant mycelium or pseudosclerotia. The conidia carried by wind or water cause secondary infection. Unlike *Fusarium*, *Verticillium* is capable of direct penetration into the roots of young plants. But entry into the roots and stems of older plants is facilitated by injuries.

Moderately cool temperatures of about 20°C and high soil moisture favor disease occurrence. The disease incidence is more during the cold months of November and December, when the temperature is about 15° - 20°C. The wilt occurs commonly on heavy clay loam soils and in soils rich in organic matter, because of greater retention of moisture around the roots and the cooling effect produced by the water.

Disease management. Refer '*Verticillium* wilt of cotton' (Page 224).

6. *Septoria* leaf spot of tomato

Septoria lycopersici

The disease is found in most of the countries of the world. In India, it is prevalent in the states of Himachal Pradesh, Jammu and Kashmir, Uttar Pradesh, Karnataka and Tamil Nadu.

Symptoms. On the leaves, spots appear as small, yellowish specks in large numbers that gradually enlarge, turn pale brown or yellowish gray and finally dark-brown. A yellow zone usually surrounds the spots. As the disease advances, the center of the spots turns gray, with a distinct, dark margin. The spots may coalesce and form large, irregular spots. Sometimes, a single spot may enlarge and cover a large area of the leaf lamina. Usually, the older leaves are infected first and then the disease gradually spreads to the upper leaves. Severely affected leaves turn yellow completely, dry and fall down. The stems and flowers are also attacked. Linear, dark-brown lesions may develop on the stem region. But the fruits are rarely attacked. Minute, black pycnidia appear as dots on the spots **(Fig.86)**.

The causal organism. The mycelium of the fungus is well-developed, branched, septate, hyaline when young and becomes slightly darker with age. The pycnidia produced by the pathogen are small, dark- brown, globose, ostiolate and sunken in the host tissues. Conidia are borne at the tips of short, hyaline phialides in succession. They are hyaline, long, slender, filiform, 3 - 9 septate, slightly curved and measure 60.0 - 120.0 x 2.0 - 4.0μ in size. When the pycnidia get wet, they swell and the conidia are exuded in long tendrils through the ostiole **(Fig.86)**.

Mode of spread and epidemiology. The conidia are spread by strong winds, splashing rain, irrigation water and agricultural implements and cattle along with soil. The fungus overseasons as mycelium and as conidia within the pycnidia on the diseased plant debris in the soil. The pathogen may also be carried through infected seeds. When a conidium falls on the surface of the host, it germinates by producing a germ tube and penetrates into the host either through the stoma or directly by penetrating the host epidermis and causes fresh infection. The disease also attacks a few solanaceous plants, such as potato, eggplant etc. and some weeds.

The fungus requires high moisture for infection and disease development, but can cause disease at a wide range of temperatures from 10° - 27°C. The disease incidence is severe during the rainy seasons. Long periods of rainfall, persistent dew and high atmospheric humidity favor the disease occurrence.

Disease management

Agronomic practices. Selection of disease-free seeds, field sanitation, crop rotation and use of resistant varieties help in minimizing the incidence of the disease.

Chemical treatment. Spraying the crop with Bordeaux mixture - 1 % or copper oxychloride - 750 gm. or mancozeb - 600 gm. or zineb - 600 gm. or captan - 450 gm. or dichloran - 450 gm. in 300 lit. of water/ acre, 3 - 4 times, at fortnightly intervals from the time initial symptoms are noticed controls the disease.

Seed treatment. Treating the seeds with captan or thiram or mancozeb at 4.0 gm./ kg. of seeds, at least 24 hours prior to sowing protects the seedlings from infection.

7. Bacterial leaf spot of tomato

Xanthomonas campestris pv. *vesicatoria*

The disease is widespread and is found in almost all the regions, wherever tomato is cultivated.

Symptoms. The disease attacks leaves, stems, petioles, flowers and fruits. On the leaves, the symptoms appear as small, irregularly circular, black, water-soaked, greasy spots, about 3.0 mm. in diameter. A yellow border usually surrounds the spots. Leaves with many lesions may turn yellow completely, become ragged and may fall off prematurely. Lesions may be formed on the stems and petioles. The lesions on the stems are larger, black and canker-like. Infection of flower parts usually results in blossom shedding. On green fruits, small, water-soaked, slightly raised spots appear, which may have greenish-white halos. The spots may enlarge and attain a diameter of 3.0 - 6.0 mm. As the disease progresses, the halos disappear and the spots become dark-brown, slightly sunken and present a scabby appearance **(Fig.87)**

The causal organism. The bacteria causing the disease are short rod-shaped, motile with a single polar flagellum and measure 1.7 x 0.9µ in size on an average. They are gram-negative and form capsules.

Mode of survival and spread. The pathogen is soil-borne and seed-borne. The bacteria can survive in the infected plant debris in the soil for a long time. They can also overseason on seeds contaminated during extraction. Secondary spread is by rain splash, irrigation water, wind and by direct contact. The bacteria can gain entry into the host only through stomata, lenticels, hydathodes or through injuries. The disease also spreads through infected seedlings. The pathogen is capable of attacking *Capsicum* species, *Solanum* species, *Nicotiana* species, pepper and many weeds.

Disease management. Refer 'Bacterial leaf spot of chillies' (Page 421).

8. Tomato mosaic

Tomato mosaic virus

The disease, which was first recorded in the United States of America in 1909, is distributed all over the world. In India, it is prevalent in all the tomato growing states.

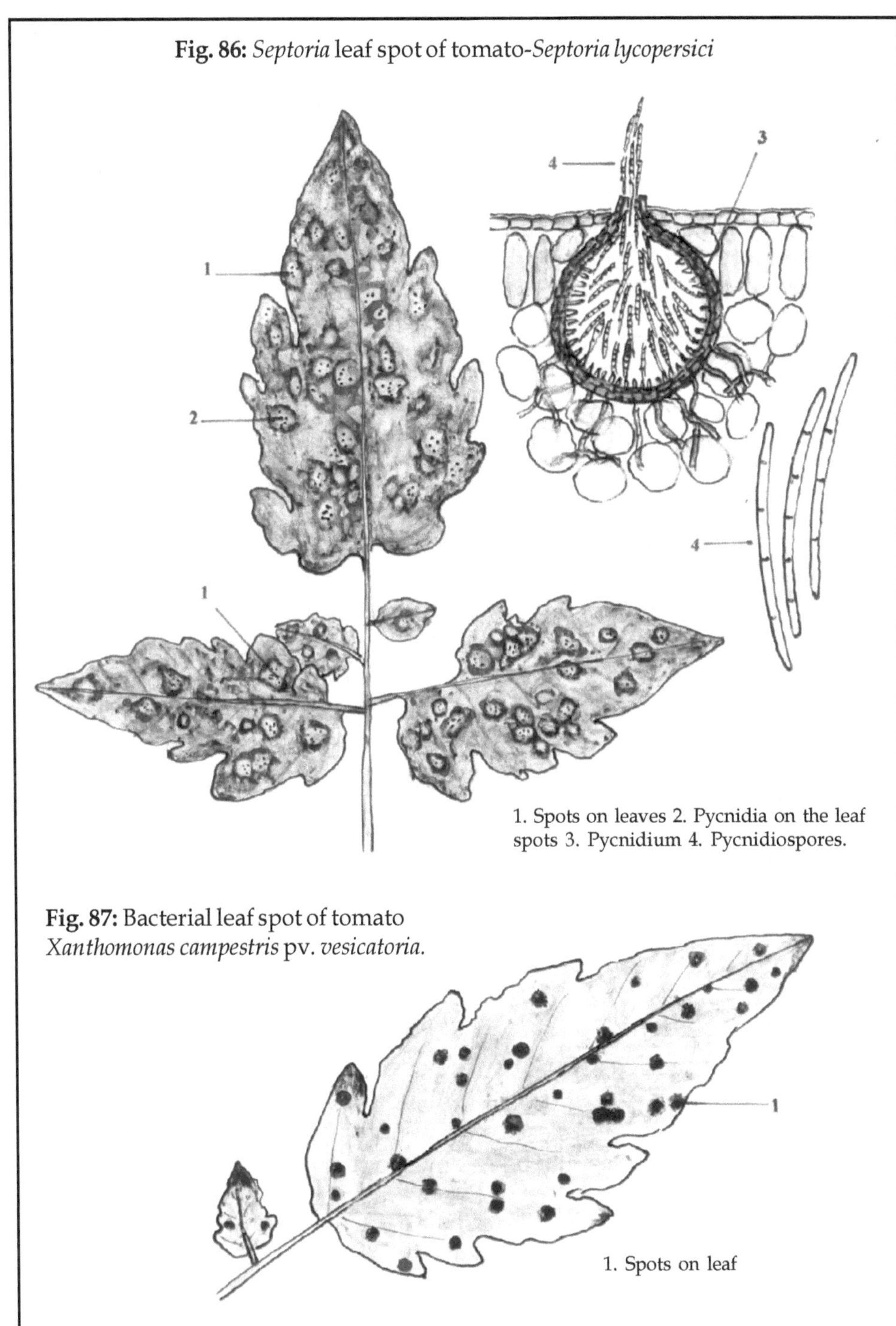

Fig. 86: *Septoria* leaf spot of tomato-*Septoria lycopersici*

1. Spots on leaves 2. Pycnidia on the leaf spots 3. Pycnidium 4. Pycnidiospores.

Fig. 87: Bacterial leaf spot of tomato *Xanthomonas campestris* pv. *vesicatoria.*

1. Spots on leaf

Symptoms. The symptoms appear as light and dark green patches, with raised dark green areas on the leaves. Young leaves become distorted and crinkled. The affected plants are stunted in growth. Often young leaves may be so distorted they become fern leaf-like or shoe string-like and the stems of young plants turn purple in color due to development of **'anthocyanin'**. When older plants are infected, mottling appears only in the new foliage, while the older leaves remain normal and look healthy. Fruit production is reduced, but the fruits appear quite normal. If infection occurs at an early stage, blemishes may appear on the fruits before they are ripe.

The causal organism. The disease is caused by *Tomato mosaic virus*, which belongs to the Genus - *Tobamovirus.* The single, rigid, rod-shaped, helical virus particles measure 300 nm in length and 15 - 18 nm in width and contain a single molecule of ssRNA coiled inside a protein capsid, with a distinct central hole.

Mode of spread. The pathogen is sap-transmissible and highly contagious. It enters the host through any kind of injury caused by mechanical means or by insects. Natural forces, such as wind, lashing rain etc., which make the leaves rub against one another may cause invisible injuries through which the virus can gain entry into the host. The virus can even be transmitted by touching a healthy leaf after touching a diseased leaf. The virus is not transmitted by insect vectors, but is transmitted through seeds. Seeds of fleshy fruits are infected and the virus is carried on the seeds, but is not found inside the endosperm. The virus can be completely destroyed, when exposed for 10 minutes to a temperature of 90°C. However, it is capable of withstanding highly adverse conditions. It can remain potent in infected plant debris for several years and is a source of primary infection. Even a dilution of 1 : 1,000,000 of the sap can cause infection. The virus has many alternate and collateral hosts, including chillies, cucurbits, tobacco, brinjal, cowpea and French bean, besides many weeds and ornamental plants.

Disease management

Agronomic practices (i) Virus-free seeds from virus-free plants should be used for sowing in the nursery (ii) All tomato plant debris in and on the soil should be removed and destroyed by burning (iii) Young plants showing symptoms of the disease should be removed and destroyed (iv) After handling diseased plants in the nursery or in the main field, healthy plants should not be touched (v) Hygienic measures, such as avoidance of tobacco products while working in tomato fields, washing of hands in soap water etc. greatly reduce the chances of spreading the disease to healthy plants (vi) Volunteer plants and weed hosts in the vicinity of tomato fields should be removed and destroyed (vii) Susceptible hosts should not be grown near about tomato fields.

Seed treatment. Soaking of seeds in trisodium phosphate ($Na_3P_2O_5$) for 10 minutes is effective in eliminating the virus from the seed surface.

Resistance. Inoculating the seedlings with inactivated virus particles or mild strains of the virus can induce resistance. Such vaccines have been developed recently.

9. Tomato spotted wilt

Tomato spotted wilt virus

The disease is worldwide in distribution and is prevalent in Africa, Asia, Australia, France, New Zealand, New South Wales, Europe, America and India. In India, the disease occurs commonly in the states of Tamil Nadu, Telungu Desam, Karnataka and Kerala.

Symptoms. Tomato plants are vulnerable to attack by this disease both in the nursery and in the main field, however young plants are more prone to attack by the virus than grown-up plants. Early symptoms appear on young leaves as thickening of the veins and development of spots or concentric rings of a brown color. Bronzing of the entire leaf lamina is another characteristic feature of the disease. Downward curling of the leaves follows this. The leaf growth is also arrested. If infection takes place at the early stages of plant growth, the bronzed leaves become necrotic and the plant may die. Sometimes, the affected plant may recover to some extent, if growth conditions are good and puts forth secondary growth. But, in such plants also mottling, consisting of indefinite yellowish areas, deformity and bronzing of leaves is seen. Sometimes, mottling and blemishes on the fruits are seen. The blemishes on fruits may be of different shapes and may consist of concentric pattern or pale areas of red or yellow. Sometimes, the fruits may be completely yellow, except for a few specks of natural red color. Seed formation in such fruits is badly affected.

The causal organism. The disease is caused by *Tomato spotted wilt virus*, which belongs to the Genus - *Tospovirus*. It is also known as *Tomato virus* 1 and *Lycopersicum virus* 3. The isometric virus particles are 85 nm in diameter and are enclosed within a lipoprotein envelope. The -ssRNA genome is composed of three molecules of RNA. The virus is highly sensitive to heat and its thermal inactivation point is 40°C.

Mode of spread. The virus is sap-transmissible and highly contagious. It can gain entry into the host even through invisible injuries caused by the breaking of epidermal hairs by gently rubbing the host surface. The virus is not transmitted through seeds or pollen, but by various species of thrips, such as *Thrips tabaci, T. palmi, Frankliniella schultzei, F. insularis, F. fusca, F. occidentalis* and *Scirtothrips dorsalis*. Adult thrips are not capable of acquiring and transmitting the virus. The virus is acquired by the larval form of the thrips and after acquisition the adults become viruliferous and transmit the virus. Minimum acquisition feeding time is 15 minutes and the incubation period lasts for 4 - 118 days till the larval form attains the adult stage. The virus has a very wide host range of at least 160 plant species. It includes numerous members of the Family - Solanaceae, such as tobacco, chillies, *Datura*, Petunia, *Physalis* etc., besides many plants from different families of Leguminaceae, Papaveraceae and Compositae. Aster, dahlia, chrysanthemum, zinnia and cineraria are well known ornamental hosts belonging to the Family - Compositae.

Disease management

Agronomic practices. Refer 'Tomato mosaic' (Page 354).

Chemical control. Spraying with endosulfan - 600 ml. or phosalone - 600 ml. or phosphamidon - 150 ml. or fenthion - 600 ml. or monocrotophos - 300 ml. in 300 lit. of water/ acre controls the vectors.

10. Tomato leaf curl

Tobacco leaf curl virus

The disease has been reported from several countries all over the world. It is more common in the winter season in India.

Symptoms. The disease is characterized by symptoms, such as dwarfing of infected plants, puckering, twisting, curling, mottling and vein clearing of leaves, excessive branching and partial or complete sterility.

The causal organism. *Tobacco leaf curl virus*, the causal agent of the disease belongs to the Genus - *Geminivirus*. The isometric particles are of size 18 - 20 nm in diameter and occur mainly in pairs. The genome consists of monopartite ssDNA.

Mode of spread. The virus is neither sap nor seed transmissible, but the seeds of fleshy infected fruits may be contaminated with the virus particles. The white fly vector - *Bemisia tabaci* transmits the virus. The vector requires a minimum acquisition feeding period of 15 - 30 minutes. About 21 hours after acquisition, the vector becomes viruliferous and can transmit the virus in a semi-persistent manner for about 2 weeks. The inoculation feeding time required is 15 - 30 minutes. White flies are very active insects and even a single viruliferous insect can transmit the virus to a number of plants within a single day. All commercial tomato varieties are found to be susceptible to the disease. Besides tomato, the virus can infect several cultivated and other hosts. The host range includes *Solanum melongena, Hibiscus esculentus, Capsicum annuum, Carica papaya, Nicotiana tabacum, Crotalaria juncia, C. mucronata, Gossypium arboreum, Solanum nigrum, Zinnia elegans, Petunia hybrida, Phlox drumondii, Euphorbia geniculata* etc.

Disease management

Agronomic practices (i) Diseased plants should be removed and destroyed (ii) Continuous cropping of tomato in the area should be avoided (iii) Alternate and collateral hosts of the virus and the vector should be removed and destroyed (iv) Yellow sticky traps should be set up in the field at the rate of 10 traps per acre to trap and kill the vectors.

Chemical control. Vector control measures should be initiated from the nursery stage onwards. At the time of sowing or a few days after sowing, the nursery should be treated with granular systemic insecticide.

For this, carbofuran 3G at 13 kg. or thiodemeton 5G at 10 kg. or phorate 10G at 5 kg./acre should be applied uniformly and irrigated. This treatment will afford protection to the seedlings from white fly infestation for a period of about 30 days. This is followed by 3 - 4 insecticidal sprays in the transplanted field at 15 days interval. Chemicals having quick knockdown effect should be used to kill the vectors, as they are capable of transmitting the virus to a number of plants within a short time. Spraying with methyl demeton - 1.0 ml. or dimethoate - 1.0 ml. or phosphamidon - 0.6 ml. or acephate - 2.0 ml. or phosalone - 2.0 ml. or ethion - 1.0 ml./ lit. of water is effective in controlling the vector. If granular insecticide is not applied to the nursery, 2 sprayings with any one of the above insecticides may be given.

Diseases of minor importance. Several other diseases are also known to attack tomato. *Corticium solani* causes 'Root rot' and 'Fruit rot'; *Leveillula taurica* causes 'Powdery mildew'; *Rhizopus nigricans* causes 'Soft rot of fruits', both in the field and in storage; *Erwinia carotovora* causes 'Soft rot of fruits' mostly in storage; 'Tomato big bud' is caused by *Mycoplasma-like organisms.*

Lady's finger or Okra or Bhendi *(Abelmoschus esculentus)*

1. Yellow vein mosaic or Vein clearing

Bhendi yellow vein mosaic virus

'Yellow vein mosaic' of bhendi was first recorded in India in 1924. It is found in all bhendi growing regions of India. It is a very serious disease and is often a limiting factor in the successful cultivation of bhendi. When the infection occurs at the early stages of crop growth, the entire crop is destroyed. The extent of disease incidence decreases with the delay in infection. In case of early and severe infection, the yield loss may go up to 90% or even more.

Symptoms. The main symptom of the disease is vein clearing, followed by veinal chlorosis of the leaves The yellow network of veins is a very conspicuous symptom. The veins and veinlets also become thickened. In case of severe incidence, the chlorosis may extend to the interveinal areas and the leaves may become completely yellow. The symptoms are more pronounced in the newly emerging leaves, which are much reduced in size, chlorotic and are wrinkled. Fruit setting is considerably reduced and the fruits, if formed are underdeveloped, malformed, fibrous and are yellowish-green in color.

The causal organism. The disease is caused by *Bhendi yellow vein mosaic virus* or *Hibiscus virus* 1. Not much is known about the nature and morphology of the virus.

Mode of survival, spread and epidemiology. . The virus is not transmitted through sap, seed or pollen, but is transmitted by the white fly vector - *Bemisia tabaci* and by the okra leafhopper - *Empoasca devastans.* However, the white flies are more efficient in transmitting the virus and the female white flies are more efficient in spreading the disease than the male insects. The acquisition-feeding period is 4 - 6 hours and the inoculation-feeding period is 15 - 20 minutes. After acquisition, the vector remains viruliferous for the rest of its life. There are several alternate hosts for the virus including *Croton sparsiflora, Malvastrum tricuspidatum* and *Agreratum* species, which are common roadside plants, besides several species of *Hibiscus* and *Abelmoschus.*

The disease incidence is more during March - June, when the temperature is high and atmospheric humidity low, which also favors multiplication and spread of the vectors.

Disease management

Agronomic practices (i) Alternate weed hosts in and around okra fields should be eradicated (ii) Diseased plants should be removed and destroyed periodically, even from the early stages of crop growth (iii) Crop rotation may be followed.

Chemical control (i) The vectors transmitting the virus may be controlled by spraying with methyl demeton - 600 ml. or dimethoate - 600 ml. or phosphamidon - 160 ml. or ethion WP - 600 gm. or acephate WP - 600 gm. in 300 lit. of water per acre. Sprayings should commence just after germination, followed by 3 or 4 sprayings, at fortnightly interval (ii) Application of granular insecticides, such as carbofuran 3G - 13 kg. or phorate 10G - 5 kg./ acre at the time of sowing, followed by one or two sprayings with any of the insecticides is also effective in controlling the vectors.

Resistant varieties. No commercial varieties of okra are found to be immune to the disease. The varieties, Arka Abhay, Arka Anamika and Parbhani Kranti are resistant to the disease. Pusa Sawani is tolerant to the disease.

Diseases of minor importance. Lady's finger is susceptible to a few more diseases of less importance. *Fusarium oxysporum* f.sp. *vasinfectum* causes 'Wilt'; *Erysiphe cichoracearum* causes 'Powdery mildew'; *Pythium aphanidermatum* and *Phytophthora palmivora* cause 'Damping off' of young plants; *Macrophomina phaseolina* causes 'Dry root rot'; *Xanthomonas campestris* pv. *esculenti* causes 'Bacterial leaf spot'.

Cucurbits

This group includes **'Cucumber'** *(Cucumis sativus)*, **'Musk melon'** *(Cucumis melo)*, **'Water melon'** *(Citrullus vulgaris)*, **'Bottle gourd'** *(Lagenaria siceraria)*, **'Bitter gourd'** *(Memordica charantia)*, **'Sponge gourd'** *(Luffa cylindrica)*, **'Ridge gourd'** *(Luffa acutangula)*, **'Snake gourd'** *(Trichosanthes anguina)*, **'Pointed gourd'** *(Trichosanthes dioica)*, **'Round gourd'** *(Citrullus vulgaris* var. *pistulosus)*, **'Ash gourd'** *(Benincasa hispida)*, **'Pumpkin'** *(Cucurbita moschata)*, **'Summer squash'** *(Cucurbita pepo)*, **'Winter squash'** *(Cucurbita maxima)* and a number of other crops, mostly having trailing habit. They all belong to the Family - Cucurbitaceae and are grown mostly during the summer season.

1. Powdery mildew of cucurbits

Erysiphe cichoracearum and *Sphaerotheca fuliginea*

Erysiphe cichoracearum causes powdery mildew of cucurbits, such as muskmelon, cucumber, squash and pumpkin. It also causes powdery mildew of potato, tobacco, lettuce, sunflower, mango, castor, snapdragon, begonia, chrysanthemum, dahlia, zinnia etc., while *Sphaerotheca fuliginea* exclusively attacks cucurbits. The disease is distributed worldwide. In India, the disease occurs in all the regions almost every year and causes extensive damage to the crops and yield loss.

Symptoms. Under excessively moist conditions, the mildew appears as small, white or dirty gray patches on the leaves and stems, but not on the fruits. These areas become powdery as they enlarge and soon cover the entire surface of leaves and stems. *Erysiphe cichoracearum* forms characteristic white, powdery coating consisting of conidiophores and conidia on the affected parts. On the other hand, the patches formed by *Spaerotheca fuliginea* are not white, but reddish-brown in color. Black pinhead-like bodies, which represent the cleistothecia of the fungus, appear late in the season on the host surface. But the ascigerous stage of the pathogen is rarely found in nature. The effect of mildew, by covering the leaves and stems with its mycelial web arrest their growth, induce chlorosis and cause reduction in the size and quality of fruits **(Fig.88).**

The causal organism. The pathogens causing the disease are obligate parasites.

Erysiphe cichoracearum. The fungus forms a tangled web of mycelium over the surface of leaves and stems, consisting of hyaline, septate hyphae, which produce rather large spherical haustoria in the epidermal cells. The superficial mycelium gives rise to a great profusion of erect conidiophores, bearing conidia in long chains. Conidia are ellipsoid to barrel-shaped and measure 25 - 45 x 14 - 26μ in size. They germinate and produce appressoria. Cleistothecia are of rare occurrence and are found only on certain parts of the diseased leaves in the cold season. They are found scattered or in groups and are globose becoming depressed or irregular in shape. They measure 90 - 135μ in diameter. Appendages on the cleistothecia are numerous, basally inserted, mycelioid, interwoven with the mycelium, hyaline to dark-brown, 1 - 4 times as long as the diameter of the cleistothecium and rarely branched. Each cleistothecium contains 10 - 15 ellipsoid, more or less stalked asci, measuring 60 - 90 x 25 - 50μ. Each ascus contains 2, very rarely 3, hyaline, oval or sub-cylindrical ascospores, measuring 20 - 28 x 12 - 20μ. **(Fig.88).**

Sphaerotheca fuliginea. The mycelium of the pathogen is hyaline, becoming reddish-brown or brownish with age. The superficial mycelium produces spherical haustoria inside the epidermal cells. Erect conidiophores are produced in profusion from the superficial mycelium. Conidia are borne singly at the end of fibrosin bodies. They are hyaline, single-celled, ellipsoid to barrel-shaped and measure 20 - 28 x 14 - 25μ in size. The conidia on germination produce a characteristic forked germ tube, distinguishing them from those of *Erysiphe cichoracearum*. Cleistothecia are dark, spherical with many mycelioid appendages and measure 90 - 180μ in diameter. Each cleistothecium usually contains only one ascus. The asci are broadly elliptic to sub-globose and 50 - 80 x 30 - 60μ in size. In each ascus, there are 6 - 10, usually eight hyaline, continuous, single-celled, ellipsoid ascospores, measuring 20 - 25 x 12 - 15μ in size **(Fig.88).**

Mode of survival, spread and epidemiology. In areas where cleistothecia are formed, they may carry over the disease from one season to the next. Cleistothecia may be formed, when sexually compatible strains of the fungus establish union in certain parts of the diseased leaves. Cleistothecia may develop on diseased plant debris left over in the field and may perpetuate the disease. It is possible that the pathogen may be harbored on some wild cucurbits during the off season and produce conidia, which may serve as primary inoculum and infect cultivated cucurbits. Secondary spread may occur through conidia carried by wind, rainsplash and by insects that attack cucurbits.

Heavy dew, cloudy and damp weather, moderate temperature range of 22° - 28°C and low light intensity favor the occurrence and development of the disease. However, even under dry conditions, the conidia can germinate and cause the disease.

At least 13 form species of *E. cichoracearum* and 84 f.sp. of *S. fuliginea* have been identified.

Disease management

Agronomic practices. Removal and destruction of diseased plant debris and destruction of weed hosts harboring the pathogen help in minimizing the occurrence of the disease.

Chemical control. Spraying the crop with wettable sulfur - 4.0 gm. or dinacap (Karathane) - 1.0 gm. or carbendazim - 1.0 gm. or benomyl - 1.0 gm. or thiophanate methyl (Topsin M) - 1.0 gm. or tridemorph (Calixin) - 1.0 gm./ lit. of water is found to be effective in controlling the disease.

Resistant varieties. In muskmelon, the varieties Diguria and Huragola have been reported to be immune to powdery mildew.

Fig. 88: Powdery mildew of cucurbits-*Erysiphe cichoracearum* and *Sphaerotheca fuliginea*

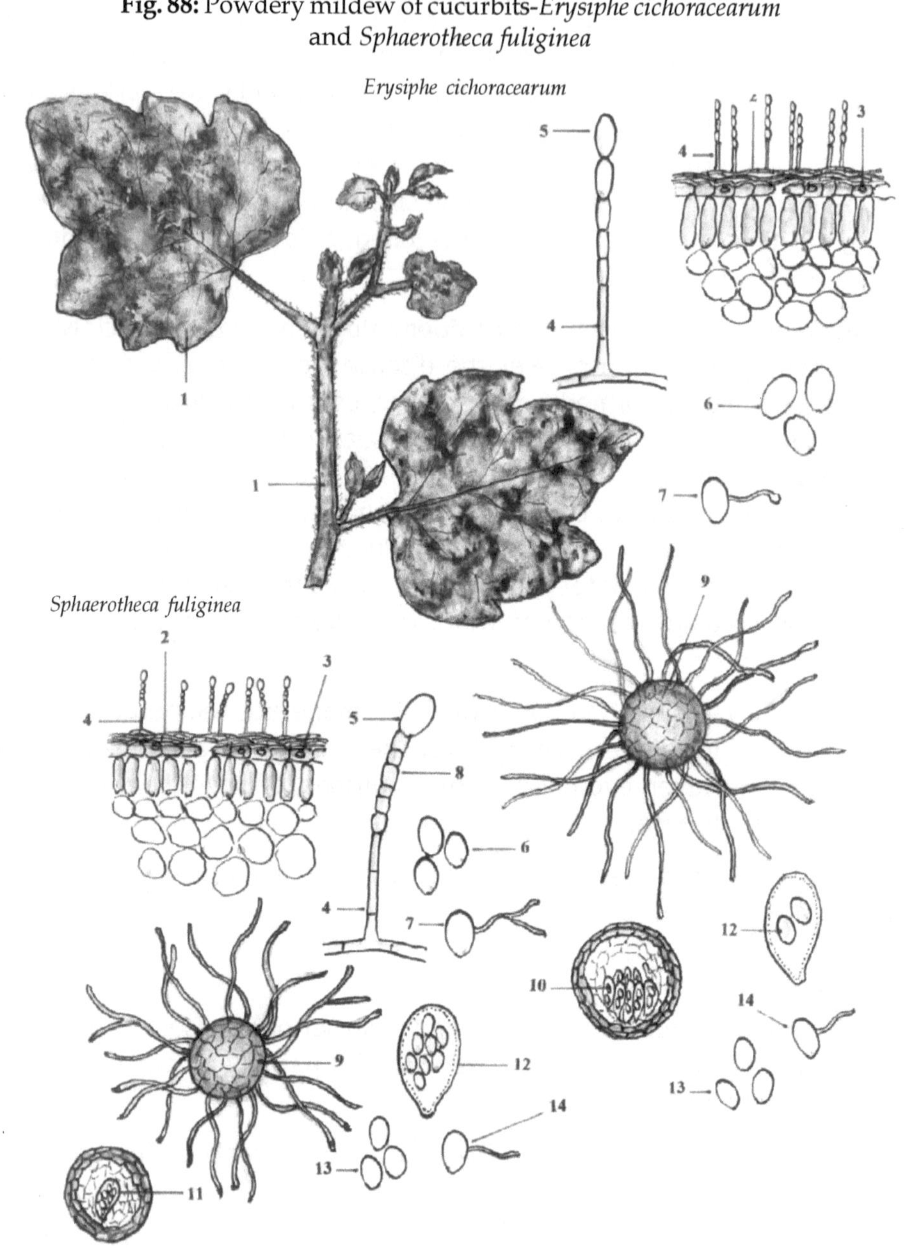

1. Powdery mildew growth on the leaf and stem 2. Superficial mycelium 3. Haustorium 4. Conidiophore 5. Conidium 6. Conidia 7. Germinating conidium 8. Fibrosin bodies 9. Cleistothecium 10. Asci inside the cleistothecium 11. Ascus inside the cleistothecium 12. Ascospores inside the ascus 13. Ascospores 14. Germinating ascospore.

2. Downy mildew of cucurbits

Pseudoperonospora cubensis

The disease, which was first recorded in Cuba in 1868, is now prevalent in all parts of the world. It occurs in most of the cultivated cucurbits, such as cucumber, muskmelon, watermelon, squash, pumpkin, gourd and several other species.

Symptoms. The symptoms appear as pale yellow, angular patches on the upper surface of the leaves, which are usually limited by the veins. The color of the patches soon deepens to brownish-yellow, as they grow older. On the corresponding lower side of these patches, a fine, purplish, downy growth appears during periods of high humidity. The growth appears scanty and is seen only under close examination. Severely affected leaves become yellow completely, dry and die. Mostly, the disease attacks leaves in the middle portion of the plants and gradually spreads to the younger leaves. Fruits are rarely affected directly. Only a few, small and misshapen fruits may be formed in the infected plants. Muskmelon fruits may be attacked and may be covered with the fungal growth.

The causal organism. The fungus causing the disease is an obligate parasite. The mycelium is hyaline, coenocytic and intercellular between the mesophyll cells of the leaves. Short, ovate or branched haustoria produced from the hyphae, penetrate the cells and absorb nourishment from the host. After a period of vegetative growth, the fungus produces sporangiophores, which arise in groups of 1 - 5 through the stomata on the under surface of the leaves. Thy are 200 - 300µ in length and 5 - 9µ in width. The lower two-third portion of the sporangiophore is unbranched, while the top one-third portion is branched 3 - 5 times, either dichotomously or in a manner intermediate between dichotomous and monochotomous branching i.e., pseudodichotomous branching. The terminal branches or sterigmata are long, subacute, slender and pointed. Sporangia are borne singly at the tips of the sterigmata. They are grayish to olivaceous-purple, ovoid to broadly elliptical, thin-walled, with an apical papilla and measure 21 - 39 x 14 - 23µ. They germinate by producing biflagellate zoospores, which are 10 - 13µ in diameter when they encyst. After a short period of rest, they germinate and cause fresh infection. Oospore production is not common.

Mode of survival, spread and epidemiology. Because of the great diversity of climate, season and cropping pattern in different parts of India and the presence of a number of wild hosts, the pathogen is able to survive from season to season and cause fresh infection. The pathogen may also perpetuate in the form of active mycelium on self-sown crops or cucurbits cultivated all through the year. In areas where oospores are formed, they may serve as an important source of primary inoculum. Secondary spread of the disease is through sporangia dispersed by wind or rainsplash. Some insects, such as the cucumber beetles may also disperse the sporangia mechanically.

Maximum sporangial production takes place in the early morning hours, under saturated atmospheric conditions and low temperatures between 18°-22°C. Sporangial production almost ceases at temperatures above 35°C. A dry period of 6 hours followed by a wet period of 6 hours is highly favorable for sporulation. Optimum temperature for germination of sporangia is 20°C. Free water on the leaf surface is essential for the zoospores to cause infection. The disease usually appears at the latter part of the rainy season and may cause extensive damage to the crops.

Disease management

Agronomic practices (i) Alternate weed hosts in and around the fields should be eradicated (ii) Severely affected plants or plant parts may be removed and destroyed (iii) Resistant varieties may be grown (iv) Proper drainage facilities should be provided.

Chemical control. Protective fungicides, such as copper oxychloride, dithiocarbamates, captafol, chlorothalonil etc. are effective only in checking the disease before its appearance. These fungicides are not effective in controlling the disease after it has established. Systemic fungicides, such as metalaxyl and fosetyl -Al. have been found to be effective in controlling the disease. Spraying with metalaxyl (Ridomil) - 1.0 gm. + mancozeb - 2.0 gm. or fosetyl -Al. (Aliette) - 1.0 gm. + mancozeb - 2.0 gm./ lit. of water is very effective in controlling the disease.

3. Fruit rot of cucurbits

Pythium aphanidermatum and *P. butleri*

'Fruit rot of cucurbits' is a very common disease of cucurbits in India. It occurs in almost all the regions, wherever cucurbits are cultivated. The occurrence of the disease is more in extended wet periods. Several species of cucurbits, including bottle gourd, sponge gourd, snake gourd, cucumber, bitter gourd, muskmelon and watermelon are attacked. In muskmelon and watermelon, sometimes 50 % or more of the fruits undergo rotting due to this disease, especially after rains.

Symptoms. The symptom appears as dark green, water-soaked lesion on the skin of the fruit during prolonged wet periods. The fruits, which are directly in contact with the soil or slightly above the ground level are more prone to attack by the fungi. On the water-soaked lesion on the surface of the fruit, a whitish, luxuriant weft of wooly mycelial growth results, while the interior fleshy portion is turned into a soft, pulpy, watery, rotten mass called **'cottony leak'**. The decaying fruit emits a bad and repulsive smell. The margin of the skin around the area covered by the wooly growth gradually becomes dark-green and water-soaked, indicating the **'killing in advance'** of the tissues below, followed by mycelial growth in that area also. Subsequently, the disease spreads to other fruits during storage and transit.

The causal organism. Mainly *Pythium aphanidermatum* and *P. butleri* are responsible for causing the disease. For the morphology of *Pythium aphanidermatum*, refer 'Damping off of seedlings of eggplant' (Page 339), **(Fig.85)**. The morphology of *Pythium butleri* is almost similar to that of *P. aphanidermatum*. However, *P. butleri* produces more copious and robust mycelium, larger, more swollen and much branched sporangia, larger zoospores and larger oospores. Some species of *Fusarium, Rhizoctonia* and *Phytophthora* also cause fruit rot of cucurbits.

Mode of survival, spread and epidemiology. The fungi, *Pythium aphanidermatum* and *P. butleri* are soil inhabiting ones and survive on dead organic matter present in the soil and cause disease, when suitable hosts are available. Secondary spread of the disease may occur through sporangia carried by irrigation and rainwater and mechanically by some insects, such as cucurbit beetles. The oospores produced by the fungi may also remain in the soil in a viable state for long periods and cause primary infection. Any kind of injury or wound on the surface of the fruit caused by either mechanical means or by insects predisposes the fruit to infection.

High soil moisture and high temperature of about 30°C favor disease occurrence and development

Disease management

Agronomic practices (i) Proper drainage facilities should be provided to prevent water stagnation in the field (ii) As far as possible, the fruits should not be allowed to come into contact with the soil (iii) During cultural operations and during storage and transit care should be taken to avoid causing injuries to the fruits (iv) The fruits should be stored in clean, moisture-free and well aerated store-houses (v) Disease affected fruits should not be stored along with healthy fruits and should be removed and destroyed.

4. Anthracnose of cucumber

Colletotrichum lagenarium

(Glomerella cingulata f.sp. *orbiculare)*

'Anthracnose of cucumber' was first reported from Italy in 1867 and later from England in 1871. Now, the disease is distributed worldwide and is prevalent in Europe, America, Australia and Asia. In India, the disease occurs in almost all the regions, wherever cucumber is grown and in some seasons it causes serious damage to the crop and yield loss. It attacks cucumbers, watermelon, muskmelon, gourds, vegetable marrow, chayote and gherkin.

Symptoms. The fungus attacks all aerial parts of the plants. Usually, the young plants are not affected. But, if young plants are attacked at the soil level, the stems are girdled and the tissues shrink and the seedlings may collapse and die. Mostly, the disease occurs somewhat late in the growing season and spreads rapidly in wet weather. Initial symptoms appear on the leaves as small, pale-green, water-soaked spots, which soon become reddish-brown in color, with a yellow zone around. The spots enlarge and coalesce to form large bronzed areas. These areas later become necrotic, develop cracks and fall out and soon the entire leaf dies. Severely affected crop presents a scorched appearance. On the upper surface of the leaf lesions, scattered, minute, pinpoint-like dots representing the acervuli of the fungus are produced in large numbers. On leaf petioles and stems oval to linear, slightly sunken, water-soaked, yellowish-brown lesions are formed. Sometimes, the stem lesions become very deep and girdle the stem completely, as a result the entire shoot may wither and die. Spots on the fruits are pale-green, almost circular and sunken and they turn black later on. On these lesions, buff to pink colored, gelatinous masses of spores are produced abundantly. On ripe fruits, the black lesions may become white, crater-like cankers on which acervuli appear as black dots. From such deep lesions on the fruits, seed infection may occur.

The causal organism. The mycelium of the fungus is septate, hyaline when young, becoming darker with age, inter- and intracellular. Acervuli are produced from sub-epidermal stromata. They are brown to black in color. The conidia are single-celled, thin-walled, hyaline, but pinkish in mass, oblong or ovate-oblong, slightly pointed at one end and are 13.0 - 19.0 x 4.0 - 6.0μ in size. Setae are numerous. They are 2 - 3 septate, brown, thick-walled and 90 - 120μ in length. The perfect stage is very rarely found in nature.

Mode of survival, spread and epidemiology. The fungus is mainly soil-borne. It can survive in the infected plant debris in the soil for prolonged periods and continues to produce conidia, which may cause primary infection.

The disease may be carried through contaminated seed externally, if the infected fruit pulp remains adhered on the seed. The inoculum may come from the various alternate hosts of the fungus also. The disease may spread through irrigation or rainwater and through soil. Secondary spread is through conidia carried by wind or rainsplash.

The optimum temperature for the germination of conidia is 22° - 27°C. High humidity of 87 - 95 % and low temperatures of about 24°C are favorable for disease occurrence and development.

Disease management

Agronomic practices (i) Diseased crop debris should be collected and destroyed by burning (ii) Crop rotation may be followed (iii) Cultivation of susceptible cucurbits or varieties in endemic areas should be avoided

Chemical treatment. Spraying with copper oxychloride - 2.5 gm. or mancozeb - 2.0 gm. or carbendazim - 1.0 gm. or chlorothalonil - 1.5 gm. or captafol - 1.5 gm./ lit. of water, at fortnightly intervals, commencing from the time the plants are well established controls the disease.

Seed treatment. Treating the seeds with captan or thiram at 4.0 gm./ kg. of seed, at least 24 hours prior to sowing will eradicate the externally seed-borne inoculum and also protects the seedlings from early infection in the field.

5. Cucumber mosaic

Cucumber mosaic virus

The disease was first reported from the United States of America in 1934 on *Cucumis sativus* and now, the disease is distributed worldwide. In India, the disease occurs in all the regions, wherever cucumbers are cultivated on a large scale intensively. *Cucumber mosaic virus* infects more different kinds of plants than any other virus. The host range includes several vegetable crops, ornamental plants, weeds and other plants. The important crops affected by the virus are cucumbers, gladioli, melon, squash, pepper, spinach, celery, tomato, beet, beans, banana and crucifers.

Symptoms. Young seedlings are usually not attacked in the field during the first few weeks. Infection starts, when the plants are about 6 weeks old and growing vigorously. The initial symptoms appear in the young developing leaves as small, greenish-yellow areas, 1.0 - 2.0 mm. in diameter and are limited by the veins. Mottling, distortion and wrinkling of the leaves follow this and the edges of the leaves begin to curl downward. The growth of the affected plant is reduced markedly and the internodes and petioles become shortened. The leaf size is also reduced considerably. Such plants produce few runners, few flowers and fruits. The older leaves of infected plants become chlorotic and necrotic areas are formed along the margins. The necrosis gradually spreads over the entire leaf area. Such leaves dry and remain hanging from the petioles or fall off, leading to defoliation of the vine. The flowers produced are also dwarfed. Fruits produced on infected plants show pale green or whitish patches, intermingled with dark-green raised areas, which later become rough and wart-like. The fruits are also deformed and they have a bitter taste.

The causal organism. The virus causing the disease belongs to the Genus - *Cucumovirus*. The viruses have tripartite genomes consisting of ssRNAs and are encapsidated in three types of protein covering. The three identical isometric particles are about 28 nm in diameter each. All the three particles are necessary for infection. The virus can easily be transmitted mechanically by sap inoculation and by insect vectors in a non-persistent manner. The virus infects and multiplies in the phloem and parenchyma cells. Drying destroys the virus.

Mode of survival and spread. The virus can be transmitted to a healthy plant by touching it, after handling a disease-infected plant. The aphid vectors, *Aphis craccivora* and *Myzus persicae* transmit the virus in a non-persistent manner. Because the virus has numerous hosts and the aphid vectors are also present throughout the year in various crops, the disease can be perpetuated easily.

Disease management

Agronomic practices (i) Alternate weed hosts belonging to species of *Commolina, Physalis. Phytolacca* etc. should be eradicated (ii) Continuous cropping of cucumbers in a particular area should be avoided (iii) After handling diseased plants, healthy plants should not be touched (iv) Resistant varieties may be grown.

Chemical control. The aphid vectors should be controlled by using insecticides, such as dimethoate, methyl demeton or phosphamidon.

Diseases of minor importance. Cucurbits are susceptible to a number of other diseases, besides the ones mentioned above. *Verticillium albo-atrum* and *V. dahliae* cause '*Verticillium* wilt'; *Erwinia tracheiphila* causes 'Bacterial wilt'; several virus diseases also occur on cucurbits.

Crucifers

The group includes '**Cabbage**' *(Brassica oleracea* var. *capitata)*, '**Cauliflower**' or '**Broccoli**' *(Brassica oleracea* var. *botrytis* sub. var. *cauliflora)*, '**Knol Khol**' or '**Kohlrabi**' *(Brassica oleracea* var. *gonglyodes)*, '**Turnip**' *(Brassica rapa)*, **Mustard** *(Brassica campestris)* and a number of other crops.

1. Club root of cabbage

Plasmodiophora brassicae

'**Club root**' or '**finger and toe disease**' attacks a wide range of cultivated and garden plants belonging to the Family - Cruciferae. The disease was first reported from Britain in 1736 and from the United States of America in 1867. However, Woronin of Russia carried out detailed investigation of the disease on cabbage in 1878 and the causal organism was named *Plasmodiophora brassicae*. Now, the disease is distributed worldwide and is considered to be a very serious one, especially on cabbage, turnip, swede, cauliflower, knol khol, mustard and radish.

Symptoms. As the name suggests, '**club root**' is essentially a disease of the root system and until the disease progresses to a considerable extent, no visible symptoms are evident on the aboveground parts. The first aboveground symptoms are unthrifty, sickly appearance of the plants, yellowing, flagging and wilting of the leaves in hot sunny days, as if suffering for want of water. When such plants are uprooted, the hypertrophied roots can be seen. The presence of the disease within the host results in increased growth of the parts affected, especially the roots, which become abnormally swollen into clubs or galls. The symptoms vary according to the rooting habit of the hosts. On cabbage, which has a slender taproot and numerous lateral roots, both the taproot and the lateral roots become transformed into spindle-shaped galls of variable thickness. On turnip and swede, in which the lateral roots occur from the main root below the bulb, galls are formed on the lateral roots or on the taproot, while the bulb is not disfigured by the formation of galls.

The gall formation within the roots, affects the vascular system, as a result absorption of water and mineral nutrients from the soil and translocation of the same to the aboveground parts of the plant is hindered. This leads to yellowing and wilting of the foliage and stunting of plants. Hypertrophied cells in the cortex and medullary rays are filled with the plasmodia and numerous minute spores produced by the plasmodia. In advanced stages of the disease, the galls become much enlarged in size, causing disruption of the outer tissues and a general decay of tissues sets in. This paves the way for the invasion of secondary saprophytic organisms, such as soft rot bacteria and the root system is converted into a black rotting mass. The spores of club root organism in the disintegrated cells are then released into the soil **(Fig.89)**.

The causal organism. The club root pathogen is an obligate parasite. The resting spores, which are released into the soil, are spherical and 2.0 - 4.0μ in diameter. The spore on germination liberates a single, naked, pear-shaped, uninucleate zoospore, with 2 unequal, whiplash-type flagella at the pointed end. After a short period of motility, it comes to rest and then changes into a myxamoeba. The myxamoeba enters the root through a hole dissolved in the epidermal cell wall and then forms a thallus in the lumen of the cell.

Mode of survival, spread and epidemiology. The resting spores, although capable of immediate germination, can remain in a viable state in the soil for long periods of 3 - 6 years, even at a depth of 30 cm. below the soil level. The spores may also be carried on the seed. The spores may be spread through soil, farm implements, feet of cattle or by the use of manure containing the crop residues. The dung of cattle fed with raw club roots may also contain the spores in a viable state. The spores may even be carried by strong winds along with the spore-laden dust. Rain and irrigation water may also carry the spores and spread the disease. Small quantities of contaminated soil adhering to the roots of transplants also serve to perpetuate the disease. The disease may also spread through spores produced from club root infected weed hosts. Infection occurs, only when the root hairs are young and full of protoplasm. Infection may also take place though wounds on the roots and stems and also through natural openings at the points of emergence of lateral roots. The presence of plasmodium in a cell, stimulates the meristematic activity of the occupied cell, as well as the cells adjacent to it, which are forced outward and ultimately become clubs or galls of hypertrophied tissues.

At soil moisture levels above 60 %, the disease occurrence is more. The optimum temperature for spore germination, infection and disease development is 27° - 30°C. The incidence is more in acidic soils.

Disease management

Agronomic practices (i) Growing cruciferous crops in infected fields should be avoided (ii) Seedlings raised in disease-free soils should be used as transplants (iii) The fields should be provided with proper drainage facilities to avoid water-logging (iv) Cruciferous weed hosts, such as wild mustard in and around the fields should be eradicated (v) Long term crop rotation without any cruciferous crop in the crop sequence may be adopted (vi) Diseased roots and other crop residues should be removed and destroyed by burning (vii) Club roots should not be fed raw to cattle (viii) Increasing the soil pH to about 7.0 by applying hydrated lime is helpful in reducing the disease incidence. A soil with pH-5 may require 1.0 ton of lime per acre and it should be applied, 6 weeks before planting (ix) Resistant varieties may be grown.

Chemical control (i) Seedbeds should be treated with soil fumigants, such as methyl bromide - 1.0 lit. or chloropicrin - 1.0 lit. or vapam - 1.3 lit./ cent of nursery area. The chemical should be injected into the soil by means of a soil-injector at a depth of 15 - 20 cm. and at 30 cm. interval, 2 - 3 weeks before sowing. Before sowing, the soil should be raked up thoroughly, so that the residual fumes may escape from the soil, otherwise the germination of the seeds will be affected (ii) Drenching the seed bed with formalin - 2 % at the rate of 10 lit./ cent of nursery area, 3 - 4 weeks prior to sowing is also effective in preventing the disease occurrence.

2. White blister or white rust of cabbage

Albugo candida (Cystopus candidus)

The disease is worldwide in distribution and is common on cruciferous plants, including cauliflower, radish, knol khol, mustard, turnip, rape and many other weeds. In India, the disease occurs in all the regions, wherever cabbage is cultivated.

Symptoms. All aboveground parts of the plant may show characteristic white, chalky pustules. The pustules are of variable size and shape on the leaves, stem and inflorescence. They are shiny white or cream-yellow in color, isolated in the beginning, slightly raised and 1.0 - 2.0 mm. in diameter. The pustules, which occur in close proximity, merge together to form large patches. Mostly, they develop in a circular arrangement around one or two central pustules. On maturity of the spores, the host epidermis is ruptured and masses of white, powdery spores are exposed. Abnormal swellings and distortions of the attacked parts occur. The inflorescence may become enormously thickened several times the normal thickness. The floral organs become very much swollen, fleshy and develop a greenish or violet color. The flowers do not drop off as usual but, remain adherent to the inflorescence. The petals and stamens become thick, green and leafy. The pistil and stamens are atrophied and the flowers become sterile. Pustules are formed usually on all the swollen parts. In the later stages, oospores are formed in the tissues inside the swellings.

In the affected leaves, the differentiation between the palisade and mesophyll tissues is lost and the mesophyll cells are enlarged 2 - 3 times the size of the normal cells. In the stem and inflorescence, the cells of the cortex region are hypertrophied. Hyperplasia of the cells also occurs, resulting in thickening and swelling. Ultimately, due to excessive uptake of nutrients from the host by the pathogen, the infected tissues of the affected parts collapse, turn brown, dry and die **(Fig.90)**.

The causal organism, mode of survival, spread, epidemiology and disease management. Refer 'White rust of mustard' (Page 189). For the life cycle of the pathogen, refer 'Life cycle of *Albugo candida*' **(Fig.53)**.

3. Downy mildew of cabbage

Peronospora parasitica

The disease is of common occurrence all over the world. In India, it is considered to be one of the most serious diseases of cabbage. The disease attacks many plants of the cruciferous family, including cauliflower, broccoli, radish, turnip and mustard.

Symptoms. The disease attacks young seedlings, senescent leaves of grown-up plants and the heads in storage. The symptoms produced by this pathogen on the stem and inflorescence almost resemble the symptoms produced by *Albugo candida*. However, *Peronospora* produces a greater deformity in the stem, while *Albugo* causes a greater deformity in the inflorescence. Further, *Peronospora* never produces any trace of violet coloration in the infected tissues as *Albugo* very often does. On the leaves, no pustules are formed as in *Albugo*, but a thin, grayish-white, downy growth occurs in scattered patches on the under surface of the leaves in cabbage, cauliflower and turnip, while in radish, such growth can be seen on the leaves, stem and inflorescence. Yellowish-white patches corresponding to the downy growth on the lower surface mark the upper surface of the leaf. In case of severe incidence, the entire leaf area is covered by the downy growth and the leaf dries, shrivels and is broken easily. In the seedlings, the entire under surface may be covered by the downy growth and oospores are formed in abundance in the cotyledons. Oospores are rarely formed inside the leaf tissues, but are formed in abundance inside the tissues of the swollen stem and inflorescence. Occasionally the roots are attacked and the tissues become black and rot near the soil surface. Oospores are formed within the tissues of the roots also **(Fig.91)**.

The causal organism, mode of survival, spread, epidemiology and disease management. Refer 'Downy mildew of mustard' (Page 192).

4. Powdery mildew of cabbage

Erysiphe polygoni

The disease is very common on a wide host range of wild and cultivated plants, which include cauliflower, broccoli, Brussels sprout, Chinese cabbage, kale, mustard, radish, rape, turnip, clover, pea and swede besides cabbage. The disease is widely spread in America, Canada, Britain, India, Australia and many other countries of the world.

Symptoms. The symptoms appear as small patches of a thin, fine meshwork of white mycelium arising at various places on the upper surface of the leaves. These white patches soon join together to cover larger areas and from the superficial mycelium, erect conidiophores, bearing conidia arise in profusion and form a whitish, powdery coating. The entire crop looks, as if dusted with a white powder from a distance. Ultimately, such leaves turn chlorotic and die.

The causal organism, mode of survival, spread, epidemiology and disease management. Refer 'Powdery mildew of pea' (Page 133), **(Fig.36)**.

Resistant varieties. The varieties Globelle, Hybelle and Sanibel are resistant to the disease.

5. Damping off of seedlings of cabbage

Pythium aphanidermatum and *P. de baryanum*

Symptoms, causal organism, mode of spread, epidemiology and disease management. Refer 'Damping off of seedlings of eggplant' (Page 339), **(Fig.85)**.

6. Black rot of cabbage

Xanthomonas campestris pv. *campestris*

The disease is very widely distributed and is found in more than 90 countries of the world, including Australia, Canada, China, Europe, Japan, Russia, Sri Lanka, the United States of America and India. In India, it was first reported on cabbage in 1928 and is now prevalent in the states of Maharashtra, West Bengal, Jammu and Kashmir, Kulu valley and in the hilly regions of Tamil Nadu. The pathogen has a very wide host range, which includes cauliflower, carrot, turnip, radish, broccoli, knol khol, mustard, potato, lady's finger, tomato, pea, egg plant, cucumber, mango and several weeds.

Symptoms. The disease mainly attacks the aboveground parts of plants of any age. Infected young seedlings show such symptoms as stunting, irregular growth and drooping of lower leaves and eventually they die. In the field, the symptoms appear as large **'V'**-shaped, chlorotic blotches at the margins of the leaves. The chlorosis progresses toward the mid rib. Some of the veins and veinlets within the chlorotic area turn black. The affected area later becomes brown and dry. The blackening of the veins spreads to the other leaves, stem and roots. Severely affected leaves dry and may fall off one after the other. The stem and the stalks of infected leaves show blackening of vascular tissues, from which yellowish droplets of bacterial ooze is exuded. Cabbage and cauliflower heads are also affected, both in the field and in storage and they become black and rot. Fleshy roots of turnip, radish and carrot are invaded and they rot. The disease, which occurs sporadically in a few plants in the beginning, spreads rapidly under favorable conditions and a large percentage of plants are affected finally **(Fig.92).**

The causal organism. The pathogenic bacteria causing the disease are rod-shaped, with rounded ends and measure 0.7 - 3.0 x 0.4 - 0.5μ in size. They are gram-negative, motile with a polar flagellum, aerobic in nature, non spore-forming and not encapsulated.

Mode of survival, spread and epidemiology. The pathogen survives in the plant residues in the soil as a saprophyte or in the seeds. In the germinating seeds, the bacteria gain entry into the cotyledons or young leaves through the stomata, hydathodes or wounds and spread intercellularly in the sub-stomatal regions and then systemically through the vascular elements to all the parts of the plant. Secondary spread of the disease in the field is mainly through irrigation or rainwater; transport of contaminated soil through implements, farm labour, cattle etc. The bacterium is capable of surviving in the infected field for several years.

The optimum temperature for the growth of the bacteria is 30°C and thermal death point is 50°C.

Fig. 89: Club root of cabbage
Plasmodiophora brassicae

1. *Spindle-shaped* galls

Fig. 90: White rust of cabbage
Albugo condida

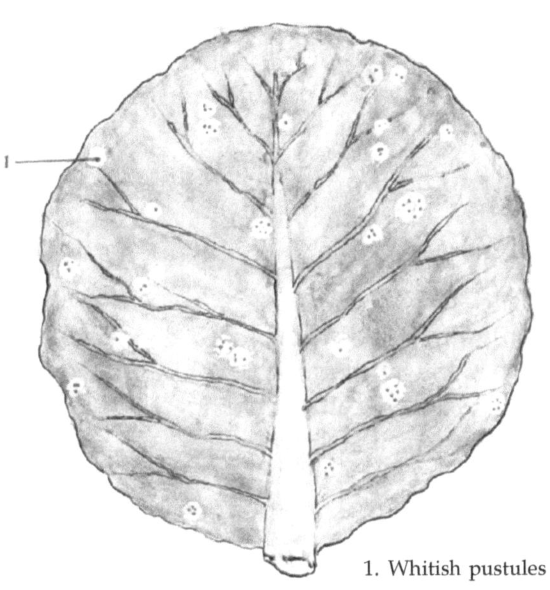

1. Whitish pustules

Fig. 91: Downy mildew of cabbage
Peronospora brassicae

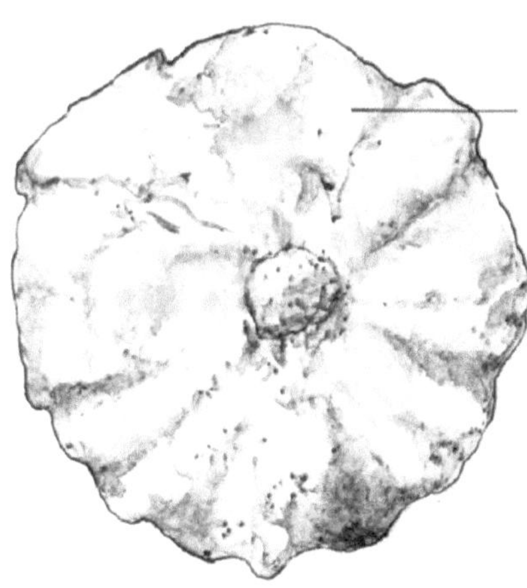

1. Grayish-white downy growth

Fig. 92: Black rot of cabbage
Xanthomonas campestris pv. *campestris*

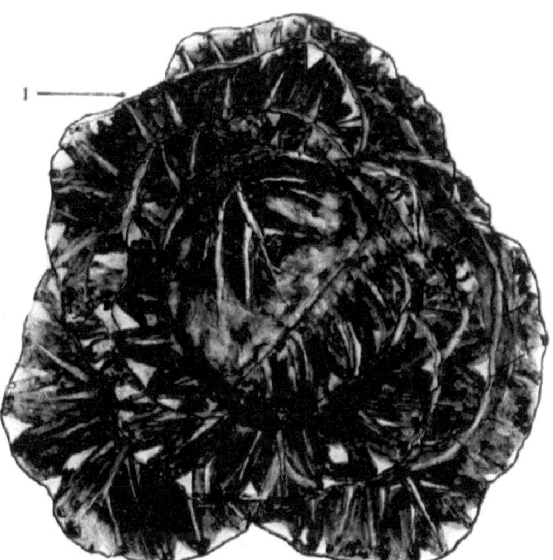

1. Infected head

Disease management

Agronomic practices (i) Disease-free seeds should be selected from disease-free crop for sowing (ii) Infected crop residues should be collected and destroyed (iii) Balanced doses of fertilizers should be applied as per recommendations. Potash induces resistance to the plants (iv) Three year crop rotation, with non-susceptible hosts may be followed (v) Resistant varieties may be grown.

Seed treatment (i) Soaking the seeds in Aureomycin - 0.1% (1 : 1000) solution for 30 minutes is effective in eliminating both the internally and externally seed-borne bacteria (ii) Hot water treatment of seeds at 50°C for a continuous period of 30 minutes eliminates the bacteria completely.

Chemical treatment (i) Drenching the seed beds with commercial formalin - 5 % at the rate of 10 lit./ cent nursery area, 3-4 weeks prior to sowing is effective in controlling the disease (ii) Drenching the seed beds with Agrimycin 100 at 600-800 ppm. (0.6-0.8 gm./ lit. of water) at the rate of 10 lit./ cent nursery area prior to sowing also controls the disease (iii) Application of bleaching powder to the soil at 5.0 kg./ acre is also effective (iv) Spraying the crop with Agrimycin 100 at 600 ppm (0.6 gm./ lit. of water), 2-3 times at fortnightly intervals gives good control of the disease (v) Spraying the crop with copper oxychloride at 2.5 gm./ lit. of water is useful in preventing the secondary spread of the disease.

7. Cauliflower mosaic

Cauliflower mosaic virus

The virus causing the disease is found in many parts of the world, but it has a limited host range and hence its distribution is also limited to certain areas only.

Symptoms. The disease usually occurs, when the plants are about half grown. The symptoms on the leaves appear as mottling, with an irregular pattern of light and dark-green color. Sometimes, a dark-green band may be formed on either side of the main veins. The disease results in reduction in size of leaves, poor and stunted growth and reduction in the size and quality of the heads. Severely affected plants at the early growth stages may die.

The causal organism. The virus causing the disease belongs to the Genus - *Caulimovirus*. The virus particles are large, icosahedral, isometric circular dsDNA. The genome consists of one molecule of dsDNA with three discontinuities, two in one strand and one in another. The genome is encapsidated in isometric particles measuring about 50 nm in diameter. The thermal inactivation point of the virus is 75° - 80°C. The viruses have a limited host range. After infection, they multiply in the parenchyma cells and pass on from one cell to another through plasmodesmata and after reaching the phloem, they multiply and spread at a much faster rate.

Mode of spread. The viruses are transmitted in nature by some species of aphids, such as *Myzus persicae* and *Brevicoryne brassicae* in a non-persistent manner. They are not transmitted through sap or seed. The viruses do not multiply in the vector and are not passed on to its progeny. The vector becomes non-viruliferous after moulting. The viruses overseason in perennial weed hosts or overlapping crops from which the aphids carry them to new crops. They infect *Brassica* sp., *Raphanus* sp. and a few other species belonging to the Family - Cruciferae.

Disease management

Agronomic practices (i) Infected plants should be rouged out from the field and destroyed (ii) Seedlings should be raised in isolated and protected areas (iii) All cruciferous weed hosts, such as wild mustard, chick weed, shepherd's purse etc. should be eradicated.

Chemical control. The aphid vectors should be controlled by using systemic insecticides, such as methyl demeton, dimethoate, phosphamidon or monocrotophos.

Resistant varieties. Cauliflower varieties, such as Ace High, Badger Ball-Head, Blue Jacket, Eastern Ball-Head, Green Boy and Green Winter have been found to be resistant.

Diseases of minor importance. Besides the above-mentioned diseases, cruciferous crops are subjected to attack by many other diseases also. *Sclerotinia sclerotiorum* causes 'Stalk rot'; *Alternaria brassisicola, A. brassicae* and *A. raphani* cause '*Alternaria* leaf spots'; *Gloeosporium concentricum* causes 'Light leaf spot' and 'Anthracnose'; *Pseudomonas solanacearum* causes 'Bacterial wilt'; *Erwinia carotovora* pv. *carotovora* causes 'Soft rot'; few virus diseases also attack crucifers.

TUBER AND ROOT CROPS

Potato *(Solanum tuberosum)*

1. Late blight of potato

Phytophthora infestans

'Late blight' is one of the most important diseases of potato and causes serious damage to the haulms and tubers in humid weather. The likely home of the disease is Mexico, where it is endemic on wild native species allied to the cultivated potato and somehow it has passed to the cultivated species of potatoes. The disease appeared with great rapidity through Holland, Northern France and Eastern Germany in 1845. During the period from 1845 to 1846, the disease swept the potato crop throughout Europe, especially Ireland and caused considerable loss of tubers. Potato being the staple food of more than half the people of Ireland, this loss caused famine, which resulted in migration of millions of people. Subsequently, the disease has spread to every potato growing country, where favorable conditions exist for the pathogen.

In India, the disease was first introduced into the hilly regions of the Nilgiris during 1870 - 1880. Soon after, it was reported from Darjeeling in the Himalayas with the introduction of European potatoes. After 1912, the disease gradually made its way from the hilly regions into the plains. Since 1943, the disease appears regularly in almost all the plains of Northern India. In South India, the disease is prevalent in the potato growing regions of Karnataka and Tamil Nadu. The disease attacks tomato crop also.

Symptoms. The disease appears mostly at the blossoming period. However, it may appear during any growth stage of the crop. The first sign of the disease aboveground is the appearance of small brown patches on the leaves, mostly on the basal leaves and in suitable weather conditions they enlarge rapidly, very often spreading over the entire leaf lamina. In dry, clear weather, successful infections are limited and the spots remain small, brown and dry, while the stems may escape infection. The dead areas of the leaf appear hard and very often the symptoms appear to be that of early blight symptoms. The infection quickly extends to the stalk and the entire crown may fall over in a rotten pulp in a few days. The blighted areas first appear as faded green patches, which soon turn to brownish-black lesions and enlarge rapidly. The symptoms mostly start at the tip or margin of the leaves and spread downward and inward, the rate of spread depending upon the weather.

Under moist conditions, the peripheral zone outside the blighted lesion is found to be pale green and merges with the brown patches. On the under surface of this pale green zone, a whitish or grayish mildew growth appears. This mildew growth consists of the aerial fructifications of the fungus. In dry weather, this downy mildew growth disappears.

Later potato tubers are also infected, while in the field and still attached to the plant or they may be infected during harvest and sometimes in storage. Affected tubers at first show purplish or brownish blotches consisting of water-soaked, dark reddish-brown tissue that may extend up to 1.5 cm. into the flesh of the tuber. Afterwards, the affected areas become firm, dry and somewhat sunken. Such lesions may be small or may cover almost the entire surface of the tuber and the rot continues to develop after the tubers are harvested. Secondary fungi and bacteria, causing soft rot and giving the rotten potatoes a putrid offensive odor may subsequently invade infected tubers. In moist atmosphere, white tufts of mycelium, sporangiophores and sporangia of the fungus appear on the surface of the tubers **(Fig.93)**.

The causal organism. The fungus causing late blight of potato is an obligate parasite. The mycelium is endophytic, consisting of hyaline, much branched, coenocytic hyphae, which are intercellular and produce single or double, club-shaped haustoria or haustoroid hyphae. During favorable periods, fructifications appear on newly infected leaves in 4 or 5 days.

The branching sporangiophores arise directly from the internal mycelium, emerging from the leaf through the stomata in small groups of about 3 - 5 in number. Sometimes, the sporangiophores emerge through or between the epidermal cells. On the tubers, they arise mostly from lenticels or aberrations in the rind. The stalk of the sporangiophore is slender, not rigid and divides at the tip into 2-4 branches of variable length up to 1.0 mm. Sporangia are formed at the tips of the branches, but growth continues just below the sporangium. In this process, the sporangium is pushed over to one side and usually falls off. At each point, where growth has been renewed, there is a small nodular swelling in the stalk and up to 9-10 such swellings may be found on a single branch. The sporangia are lemon-shaped, multinucleate, thin-walled, hyaline and measure 28-32 x 16-24µ in size. In the presence of water, the apex becomes papillate and 3-8 zoospores are liberated. The zoospores are reniform and biflagellate. One flagellum possesses fine hairs on it. After actively swimming in the film of water on the host surface for a few minutes, the zoospores lose their flagella and become spherical. After a small resting period, they germinate by a germ tube and enter the host, either by direct penetration of the epidermis or through the stomata. Otherwise, the sporangia germinate by issuing germ tubes and thus behave as conidia.

Sexual reproduction is very rare in nature. The oogonia are yellow or hyaline, pear-shaped or spherical and 31-43µ in diameter. They are fertilized by club-shaped antheridia, borne on a hypha, other than the one forming the oogonium. Sometimes, oospores are produced without fertilization. The oospores have a thick, smooth, hyaline wall and measure 24-35µ in diameter. The oospore germinates by a germ tube that produces a secondary sporangium or germinates directly into the mycelium **(Fig.93).**

More than 14 different pathogenic races of the fungus are found in India.

Mode of survival, spread and epidemiology. The pathogen overwinters as mycelium in infected potato tubers left over in the field after harvest or in tubers stored and used as seed in the ensuing season. The mycelium spreads into shoots produced from infected tubers, causing discoloration and collapse of the cells. When the mycelium reaches the aerial parts of the plant, it produces sporangiophores, which emerge through the stomata of the stems and leaves and produce sporangia. The sporangia, when ripe, get detached and drift off or are dispersed by rain. When they land on wet potato leaves or stems, they germinate and cause new infections. A few days after infection, new sporangiophores emerge and produce numerous sporangia, which are spread by wind and infect new plants. In cool, moist weather, new crop of sporangia is produced within 4 days from infection. Thus, several generations of sporangia may be produced within a single cropping season.

In the second phase of the disease, tuber infection takes place in the field. During wet weather, sporangia are washed down from the leaves and stems and are carried to the soil. The zoospores emerging from the sporangia, germinate and penetrate the tubers through lenticels or through wounds. In the tuber, the mycelium grows mostly between the cells and sends haustoria into the cells. Tubers contaminated with living sporangia present in the soil or on diseased foliage at the time of harvest may also get infected. Most of the blighted tubers rot in the field or during storage.

The development of late blight epidemics depends greatly on the prevailing humidity and temperature during the different stages of the life cycle of the fungus. The fungus grows and sporulates most abundantly at relative humidity near 100 % and at temperatures between 15° and 25°C. Temperatures above 30°C check the growth and sporulation of the fungus in the field, but do not kill it and the fungus can start to sporulate again, when favorable conditions return. Sporulation occurs at temperatures ranging from 9°-26°C. The optimum temperature for germination of sporangia by zoospores is 12°-13°C and by a germ tube 24°C. The mycelium of the fungus grows best at 16°-18°C.

Disease management

Agronomic practices (i) Selection of seed tubers from disease affected fields should be avoided. Tubers suspected to be infected should also be discarded (ii) The diseased plant parts lying on the field should be removed and destroyed At the time of harvest, care should be taken to avoid contact of tubers with the aerial parts of the plants (iii) Harvesting of a diseased crop should be delayed until the plants are fully matured. The plants should be allowed to dry completely in the field before harvesting. This will kill all the spores present on the foliage and thus helps to avoid infection of the tubers during harvest (iv) Harvesting should be done, when the weather is dry and only good tubers without any signs of the disease should be collected first and stored in store-houses. Care should be taken not to injure the tubers at the time of harvest (v) The tubers should be stored in cool, dry, well ventilated store-houses. Temperatures above 30°C are best suited for storage (vi) The store-houses should be inspected periodically and tubers showing rotting symptoms should be removed and destroyed (vii) At the time of earthing up, ridging to a height of 10 - 15 cm. around the stem region is advocated. This will prevent the spores washed down from the foliage reaching the tubers, thus help to prevent tuber infection (viii) All volunteer potato plants in and around the fields should be removed and destroyed.

Chemical control. Fungicides used for late blight control include dithiocarbamates, metalaxyl, a combination of metalaxyl and dithiocarbamate or chlorothalonil, organo tin compounds and several copper fungicides. In areas, where late blight is common, it is necessary to start the spraying operations sufficiently in advance of the usual time of appearance of the disease. The first spray should be given, when the crop is about 6 weeks old, followed by two to three more sprayings, at 15 - 21 days interval. Bordeaux mixture- 1 % or copper oxychloride (Fytolan, Blue copper or Blitox-50) at 750 - 900 gm. or dithiocarbamate (Zineb or Mancozeb) at 600 - 750 gm. or organo tin compounds (Brestan or Du-ter) at 375 - 450 gm. or metalaxyl (Ridomil) at 300 - 350 gm. or chlorothalonil (Daconil, Bravo or Kavach) at 375 - 450 gm. in 300 - 350 lit. of water per acre have been found to be effective in controlling the disease.

Seed treatment. Before storage, seed tubers should be dipped in a solution of mercuric chloride - 1:1000 for 90 minutes. This will prevent secondary infection by rot-causing organisms.

Resistant varieties. Because of the presence of many pathogenic races of this fungus, it is very difficult to evolve varieties resistant to all the races of this pathogen. However, many potato varieties have field resistance to this pathogen. Kufri Naveen, Kufri Jeevan, Kufri Alankar, Kufri Khasi Garo and Kufri Moti show resistance to this disease.

2. Early blight of potato

Alternaria solani

'Early blight', which occurs much earlier than the late blight is one of the common and destructive diseases of potato. It is found in all countries, where potato is grown. In India, the disease is prevalent in the hilly regions, as well as in the plains and causes heavy losses, both in the summer and winter seasons. When the disease occurs in a severe form, the loss in yield may go up to 40 per cent. Besides potato, the fungus attacks tomato and chillies also.

Symptoms. The first symptoms of the disease appear as small, isolated, brown spots, scattered on the leaflets. The older leaves are first attacked and the disease then proceeds up. As the spots enlarge, the central tissues become necrotic and concentric rings are formed as in a target-board, which is very characteristic of the disease. A narrow chlorotic zone surrounds the necrotic area. As the spots increase in size, the chlorotic zone also expands simultaneously. The chlorotic zone is formed due to the production of a toxin, **'alternaric acid'.** When these leaf spots extend to the larger veins, the chlorotic zone extends much beyond the necrotic spots along the veins due to translocation of the toxin to a larger extent through the veins. A few to many spots may develop in a leaflet and occupy major portion of the lamina, when the disease becomes severe. In dry weather, the spots become hard and shrink and the leaves curl. When the weather is moist, the spots are covered with a dense, greenish-blue growth of the fungus. The spots coalesce and become big, rotting patches under such conditions. In case of severe attack, the leaves shrivel, dry and fall down. The fungus also attacks the potato stems and brown to black necrotic lesions are formed on the stem and branches. These lesions may enlarge and girdle the branches or stem leading to collapse of the affected branches or the entire plant. In the later stages of crop growth, the tubers are also affected and they begin to rot. The number of tubers and the tuber size are reduced as a result of infection **(Fig.94).**

The causal organism. The mycelium of the fungus is septate, branched, light-brown, which becomes darker with age. The hyphae are intercellular at first and later become intracellular. Conidiophores emerge through the stomata from the necrotic zones of the spots. They are short, septate and dark colored. Conidia are produced singly from a bud at the apical cell of the conidiophore. Conidia are beaked, muriform and dark colored, with 5 - 10 transverse septa and a few longitudinal septa. They measure 120 - 296 x 12 - 20µ in size. The conidia germinate in moist weather by putting forth germ tubes. From each conidium 5 - 10 germ tubes may arise and enter the host, either through stomata or by penetrating the host epidermis **(Fig.94).**

In India, 3 different strains of the fungus, differing in morphology and physiology have been identified

Mode of survival, spread and epidemiology. The mycelium of the fungus remains viable in dry infected leaves for a year or more. The conidia have also been found to be in a viable state in the infected plant debris in the fields for about 17 months and cause primary infection in the ensuing crop. Mycelium and conidia present in the affected tubers also cause primary infection.

The lower leaves near the ground level are infected first through conidia found on the soil. Secondary spread occurs through conidia produced on the primary spots. Wind, water and some insects disperse the conidia. The conidia produced from alternate hosts of this fungus also serve as primary inoculum.

The optimum temperature range for germination of conidia is 28° - 30°C and at these temperatures the conidia germinate in 35 - 45 minutes. High atmospheric humidity, frequent rains, followed by warm and dry weather are conducive for the rapid development and spread of the disease. Continuous dry spell is unfavorable for the development and spread of the disease.

Disease management

Agronomic practices (i) Disease-free seed tubers should be used for planting (ii) Dead haulms and other plant debris should be collected from the fields after harvest and destroyed by burning (iii) The crops should be maintained properly, so as to maintain the vigor of the plants (iv) Wherever possible, crop rotation may be followed (v) Growing of alternate hosts, such as tomatoes and chillies near about potato fields may be avoided.

Chemical control. As the disease appears much earlier than the late blight, control measures should be taken up sufficiently earlier. The first spraying should be given 30 days after planting and subsequently 3 - 4 sprayings should be given at an interval of 7 - 14 days. The chemicals recommended for the control of late blight of potato are effective for the control of this disease also.

Fig. 93: Late blight of potato-*Phytophthora infestans*

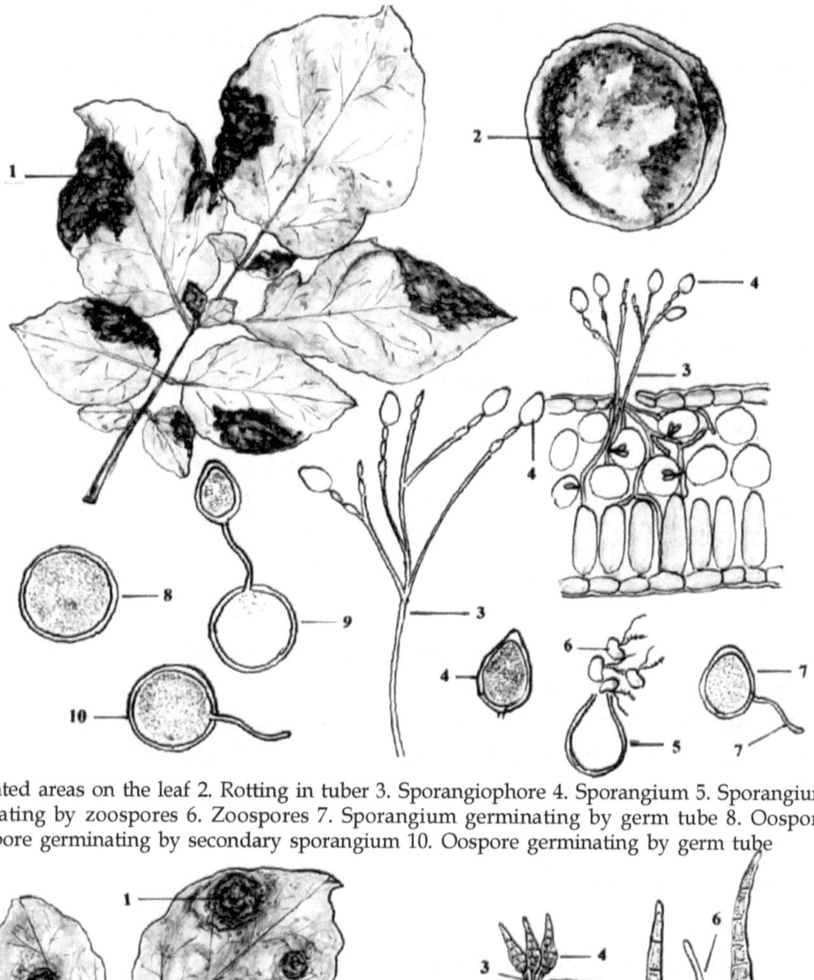

1. Blighted areas on the leaf 2. Rotting in tuber 3. Sporangiophore 4. Sporangium 5. Sporangium germinating by zoospores 6. Zoospores 7. Sporangium germinating by germ tube 8. Oospore 9. Oospore germinating by secondary sporangium 10. Oospore germinating by germ tube

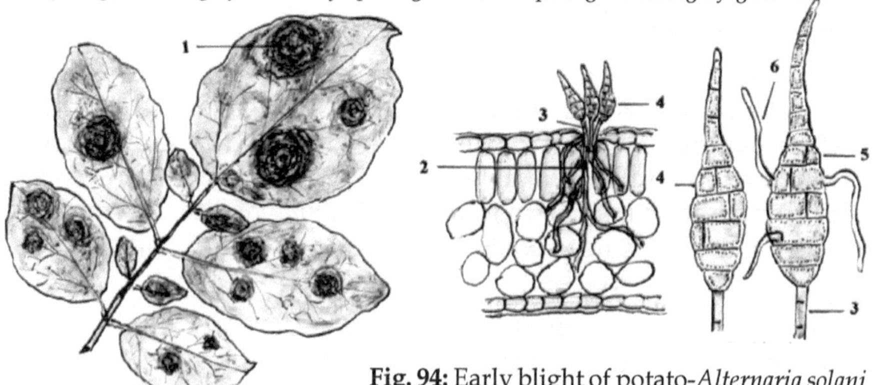

Fig. 94: Early blight of potato-*Alternaria solani*

1. Spots on leaves 2. Inter and intracellular mycelium 3. Conidiophore 4. Conidium 5. Germinating conidium 6. Germ tube

3. Black scurf and *Rhizoctonia* stem canker of potato

Rhizoctonia solani

(Thanatephorus cucumeris)

This disease was first reported from Germany in 1858 and from India in 1912. The disease occurs throughout the world and in India, it is found in all potato growing areas, both in the hilly tracts and in the plains. Major damage to the crop is caused in the stem canker phase of the disease. The black scurf of the tubers adversely affects the quality and marketability of the potatoes and the affected tubers become unfit for seed purposes also.

Symptoms. There are two distinct phases of the disease viz., **'the stem canker and blight phase'** and **'the black scurf phase'**. In 'the stem canker phase', when the tuber sprouts, the growing tip is killed before emergence. The dormant buds near the base then sprout to grow, but these may also be attacked and killed, until there is no further emergence. If conditions become unfavorable to the pathogen, further infection may stop and the lateral buds may sprout and grow. But, such plants become stunted, yellowish and the yield from these plants is badly affected. The sprouts, which escape infection, continue to grow and as the tissues harden, infection may be less. But, even on growing plants, sunken or shallow, brown cankers are formed due to the infection. The cankers may girdle the stem region and cause damage to the stem, as a result normal downward translocation of carbohydrates is impaired and carbohydrates tend to accumulate in the top portions. This leads to excessive formation of **'anthocyanin'** and the leaves turn purple. Moreover, buds near the base of the stem may swell abnormally and give rise to aerial tubers. The uppermost leaves tend to become bunchy and often the leaf edges may role abnormally.

In **'the black scurf phase'**, newly formed tubers show brown, discolored, crusty areas extending from the affected stolons. These areas may consist of a mere russeting of the skin or a more severe corrosion and narrow cavities of irregular shape. Such cavities may contain the brown mycelium of the fungus. As the tubers mature, sclerotia begin to appear and before harvest, the tubers develop a slow type of rot starting from the heel end. In storage, decay of tuber sets in, in the form of a dry rot. Sclerotia are typically formed on the tubers and less frequently on the affected stems or roots. They are round to irregular in shape, black, and closely adpressed to the skin and vary in size from a pinhead to the size of a pea. The sclerotia are formed at various spots on the tubers, especially on the underside to a larger extent. They are formed by small groups of white, cottony hyphae, composed of closely interwoven filaments. These groups of hyphae gradually darken and harden to form the black sclerotial bodies **(Fig.95).**

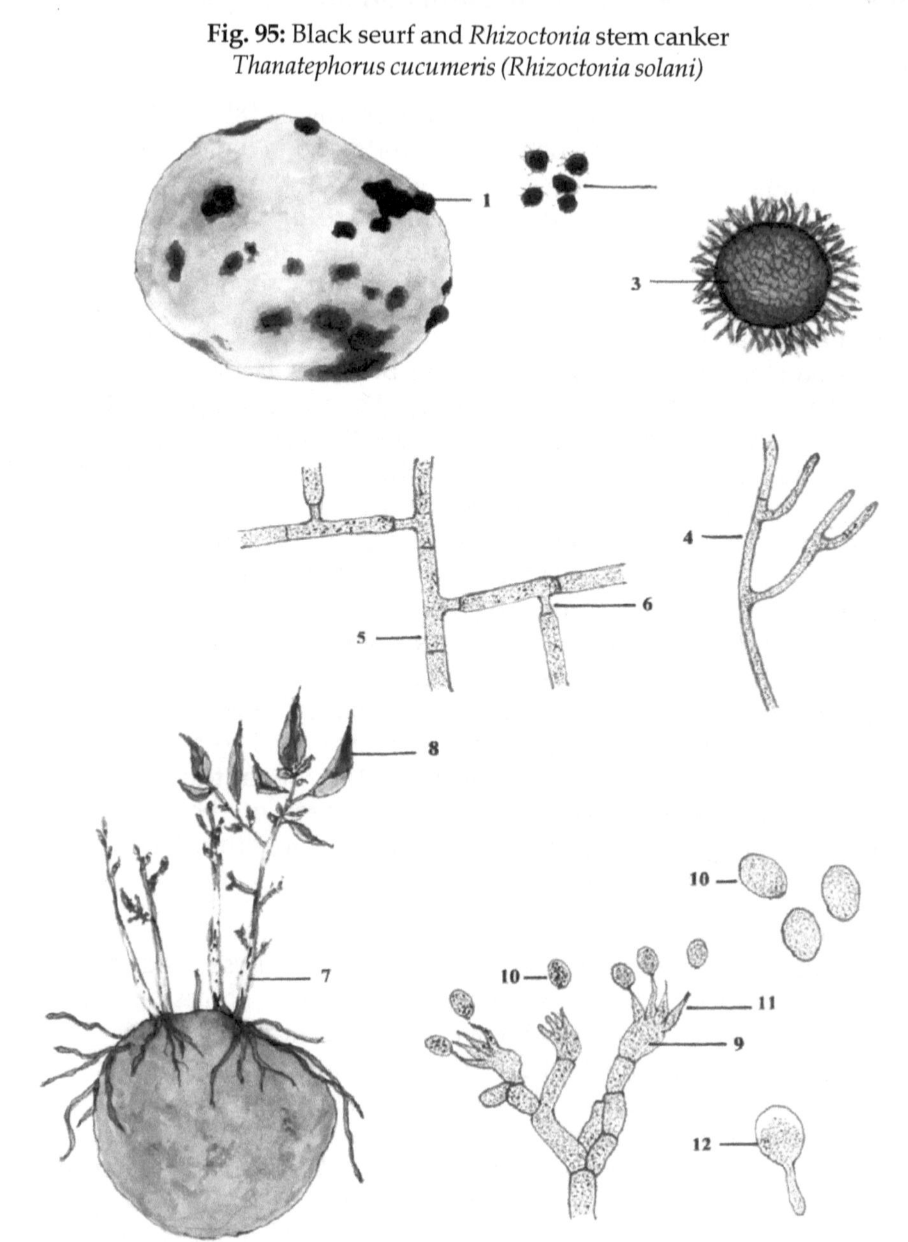

Fig. 95: Black seurf and *Rhizoctonia* stem canker
Thanatephorus cucumeris (Rhizoctonia solani)

1. Sclerotia on the surface of potato tuber 2. Sclerotia 3. Sclerotium 4. Young hyphae 5. Older mycelium 6. Constriction at the junction of branching 7. Stem canker 8. Abnormal rolling of uppermost leaves 9. Basidium 10. Basidiospore 11. Sterigma 12. Germinating basidiospore producing mycelium.

The causal organism. The mycelial and sclerotial stage of the fungus, *Rhizoctonia solani* is commonly encountered in India. The mycelium appears white in the beginning, but later turns light-brown, which is a distinctive feature of the fungus. The septate mycelium consists of multinucleate cells and produces branches that grow at approximately right angles to the main hypha and are slightly constricted at each junction and have a cross wall near the junction. The branching characteristics are important for identification of the fungus as *Rhizoctonia.* The mycelium inside the host tissues is slender and hyaline, the cells longer and is quite different from the sclerotial forming superficial mycelium, which is dark- colored, short-celled, stout and profusely branched. The sclerotia of the fungus are quite distinct from sclerotia produced by other fungi in that the sclerotium of *Rhizoctonia* has a uniform texture throughout and there is no differentiation of the sclerotial tissue into a rind and internal medulla. Under certain conditions, the fungus produces sclerotia-like tufts of short, broad cells that function as chlamydospores or eventually, the tufts develop into small, loosely formed brown to black sclerotia. No other spores are produced in the asexual stage of the fungus.

The basidial stage of the multinucleate *Rhizoctonia solani* is *Thanatephorus cucumeris.* This is an inconspicuous, mostly saprophytic stage of the pathogen. The perfect stage is formed under high humidity and appears as a thin, mildew-like growth on the soil, leaves and infected stems just above the ground level. The basidia are borne on small, imperfectly symmetrical cymes. They are clavate, obpyriform or barrel-shaped, which bear four, horn-shaped sterigmata. Each sterigma bears one basidiospore. The basidiospores are ellipsoid or obellipsoid, hyaline, truncate and measure 9.0 - 14.0 x 6.0 - 8.0 μ in size. They germinate to produce secondary basidiospores or primary mycelium **(Fig.95).**

The basidial stage is extremely rare in nature and therefore of little value in identifying the fungus. So, this fungus continues to be considered as **'Mycelia Sterilia'** and referred to by the name ***Rhizoctonia.***

Mode of survival, spread and epidemiology. The pathogen overwinters usually as mycelium or sclerotia in the soil and in or on infected plant parts or propagative material, such as potato tubers. The pathogen is present mostly in soils and once established in a field, remains there indefinitely. The fungus possesses considerable degree of resistance to the metabolic products of other soil microorganisms.

The fungus spreads through rain, irrigation or floodwater, through tools and anything else that carries contaminated soil and through infected or contaminated propagative materials. The rejuvenated mycelium in the soil or the mycelium from germinating sclerotia, attacks the tubers at the eyes and proceeds to penetrate the young sprout at the apex or a little lower down to form brown, sunken lesions. After entry into the host, the epidermal cells are rapidly destroyed and soon the fungus makes its way into the meristem of the young shoot after dissolving the tissues underlying the epidermis.

There are many races of the pathogen and for most of the races the optimum temperature for infection is 15 - 18°C, but some races are most active at much higher temperatures, up to 35°C. The disease is more severe in soils that are moderately wet, than in soils that are water-logged or dry. The fungus thrives best in acid soils. Infection of young plants is most severe, especially when plant growth is slow because of adverse environmental conditions for the plant.

Disease management

Agronomic practices. Control of *Rhizoctonia* disease is rather very difficult. Proper soil and water management and use of disease free seed material will help to minimize the incidence of the disease to a large extent. (i) The disease is more severe in moderately wet soils. So, proper drainage facilities should be provided to keep the soil moisture level low. Wet and ill-drained areas should be avoided. (ii) Disease-free seed material should be used for planting. (iii) Shallow planting encourages quick growth and avoids too much damage to sprouts. (iv) Soil amendment with neem cake (40 kg./ ac.) or margosa cake (40 kg./ ac.) or wood sawdust (40 kg./ ac.) has been reported to give good control of the disease. The amendments are to be given, 3 weeks before planting.

Seed treatment. Steeping the seed tubers in a solution of 1:1000 mercuric chloride for 90 minutes or in an acidulated mercuric chloride solution of 2:2:1000 mercuric chloride, commercial hydrochloric acid and water for 5 minutes kills the sclerotia adhering to the tubers.

Soil treatment. Soil application of Brassicol (PCNB) - 2000 gm. in 1,000 liter of water / ac. (2.0 gm./ liter of water) has been found very effective in controlling the disease.

Biological control. The fungus can be suppressed by soil application of its antagonists, such as *Trichoderma* species.

4. Wart disease of potato

Synchytrium endobioticum

'Wart' or **'black wart'** disease was first reported from Hungary in 1896. Now, it is widely distributed in several countries of the world, including the United States of America, Britain, Canada, South Africa, Peru, South America, India etc. The disease was introduced to India from Denmark in 1953 along with potato tubers. Besides potato, the disease attacks several other species of *Solanum* and also tomato.

Symptoms. The disease affects mainly the tubers during the growing season. No visible signs of wart disease are seen on the green shoots of plants. But, typical symptoms of wart disease appear on the underground tubers. On the tubers, characteristic brown or brownish-black, cauliflower-like lumps of variable sizes and shapes grow, chiefly out of the eyes of the tubers and in case of severe infection wart appears all over the tubers and appear as dirty protuberances of cauliflower-like outgrowths. All the tubers in a stool may not be affected, but some tubers, which look normal and healthy may show tiny warts at the eyes, which are not easily detectable. The disease also attacks the underground stolons and produce warts. The infected tubers may rot due to attack by secondary saprophytic soil microbes. The warts consist of distorted outgrowths of proliferated, intricately branched, cauliflower-like structures. They are produced as a result of hypertrophy of epidermal cells surrounding the infected cell **(Fig.96)**.

The causal organism. The fungus causing the disease is an obligate parasite. It belongs to Family - Olpidiaceae, Order - Chytridiales of Class - Chytridiomycetes. The life cycle of the organism is very simple and throughout the life cycle, no mycelium is developed at all. The organism starts its parasitic life as uniciliated zoospores formed in sporangia inside the warted tissues, which are released into the soil in large numbers. Later, there is a resting stage to tide the organism over the winter, at the end of which zoospores are again formed from resting sporangia and released into the soil to start fresh infection.

Mode of survival, spread and epidemiology. In the absence of potato crop, the resting sporangia may remain viable in the soil for many years, sometimes as long as even up to 12 years. The resting sporangia may be carried from one place to another through irrigation or rain water, through contaminated soil adhering to seed tubers, farm implements, feet of cattle and birds and even by man. It is also carried through infected seed tubers. If warted tubers are eaten raw by cattle, the resting sporangia remain unharmed and are still in a viable state, even after passing through the intestine. They also retain their viability in compost manure heaps into which warted tubers have been dumped.

The resting sporangia are very resistant to extremes of temperature, either high or low. They can withstand temperature of boiling water for 8 - 12 minutes. In the soil, they remain in a viable state, even up to a depth of 20 cm. A very high degree of soil moisture is necessary to bring about infection by the resting sporangia. Zoospore emergence is highest at very high humidity of 90 - 100 % and low temperatures between 14° - 24°C. A period of flooding, followed by draining and aeration of the soil is favorable for disease occurrence. The disease incidence is low at high degrees of alkalinity.

Disease management

Agronomic practices (i) It is very difficult to eradicate the disease, once it is introduced to a field. So, strict quarantine measures should be taken to prevent introduction of the disease to a new area or field through infected seed tubers (ii) Field sanitary measures should be adopted and infected tubers or left over tubers, debris etc. should be collected and destroyed by burning.

Resistant varieties. A number of varieties resistant to wart disease have been introduced. Varieties, such as Adina, Ackersegen, Blanki, Capella, Mira, Pimpernel, Ronda, Ultimus and Voran are reported to be immune to wart disease. Kufri Kanchan, Kufri Sherpa, Kufri Jyoti etc. have been found to be resistant.

5. Powdery scab of potato

Spongospora subterranea

The disease was first reported from Germany in 1841. The appearance of the disease has been since reported from several countries of the world, including Africa, America, Canada, Australia, New Zealand, Tasmania, Kenya, Cyprus, Hawaii, Netherlands, Russia and India. In India, the disease occurs in the upper hilly regions of Himachal Pradesh, Sikkim, Uttar Pradesh, West Bengal and Tamil Nadu.

Symptoms. The disease attacks all underground parts of the plant, such as the stem, stolons, tubers and roots. Typical symptoms appear only on the tubers, on which unsightly, brown blisters, discharging dry masses of powdery spores appear. The disease initially appears during tuber formation as small, circular, light-brown spots, surrounded by a translucent halo at the apical end of tubers. The spots spread over the greater part of the infected tubers, as the tubers enlarge in size. The spots soon become raised to form small pimple- or blister-like lesions, which are quite smooth in the early stages of development. Later, the skin ruptures and forms a frill around the margin and the blister becomes a shallow crater-like cavity, filled with a dry, powdery mass of spores, which are dispersed into the air. This is the first stage of the disease known as the **'powdery scab stage'** and at this primary stage there is very little destruction of the fleshy part of the tubers. Sometimes, only this primary stage is present.

If the soil is continuously wet during the growing season, a second stage known as the **'canker stage'** follows. This stage develops from the same primary lesions and large areas of the tubers may be covered with corroded, cankerous patches and the tubers become deformed. These cankers are quite different from the blisters formed in the primary stage.

If the soil continues to remain wet before the time of harvest or if the infected tubers are stored under too moist conditions, a third stage of the disease known as the **'wart stage'** occurs. At this stage, warty outgrowths develop from the tubers. The warts have rather a smooth contour and differ from the rough, rugose, cauliflower-like warts produced by *Synchytrium endobioticum*. These warts eventually collapse and release spore balls into the soil. Such collapsed warts resemble those of the canker stage.

On the roots, the disease causes small, warty outgrowths that appear like nodules **(Fig.97).**

The causal organism. The fungus causing the disease is an obligate parasite. It belongs to the Family - Plasmodiophoraceae, Order - Plasmodiophorales of Class - Plasmopodiophoromycetes. The fungus does not produce any mycelium during its entire life cycle. The naked plasmodium emerging from the zoospore, enters the potato tuber through the lenticel or through injury and invades the host cell. After entry into the host cell, the plasmodium is capable of penetrating and passing between the epidermal cells by solvent action of the middle lamella.

Mode of survival, spread and epidemiology. The disease does not attack any other crop, except potato under field conditions. The spore balls formed in the cankers on tubers and in the soil remain in a viable state for long periods of 5 years or more. The encysted myxamoebae can also tide over adverse conditions and persist in the soil for long periods. In the absence of the susceptible host, the plasmodia can exist saprophytically in the soil. If infected potatoes are fed raw to pigs, the spores can withstand digestion and find their way into the pig manure or compost.

The spores may be carried from one place to another through irrigation or rainwater, through soil adhering to the tubers, farm implements or feet of cattle and through infected tubers. High soil moisture level of 90 - 100 % and low temperatures below 18°C favor the disease occurrence. Generally, prolonged rainfall, followed by cool and damp weather are conducive for the development of the disease. The disease incidence is higher in heavy clay soils than in sandy loam soils.

Disease management

Agronomic practices (i) Strict quarantine measures should be enforced to prevent introduction of the disease to new areas through infected seed tubers or contaminated soil (ii) Infected tubers should not be used for planting (iii) Infected tubers should not be dumped into manure or compost heaps (iv) Only after boiling the infected tubers in water, they may be fed to pigs or cattle (v) Proper drainage facilities should be provided to make the soil dry and well aerated.

Chemical treatment. Treating seed tubers in mercuric chloride - 0.1 % solution for 1½ hours or in formaldehyde - 5 % solution for 3 hours prior to planting helps to eliminate the spores on the tubers.

Resistant varieties. The varieties, Arran Chief, Arran Banner and Bintje are resistant to the disease to some extent.

6. Common scab of potato

Streptomyces scabies

The disease was first recorded in Britain in1825. It is widely distributed in many European countries, Britain, the United States of America and India. In India, it is prevalent in the states of Himachal Pradesh, Bihar, Uttar Pradesh, Meghalaya, Punjab, West Bengal, Maharashtra and Tami Nadu.

Symptoms. Several types of scab have been identified. While **'the shallow scab'** or **'common scab'** is the most common one, **'the pitted scab'** or **'deep scab'** is less common. The pathogen attacks growing tubers at a very early stage of development. The common or shallow scab is characterized by the formation of concentric wrinkled layers of the skin around a central depression, while the pitted scab is characterized by the formation of deep depressions or pits bordered by torn skin, assuming a severe and unsightly form. The tissues around the interior of the pits become corky and dark. Saprophytic bacteria and various other soil microbes attack such infected tubers and big cankers are formed in due course. Small nodular structures may be formed on the roots and stolons and brown areas may be found on the rootlets. The scab does not affect matured tubers **(Fig.98)**.

The causal organism. The organism belongs to Family - Streptomycetaceae, Order - Actinomycetales of Class - Thallobacteria. The thallus consists of a mass of simple or branched filaments, which are non-motile, very thin, 0.5 - 1.0µ in diameter and septated at irregular intervals. There is a characteristic spiral twist at one end of the filament or branch and sporulation is confined to this region. The spores, which are abstricted in succession at the apex are roughly cylindrical and measure 1.0 - 2.0 x 0.6 - 0.7µ in size. No endospores are formed. The organism is aerobic in nature. The spores germinate by means of one or two germ tubes **(Fig.98)**.

Mode of survival, spread and epidemiology. The organism is a soil inhabitant and can continue its life in the soil indefinitely as a saprophyte. The spores can withstand adverse climatic conditions. The pathogen attacks young tubers at very early stages of development and infection continues, as long as the tubers are growing. Penetration of the pathogen is by way of stomata or young lenticels or through wounds. The presence of the organism in the lenticel, stimulates the cells of the lenticel meristem, which divide more rapidly and the infected areas increase in size radially.

Dry, gravelly soils, with alkaline reaction are favorable for the disease development, while acid reaction greatly reduces the disease occurrence. The disease is inhibited in clay soils and in soils with high moisture content. At higher temperatures, above 25°C, the disease incidence is more. Besides potato, the disease attacks cabbage, carrot, eggplant, onion, radish, turnip, spinach etc.

Disease management

Agronomic practices (i) Strict quarantine measures should be enforced to prevent introduction of the disease to new areas (ii) Only disease-free tubers should be used as seed material (iii) Infected tubers should not be dumped into manure heaps (iv) Pigs or cattle should not be fed with raw infected tubers (v) Proper irrigation practices should be followed to prevent drying of soil, from the time of tuber formation, till the time of tuber maturity.

Chemical treatment. Treatment of seed tubers in mercuric chloride - 0.1 % solution for 1½ hours or in formaldehyde - 5 % solution for 3 hours before sowing helps to eliminate the seed-borne pathogen.

Resistant varieties. The commercial varieties Menominee, Russet, Rural and Sebago possess high degree of resistance to scab.

7. Bacterial brown rot and wilt of potato

Pseudomonas (Ralstonia) solanacearum

This **'vascular disease'** of potato is widespread in the tropical, sub-tropical and warm temperate regions of Asia, Africa, Australia, Europe, America, West Indies, Indonesia, Peru and India. In India, the disease was first recorded in 1892 and is prevalent in the states of Karnataka, Assam, Punjab, Bihar, Madhya Pradesh, Uttar Pradesh, Maharashtra, West Bengal and Tamil Nadu. The pathogen has a very wide host range and attacks tomato, chillies, eggplant, castor, groundnut, papaya, cabbage, radish, banana, ginger and a large number of other cultivated and wild plants. Losses due to the disease may go up to 75 % in some cases. The disease is also known as **'bangle blight'**, **'bangdi'**, **'ghera'**, **'utka'** and **'rassa'** in different parts of the country

Symptoms. The disease affects both the above- and underground parts of the plants. The characteristic symptoms of the disease are yellowing of the foliage and sudden wilting, followed by collapse of the infected plants completely. The name **'brown rot'** indicates the browning of the xylem in the vascular bundles. This browning is visible as dark patches or streaks on the surface of infected stems. If the infected stem is cut across and squeezed, creamy white bacterial ooze comes out of the vascular ring. Potato tubers from diseased plants may or may not show any external symptoms. But in advanced stages of the disease, brown discoloration may be seen through the skin. A bacterial mass oozes out of the eyes on the tubers. Cross section of the tuber shows brown discoloration of the vascular bundles as a ring and if such a tuber is squeezed gently, cream-like drops of sticky fluid containing bacteria are exuded from the vascular bundles. The bacterium is mainly confined to the vascular region, but in advanced stages it may invade the cortex and pith regions and cause discoloration and rotting of the tissues. In wet soils, the tubers may start rotting and bacterial mass oozes out from the eyes as a slimy exudate **(Fig.99).**

The causal organism. The bacterium causing this disease is a gram-negative, short rod, 1.5 x 0.5μ in size and motile by a single or sometimes many polar flagella. It is aerobic in nature, non-spore-forming and not encapsulated. The optimum temperature for the growth of the bacterium is 35° - 37°C. The pathogen is known to produce a toxin **'polysaccharide'**, which causes disintegration of the cell membrane and is partly responsible for causing the wilt **(Fig.99).**

Five races of the species have been identified.

Mode of survival, spread and epidemiology. The pathogen is both soil-borne and seed-borne. The pathogen survives in the soil on diseased plant residues and leftover diseased potato tubers and initiates infection. In cultivated soils, survival of the pathogen has been reported for a period of more than 2½ years. Potato tubers carry the infection in three ways viz., through vascular tissues, on the tuber surface and in the lenticels. The latent infection through the tubers is the most important method of transmission of the disease. Cutting knives used for preparing the seed materials, spread the infection from diseased to healthy tubers and thereafter to the fields. The pathogen has numerous alternate hosts and so, there is a steady supply of inoculum all through the year. In some regions, potato is cultivated throughout the year and so, the bacteria continue to infect the crop. The bacteria are also spread through irrigation and rainwater and through soil. Infection takes place through wounds caused during cultural operations, by insects and by nematodes. Potato tubers are also infected by means of stolons.

The disease develops more rapidly at 37°C. High temperatures and high soil moisture levels of 50 - 100 % favor disease occurrence. The incidence of the disease is severe in acidic and alkaline soils and in sandy loam soils. The disease severity decreases with increase in the age of the plants.

Disease management

Agronomic practices (i) Infected plant residues should be removed and destroyed (ii) Disease-free seed materials should be used for planting. The planting materials should be procured from areas, where the disease is not endemic (iii) Crop rotation with wheat, maize, sunnhemp, finger millet, cabbage, onion, garlic, soybean etc. for a period of three years helps to reduce the incidence of the disease (iv) The knives used for cutting seed tubers should be properly disinfected by dipping them frequently in formalin or lysol solution (v) Rain or irrigation water should not be allowed to flow from infected to healthy fields (vi) After harvest of the crop, the field should be ploughed to expose the soil to the summer heat, so as to kill the soil-borne inoculum (vii) Application of high doses of nitrogenous fertilizer has been found to reduce the disease incidence.

Seed treatment. The seed tubers should be treated in Streptocycline - 0.02 % solution for 30 minutes prior to planting.

8. Leaf roll of potato

Potato leaf roll virus

Virus diseases of potato are of great economic importance all over the world and account for greater losses than many of the other diseases. Nineteen virus diseases have been reported on potato. However, in India only a few of them occur and cause extensive damage to the crop. Among the virus diseases that attack potato, **'leaf roll'** is of great economic importance and is the main reason for huge yield losses and potato degeneration in the country

Symptoms. In many varieties, the margins or tips of leaves of infected plants become yellow. The leaves roll up characteristically along the margins like a boat, with the midrib at the bottom. On plants from infected seed tubers, such rolling of leaves starts in the lower leaves and progresses upward throughout the plant. In the case of secondary infection on the standing plants, the rolling of leaves starts in the upper part of the plant and progresses downward. In some varieties, individual leaflets become more erect, giving the infected plant an upright appearance. The infected leaves turn leathery, brittle and pink- or brown- pigmented. Infected plants remain stunted, chlorotic and make a rattling noise, when the leaves strike against each other. Often, the tuber bearing stolons are shorter and the tubers appear to be attached directly to the underground stem. Tuber formation is considerably reduced and the tuber size also becomes small, resulting in poor yield and quality. Phloem necrosis and accumulation of starch occur.

The causal organism. *'Potato leaf roll virus'* or *'Solanum virus* 14' or *'Potato virus* 1' causes the disease. It belongs to the Genus - *Luteovirus.* The genome consists of monopartite ssRNA within isometric particles of about 25 nm diameter. Four distinct strains of the virus have been identified.

Mode of spread. The virus is transmitted only through infected seed tubers and by some aphid vectors. The main vector transmitting the virus is *Myzus persicae*, which transmits the virus in a persistent manner. The vector requires several hours of acquisition feeding on infected plants and several hours of incubation afterwards before it can become viruliferous. The virus persists in the body of the vector for many days, even if it happens to feed on many immune hosts. The host range of the virus includes *Datura stramonium, D. tatula, Physalis angulata* and *Lycopersicon esculentum.*

9. Mild mosaic or latent mosaic of potato

Potato virus X

Several types of mosaic diseases affect potato crop. Of these **'mild mosaic'**, **'severe mosaic'**, **'super mild mosaic'**, **'rugose mosaic'** and **'crinkle'** are most important. Mild mosaic is one such disease, which causes degeneration in potato crop, but it is not as severe as leaf roll or some other virus diseases in India. The disease occurs worldwide and causes heavy losses.

Symptoms. The disease attacks almost all cultivated varieties of potato. Many varieties exhibit very slight or negligible visible symptoms on the infected plants and hence it is called **'latent mosaic'**. The affected plants may show interveinal mottling or mosaic. However, under favorable growth conditions of the host, the symptoms are masked, thus making detection of the disease very difficult. Further, at higher temperatures above 21 °C, the mosaic symptoms are almost completely masked. Slight stunting of infected plants or deformation of leaves may be seen. In some varieties viz., Key Edward, Arran, Crest, Epicure etc. top necrosis may appear.

The causal organism. The virus causing this disease is called *'Potato virus X'*, *'Potato mottle virus'*, *'Latent potato mosaic virus'* or *'Solanum virus* 1'. It belongs to the Genus - *Potexvirus*. The virus particles are flexuous rods with helical symmetry, 470 - 580 nm in length and 13 nm in width. The genome is a single ssRNA.

Mode of spread. The disease is sap transmissible. In the field the disease spreads by contact between healthy and diseased plants. The virus can also be transmitted by core grafting and by the dodder, *Cuscuta campestris*. The virus is not transmitted by any insect vectors. However, diseased seed tubers mainly perpetuate the disease. Besides potato, the virus infects *Datura* sp., *Solanum nigrum* and tobacco.

10. Severe mosaic or vein banding of potato

Potato virus Y

The disease is next in importance to 'leaf roll'. It is distributed worldwide, including Britain, Australia, the United States of America and India.

Symptoms. The symptoms vary very widely depending upon the susceptibility of the host plant and strain of the virus. However, the symptoms are more severe than in the case of mild mosaic. On susceptible varieties, local lesions, consisting of black, necrotic spots appear on the underside of the leaves. Later, the uppermost leaves develop a wrinkled appearance, turn yellow or mottled, while the lower leaves collapse and hang on to the stem or fall off. Black, necrotic spots appear on the underside of the veins of leaves in the middle portion of the plant. The necrosis gradually spreads into the petioles and finally into the stem causing brown, longitudinal streaks. This is followed by leaf fall from the lowermost upwards, which is often called **'leaf drop streak'**. In the case of systemic infection from infected seed tubers, the symptoms are more distinctly pronounced. Such infected plants are stunted and brittle and the leaves exhibit severe mottling and deformation. The diseased plants produce fewer and undersized tubers.

The causal organism. *Potato virus Y* or *Solanum virus* 2 causes the disease. It belongs to the Genus - *Potyvirus.* The flexuous, rod-shaped particles measure 680 - 970 nm in length and 11 nm in width and contain a single ssRNA genome.

Mode of spread. The virus is transmitted by sap, as well as by insects. Several aphids including *Myzus persicae, Aphis gossypii, A. rhamni, A. fabae* etc. transmit the virus. Of these, *Myzus persicae* is the most efficient vector. The virus is transmitted in a non-persistent manner by the insects. It has a wide range of hosts, including *Nicotiana tabacum, Datura stramonium, Solanum nigrum, S. demissum, Capsicum annuum, Lycopersicon esculentum* etc.

11. Super mild mosaic of potato

Potato virus A

Symptoms. The symptoms of the disease vary widely with the variety of the host and the strain of the virus. The symptoms are slightly more distinct than those of mild mosaic. In some varieties, mild mosaic is caused, while in others severe necrosis may occur and the affected plants may even be killed. In such severely affected plants, necrosis may occur in the phloem vessels of the tubers and the eye-buds are destroyed. This is followed by rotting of the tubers.

The causal organism. The disease is caused by *Potato virus A* or *Solanum virus* 3. It is another member of the Genus - *Potyvirus.* The virus particles are about 730 nm in length and 11 nm in width.

Mode of spread. The disease is transmitted through infected seed tubers and by aphid vectors. *Myzus persicae, M. circumflexus* and *Aphis rhamni* are the main vectors and they transmit the disease in a non-persistent manner.

12. Rugose mosaic of potato

Potato virus X and *Potato virus Y*

Symptoms. The disease is more severe than 'mild mosaic' caused by *Potato virus X* and 'severe mosaic' caused by *Potato virus Y*. It is characterized by severe mottling of leaves. Further, the leaves become wrinkled, puckered and rough and are markedly reduced in size. The margins of leaves roll downward and the plants are very much stunted. The lower leaves become abnormally hairy and develop black necrotic veins. In case of severe incidence, the plants may even die. Tuber production and the size of tubers are very much reduced.

The causal organism. The disease is caused by a combination of *Potato virus X* and *Potato virus Y* and both the virus components are necessary to cause the disease.

Mode of spread. *Potato virus X* is transmitted by sap and *Potato virus Y* by aphid vectors. The disease is also transmitted through infected seed tubers.

13. Crinkle virus disease of potato

Potato virus X and *Potato virus A*

Symptoms. Affected plants show marked downward curling and puckering of leaves. The leaves become hard and brittle. Slight mosaic symptoms also appear on the leaves. The plants become stunted, bushy and chlorotic.

The causal organism. The disease is caused by a combination of *Potato virus X* and *Potato virus A*. Both the virus components are necessary to cause the disease.

Mode of spread. *Potato virus X* is transmitted by sap and *Potato virus A* by aphid vectors. The disease is also transmitted through infected seed tubers.

Management of virus diseases of potato

Agronomic practices (i) The main source of virus diseases in potato crop is infected seed tubers. So, care should be taken in selecting disease-free tubers for seed purposes. Seed tubers should be obtained from disease-free crops. Certified disease-free tubers may be obtained from reliable sources. Healthy, well-formed tubers of size about 5.0 cm. in diameter should be selected for seed purposes (ii) Early planting may help to ward off the disease incidence to some extent by way of disease escape (iii) Plants showing symptoms of the diseases, as well as weak and sickly plants should be removed and destroyed by burning (iv) Alternate hosts of the viruses should not be allowed to grow in the vicinity of potato fields.

Chemical control. In the case of virus diseases that are transmitted by insect vectors, suitable insecticides should be used to eliminate them. (i) Application of systemic granular insecticides in the planting furrows at the time of planting helps to prevent the infestation of aphids, which transmit the viruses, for a period of about 40 - 50 days. For this, aldecarb 10G - 4 kg. or phorate 10G - 4 kg. or carbofuran 3G - 13 kg. or disulfotan 5G - 8 kg./ acre, mixed with 10 kg. of sand is applied uniformly in the planting furrows. Following this, spraying with dimethoate - 600 ml. in 300 lit. of water/ acre on the 50th day protects the crop from aphid infestation for another 15 days (ii) If granular insecticide is not applied at the time of planting, 4 - 5 sprayings, with dimethoate - 600 ml. or methyl demeton - 450 ml. or phosphamidon - 150 ml. in 300 lit. of water per acre, at 15 days interval, commencing from the 15th day of planting effectively protects the crop from aphid infestation. Instead of repeatedly applying the same insecticide every time, different insecticides suggested may be used each time for better results.

Diseases of minor importance. Besides the diseases discussed above in detail, potato is subjected to attack by several other diseases, some of which may become very serious during some seasons. '*Sclerotium* rot' is caused by *Sclerotium rolfsii*; 'Charcoal rot' is caused by *Macrophomina phaseolina*; '*Verticillium* wilt' is caused by *Verticillium albo-atrum*; '*Fusarium* wilt' is caused by *Fusarium oxysporum* and *F. solani*; 'Powdery mildew' is caused by *Erysiphe cichoracearun*; 'Soft rot' is caused by *Erwinia carotovora* pv. *carotovora; Erwinia carotovara* pv. *atroseptica* causes 'Soft rot' and 'Black leg'; 'Pyllody' and 'Witches' broom' are caused by *Mycoplasma-like* organisms.

Carrot *(Daucus carota)*

1. Bacterial soft rot of carrot

Erwinia carotovora pv. *carotovora*

The disease occurs all over the world, wherever carrot is grown. It occurs most commonly on fleshy storage organs of vegetables and root crops, annual ornamentals, fleshy fruits and succulent stems, stalks and leaves. The organism attacks the plant parts in the filed, in transit and especially in storage. Besides carrot, the disease attacks potato, radish, turnip, tomato, cabbage, onion and many other crops.

Symptoms. The soft rot symptom appears as small, water-soaked lesions, which enlarge rapidly in diameter and in depth. The affected areas become soft and pulpy, while the surface of the lesions becomes discolored and depressed. The tissues within the affected region become cream colored and slimy, disintegrating into a soft, pulpy mass of disorganized cells. Sometimes, cracks may develop on the surface of the affected portions and the slimy mass exudes to the surface and turns brown. An infected carrot may become a soft, watery, decayed mass within 3-5 days. When other saprophytic bacteria and fungi invade soft rot affected carrots, the tissues decompose and emit a foul odor. When the crop is affected in the field, the lower succulent parts may also be affected and they become watery, turn black and shrivel **(Fig.100).**

The causal organism. The bacteria causing the disease are rod-shaped, occurring singly, in pairs or in chains. They do not form spores or capsules. They are gram-negative, facultative anaerobic, with 2 - 8 peritrichous flagella and measure 1.5-5.0 x 0.5-0.9µ in size **(Fig.100).**

Mode of survival, spread and epidemiology. The soft rot bacteria may survive in the tissues of infected carrots, diseased plant debris and in the soil. The pathogen can lead a saprophytic life in the soil and can continue to live in the soil for 20 years or even more. Direct contact, hands, agricultural implements, soil, water and insects spread the bacteria. They enter the plant tissues primarily through wounds. Within the tissues they multiply profusely in the intercellular spaces, where they produce large quantities of **'pectolytic enzymes'** and dissolve the middle lamella. The cells are thus disintegrated, macerated, softened and transformed into a slimy mass containing innumerable bacteria that swim about in the liquid substance.

The optimum temperature range for the growth of the bacteria is 23°-27°C. High soil moisture and low temperatures favor the occurrence of the disease. Application of heavy doses of nitrogenous fertilizer increases the disease incidence.

Disease management

Agronomic practices (i) At the time of harvest, diseased carrots should not be left behind in the field. The diseased carrots and other plant debris should be collected and burnt (ii) The bacteria causing the disease enter the plant tissues mainly through injuries. So, at the time of cultural operations, when the crop is in the field and at the time of harvest, care should be taken to avoid causing injuries to the carrots (iii) The plants showing symptoms of the disease should be uprooted and destroyed (iv) Diseased carrots should

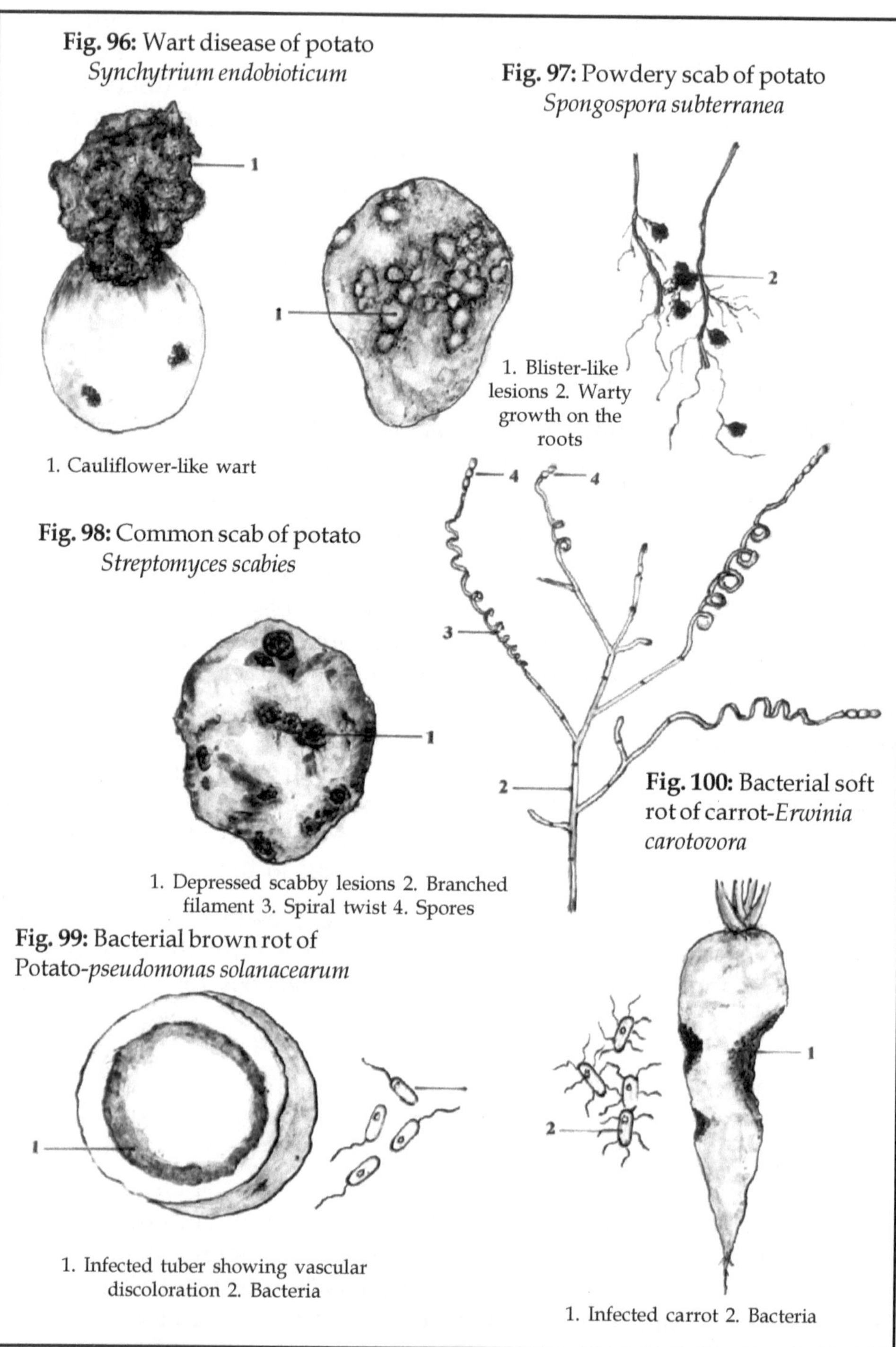

Fig. 96: Wart disease of potato *Synchytrium endobioticum*

1. Cauliflower-like wart

Fig. 97: Powdery scab of potato *Spongospora subterranea*

1. Blister-like lesions 2. Warty growth on the roots

Fig. 98: Common scab of potato *Streptomyces scabies*

1. Depressed scabby lesions 2. Branched filament 3. Spiral twist 4. Spores

Fig. 99: Bacterial brown rot of Potato-*pseudomonas solanacearum*

1. Infected tuber showing vascular discoloration 2. Bacteria

Fig. 100: Bacterial soft rot of carrot-*Erwinia carotovora*

1. Infected carrot 2. Bacteria

not be stored along with healthy carrots (v) Irrigation through diseased field to other fields should be avoided (vi) Application of excessive doses of nitrogenous fertilizer should be avoided (vii) The store-houses should be kept clean, dry and well ventilated (viii) The carrots should not be stored, when they are too moist or wet (ix) The carrots should be stored in cold-storage houses at temperatures just above 0°C.

Chemical treatment. The carrots may be dipped in 1 : 500 solution of sodium hypochlorite before packing and transporting.

Diseases of minor importance. Besides the above-mentioned disease, a few other diseases also occur commonly on carrot. *Erysiphe polygoni* causes 'Powdery mildew'; *Cercospora carotae* causes '*Cercospora* leaf blight'; *Alternaria porri* causes '*Alternaria* leaf blight'.

Radish *(Raphanus sativus)*

Many of the diseases that attack cruciferous crops, attack radish also. *Pythium de baryanum* and *P. aphanidermatum* cause 'Damping off of seedlings'; *Alternaria brassicae* causes 'Leaf spot and leaf blight'; *Albugo candida* causes 'White rust'; *Peronospora parasitica* causes 'Downy mildew'; *Erwinia carotovora* pv. *carotovora* causes 'Soft rot'; *Macrophomina phaseolina* causes 'Root and stem rot'.

Tapioca or Cassava *(Manihot esculenta / M. utilissima)*

1. Brown leaf spot of tapioca

Cercospora henningsii

(Mycosphaerella manihotis)

The disease was first reported in 1904 from Kerala in India and it occurs commonly in Kerala, Tamil Nadu and other states, wherever tapioca is cultivated extensively. It is one of the most serious diseases of tapioca and causes considerable damage to the crop and yield loss.

Symptoms. The symptoms appear as small, circular to irregular spots, 3.0 - 10.0 mm. in diameter on both the surfaces of leaves. The spots on the upper surface appear uniformly brown with a distinct dark margin. On the under surface, the margins of the spots are not quite distinct and the center of the spots turn grayish later on, when the fungus sporulates. In a single leaf, about 20 such spots may appear. The older leaves are more prone to attack by the disease than the younger leaves. As the disease advances, the spots enlarge in size and are limited by the larger veins, as a result the spots become angular in shape. Small veins within the spots turn black. The spots may coalesce and large areas of the leaves get blighted. Severely affected leaves become yellow, dry and fall off prematurely. Affected plants are stunted in growth.

The causal organism. The mycelium of the pathogen is septate, inter- and intracellular. Before fructifications are formed, the hyphae aggregate beneath the host epidermis of the spots and form stromata. From the stromata, conidiophores emerge in dense fascicles through the stomata. The conidiophores are erect, tubular, up to 100µ in length, septate, light-brown in color, geniculate with scars at every point from where the matured conidia have been dislodged. The conidia are long, straight or slightly curved, pale olive- brown in color, thin-walled, 2 - 8 septate and slightly narrowing towards the apex and with rounded ends. They measure 30.0 - 50.0 x 4.0 - 6.0µ in size.

Perithecia, which may be formed later, are found scattered in the necrotic areas on the upper surface of the leaves. They are ostiolate, dark colored and about 100µ in diameter. The asci are sub-sessile, elongated, clavate, 8-spored and are 72 - 77 x 10 - 13µ in size. The ascospores are hyaline, ovoid, and single-septate, slightly constricted at the septum and are 17.0 - 22.0 x 5.2 - 6.8µ in size.

Mode of survival, spread and epidemiology. The disease is primarily air-borne. Infected leaves fallen on the ground continue to produce conidia, which are carried through wind and rain. Further, tapioca is a long duration crop and remains in the field for a period of 6 - 10 months and the crop may be present all through the year in different localities. The spores produced from these crops may provide enough inoculum to perpetuate the disease. The conidia remain in a viable state for 26 - 28 days under field conditions. The pathogen also attacks *Manihot glaziovii* and sweet potato.

High atmospheric humidity of 98 - 100 % and moderate temperatures of about 27°C are favorable for the production of conidia and disease development. The disease incidence is more during the periods of the monsoon rains. Application of heavy doses of nitrogenous fertilizer increases the disease severity.

Disease management

Agronomic practices (i) Severely affected leaves from the plants, as well as the leaves fallen on the ground should be collected and destroyed by burning (ii) Balanced fertilizer application helps to reduce the severity of the disease.

Chemical control. Spraying with Bordeaux mixture - 1 % or copper oxychloride - 2.5 gm. or wettable sulfur - 4.0 gm. or clorothalonil - 1.5 gm. or thiophanate methyl - 1.0 gm./ lit. of water controls the disease effectively. The sprayings should commence, when the crop is about 3 months old, followed by 2 - 3 more sprayings, at 3 weeks interval. A quantity of 300 - 350 liters of spray fluid may be required to cover one acre of the crop with a high volume sprayer.

2. Anthracnose of cassava

Colletotrichum manihotis (Gloeosporium manihotis)

(Glomerella manihotis)

'Anthracnose', **'die-back'** or **'wither tip'** occurs in all tapioca growing states of India, including Tamil Nadu, Kerala and Uttar Pradesh. It occurs sporadically in some regions during some seasons, but under favorable conditions may become very serious and cause considerable damage to the crop and yield loss.

Symptoms. The initial symptoms of the disease appear as small, brown, circular to irregular spots on the leaves, which enlarge gradually. The spots may coalesce to form large, irregular spots. Severely affected leaves crinkle, turn yellow, dry and fall off. Light to dark-brown, elongated, slightly sunken lesions may develop on young, tender shoots and petioles. The lesions may become black and necrotic and the parts above them may wither and dry, leading to tip drying. As the infection spreads, the basal portion of the petioles shrivels, the leaves turn yellow and dry. In advanced stages of infection, tip drying may be seen in many of the shoots. On the central portions of the necrotic lesions on the leaves, stems and petioles, scattered, dark-brown to black, erumpent, pin point-like dots, representing the acervuli of the fungus are produced in large numbers.

The causal organism. The mycelium of the fungus is hyaline when young and becoming darker with age, branched, septate, inter- and intracellular. Prior to sporulation, the hyphae aggregate beneath the epidermis of the spots and produce acervuli. The conidiophores, which arise from the base of the acervuli as palisade cells in large numbers, are erect, minute, cylindrical, short and hyaline. Conidia are produced from the tips of the conidiophores in succession. They are single-celled, hyaline to olive brown in color and pale-pink in mass. They are oblong to cylindrical, with rounded ends and measure 12.0 - 18.1 x 3.4 - 4.3µ in size. Setae are produced sparingly and are long, tapering towards the tip, spiny, unbranched, septate, dark-brown in color and are 29 - 58µ in length.

In the perfect stage, the fungus produces perithecia. They are sub-globose to pear-shaped with a short beak, ostiolate, dark-brown in color and measure 92.0 - 153.3μ in diameter. The 8-spored asci present inside the perithecia are long, cylindrical to clavate, sub-sessile and measure 53.4 - 80.1 x 9.0 - 13.4μ in size. The ascospores, which are arranged in a single row inside the ascus are hyaline, single-celled, smooth, oval to obovate and are 13.8 - 17.2 x 5.2 - 6.0μ in size. Paraphyses are absent.

Mode of survival, spread and epidemiology. Cuttings from diseased plants are the main source of primary inoculum. Secondary spread is by air-borne conidia produced in the acervuli. Tapioca is a long duration crop and the crop may be present throughout the year in different localities. The conidia produced from the standing crops may provide enough inoculum to cause fresh infection. Conidia may also be produced from infected leaves and other plant parts fallen to the ground.

Long periods of rain, high atmospheric humidity, water-logging, moderate temperature and poor crop growth and susceptibility of the varieties predispose the crops to the disease attack.

Disease management

Agronomic practices (i) Disease-free planting material should be used (ii) Planting during periods of heavy rainfall should be avoided (iii) Severely affected plant parts may be pruned and destroyed to reduce the inoculum potential (iv) The crop should be maintained properly by providing sufficient irrigation, adequate drainage facilities and by the application of balanced dose of fertilizers.

Chemical control. Spraying with Bordeaux mixture - 1 % or copper oxychloride - 750 gm. or carbendazim - 300 gm. in 300 lit. of water per acre controls the disease.

3. Sett rot or stem rot of cassava

Diplodia natalensis

The disease is widespread and is found in all tapioca growing regions and results in mortality of seedlings and gappiness in the fields.

Symptoms. The disease occurs both in the stems stored for seed purposes, as well as in the setts after planting in the fields. The symptoms appear as dark-brown blisters on the epidermis. The infection spreads deeper into the tissues, which become dark-brown or black and rot. Black discoloration of the vascular strands is also seen. From the setts, the infection spreads to the growing shoot resulting in the rotting of the stems of young plants and eventual death of the plants. When the blisters rupture, black, pinpoint-like dots, representing the pycnidia of the fungus are seen in large numbers.

The causal organism. The mycelium of the fungus is hyaline, septate, inter- and intracellular. The hyphae aggregate beneath the epidermis of the infected area and form pycnidia. They are dark-brown to black in color, globose and erumpent. The pycnidia vary in size considerably and are 130 - 340μ in diameter. Conidia are produced in succession from the tip of conidiogenous cells lining the inside of the pycnidia. The conidia are brown in color and are medially single-septate and slightly constricted at the septum. They have longitudinal striations and measure 20 - 30 x 10 - 15μ in size. The perfect stage of the fungus is *Physalospora rhodina,* which is found very rarely in nature.

Mode of survival, spread and epidemiology. The disease is mainly transmitted through infected setts. The fungus gains entry into the setts through wounds and cut ends, both in storage and after planting. Secondary spread is through conidia carried by wind and rainwater.

High soil moisture at the time of planting and low to moderate temperatures favor the disease occurrence.

Disease management

Agronomic practices (i) Cuttings for planting should be selected from disease-free, healthy plants. (ii) Care should be taken to avoid injuries to the cuttings (iii) Planting should not be taken up when the soil is too moist (iv) Stems for planting should not be stored under moist conditions.

Chemical control. Dipping the setts (cuttings) prior to planting in a solution of carbendazim - 1.0 gm. or carboxin - 1.0 gm. or chloroneb - 1.0 gm./ lit. of water for 15 minutes is effective in eliminating the inoculum present in the setts and also gives protection against fresh infection. However, treating the setts in a mixture of carbendazim - 1.0 gm. + mancozeb - 2.0 gm. or chloroneb - 1.0 gm. + captan - 2.0 gm./ lit. of water is found to be more effective in giving wider control.

4. Cassava mosaic

Cassava latent mosaic virus

'Mosaic of cassava' is a most serious and destructive virus disease. In India, the disease was first reported from Tamil Nadu in 1966. Now, it is commonly found in all the tapioca growing regions of Tamil Nadu, especially in the districts of Salem and Kanyakumari.

Symptoms. The disease is characterized by severe mosaic symptoms. Large, yellow and chlorotic patches are found fading into the green areas of leaves. The symptoms are more pronounced in the younger leaves, which are misshapen, variously twisted and wrinkled and are reduced in size. Affected plants are stunted in growth. Plants growing from setts obtained from diseased plants are very much stunted with thinner stems and less of leaf canopy. Repeated use of setts from diseased plants results in heavy reduction in yield, besides tuber splitting occasionally.

The causal organism. The disease is caused by *Cassava latent mosaic virus*, which belongs to the Genus - *Geminivirus*. The morphology of the virus particles is unique in that their isometric particles of size 18 - 20 nm diameter occur in pairs. The genome of the virus particles consists of ssDNA molecule.

Mode of spread and epidemiology. The disease is primarily perpetuated through cuttings or setts from infected plants. Secondary spread of the disease is by the white fly vector, *Bemisia tabaci*. The acquisition feeding time is 4 hours on young diseased leaves and the insects become viruliferous only after an incubation period of 4 hours. Afterwards, the vectors are capable of transmitting the virus in a persistent manner. The virus is retained in the vector even after moulting, but is not passed on to its progeny. Increase in the number of white flies, increases the disease incidence however, a single white fly can transmit the virus to a large number of plants in a single day. The virus is not transmitted through sap. When environmental conditions are not favorable for the multiplication of the virus inside the host, typical symptoms may not be manifested and the plants remain as symptomless carriers of the virus. However, if cuttings from such plants are planted, they produce the disease. *Cassava* varieties, with green petioles and soft leaves are preferred by the vectors and are more prone to attack by the disease compared to varieties with red or reddish-green petioles and coarse leaves.

Conditions that favor the rapid multiplication of white flies, such as moderate temperature and high humidity favor the development of the disease also. Besides *Manihot utilissima*, the disease occurs in *Manihot dulcis, M. glaziovii*, cucumber etc.

Disease management

Agronomic practices (i) Care should be taken to select cuttings from disease-free, healthy plants for seed purposes. Cuttings should not be taken from plants showing any symptoms of the disease on the leaves (ii) Plants showing symptoms of the disease in the early stages of growth should be uprooted and destroyed (iii) Seed materials may be obtained from disease-free nursery crop (iv) Application of excessive doses of nitrogenous fertilizer should be avoided (v) Yellow sticky traps may be set up in the fields at the rate of 10 traps per acre to attract and kill the white flies.

Chemical control. Chemicals having quick knockdown effect should be used to control the vectors, as they can transmit the virus to a number of pants within a short time. Spraying with methyl demeton - 300 ml. or dimethoate - 300 ml. or phosphamidon - 170 ml. or acephate - 600 gm. or phosalone - 600 ml. or ethion - 300 ml. in 300 lit. of water per acre is effective in controlling the vectors.

Resistant varieties. Varieties, such as H-43, H-86, H-97, H-165, H-226, S-856, S-2304 and S-2371 have been found to be fairly resistant to the disease.

Minor diseases. Besides the diseases discussed above in detail, *Cassava* is subjected to attack by several other diseases, some of which may assume serious proportions during some seasons. *Xanthomonas axonopodis* pv. *manihotis* causes 'Bacterial blight'; *Sclerotium rolfsii* causes '*Sclerotium* root rot'; several fungi, such as *Cercospora manihotae, Phyllosticta manihotae, Alternaria palandui, Drechslera rostrata* and *Phomopsis manihotis* cause characteristic 'Leaf spot diseases'.

Sweet potato *(Ipomoea batatas / Convolvulus batatas)*

1. *Cercospora* leaf spot of sweet potato

Cercospora batatae (C. ipomoae) and

Cercospora bataticola (Phaeoisariopsis bataticola)

The disease was first recorded in India from Bihar in 1976. It is widely distributed and is commonly found in the states of Telungu Desam, Maharashtra, Bihar and Tamil Nadu. It is one of the most serious and destructive diseases of sweet potato and causes considerable damage to the crop and yield loss.

Symptoms. The disease is primarily a foliar disease. Symptoms appear as small, brown, roughly circular spots, about 3.0 - 5.0 mm. in diameter, with reddish-purple margin. Later, the center of the spots become ashy-gray, thin, papery and brittle and may drop off, leaving a ragged hole. If many spots develop on the leaf, the spots may coalesce to form large blighted areas, covering major portions of the leaf lamina. In humid and damp weather, the affected area is covered with an ashen-gray, mouldy growth, representing the conidiophores and conidia. In case of severe infection, the leaves turn yellow, dry and fall off prematurely. Similar elongated lesions may be produced on the stem region and leaf petioles also.

The causal organism. The mycelium of the fungus is septate, hyaline or pale olive in color, inter- and intracellular. The hyphae aggregate beneath the epidermis of the spots prior to fructification and form stromata. From the stromata, conidiophores arise out of the surface in fascicles through the stomata and form conidia in succession on new growing tips. The conidia are easily detached and often blown over long distances by wind. In ***Cercospora batatae***, the conidiophores emerge through the stomata in groups of 1 - 7. They are septate, dark-brown in color, geniculate and are 120 - 130µ in length. The conidia are long, slender, hyaline to pale olive in color, multiseptate, straight or slightly curved, slightly tapering towards the apex and measure 160.0 - 175.0 x 3.0 - 4.0µ. In ***Cercospora bataticola***, the conidiophores, which arise in clusters through the stomata are septate, dark-brown, geniculate and are 80 - 140µ in length. The conidia are slender, rather short, cylindrical, pale olivaceous in color, multiseptate and are 50.0 - 120.0 x 3.0 - 5.0µ in size. The conidia germinate by issuing one or more germ tubes.

Mode of survival, spread and epidemiology. The pathogens continue to produce conidia from infected, fallen leaves and other plant parts. The fungi overseason in the infected leaves or stems as stromata and cause primary infection. Secondary spread is through conidia carried by wind or rain.

High temperature and high atmospheric humidity favor disease occurrence and development. The coinidia require water for germination and infection, which is provided by rain or heavy dew.

Disease management

Agronomic practices. Removal and destruction of diseased crop debris help to reduce the inoculum potential.

Chemical control. Spraying with Bordeaux mixture - 1 % or copper oxychloride - 2.5 gm. or mancozeb - 2.0 gm. or carbendazim - 1.0 gm. or chlorothalonil - 1.25 gm./ lit. of water is effective in controlling the disease. Sprayings should be repeated 2 or 3 times, at fortnightly intervals.

2. Black rot or mouldy rot of sweet potato

Ceratocystis (Ceratostomella) fimbriata

The disease is found in many of the sweet potato growing countries of the world, including Africa, America, India and many other Asian countries. It is primarily a storage rot disease of tubers, but also occurs in the field and affects the tubers in the standing crop.

Symptoms. In the field, the pathogen affects all underground parts of the plant. Infection by the pathogen leads to production of black cankers on the underground parts and grayish-black, circular, sunken spots in the fleshy tubers. Infection gradually spreads to a short distance up the stem, causing blackening of the stem above the soil surface. This is commonly known as **'black shank'**. The foliage of infected plants turn yellow and the plant appears sickly. The spots on the fleshy portions enlarge and may soon cover almost the entire tubers. A mouldy growth, consisting of mycelium and spores appear in the central parts of the lesions. The tissues below the spots rot. If infected tubers are stored, the rot spreads rapidly and the tubers become unfit for consumption or for industrial use. Infected tubers when cooked have an unpleasant and acrid taste.

The causal organism. The mycelium of the fungus is hyaline or light greenish in color, septate and branched. In the asexual phase, the fungus produces both microconidia and macroconidia or chlamydospores. Microconidia are produced endogenously in succession from conidiophores, which are 50 - 100μ in length. They are forced out of the conidiophores in long chains and are held together in a drop of mucus. They are thin-walled, single-celled, oblong to cylindrical, hyaline at first, becoming black later on and measure 11.0 - 25.0 x 2.5 - 5.0μ. Macroconidia are produced singly or in chains on lateral conidiophores. They are oval to obovate, single-celled, thick-walled, olive brown in color and measure 9.0 - 18.0 x 6.0 - 13.0μ in size.

In the sexual phase, the fungus produces perithecia. They are flask-shaped, ostiolate, brown in color and are 140 - 220μ in diameter with a globose bottom and long, narrow neck measuring 350 - 900 x 20 - 30μ. The tip of the neck is shredded and feathery. Asci are formed at the inside base of the perithecium. The 8-spored asci are club-shaped and thin-walled. The ascospores are liberated from the asci, while still inside the perithecium and the ascospores emerge through the ostiole and are held together in a viscous fluid at the ostiole. They are oval to elliptical, hyaline, thin-walled, single-celled and are 4.5 - 8.0 x 2.5 - 5.5μ in size.

Mode of survival, spread and epidemiology. The macroconidia, which behave as chlamydospores remain viable for long periods in the soil and in the diseased plant debris, as well as in the leftover tubers. The microconidia cannot survive in the soil for long. The fungus is also capable of leading a saprophytic life in the soil in the absence of the host. The perithecia can also persist in the soil for prolonged periods and cause fresh infection. The fungus is a weak parasite and can gain entry into the host only through bruises, wounds etc. After entry into the host, the disease spreads very rapidly. If infected tubers are stored, then also the disease spreads rapidly and very soon the entire tuber gets rotten. Plants may get infected in the seedbeds or in the field.

The optimum temperature for the fungal growth and infection is 23° - 28°C. High soil moisture and warm weather conditions favor occurrence and development of the disease.

Disease management

Agronomic practices (i) Using disease-free planting stock is the best method of control of the disease. Healthy seed sweet potatoes can be grown from cuttings from healthy, disease-free plants. The cuttings should be planted in soils free from the black rot pathogen. (ii) Crop rotation for 2 - 3 years with crops, which are not susceptible to black rot should be followed (iii) Cattle manure obtained from animals fed with raw, infected sweet potatoes should not be applied to sweet potato crop, as the dung may contain spores of the fungus in a viable state (iv) Varieties resistant to black rot should be grown.

Chemical control. Dipping seed sweet potatoes in a solution of thiobendazole - 1.0 gm. or carbendazim - 1.0 gm./ lit. of water gives good control of the disease.

3. Charcoal rot of sweet potato

Macrophomina phaseolina

The disease, which is primarily a storage rot of sweet potato, is of common occurrence wherever the crop is grown and it causes considerable damage to the tubers in storage.

Symptoms. The pathogen causes slow storage rot of the tubers. The tissues of affected tubers become chocolate-brown in color, which later turn dark-brown or dark red and ultimately black. The decay, which is soft and spongy in the beginning, becomes hard later on and the tubers look mummified. Black, minute sclerotia of the fungus are seen on the surface of the affected areas along with a dense growth of white to gray mycelium.

Causal organism mode of survival, spread and epidemiology. Refer 'Charcoal rot of soybean' (Page 153), **(Fig.43)**. The pathogen attacks several other crops, including potato, maize, soybean, redgram, chickpea, cowpea, black gram, green gram and bean.

Disease management

Agronomic practices (i) Care should be taken to avoid causing injuries to the tubers during cultural operations and at the time of harvesting and storing (ii) At the time of harvest, diseased tubers should not be left behind in the field (iii) Diseased tubers should not be stored along with sound tubers. Cut and injured tubers should be sorted out and removed (iv) Store-houses should be kept clean, dry and well aerated (v) The tubers should not be stored when they are too moist or wet.

Chemical treatment (i) Seed tubers may be dipped in 1 : 1000 solution of mercuric chloride before planting (ii) Tubers may be dipped in 1 : 500 solution of sodium hypochlorite before storing.

4. *Rhizopus* soft rot of sweet potato

Rhizopus nigricans

Symptoms. The pathogen causes storage rot of tubers. The symptoms appear as small, soft, water-soaked, rotten patches on the tubers. The tissues inside become soft, pulpy and stringy. The disease progresses rapidly and within a few days the infected tubers rot completely. When the outer skin is broken, the inner tissues show brown discoloration and amber colored liquid oozes out from the rotten tissues and an unpleasant odor is also emitted. The surface of the lesions is covered with a whitish, fine, cottony mycelial growth, which turns black when spores are formed abundantly.

The causal organism. Refer 'Head rot of sunflower' (Page 196), **(Fig.56)**.

Mode of survival, spread and epidemiology. The fungus is a weak facultative parasite and can lead a saprophytic life in the diseased crop debris in the soil and continue to produce sporangiospores. The spores are present in the atmosphere throughout the year and can cause infection at any time, when conditions are favorable. The organism can gain entry into the tubers only through wounds caused by mechanical means or by insects, nematodes etc.

Moderate temperature, high atmospheric humidity and moist conditions favor infection and development of the disease. Besides sweet potato, the fungus attacks potato, tomato, apple, sunflower etc.

Disease management

Agronomic practices (i) Care should be taken to avoid causing injuries to the tubers during cultural operations and at the time of harvesting and storing (ii) Diseased tubers should not be stored along with good tubers. Cut and injured tubers should be sorted out and removed before storing (iii) Store-houses should be kept clean, dry and well aerated (iv) The tubers should not be stored when they are too moist.

Chemical treatment. The tubers may be dipped in 1 : 500 solution of sodium hypochlorite before storing.

Diseases of minor importance. A few other diseases besides the ones discussed above also attack sweet potato. *Fusarium oxysporum* f.sp. *batatas* causes 'Vascular wilt' and 'Stem rot'; *Streptomyces ipomoeae* causes 'Soil rot' or 'Pox', which affects the tubers; *Sclerotium rolfsii* causes 'Collar rot'; specific 'Leaf spots' are caused by *Alternaria capsici*, *Helminthosporium euphorbiae*, *Septoria bataticola* and *Phyllosticta batatas*; few virus diseases also occur on this crop.

Coco yam *(Colocasia antiquorum / C. esculenta* var. *antiquorum)* Yam *(Dioscoria alata / D. globosa)* and Elephant's foot yam *(Amorphophallus campanulatus)*

1. *Phytophthora* blight of yam

Phytophthora colocasiae

The disease was first recorded in Java in 1900. It was first reported in India in 1907 and now, it is prevalent in the states of Assam, Orissa, Himachal Pradesh, Punjab, Haryana, Kerala and Tamil Nadu. The yield loss due to this disease may go beyond 50 % during some seasons.

Symptoms. The initial symptoms of the disease appear as small, dark, circular specks, which enlarge rapidly in a centrifugal manner and become large, circular or irregular spots. Several such spots may develop on the leaf and often the entire leaf lamina gets affected. Drops of clear, yellow-colored liquid ooze out from the surface of the spots in the initial stages. As the disease progresses, the central portions of the spots dry and drop off leaving holes in the leaf. The periphery of the spots shows a zonate pattern of different shades of brown, green and yellow. When conditions are favorable for the pathogen, all the plants in the field are blighted within a few days. As the spots increase in size, sporangia are produced towards the peripheral region in a zonate manner. Under favorable conditions of temperature and humidity, the infection spreads to the petioles, which become softened and weak, as a result the leaves break and fall off. The infection may then proceed down into the stem and ultimately into the corm. Such infected corms may rot completely. The pathogen infects the inflorescence also.

The causal organism. The mycelium of the fungus is coenocytic, profusely branched and intercellular. Long, unbranched, slender haustoria are sent into the cells. The hyphae do not enter the fibro-vascular bundles of the leaf or petiole, but in the corms they penetrate all the cells, including the vascular bundles. As a result of infection, the cells are destroyed and they become a brown, rotten mass.

From the internal hyphae, sporangiophores grow out to the surface on both the surfaces of leaves, petioles and inflorescence. They are slender, unbranched and narrow at the tip and up to 50µ in length. They bear single, ellipsoid, ovoid or pear-shaped sporangia, which are 38 - 60 x 18 - 26µ in size. They are semi-papillate and pedicellate. Each sporangium may contain as many as 12, reniform, biflagellate zoospores. Sometimes, the sporangia germinate directly by means of a germ tube. The fungus also produces chlamydospores, which are thick-walled, yellow and are 26 - 30µ in diameter. In the sexual phase, the fungus produces oospores, which are spherical, 20 - 28µ in diameter and lie free within the oogonia. The oospores also produce zoosporangia and zoospores.

Mode of survival, spread and epidemiology. The disease is perpetuated through infected corms, chlamydospores and oospores persisting in the soil. The pathogen cannot survive in the free state in the soil for a long time. Infected corms used as seed material are the major source of primary inoculum. Self-sown host plants may also play an important role in the recurrence of the disease. Secondary spread of the disease is by zoospores produced by sporangia or zoosporangia and chlamydospores carried by rain and irrigation water and by rainsplashes.

Sporangial production is maximum when the temperature is 21°C and humidity is 100 %. Germination of sporangia by zoospore production is optimum at 20° - 21°C. Relatively moderate temperatures of 21° - 30°C, high relative humidity of 90 - 100 % and free water on the host surface favor disease occurrence and development.

Disease management

Agronomic practices (i) Disease-free seed corms should be used for planting (ii) Removal and destruction of diseased crop debris from the field and destruction of badly affected plants help to minimize the inoculum potential (iii) Crop rotation may be followed (iv) Resistant varieties may be grown.

Chemical control. Spraying with copper oxychloride - 2.5 gm. or metalaxyl - 1.0 gm. or captafol - 1.5 gm. or chloroneb - 1.0 gm. or mancozeb - 2.0 gm./ lit. of water at 15 days interval have been found to be effective in controlling the disease.

2. Anthracnose of yam

Colletotrichum gloeosporioides

(Glomerella cingulata)

The disease occurs in all the states of India, wherever yam is cultivated. It is more prevalent in Rajasthan, Kerala, Tamil Nadu and parts of North-East India.

Symptoms. The symptoms may vary slightly depending upon the Genera and species of yam. But in general the pathogen attacks the leaves, petioles and stems. On the leaves, the symptoms appear as small, brown, slightly sunken spots, about 3.0 - 5.0 mm. in diameter, sometimes with a yellow border. The lower leaves are mostly affected first. The spots enlarge and may reach a diameter of about 20 mm. As the spots enlarge, acervuli are produced in concentric circles as brown to black, pinpoint-like dots. Sometimes, the center of the spots becomes necrotic and falls off leaving ragged shot holes. Severely affected leaves wither and dry completely. The disease spreads to the petioles and stems and linear, brown, sunken lesions are produced. Large number of acervuli are formed on the lesions on petioles and stems. Severely affected plants become yellow, dry and die.

The causal organism, mode of survival, spread, epidemiology and disease management. Refer 'Anthracnose of mango' (Page 283), **(Fig.73)**.

Minor diseases. A few other diseases are also known to attack yams besides the ones discussed above. *Sclerotium rolfsii* causes 'Collar rot' and 'Wilt'; *Erwinia carotovora* causes 'Corm rot'.

SPICES AND CONDIMENTS

Chillies *(Capsicum annuum)*

1. Damping off of seedlings of chillies

Pythium aphanidermatum and *P. de baryanum*

Symptoms, causalorganism, mode of survival, spread, epidemiology and disease management. **Refer 'Damping off of seedlings of eggplant'** (Page 339), **(Fig. 85).**

2. Die-back and fruit rot of chillies

Colletotrichum capsici

The disease is widely distributed in the tropical and sub-tropical regions and is prevalent in Africa, America and many Asian countries. It was first reported in South India in 1914 on young twigs and ripening fruits of chillies. In India, the disease occurs commonly in the states of Assam, Bihar, Haryana, Maharashtra, Punjab, West Bengal and Tamil Nadu. It is found to occur in all the chillies growing regions of Tamil Nadu.

Symptoms. The disease is characterized by necrosis of tender twigs from the tip backwards. The symptoms may appear in a single branch or in a few branches or the entire plant may be affected and start drying from the tips. The necrotic twigs are water-soaked and become brown in color. Within a few days the dead areas turn grayish-white or straw-colored. On these areas, numerous, black, scattered, pinpoint-like dots, representing the acervuli of the fungus are found. Sometimes, the necrotic areas are demarcated from the healthy areas by a dark-brown to black, irregular band. When one or a few branches are affected, only a few and poor quality fruits may be formed on the unaffected branches.

The fungus also affects the ripening fruits, which start turning red. In the beginning, one or two, small, black, circular spots appear on the skin of the fruit. The spots enlarge gradually in the direction of the long axis of the fruit and become somewhat elliptical. The spots, which are slightly sunken, get diffused and become dirty-gray in color or they may be surrounded by a thick and sharp, black margin, while the central area becomes grayish-black or straw-colored. Sometimes, the entire surface of the fruit turns straw-colored or dirty-white from its normal red color. On the discolored area, numerous, black, scattered dots, representing the acervuli are formed, often in concentric zones. When such a diseased fruit is broken open, the inner surface of the skin is found covered with whitish-gray mycelial growth. The seeds are also covered by a mat of mycelium and are rust colored.

Affected fruits drop off easily. The disease may affect the ripe fruits in storage also.

The causal organism. The mycelium of the fungus is septate, hyaline, inter- and intracellular. The hyphae aggregate below the epidermis prior to fructification and form stromata, from which acervuli are formed. Acervuli are inverted saucer-shaped and measure 70 - 120µ in diameter. Short, aseptate, unbranched conidiophores arise in large numbers from the base of the acervuli. Conidia are produced in succession, singly at the tip of the conidiophores. Conidia are hyaline, single-celled, slightly curved, with slightly narrowing ends and are 17.0 - 28.0 x 3.0 - 4.0µ in size. Setae are found interspersed with the conidiophores. These sterile organs are long, dark-brown in color, multiseptate, hard and needle-like. They measure up to 150µ in length. Due to the pressure exerted by the continuous production of conidia in large numbers, the host epidermis is ruptured and the spores are exposed. The spores appear pinkish and slimy in mass. The conidia after dispersal germinate on the host surface in the presence of water within 4 hours. The germ tube produced by the conidium forms an appressorium, from which an infection hypha is produced. The infection hypha penetrates into host through the stoma and causes fresh infection. The pathogen produces a **'thermostable toxin'** inside the host that causes injury to the protoplasmic content of the host cells. This toxic substance is also responsible for the drying of the twigs and rotting of the fruits **(Fig.108).**

Mode of survival, spread and epidemiology. The pathogen survives in the diseased plant debris in the soil for a long time, sometimes even up to a period of 12 months and continues to produce conidia. Seeds from infected fruits also carry the primary inoculum. Secondary spread of the disease is mainly through wind-borne conidia.

The disease incidence is more after the rainy season, followed by prolonged periods of dewfall. Temperatures ranging from 28° - 30°C and atmospheric humidity above 95 % and dewfall are favorable for spore germination and disease development. The pathogen also attacks marigold, pothos, coriander, turmeric and snap melon (*Cucumis melo* var. *momordica*).

Disease management

Agronomic practices (i) Seeds should be selected from disease-free, spotless and well formed fruits (ii) Diseased plant debris and fallen fruits should be collected periodically and destroyed by burning (iii) Affected branches may be cut and destroyed (iv) Alternate hosts should be eradicated (v) Disease affected fruits should be sorted out and stored separately. They should not be mixed with healthy fruits during drying and storage.

Chemical control. Spraying the crop with ziram - 600 gm. or zineb - 600 gm. or mancozeb - 600 gm. or wettable sulfur - 1200 gm. or copper oxychloride - 750 gm. in 300 lit. of water / acre, 3 - 4 times, at fortnightly intervals commencing from the time the plants are 60 days old controls the disease.

Seed treatment. Treating the seeds with captan - 4 gm. or thiram - 4 gm. or zineb - 4 gm./ kg. of seeds, at least 24 hours prior to sowing is effective in eliminating the seed-borne inoculum.

Resistant varieties. The varieties Bengal Green, Lorai, Perennial, H-1, H-4 and H-6 have been found to be resistant to the disease.

3. Leaf spot of chillies

Alternaria solani

Symptoms, causal organism, mode of survival, spread, epidemiology and disease management. Refer 'Early blight of potato' (Page 384), **(Fig.94).**

4. Bacterial leaf spot of chillies

Xanthomonas campestris pv. *vesicatoria*

The disease was first reported from the United States of America in 1912 and now it is prevalent in many countries of the world. In India, the disease occurs very commonly in all the regions, wherever chillies are cultivated on a large scale.

Symptoms. Initial symptoms appear as small, dark-brown or black, circular or irregularly circular, water-soaked spots. As the spots enlarge in size, the center becomes lighter and the spots are surrounded by a black margin. The spots may coalesce to form larger, irregular spots. Severely affected leaves turn yellow, wilt and drop off prematurely. Linear, dark-brown to black lesions may develop on the petioles and stem region. Stem infection may often lead to cankerous growth and the branches wilt and die. On the green fruits, small, circular, black, water-soaked spots, with a light yellow border are formed scattered on the surface. Slimy droplets of bacterial ooze come out from the spots. The affected fruits may drop off prematurely.

The causal organism. The bacteria causing the disease are rod-shaped, with rounded ends and measure 1.5 - 2.0 x 0.5 - 0.75μ in size and are motile with a single polar flagellum. They are aerobic in nature and are non-spore forming.

Mode of survival, spread and epidemiology. The bacteria can survive in the diseased crop debris and fallen diseased fruits in the soil for a very long time. They are also carried through seeds. When the seeds germinate, the bacteria present in the seeds or in the soil infect the germinating seedlings through the stomata on the cotyledons and cause primary infection. The bacteria may also gain entry into the leaves through the stomata, hydathodes, injuries caused by insects and mechanical means. Infected seedlings also carry the disease to the main field. Secondary spread is caused by rainsplash, strong winds and by some insects mechanically.

Plants at the age of 40 - 50 days are more prone to attack by the bacteria in the main field. Moderate temperatures ranging from 24° - 28°C, high atmospheric humidity and intermittent rainfall favor infection and development of the disease. The bacteria also attack tomato and a few other species of *Capsicum*.

Disease management

Agronomic practices (i) Seeds should be selected from disease-free, good quality fruits (ii) The nursery should be set up in disease-free area (iii) Field sanitation measures should be adopted (iv) Alternate hosts should be eradicated (v) The crop should be maintained properly by supplying recommended doses of fertilizers and by providing adequate irrigation, so that the plants remain healthy and strong (vi) Disease-free and healthy seedlings should be used as transplants.

Seed treatment (i) Seeds should be treated with captan or thiram at 4.0 gm./ kg. of seeds, at least 24 hours prior to sowing (ii) Hot water treatment of seeds at 50°C for a continuous period of 25 minutes eliminates the bacterial inoculum in the seeds.

Chemical control. Spraying the crop with Agrimycin 100 at 800 - 1000 ppm (0.8 - 1.0 gm./ lit. of water) or Streptocycline at 20 ppm (0.2 gm./ lit. of water) at fortnightly intervals during the monsoon period is effective in controlling the disease.

5. Mosaic disease of chillies

Tobacco mosaic virus

Symptoms, causal organism, mode of spread, epidemiology and disease management. Refer 'Mosaic disease of tobacco' (Page 245).

Minor diseases. Several other diseases also attack chillies and some of them may cause serious damage to the crop during some seasons. *Cercospora* *capsici* causes '*Cercospora* leaf spot'; *Fusarium solani* causes '*Fusarium* wilt'; *Sclerotium rolfsii* causes 'Root rot'; 'Powdery mildew' is caused by *Leveillula taurica; Peronospora tabacina* causes 'Downy mildew'; *Erwinia carotovora* pv. *carotovora* causes 'Bacterial soft rot of fruits'; several virus diseases also attack chillies.

Onion *(Allium cepa)*

1. Downy mildew of onion

Peronospora destructor

The disease attacks all kinds of onions and garlic. It was first reported in England in 1841 and later in America in 1884. Now, the disease occurs in all the countries of the world, where onion is grown. The disease is found in all the states of India.

Symptoms. Plants at all growth stages are attacked, either by primary infection systemically from diseased bulbs or by secondary local infection through spores produced from primary infection. In the case of primary infection from bulbs already infected with the hibernating mycelium, one or more leaves or sometimes the whole plant turns yellow and glazed. In moist and humid conditions, the infected leaves gradually collapse and wither from tip downward. Meanwhile, the fungus inside the leaf tissues produces sporangiophores, which emerge through the stomata. Sporangia are borne in large numbers on the sporangiophores, which are dispersed by wind to cause secondary infection. Thus, infected bulbs serve as the main source of primary infection. The fungus keeps pace with the growing shoot and infects all parts of the plant, such as leaves, stem, inflorescence axis, floral stalks and flowers, but the roots are not affected. The lesions may completely girdle the pedicels and seed setting is adversely affected **(Fig.101).**

Secondary infection occurs through wind-borne sporangia. The sporangium deposited on the host leaf germinates by producing a germ tube and enters the host through the stoma. Lesions develop in about 14 days after successful infection. Initial symptoms appear as light-colored spots on the leaves, which turn yellow and then white. The lesions are somewhat sunken due to collapse of the tissues. Under humid and moist conditions, the spots increase in size and number and the green foliage presents a white, flecked appearance. The affected leaves lose their turgidity, collapse and shrivel, becoming very thin and gray. The mildew then appears on the affected spots to form a purplish growth. From the point of infection, the fungus travels between the cells, chiefly downwards and reaches the tissues of the bulb. In the bulb the fungus collects near the short stem and invades the developing leaves. The fungus may even invade the leaf primordia at the center of the stem. The withering of the foliage prevents normal development of the bulbs, which remain small and deformed. Their keeping quality is adversely affected and they become soft in storage and sprout prematurely.

The causal organism. The mycelium is intercellular and coenocytic. Haustoria are mostly simple, narrow tubes, but sometimes branched. The sporangiophores are non-septate, long, stout tubes that divide into four or more branches at the top. The branches terminate in a pair of short sterigmata, each bearing a lemon-shaped sporangium. The sporangia measure 43.5 - 69.0 x 22.5 - 30.0μ in size. After dispersal of the sporangia, a fresh crop of sporangia is formed. The sporangium germinates by issuing a germ tube and causes fresh infection. Thick-walled oospores are formed later in the intercellular spaces of tissues in the leaves, flowering shoots and rarely on the walls of capsules and sometimes adhering to the seeds. In the leaves, they are found in parallel rows mostly along the veins. They are colorless in the beginning, but turn red later on, round and measure 17 - 37μ in diameter **(Fig.101).**

Mode of survival, spread and epidemiology. The perennating mycelium in the bulbs is the chief source of carry over of the disease. The disease may also be propagated through oospores adhering to the testa of seeds. The oospores present in the crop residue in the soil may also cause primary infection. Sporangia, which are mostly wind-borne, cause secondary infection.

Under favorable conditions of high humidity, sporangia may be produced continuously. Day temperatures between 13° - 18°C and night temperatures between 5° - 13°C. are conducive for the development of sporangia. The optimum temperature for germination of sporangia is about 10°C. Prolonged rainy or damp, muggy weather conditions favour the spread of the disease. Sporangia are not formed when the leaves are dry. Dew is more important for the production of sporangia than actual rain.

Disease management

Agronomic practices (i) Disease-free bulbs should be used for planting (ii) At the time of planting and in the early stages of crop growth, plants showing any signs of the disease, such as softness of the bulbs or yellowing of the young leaves should be removed and destroyed.

Chemical control. Spraying with mancozeb - 600 gm. or Cuman L - 600 ml. or captafol (Difolatan or Difosan) - 375 gm. or metalaxyl (Ridomil or Acylon) - 300 gm. in 300 lit. of water per acre, 20 days after planting, followed by one or two more sprayings, at 10 - 12 days interval controls the disease.

Fig. 101: Downy mildew of onion-*Peronospora destructor*

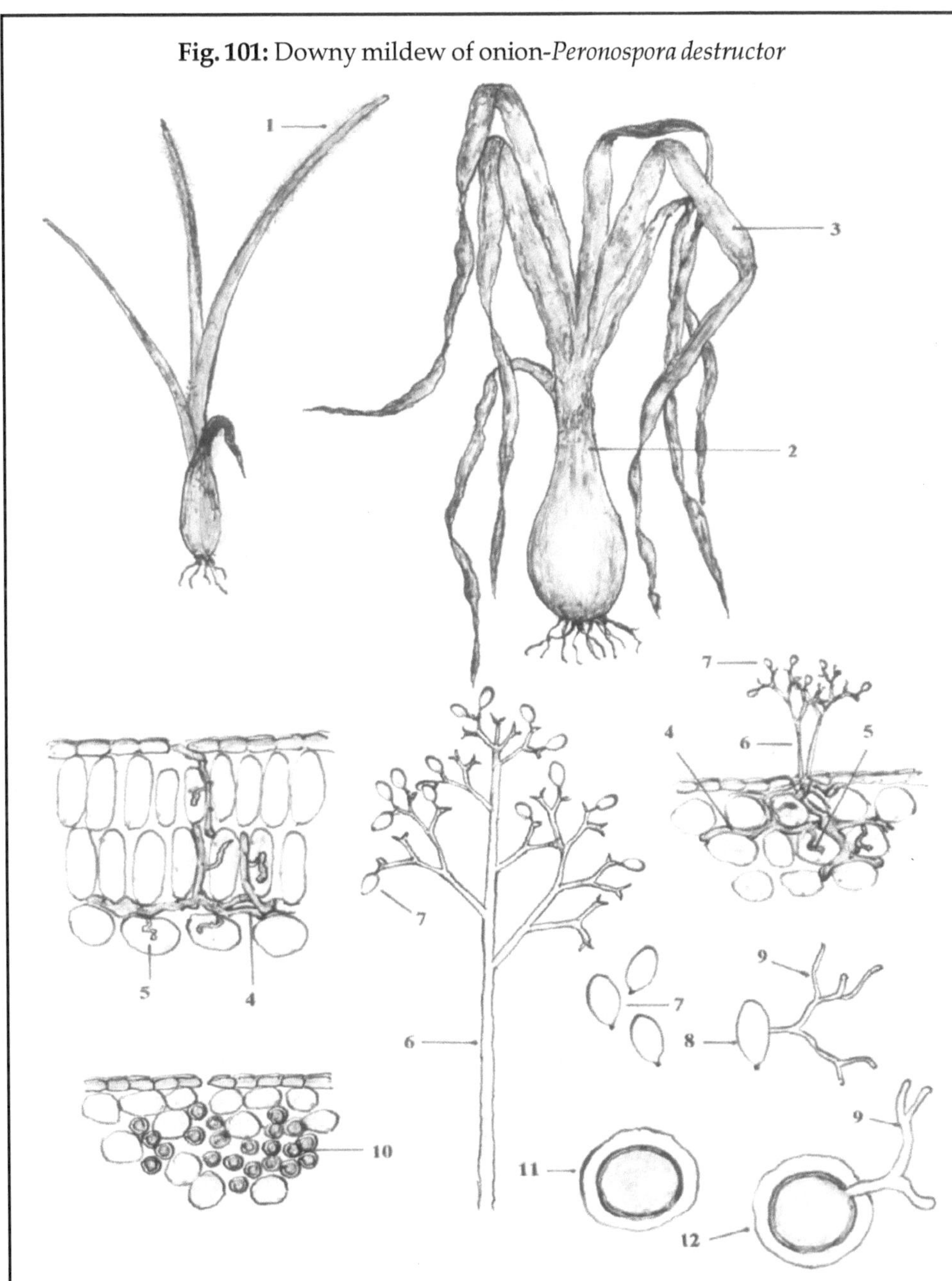

1. Downy mildew growth (fructifications) on leaves of young plant 2. Young plant destroyed by downy mildew 3. Lesions on the leaves 4. Intercellular coenocytic mycelium 5. Haustorium 6. Sporangiophere 7. Sporangia 8. Germinating sporangium 9. Mycelium 10. Oospores inside the leaf tissues 11. Oospore 12. Germinating oospore.

2. Smut of onion

Urocystis cepulae

The disease first occurred in America in 1860 and was later reported from France and Germany in 1879. Now, the disease is prevalent throughout Europe, Canada, New Zealand and India. In India, the disease is found in the states of Telungu Desam, Karnataka and Tamil Nadu. The disease attacks all kinds of onions, garlic, leeks and a few other crops.

Symptoms. The disease attacks only seedlings and young plants. Soon after emergence, seedlings show one or more long, silvery streaks below the epidermis of the cotyledon and sometimes on the succeeding leaves. The affected leaves are somewhat thicker and usually bend over. When the silvery epidermis over the blister-like lesions is broken, sooty, black masses of spores are released from the affected leaf tissues below. Infected, 3 - 5 weeks old young seedlings may die. If the seedlings survive longer, then the newly formed leaves are attacked and they develop stripes and blisters. The disease usually keeps pace with the growth of the surviving plants and develops silvery lesions and blisters on all the new leaves. Finally, smut blisters may be formed on the outer, as well as inner scales, as a result the bulbs become shrunken and reduced in size. The smutted bulbs do not rot in storage, but can be easily differentiated by the silvery stripes at the base of the scales. When the lesions break, the spores fall into the soil and remain in a viable state for several years **(Fig.102).**

The causal organism. The fungus produces black, powdery spore masses. Each spore is not merely a single cell, but a spore ball consisting of a true, fertile, dark spore surrounded by a close fitting covering of as many as 20 sterile, hyaline spores. Rarely two fertile spores are found within a spore ball. The spore balls measure 15 - 20µ in diameter, while the central fertile spores measure 11 - 15µ in diameter.

A true or brand spore germinates to produce a single (rarely two) hypha, which comes out between the sterile spores of the spore ball and forms a short, hemispherical tube, the promycelium or basidium. The basidium gives rise to many, septate and filamentous branches consisting of uninucleate cells. These filaments may either bear a number of lateral sporidia or they may branch to form a weft of mycelium without forming any sporidia. The sporidia, if formed may bud freely. The filamentous branches arising from the basidium may separate into individual cells, which again bud and function as sporidia. The mycelium or sporidia produced by the true spores exist in the soil as saprophytes for many years in the absence of susceptible hosts. In the presence of a susceptible host, the fungus becomes parasitic and attacks the emerging seedlings through the cotyledons or the collar region. The penetrating hypha bores its way directly into an epidermal cell at infection, passes between the mesophyll cells, establishes itself in the cotyledon and then makes its way into the growing point of the shoot, and ultimately infects the leaf primordium. From the time of infection, in about 3-4 weeks smut spores are formed. The haploid mycelium becomes diploid prior to spore formation through somatogamy and only from the binucleate mycelium, true smut spores are formed **(Fig.102).**

Mode of survival, spread and epidemiology. Onion smut is carried from place to place through contaminated soil and diseased plants. The disease is mainly distributed over long distances through infected bulbs or transplants. The spores in the soil may persist for many years. Smut spores may also be carried by wind, irrigation water, farm implements and cattle, and by some insects. Seeds do not carry the disease.

The optimum temperature range for smut development is 15°-20°C. The temperatures prevailing in the upper layer of the soil have an important bearing on the incidence of the disease. Soil temperatures of 20°-25°C favor disease occurrence.

Disease management

Agronomic practices (i) Infected bulbs and transplants should not be used for planting (ii) All infected crop debris should be collected from the field and burnt (iii) Long term rotation with crops not susceptible to this pathogen may be adopted (iv) Resistant varieties may be grown.

Seed treatment. Treating the seeds with a mixture of carboxin - 2 gm. + thiram or captan - 4 gm. / kg. of seed, prior to sowing protects the emerging seedlings from soil-borne infection.

3. White rot or dry rot of onion

Sclerotium cepivorum

The disease attacks onion, garlic, shallots and leeks. It is very common in regions of cool climate. The disease, which was first reported from England in 1841, is now prevalent in Italy, France, Syria, Holland, Germany, Egypt, Australia and India.

Symptoms. The disease attacks the hosts during the active growth stage and continues to attack the bulbs after harvest in storage. Early symptoms are noticed in older leaves, when they turn yellow from the tips downwards. The die-back may be severe particularly when the plants are young, resulting in complete wilting of all the leaves. The infection starts from the roots and then the bases of the bulb scales are attacked. In the later stages, the bulbs are directly infected by the mycelium. Due to extensive rotting of the root system, the diseased plants can easily be pulled out from the soil and the bulbs usually contain abundant white mycelium in the form of felt. On the mycelial weft, tiny, black, hard sclerotial bodies are found in great numbers and appear as a black crust **(Fig.103).**

The causal organism. The fungus is a soil-inhabiting organism. It does not produce any kind of spores except sclerotia (microsclerotia). The sclerotia measure 0.28 - 0.60 mm. in diameter. On disintegration of the diseased bulbs, the microsclerotia are liberated into the soil.

Mode of survival, spread and epidemiology. The pathogen survives in the soil in the form of minute, black sclerotia (microsclerotia). The microsclerotia remain viable in the soil for 8 - 10 years. They are highly resistant to adverse soil environment and are the principal means of carry over of the disease from one season to the next. The fungus has several strains.

The disease is favored by cool weather, low soil temperatures between 15° - 18°C and moderate soil moisture. The disease is almost absent at soil temperatures above 26°C and soil moisture content above 50 per cent.

Disease management

Agronomic practices (i) The diseased bulbs should be removed from the field along with the adherent soil and destroyed by burning, so as to prevent dispersal of the sclerotia into the soil (ii) Irrigation water should not be allowed to pass from infected fields to other fields (iii) Healthy planting materials should be used (iv) Wild onion *(Allium vineale)*, which is a carrier host of the pathogen should be destroyed (v) Application of organic manure reduces the disease occurrence, as it encourages the growth of antagonistic organisms present in the soil, which serve to destroy the sclerotia.

Seed treatment. Seed treatment with carbendazim or thiophanate methyl (Topsin M) at 2.0 gm./ kg. of seed provides certain amount of protection to the seedlings from soil-borne infection.

4. Bacterial soft rot of Onion

Erwinia carotovora pv. *carotovora*

The disease occurs in all countries, where onion is grown. The pathogen attacks the bulbs in the field, as well as in transit and storage.

Fig. 102: Smut of onion
Urocysits cepulae

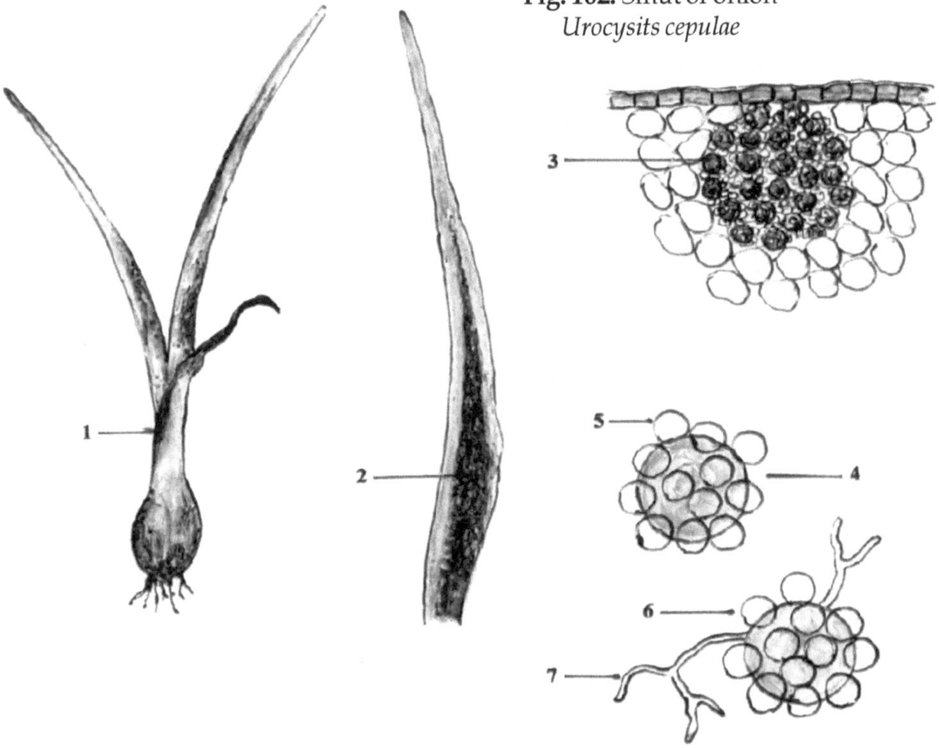

1. Smut lesions on the base of a seedling 2. Smut lesion on a leaf 3. Smut sori inside the leaf tissue 4. Smut spore ball 5. Sterile cells 6. Germinating fertile smut spore 7. Mycelium.

Fig. 103: White rot or Dry rot of Onion
Sclerotium cepivorum

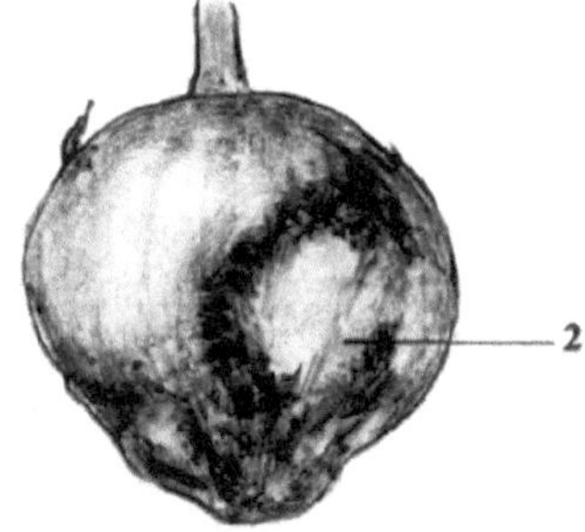

1. Sclerotium near the base of a bulb 2. Bulb showing basal rot

Symptoms. The initial symptoms appear as small, irregular, water-soaked lesions near the points of infection. The infection, which starts at the collar region or through senescent leaves, spreads rapidly and progresses downward into the bulb and very soon affect all the tissues of the bulb. The affected areas become soft and mushy. The outer scales become completely rotten and slimy ooze is seen on the surface. When the infected bulb is pressed slightly, a turbid, slimy liquid comes out. Bulbs infected by the soft rot bacteria gives off a repulsive odor.

The causal organism. The bacteria causing this disease are rod-shaped and measure 1.5 - 5.0 x 0.9µ in size. They occur, either singly or in pairs or in long chains and have peritrichous flagella numbering 2 - 8. They are non-spore forming.

Mode of survival, spread and epidemiology. The soil-inhabiting bacteria are facultative parasites and can survive in the soil up to 20 years in the diseased plant parts and in dead organic matter present in the soil as a saprophyte. When conditions are favorable for the bacteria and when suitable hosts are present they become parasitic and attack the hosts. They are primarily wound parasites and can enter the host only through injuries or bruises caused during cultural operations or while handling during transit or storage. Once the bacteria enter the host tissue, they multiply enormously within a short period. They produce certain enzymes, which dissolve the cell walls and the tissues become a mushy mass of disintegrated, rotten cells. They are spread by direct contact, through agricultural implements, soil, water and insects, and by hands.

The bacteria can grow and are active over a wide range of temperatures from 5° - 35°C and moisture levels. However, temperatures between 28° - 35°C and high soil moisture favor the occurrence and spread of the disease. Besides onion, the pathogen can attack several other crops, such as carrot, turnip, radish, potatoes etc.

Disease management

Agronomic practices (i) The pathogen is a typical wound parasite. So, precautionary measures should be taken to avoid causing possible damages, such as injuries or bruises during cultural operations or while handling during transit or storage (ii) Diseased bulbs should be sorted out at the time of harvest and before storing (iii) Affected bulbs should not be allowed to be mixed with sound bulbs (iv) Diseased plant debris should be removed from the field and destroyed (v) Irrigation water should not be allowed to flow from diseased fields to other fields (vi) The store-houses should be kept clean, moisture-free and well-ventilated (vii) The bulbs should be sufficiently dry before being stored (viii) It is better to store the bulbs in air conditioned ware-houses.

5. Onion yellow dwarf

Onion yellow dwarf virus

The disease, which was first reported from Virginia in 1916, occurs in several countries of the world including Argentina, Austria, Australia, Canada, Czechoslovakia, Denmark, England, New Zealand, Russia, the United States of America and India. The disease may cause severe damage to the crop and appreciable loss in yield during some seasons.

Symptoms. The symptoms first appear as irregular, yellow streaks, mostly on the upper leaves, which become completely yellow. The leaves are crinkled, flattened and they curl downwards. Eventually the leaves collapse and die. The flower stalks of infected plants show yellow stripes and are twisted and dwarfed. Flower clusters become small, with fewer number of flowers. Seed formation is considerably reduced. The bulbs are under-developed and small. The affected plants are very much stunted in growth.

The causal organism. The virus causing the disease belongs to the Genus - *Potyvirus.* The virus particles are flexuous rods, measuring 680 - 900 nm in length and 11 nm in width. The genome consists of a single ssRNA molecule.

Mode of spread. The virus is transmitted through sap, pollen and seed, as well as by insect vectors. The efficient vectors are *Aphis fabae, Rhopalosiphum maidis* and *R. padi.* It is also transmitted by soil-inhabiting nematodes, mites and through sap adhering to knives.

Disease management

Agronomic practices (i) Disease-free seed stock should be used for planting (ii) Affected plants may be rouged out and destroyed.

Chemical control. The insect vectors may be controlled by the use of systemic insecticides, such as methyl demeton, dimethoate, phosphamidon or monocrotophos.

Diseases of minor importance. In addition to the above-mentioned diseases, many other diseases attack onion. 'Blast' and 'Neck rot', caused by *Botrytis allii, B. squamosa* and *B. hyssoidea* attack the leaves and bulbs; *Alternaria porri* causes 'Leaf scald disease'; 'Basal stem rot' or 'Root rot' or 'Wilt' is caused by *Fusarium oxysporum* f.sp. *cepae*; *Pythium* *aphanidermatum*, *P. de baryanum* and *P. ultimum* cause 'pre-emergence and post-emergence damping off' and 'root rot' in the later stages of crop growth; 'Black mould' is caused by *Aspergillus niger* and is very destructive in storage; *Macrophomina phaseolina* causes 'Bulb rot' in storage.

Garlic *(Allium sativum)*

Most of the diseases that occur on onion, occur on garlic also. 'Downy mildew of onion' caused by *Peronospora destructor* attacks garlic leaves; 'Smut of onion' caused by *Urocystis cepulae* affects the leaves and bulbs of garlic; 'White rot' or 'Dry rot of onion' caused by *Sclerotium cepivorum* affects the bulbs; *Botrytis allii* and *Fusarium oxysporum* causes 'Neck rot' and 'Bulb rot'; *Macrophomina phaseolina* causes 'Bulb rot'; *Aspergillus niger, A. repens* and *A. alliaceous* cause 'Black mould' of bulbs.

Pepper *(Piper nigrum)*

1. Foot rot and quick wilt of pepper

Phytophthora capsici

The disease was reported from Kerala in 1902 and now, it is common in all the major pepper growing regions. It causes severe damage to pepper crop in Kerala and Tamil Nadu.

Symptoms. The disease affects all the parts of the plants. The pathogen causes local lesions on the leaves, spikes and berries. On the leaves, large, brownish, water-soaked, circular to irregularly circular spots, with fimbriate margins are produced, starting from the leaf margins or in the center of the leaves. The spots enlarge rapidly under favorable conditions and become big patches covering large areas of the leaves. Often, concentric, dark-brown and light-brown zonations are seen in the spots. The under surface of the spots may be covered with a delicate, downy growth of the fungus. Severely affected leaves turn yellow, wilt and fall off prematurely. Small, oval, dark-brown to black, slightly sunken spots develop on the spikes and the fruits.

The aerial branches of the vines may be infected at any point. The branches at the points of infection become dark-brown and often the branches are girdled. At these points, rotting sets in, which extends both upwards and downwards, as a result the branches wilt and die from the tips downwards. This is referred to as the **'die-back'** phase of the disease.

The basal part of the stem at the ground level is also affected leading to rotting of the collar or foot region. The affected collar region becomes dark colored and water-soaked. This is known as the **'foot rot'** phase and is the most destructive phase of the disease, as the infected vines are killed within 2 - 3 weeks. In a single vine, one or more runner shoots may be affected and die, while the others remain healthy, leading to partial wilting and death. Sometimes, only one side of the stem alone is affected and in that case the branches arising from that side of the infected stem alone may wilt and die. The rotten tissues within the affected region emit a foul odor. From the collar region, the infection may spread down to the underground stem and may go still further into the roots

Usually the infection starts from the feeder roots and then spreads to the lateral roots and finally to the main roots. The main roots may be infected directly also. When the root system is thus affected, the plant may wilt suddenly even in the absence of collar infection. This is referred to as **'quick wilt'**. Actually the wilting of plants is caused due to the rotting and disintegration of the tissues of the root system and the collar region. The vascular elements are not affected as in the case of *Fusarium* wilt. The extent of damage to the vine or a part of the vine depends upon the extent of rotting of the roots or the collar region.

The causal organism. The causal organism, *Phytophthora capsici* is a **'morphological form' (MF-4)** of *Phytophthora palmivora*. The mycelium of the fungus is coenocytic, hyaline and intercellular. The hyphae send haustoria into the cells. Sporangiophores arise from the endophytic hyphae, which are slightly different from the normal hyphae. The sporangiophores may be simple or may have one or few branches that bear terminal sporangia singly. The sporangia are inverted pear-shaped, rarely spherical, hyaline, thin-walled with an apical papilla, tapering slightly towards the stalk, deciduous, pedicellate and are 38 - 72 x 33 - 42µ in size. They germinate in water and release large zoospores through the papilla. The zoospores after swimming in free water for a short while, come to rest and encyst. The encysted zoospores measure 8.0 - 10.0µ in diameter. The fungus in its sexual phase produces oospores in the affected tissues. They are spherical, thick-walled and are 35 - 45µ in diameter. They germinate by producing secondary zoosporangia at the end of a short stalk. The fungus also produces thick-walled, terminal or intercalary chlamydospores, measuring 30 - 55µ in diameter.

Mode of survival, spread and epidemiology. Pepper being a perennial plant, the pathogen may continue to survive in the live plant tissues all through the year. Infected plant debris in the soil in the plantation serves as the main source of inoculum. However, the fungus cannot survive in the soil for a long time saprophytically, as it is not capable of competing with the other soil microorganisms. The oospores and chlamydospores found inside the dead tissues in the soil may also perpetuate the disease. When moist and wet conditions, as well as low temperatures prevail, the pathogen becomes active and produces sporangia and zoospores, which are carried to the upper leaves, one above the other through rainsplash and the disease spreads in a ladder-like manner. Secondary spread of the disease occurs through soil and water.

Generally, heavy rainfall, damp and cloudy weather conditions, short day periods and low light intensity favor the disease occurrence. In areas where the annual rainfall is above 300 cm., the incidence of the disease is more. High relative humidity of 85 - 99 %, low temperatures of 20° - 24°C and shorter sunshine hours of 2.8 - 3.5 hours per day are highly conducive for the disease occurrence and development. *Phytophthora capsici* attacks *Piper attenuatum, P. betle, P. longum* and *Areca catechu*. It also attacks leaves of rubber, castor, pods of cocoa, capsules of cardamom, fruits of citrus and arecanut.

Disease management

Agronomic practices (i) Foot rot is mainly a soil-borne disease, So movement of soil from infected gardens to disease-free gardens should be avoided (ii) Nursery stocks raised in disease-free nurseries under protected conditions should be used for planting in disease-free gardens (iii) Seedlings raised in infected fields should not be used for planting in new gardens, as the roots may carry contaminated soil (iv) Diseased plant debris should be removed and destroyed (v) Wilted shoots and branches should be cut off and destroyed. Afterwards Bordeaux paste should be applied to the cut ends. If the vine is completely affected, the entire plant is uprooted and destroyed and the pathogen present in the soil in the pit should be killed by burning trash (vi) The vines should be trained properly, so that the leaves may not come in contact with the soil (vii) Dense growth of live standard trees should be trimmed before the onset of the monsoons to let in more sunlight and air to reach the lower region of the plants. This will help to reduce build-up of humidity near the base of the plants (viii) Adequate drainage facilities should be provided to avoid water stagnation (ix) Care should be taken to avoid causing injuries to the root system and the collar region during cultural operations (x) Organic amendment with neem cake at 1 - 2 kg./ vine, once in 6 months during the monsoon periods, helps in suppressing the growth of the pathogen by encouraging the multiplication of soil-inhabiting antagonistic microorganisms (xi) Plantings should be avoided during periods of heavy rains (xii) Application of excessive doses of nitrogenous fertilizer should be avoided.

Chemical control (i) Soil drenching around the stem of vines with Bordeaux mixture - 1 % or copper oxychloride - 2.5 gm./ lit. of water at the rate of 3 - 4 lit./ vine, 4 times, at monthly intervals and foliar spraying with the same chemicals at the same concentrations, 8 times, at fortnightly intervals starting from the time just prior to onset of the South-west monsoon controls the disease (ii) Soil drenching with metalaxyl - 1.0 gm. or chlorothalonil - 1.25 gm./ lit. of water is also effective in controlling the disease (iii) Foliar spraying with metalaxyl - 1.0 gm. or Fosetyl-Al - 1.0 gm./ lit. of water, 3 times at monthly intervals also gives good control.

Biological control. Dipping the transplants in a spore suspension of *Trichoderma viride* before planting or drenching the soil with the spore suspension is found to be very effective in reducing the soil inoculum.

2. Anthracnose of pepper

Colletotrichum gloeosporioides

(Glomerella cingulata)

Symptoms. The disease affects all aboveground parts of the plants. On the leaves, many, small, circular or irregularly circular, brown spots appear on the upper surface. Under humid conditions, the spots enlarge and form irregular, gray, necrotic areas. Concentric rings of acervuli appear on the necrotic lesions. Eventually, the tissues in the necrotic areas dry up and break. Young leaves are more prone to attack by the disease and the affected leaves become crinkled. Severely affected leaves may fall off. On the stem, the fungus infects the tip region and black, sunken, linear, necrotic lesions appear, which may extend down to some distance. The lesions may spread and girdle the affected area and in due course the branch starts drying from the tip downwards. On the dried, dead areas, black colored acervuli can be seen. On old vines, the infection starts at the place where the branches arise and the branches above the point of infection may dry and die. The pathogen also attacks the spikes and the berries and causes direct loss in yield. The infection starts at the leaf axil and spreads to the spikes, as a result the tissues rot and the spikes dry, die and fall down. Berries in the fallen spikes are ill filled. Individual berries may also be attacked. On the berries, dirty-brown, sunken lesions develop from the upper portion and progress downwards. The affected berries, which dry up, are ill filled and chaffy. On the dried spikes and berries, acervuli are formed. The pathogen attacks many other crops including mango, cashew, yam etc.

The causal organism, mode of survival, spread, epidemiology and disease management. Refer 'Anthracnose of mango' (Page 283), **(Fig.73).**

Minor diseases. Pepper is subjected to attack by a few more diseases. *Sclerotium rolfsii* causes 'Basal wilt'; *Ganoderma lucidum* causes 'Red root rot'; *Cercospora piperina* causes '*Cercospora* leaf spot'; *Xanthomonas campestris* pv. *vesicatoria* causes 'Bacterial leaf spot'; *Cucumber mosaic virus, Green mottle virus* and *Ring spot virus* cause 'virus diseases'.

Fenugreek *(Trigonella foenum - graecum)*

1. Powdery mildew of fenugreek

Erysiphe polygoni and *Leveillula taurica*

Symptoms, causal organism, mode of survival, spread, epidemiology and disease management. *Erysiphe polygoni* - Refer 'Powdery mildew of pea' (Page 131), **(Fig.36).** *Leveillula taurica* - Refer 'Powdery mildew of pigeon pea' (Page 123).

Diseases of minor importance. A few other diseases, which are not of much economic importance also attack fenugreek occasionally. *Rhizoctonia solani* causes 'Collar rot'; *Cercospora traversiana* causes 'Leaf spot'; *Alternaria alternata* causes '*Alternaria* root rot'; *Peronospora trigonella (P. trifoliorum)* causes 'Downy mildew'.

Ginger *(Zingiber officinale)*

1. Rhizome rot or soft rot of ginger

Pythium species

The disease, which was first reported from India in 1907, is prevalent in all ginger growing states, such as Karnataka, Kerala, Tamil Nadu, Gujarat and West Bengal. The disease is more severe in parts of South India. In some of the low-lying and water-logged areas total loss occurs, especially when the crop is infected at the early stages of crop growth. The disease is not only confined to the crop in the field, but the rhizomes are also attacked during storage and may cause considerable damage.

Symptoms. The leaves of infected plants become pale green in color initially. Yellowing of the leaves starts from below and spreads upwards. Eventually, the leaves wither and die. The basal portion of the plant becomes watery and soft. The rhizomes start rotting and become a decaying, pulpy mass of tissues covered by a fairly tough rind. But the fibrovascular strands are not affected. The roots are also affected and they become rotten. The rotten rhizomes emit a putrefying odor. In case of systemic infection from the rhizomes, the young shoots show damping off symptoms. If infection occurs at a later stage, the pseudostem withers and dies, which results in a poor crop **(Fig. 104).**

The causal organism. Several species of *Pythium* viz., *Pythium aphanidermatum, P. ultimum, P. myriotylum, P. butleri, P. graminicola, P. deiense, P. pleroticum, P. vexans, P. monospermum* and *P. gracile* have been reported to be associated with the disease. However, the pathogenicity of *P. aphanidermatum, P. ultimum* and *P. myriotylum* has been proved. The morphology of *P. aphanidermatum* and *P. ultimum* have been described under damping off of seedlings of eggplant (Page 339) and that of *P. myriotylum* under damping off of seedlings of tobacco (Page 241).

Mode of survival, spread and epidemiology. The disease is both soil-borne and seed-borne. The decaying diseased plant tissues lying in the soil, harbor the pathogens, which carry on a saprophytic life for a long time and cause infection in the succeeding crop. The oospores of the fungi remain in a viable state in the soil for a long time and perpetuate the disease. The disease is also carried through infected rhizomes used for seed. The rhizomes may contain the mycelium, as well as the spores of the fungus. The disease is carried over long distances in ginger rhizomes. Secondary spread is through water and soil carried from field to field by cattle, agricultural implements etc.

High soil moisture and temperatures ranging from 28° - 35°C favor the occurrence and spread of the disease. Water-logging predisposes the plants to infection. Presence of abundant decomposing organic matter in the soil is another factor that contributes to the severity of the disease.

Disease management

Agronomic practices. Use of disease-free rhizomes for planting, removal and destruction of diseased plant debris from the fields, avoiding low-lying, water-logged areas for raising ginger crop, providing adequate drainage facilities, crop rotation, growing of resistant varieties etc., go a long way in controlling the disease outbreak.

Chemical treatment (i) Seed treatment in a fungicidal solution of copper oxychloride - 2.5 gm. or mancozeb - 2.0 gm./ lit. of water for 30 minutes helps in eradicating externally seed-borne inoculum. However, deep-seated mycelium and spores inside the rhizomes cannot be eliminated by this method (ii) Treating the rhizomes in metalaxyl (Ridomil) - 1.0 gm. or chlorothalonil - 1.5 gm./ lit. of water is more effective in controlling seed-borne inoculum (iii) Drenching the soil with difolatan - 1.5 gm. or copper oxychloride - 2.5

gm./ lit of water effectively checks the disease. A quantity of 800 - 1000 liters of fungicidal fluid may be required to drench one acre of field.

Heat therapy. Treating the seed rhizomes in aerated steam at 40°C for 1 hour eliminates both externally and internally seed-borne inoculum.

2. Leaf spot of ginger

Colletotrichum zingiberis
(Glomerella cingulata)

The disease was first reported in India from Telungu Desam in 1923. Now it is widely prevalent in all the ginger growing regions of India.

Symptoms. Initial symptoms appear as small, circular to oval, pale yellow spots. The spots enlarge more along the length of the leaves and turn brown. The spots may coalesce to form irregular, brown patches. In the later stages, the central portion of the spots may become necrotic, the tissues crack, dry and fall off leaving ragged holes. Severely affected leaves turn yellow completely, dry and die. Linear, brown lesions are also formed on the leaf sheaths and scales. Minute, dark colored, pinpoint-like dots representing the acervuli of the fungus are formed on the spots in an irregularly concentric manner.

The causal organism. The mycelium of the fungus is septate, hyaline, inter- and intracellular. The acervuli, which are formed from the sub-epidermal stromata are inverted saucer-shaped and are 50 - 140μ in diameter. Conidia are borne singly at the tip of short, minute, hyaline conidiophores, which are formed at the base of the acervuli. The conidia are produced in succession in large numbers and they are sub-fusoid, slightly curved with blunt or rounded ends, hyaline and single-celled. Numerous, sterile setae are found interspersed among the conidiophores. They are long, erect, narrowing towards the tip, multiseptate, dark- brown in color and are 85 - 170μ in length. After the maturity of the spores, the host epidermis is ruptured and the spores are exposed. The perfect stage of the fungus is very rarely found in nature.

Mode of survival, spread and epidemiology. The disease is primarily air-borne. The fungus survives in the diseased plant debris in the soil for long periods and continues to produce conidia. Secondary spread of the disease is through air-borne and water-borne conidia.

The disease incidence is more during the monsoon seasons, when there is continuous rains for several days. Moist and damp weather conditions and moderate temperatures are conducive for the occurrence and development of the disease.

Disease management

Agronomic practices. Diseased crop debris in the field should be collected and destroyed.

Chemical control. Spraying the crop with Bordeaux mixture - 1 % or copper oxychloride - 750 gm. or ziram - 600 gm. or zineb - 600 gm. or mancozeb - 600 gm. in 300 lit. of water / acre is effective in controlling the disease. The first spraying should be given, as soon as the symptoms of the disease are noticed, followed by 2 or 3 more sprayings at fortnightly intervals.

3. *Phyllosticta* leaf spot of ginger

Phyllosticta zingiberi

The disease was first reported in India from Telungu Desam in 1942. It is prevalent in most of the ginger growing states of Kerala, Karnataka, Telungu Desam, Tamil Nadu, West Bengal, Bihar, Himachal Pradesh, Uttar Pradesh and Maharashtra.

Symptoms. The disease appears as oval to irregular, water-soaked spots on the leaves in large numbers. The leaves become whitish and papery. Numerous, black, pinpoint-like dots, representing the pycnidia of the fungus are formed on the upper surface of these leaves **(Fig.105).**

The causal organism. The pycnidia produced by the fungus are small, dark-colored, globose, leathery, ostiolate and are sunken in the leaf tissue. They measure 94µ in diameter. Pycnidiospores are borne at the tip of very short, hyaline phialides, lined on the inside of the pycnidia. The pycnidiospores are produced in succession in large numbers. They are hyaline, single-celled, oblong, with rounded ends and measure 4.4 x 1.8µ in size on an average **(Fig.105).**

Mode of survival, spread and epidemiology. The disease is primarily air-borne. The fungus can survive in the diseased plant debris in the soil for more than 12 months and continues to produce pycnidiospores. Secondary spread of the disease is through pycnidiospores carried by wind and rain splash.

Moist, humid weather conditions and moderate temperatures favor the disease occurrence and development. The disease incidence is more during the monsoon periods.

Disease management

Agronomic practices. Diseased crop debris should be collected from the field and destroyed.

Chemical control. Spraying the crop with Bordeaux mixture - 1 % or copper oxychloride - 750 gm. or mancozeb - 600 gm. in 300 lit. of water / acre controls the disease.

4. Bacterial wilt of ginger

Pseudomonas (Ralstonia) solanacearum

The disease occurs in Jamaica, Nigeria, Brazil, Indonesia, the Philippines and other tropical and sub-tropical countries of the world. In India, it is found in the states of Bihar, Himachal Pradesh, Kerala and Tamil Nadu. The disease was first reported in Tamil Nadu in 1914.

Symptoms. The disease attacks all the parts of the plants. The characteristic symptoms of the disease are, sudden wilting and death of the affected plants. The aboveground symptoms appear as linear, dark, water-soaked streaks or patches on the collar region of the pseudostem, followed by yellowing, curling, drooping and wilting starting from the upper leaves. The leaf sheaths and ligules also become yellowish. The pseudostem comes off easily when pulled gently. Eventually, the affected plants die within a short time. If such a wilted plant is uprooted, the rhizome shows translucent, water-soaked patches, while the inner tissues are dark-colored and show varying degrees of rotting. The outer skin of the rhizome appears hard, but the inner tissues are soft and pulpy and when slightly pressed, bacterial ooze comes out as a milky slime. The affected pseudostems show dark vascular streaks and bacterial ooze comes out from the cut ends also.

The causal organism, epidemiology and disease management. Refer 'Bacterial brown rot and wilt of potato' **(Page 396).**

Diseases of minor importance. Besides the above-mentioned diseases, a few more diseases also attack ginger. *Sclerotium rolfsii* causes '*Sclerotium* rot'; *Pyricularia zingiberi* causes '*Pyricularia* leaf spot'; several fungi belonging to different species of fungi cause 'Storage rot'; *Wheat streak mosaic virus* causes 'Mosaic'

Coriander *(Coriandrum sativum)*

1. Powdery mildew of coriander

Erysiphe polygoni

The disease is found in many countries, where coriander is cultivated. In India, it is prevalent in the states of Telungu Desam, Gujarat, Rajasthan and Tamil Nadu. When conditions are favorable for the disease, it may cause serious damage to the crop and loss in yield.

Symptoms. The disease attacks the foliage and causes severe damage to the crop. Usually, it appears in the later stages of crop growth and becomes serious when the plants are in the flowering and seeding stages. Initially, small, white, circular to irregular, powdery patches appear on both the surfaces of the leaves. The disease soon spreads to all green parts of the plants. The powdery patches enlarge rapidly and often cover the entire area of the leaves, young branches, petioles and the umbels. The plants appear as if covered by a grayish-white, powdery coating from a distance. The leaves are also reduced in size and are distorted. Eventually, the leaves turn yellow, wither and dry prematurely. Sometimes, the seeds are also attacked. The whitish powdery coating consists of conidiophores and conidia produced in profusion from the superficial mycelium of the fungus. Later in the season, numerous, small, black, pinhead-like cleistothecia are formed, which are embedded in the mycelial weft. Due to excessive transpiration of the plants due to the fungal invasion, increase in the rate of respiration and reduction in the photosynthetic activity, the infected plants become very much stunted

The causal organism, mode of survival, spread, epidemiology and disease management. Refer 'Powdery mildew of pea' (Page 133), **(Fig.36).**

2. Stem gall of coriander

Protomyces macrosporus

It is a common and widespread disease of coriander and is found in many coriander growing regions of India, especially in the states of Uttar Pradesh, Telungu Desam and Tamil Nadu.

Symptoms. The symptoms appear as tumor-like swellings or galls on the leaf veins, leaf stalks, peduncles, stems and on fruits. The galls on the veins give a swollen, hanging appearance to the leaves. The tumors are elongated and may be 10.0 - 13.0 x 3.0 - 4.0 mm. in size. They look glossy in the beginning, but later rupture and become rough and appear corky. When the stems remain succulent for longer periods due to excessive soil moisture and particularly under shaded conditions, numerous tumors may be formed on them. In case of severe attack, the plants may be killed prematurely **(Fig.106).**

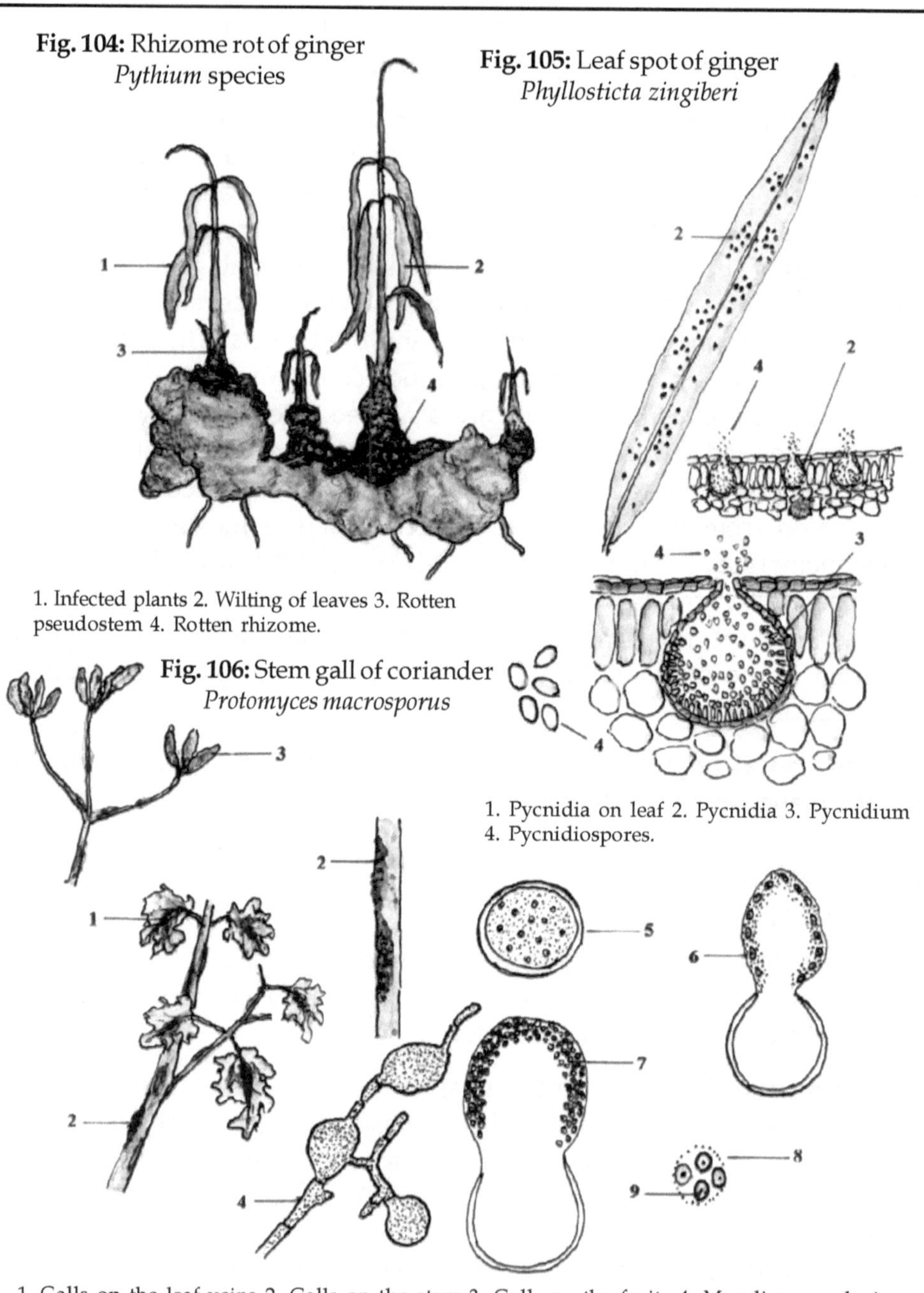

Fig. 104: Rhizome rot of ginger
Pythium species

1. Infected plants 2. Wilting of leaves 3. Rotten pseudostem 4. Rotten rhizome.

Fig. 105: Leaf spot of ginger
Phyllosticta zingiberi

1. Pycnidia on leaf 2. Pycnidia 3. Pycnidium 4. Pycnidiospores.

Fig. 106: Stem gall of coriander
Protomyces macrosporus

1. Galls on the leaf veins 2. Galls on the stem 3. Galls on the fruits 4. Mycelium producing intercalary and terminal resting spores 5. Resting spore 6. Protoplasm segmenting to form uninucleate spore mother cells inside the sac 7. Mature endospores 8. Ascospores inside a wall-less ascus 9. Ascospores.

The causal organism. The pathogen is found only in the tumours. The hyphae are intercellular, closely septate, branches irregularly, quite broad and are diploid. Some of the cells in the hyphae swell and form elliptical or globose bodies, that develop into chlamydospores. The chlamydospores or resting spores are multinucleate and the nuclei are diploid. They are surrounded by a thick, hyaline exospore and they measure 50 - 60µ in diameter. A resting spore germinates in water and the inner wall is pushed through the outer wall and forms an elongated, more or less cylindrical sac or vesicle. The protoplasm from the spore passes into the sac and the multinucleate protoplasm is displaced to the periphery of the sac. Each of the diploid nucleus develops into an ascus. Meiotic divisions of the nuclei take place in the sac. The four haploid nuclei formed, following meiotic divisions of each of the diploid nucleus become four ascospores inside a wall-less ascus. Thus, the large number of wall-less asci produced in the sac, makes it a **'compound spore sac'** or **'synascus'**. On maturity, the ascospores separate and are forcibly ejected from the spore sac in a single mass. After liberation, the uninucleate, haploid ascospores reproduce further by budding and they may conjugate in pairs. The conjugation spore (zygote) germinates and forms a diploid mycelium that penetrates into the host by producing a germ tube and causes fresh infection **(Fig.106).**

Mode of survival, spread and epidemiology. The chlamydospores present in the plant debris in the soil can survive in the soil for a number of years, but their survival capacity decreases with age. The disease is also perpetuated through chlamydospores present in the seeds.

High soil moisture, low night temperature and dewfall favor the disease occurrence. The disease severity is more in shady places.

Disease management

Agronomic practices. Use of disease-free, healthy seeds, field sanitation, provision of adequate drainage facilities and crop rotation helps to reduce the inoculum potential.

Seed treatment. Treating the seeds with thiram at 4.0 gm./ kg. of seed, atleast 24 hours prior to sowing is effective in eradicating the external seed-borne inoculum.

Chemical treatment. Spraying the crop with thiram - 600 gm. or captafol - 450 gm. or carboxin - 300 gm. in 300 lit. of water / acre has been found to be effective in controlling the disease.

Resistant varieties. Delhi Local, Kota Local, NP-92, NP-95, P-107, Rajendra Swathi, Shujalpur Local and Synthetic North have been found to be resistant to the disease to some extent.

Minor diseases. A few other diseases also attack coriander. *Fusarium oxysporum* f.sp. *corianderii* causes 'Wilt'; *Curvularia pallescens* causes 'Root rot'; *Glomerella cingulata (=Colletotrichum gloeosporioides)* causes 'Anthracnose'; *Sclerotinia sclerotiorum* causes 'Stem rot'.

Turmeric *(Curcuma longa / C. domestica)*

1. Rhizome and root rot of turmeric

Pythium aphanidermatum and *P. graminicolum*

The disease is found in Telungu Desam, Tamil Nadu, Karnataka and Kerala. In Tamil Nadu, it is widely distributed in Coimbatore, Erode, Salem and Trichirapalli districts.

Symptoms. Initially, the leaves of infected plants turn yellow and start drying along the margins. Eventually, the leaves dry completely and die. The basal portion of the shoot becomes watery and soft. The root system is almost completely affected. It is very much reduced and the roots rot. As the disease progresses the rhizomes are affected and they rot and turn into a decaying mass of soft, pulpy tissues. The affected areas turn dark-brown to black in color. The development of rhizomes is badly affected. Usually, the disease appears in patches in the field.

The causal organism. The disease is caused by *Pythium* aphanidermatum and *P. graminicolum*. The morphology of *P. aphanidermatum* has been described under 'Damping off of seedlings of eggplant' (Page 339).

The mycelium of *P. graminicolum* is hyaline, coenocytic, granular, irregularly branched, inter- and intracellular. Sporangia are actually enlarged hyphae and are sometimes irregularly lobed. They may be terminal or intercalary. Each sporangium contains 15 - 48 zoospores, which are reniform, with 2 lateral flagella. They swarm about in free water for a short while, come to rest and encyst. The encysted zoospores measure 8.0 - 11.0μ in diameter. Oospores are plerotic, spherical, single, smooth, thick-walled, hyaline to light-brown and measure 15.4 - 35.3μ in diameter.

Mode of survival, spread, epidemiology and disease management. Refer 'Rhizome and root rot of ginger' (Page 436).

2. Leaf spot of turmeric

Taphrina maculans

This is quite a common disease of turmeric and is found in almost all the turmeric growing regions. It was first reported in India in 1911. It is widely prevalent in the states of Telungu Desam, Bihar, Gujarat, Maharashtra, Orissa, Uttar Pradesh, West Bengal and Tamil Nadu.

Symptoms. The disease usually attacks all the plants in a field and all the leaves in a plant are affected. The symptoms appear as numerous, circular, light-yellow to dark-yellow spots, 1.0 - 2.0 mm. in diameter on both the surfaces of leaves, but are more prominent on the upper surface. The affected leaves appear reddish-brown instead of the normal green color. The spots may coalesce to form large, irregular patches and soon the entire leaf turns yellow and dries. Excessive spotting on the leaves destroys the green tissues of the leaf, as a result the plant becomes weak and the yield is reduced drastically. However, the affected plants are not killed. The pathogen does not induce hypertrophy and hyperplasia of the tissues and so there is no deformation, wrinkling or curling of the leaves as in the case of peach leaf curl, caused by *Taphrina deformans* **(Fig.107).**

The causal organism. The mycelium of the fungus develops in-between the cuticle and epidermis or inside the epidermal cells. Sometimes, the hyphae may grow down intercellularly to the next layer beneath the epidermis, but do not penetrate deeper. Branched or lobed haustoria are sent into the cells. In the fully developed spots, a more or less continuous layer of hyphae occupies the central portions. The outermost cells of this layer develop into cylindrical or club-shaped, thin-walled projections, which rupture the host epidermis and become asci. All the outer cells are ascigerous, but the asci mature at different intervals and the matured ones are found in small groups. Basal cells are present below each ascus. Asci are produced on both surfaces of leaves. The asci measure 20.0 - 30.0 x 6.0 - 10.0μ. Each ascus usually contains 8 ascospores. They are ovoid, hyaline, single-celled and are 4.0 - 7.0 x 2.0 - 3.0μ in size. The ascospores multiply within the ascus by budding and produce many sprout conidia. When the ascus wall is ruptured, the spores are released. They fall on the host surface and germinate by producing a germ tube and cause fresh infection **(Fig.107).**

Mode of survival, spread and epidemiology. The asci present in the dried, diseased leaves in the soil may perpetuate the disease from one season to the next. Secondary spread may occur through wind-borne ascospores or sprout conidia.

The incidence of the disease is more during the monsoon periods. A few other species of *Curcuma* and *Zingiber* are also attacked by the pathogen and may serve as alternate hosts.

Disease management

Agronomic practices (i) Removal and destruction of diseased plant debris from the field may help to reduce the inoculum potential (ii) Resistant varieties may be grown.

Chemical control. Spraying the crop with Bordeaux mixture - 1 % or copper oxychloride - 750 gm. or mancozeb - 600 gm. or zineb - 600 gm. in 300 lit. of water/ acre is effective in reducing the disease incidence.

3. *Colletotrichum* leaf spot of turmeric

Colletotrichum capsici

The disease is one of the most destructive one among the diseases of turmeric and causes considerable loss in yield of rhizomes. It is found in many countries of the world, including India and Sri Lanka. The disease was first reported in Tamil Nadu in 1917. In Tamil Nadu of India, it is widely prevalent in almost all the turmeric growing districts, such as Coimbatore, Tiruchirapalli, Salem and North Arcot.

Symptoms. The disease attacks plants at all stages of growth. The symptoms appear as small, elliptical or oblong, brown-colored spots of varying sizes on the upper surface of leaves, but may be formed on the under surface also. The spots enlarge gradually and may reach 2.0 - 3.0 cm. in length and about 1.0 cm. in width. The spots are marked with light-brown and dark-brown zonations. Two or more spots may coalesce to from irregular patches and may cover large areas of the leaf lamina. As the disease progresses, the center of the spots become grayish-white and thin, and numerous, black, pinpoint-like dots, which are the acervuli of the fungus are produced in concentric circles. Around the spots, yellow halo is formed, which is due to the production of certain toxins by the pathogen. Severely affected leaves turn completely yellow, dry and die. Spots may be formed on the leaf sheaths. The scales covering the rhizomes are also attacked and black, necrotic spots or patches may be formed **(Fig.108).**

The causal organism, mode of survival, spread, epidemiology and disease management. Refer 'Die-back and fruit rot of chillies' (Page 419)

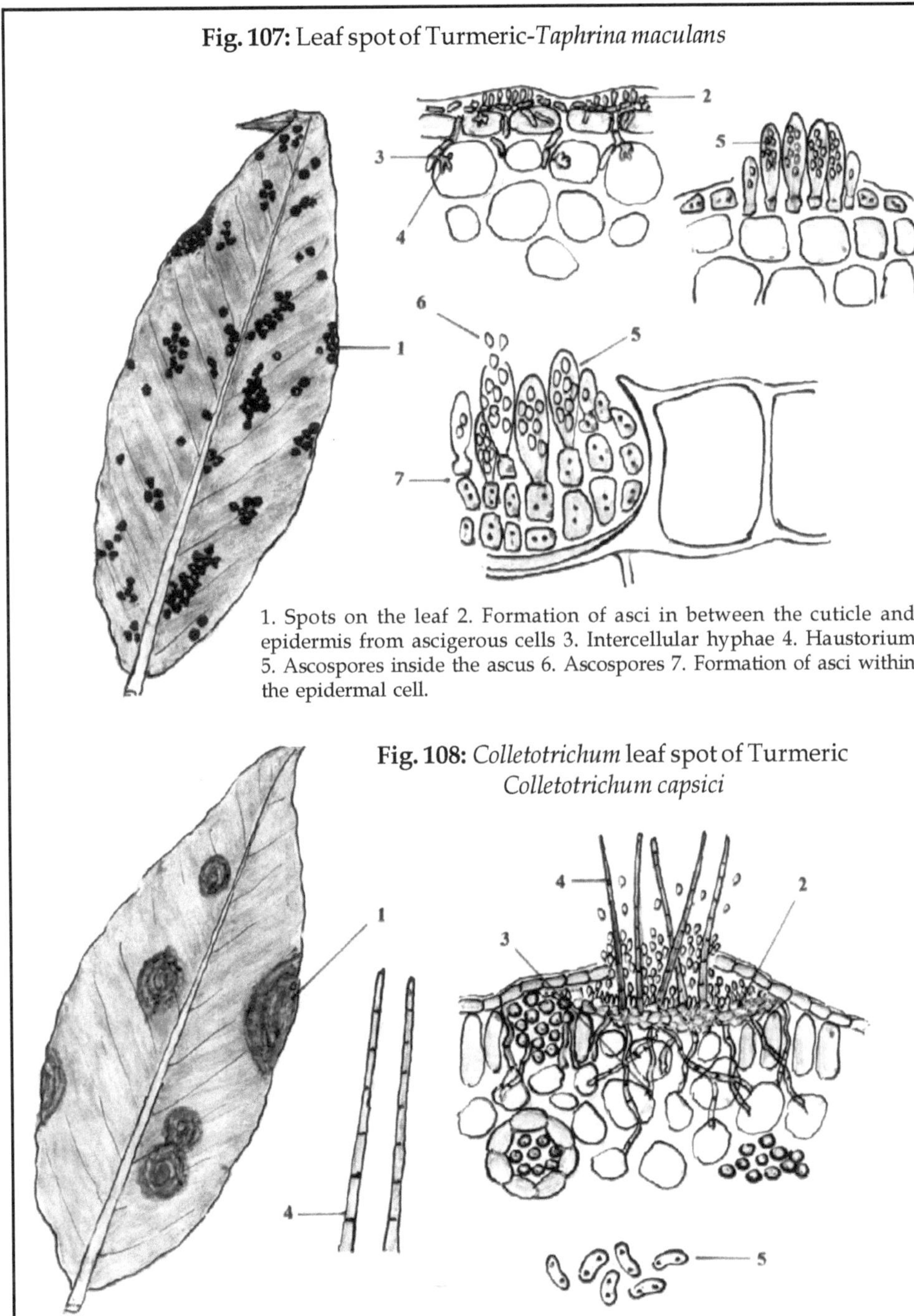

Fig. 107: Leaf spot of Turmeric-*Taphrina maculans*

1. Spots on the leaf 2. Formation of asci in between the cuticle and epidermis from ascigerous cells 3. Intercellular hyphae 4. Haustorium 5. Ascospores inside the ascus 6. Ascospores 7. Formation of asci within the epidermal cell.

Fig. 108: *Colletotrichum* leaf spot of Turmeric *Colletotrichum capsici*

1. Spots with concentric zonation on the leaf 2. Acervulus 3. Hyphae 4. Setae 5. Conidia

Diseases of minor importance. A few other diseases are also known to attack turmeric, but they are of less importance. *Rhizoctonia bataticola* causes 'Dry rot of rhizomes'; *Cercospora curcumae* causes '*Cercospora* leaf spot'.

Cardamom *(Elettaria cardamomum)*

1. Marble mosaic or 'Katte' disease of cardamom

Cardamom mosaic virus

The disease was first reported from the United States of America and it is now prevalent in many countries of the world. In India, the disease is found on a large scale in Kerala, Karnataka and Tamil Nadu.

Symptoms. The disease attacks the crop at all growth stages. The initial symptoms appear on the young leaves. Pale green, interrupted streaks, running parallel to the veins from the midrib to the leaf margin are seen on the entire leaf lamina. The leaves become chlorotic and mosaic symptoms also appear. On young shoots, such streaks may be seen on the leaf sheaths also. Newly formed leaves become small and the affected plants become very much stunted. When one or two year old plants are infected, flower and capsule formation is almost completely stopped. When fully-grown plants are affected, the yield is drastically reduced. The affected clumps deteriorate rapidly. The rhizomes shrivel and eventually the plants die.

The causal organism. The virus causing the disease is a flexuous rod and measure 700 - 720 nm long.

Mode of spread. The disease is transmitted mainly through infected rhizomes used as seed material. The virus is also transmitted through sap and seed. Secondary spread of the disease is through aphid vectors. The cardamom aphid, *Pentalonia nigronervosa* f. sp. *caladii* and the banana aphid, *Pentalonia nigronervosa* f. sp. *typica* are the vectors, of which the former is more efficient. The banana aphid can also infest, multiply and live underneath the leaf sheaths of cardamom plants during the winter months. Both the nymphs and adults can transmit the virus in a non-persistent manner. The acquisition feeding time is about 10 minutes. After an incubation period of about 90 minutes, the insects become viruliferous and they can transmit the virus only for about 24 hours. Some of the common weed plants found in cardamom fields, such as *Caladium* species, *Colocacia* species, *Alpinia* species etc. serve as alternate hosts for the virus.

Disease management

Agronomic practices (i) Rhizomes for planting should be selected from disease-free fields (ii) Instead of planting rhizomes directly, seedlings may be raised in disease-free nurseries under protected conditions and healthy seedlings may be selected as transplants (iii) Diseased plants should be uprooted along with the rhizomes and destroyed (iv) Alternate hosts in and around the fields should be eradicated.

Chemical control. Secondary spread of the virus by the insect vectors can be controlled by periodical application of suitable insecticides. Spraying with dimethoate - 2.0 ml. or methyl demeton - 1.5 ml. or phosphamidon - 0.6 ml./ lit. of water controls the vector.

Minor diseases. A few other diseases have also been reported on cardamom. 'Damping off of seedlings' is caused by a few species of *Pythium*; 'Capsule rot' is caused by a few species of *Phytophthora*; *Phyllosticta elittariae, Sphaceloma cardamomi* and *Cercospora zingiberi* cause 'Leaf spots'.

Clove *(Syzygium aromaticum)*

1. *Colletotrichum* leaf spot of clove

Colletotrichum gloeosporioides

(Glomerella cingulata)

Symptoms. The pathogen affects the crop at all growth stages. On the leaves, the symptoms appear as small, brown specks, scattered all over the leaf lamina. The spots gradually enlarge to form circular to oval, distinct spots, with gray center and dark margin. The spots may coalesce to form large, necrotic patches. The infection then spreads to the petioles resulting in premature defoliation. Linear, brown, necrotic lesions may develop on the branches and twigs also. On older plants, during periods of heavy and continuous rainfall, the flower buds and flowers are attacked. The affected flower buds and flowers become black, shrivel and are shed. Young flower buds are more vulnerable to attack by the fungus.

The causal organism, mode of spread, epidemiology and disease management. Refer 'Anthracnose of mango' (Page 283), **(Fig.73).**

Minor diseases. Besides the disease mentioned above, a few more diseases attack clove. *Leucostoma (Valsa) eugeniae* causes 'Canker'and 'Sudden death'; *Cryphonectria (Endothia) eugeniae* causes 'Die-back'.

Nutmeg *(Myristica fragrans)*

Nutmeg is attacked by a few diseases, which are of economic importance. *Diplodia natalensis* causes 'Die-back' and 'Fruit rot'; *Colletotrichum gloeosporioides* causes 'Leaf spot'.

Cinnamon *(Cinnamomum zeylanicum)*

A few diseases are known to attack cinnamon and sometimes may cause appreciable damage to the crop. *Phytophthora cinnamomi* causes 'Bark canker'; *Colletotrichum gloeosporioides* causes 'Leaf spot' and 'Die-back'

PLANTATION CROPS

Coconut *(Cocos nucifera)*

1. Basal stem rot

Ganoderma lucidum (Polyporus lucidus)

'Basal stem rot' or **'bole rot'** of coconut, commonly known as **'Thanjavur (Tanjore) wilt'** in Tamil Nadu, **'Anabe roga'** in Karnataka and **'*Ganoderma* wilt'** in Andhra Pradesh is a very serious disease of coconut palms and is found scattered in Tamil Nadu in the districts of Thanjavur, Kanyakumari, Tirunelveli, South Arcot, North Arcot, Coimbatore and Salem and causes death of grown up trees. The disease is also found to attack arecanut, palmyrah palms and several other plants. The disease has been reported from Malaya, New Guinea, Philippines and Sri Lanka. In India, besides Tamil Nadu, the disease occurs in parts of Andhra Pradesh, Gujarat, Karnataka, Kerala, Maharashtra and Orissa.

Symptoms. Typical disease symptoms can be noticed on the stem region, the crown region, inflorescence and nuts, as well as the root region.

On the stem region, visible symptoms are found on the basal part of the stem. Numerous, small cracks develop on the basal part of the stem, just above the soil level and reddish-brown, viscous fluid exudes through these cracks continuously. Gradually, this type of bleeding extends from the base, up to a height of about 3 meters. The tissues around the bleeding patches become red or reddish-brown and are rotten. The discoloration may reach the inner tissues up to the core of the stem and even the vascular regions show such discoloration and rotting. In advanced stages of infection, when the trees are in a dying state or already dead, the fructifications (basidiocarps) of the pathogen appear on the basal stem portion, just above the ground level.

On the crown region, the outer whorls of the leaves become yellow and brown, then start drooping. As the disease advances, the remaining leaves also droop down in quick succession, while the spindle leaf and a few young leaves alone stand erect without unfurling. New leaves may not be produced or there is much delay in the production of new leaves. Soon, the hanging outer leaves fall off. When strong winds blow, the entire crown is blown off, leaving the decapitated stem.

In the diseased plants, when the leaves droop down, the coconut bunches also hang down. The nuts are much reduced in size and the quality of the kernel is also very poor. Development of inflorescence is arrested and button shedding is common. Most of the coconut trees bear profusely during the initial stages of disease development.

Decay and death of finer roots result during the initial stages of disease development. As the disease advances, most of the roots show red or reddish-brown discoloration and rotting. Production of new roots is also very poor.

In some cases, affected coconut trees may exhibit all other wilt symptoms except the stem-bleeding symptom. Trees 10 to 30 years of age are more vulnerable to attack by this pathogen. In advanced stages of infection, the stem is attacked by bark and stem borers, such as *Xyleborus parvulus, X. fornicatus* and *Dicalandra stigmaticollis* and wood dust along with faecal matter of the insects are pushed out through the small holes made by them. The infected trees may die within a period of 6 months to 2 years **(Fig.109)**.

Causal organism. The mycelium of the pathogen attacks both the parenchymatous tissues and vascular tissues, as a result translocation of nutrients through the stem is affected. Further, as the vascular tissues are attacked, there is continuous bleeding, which results in the rapid death of the affected trees.

The mycelium of the fungus is hyaline, thin-walled, much branched and produce frequent clamp connections. The hyphae produce chlamydospores, which are ellipsoid, slightly thick-walled and may be terminal or intercalary, sometimes formed in chains. The sporocarps (basidiocarps), which are perennial and bracket-shaped, are reddish-brown, cork-like in the beginning, becoming hard and woody later on. The lateral stem and the upper surface of the basidiocarps are coated with a hard, reddish, shiny substance resembling shellac. The whitish under surface of the basidiocarp consists of numerous, tiny, round pores and the basidia typically line the inner surface of the pores or tubes. At the apical end of each basidium, two or four basidiospores are borne on minute sterigmata. The spores are slightly thick-walled, hyaline or light-brown **(Fig.109)**.

Fig. 109: Basal stem rot of coconut-*Ganoderma lucidum*

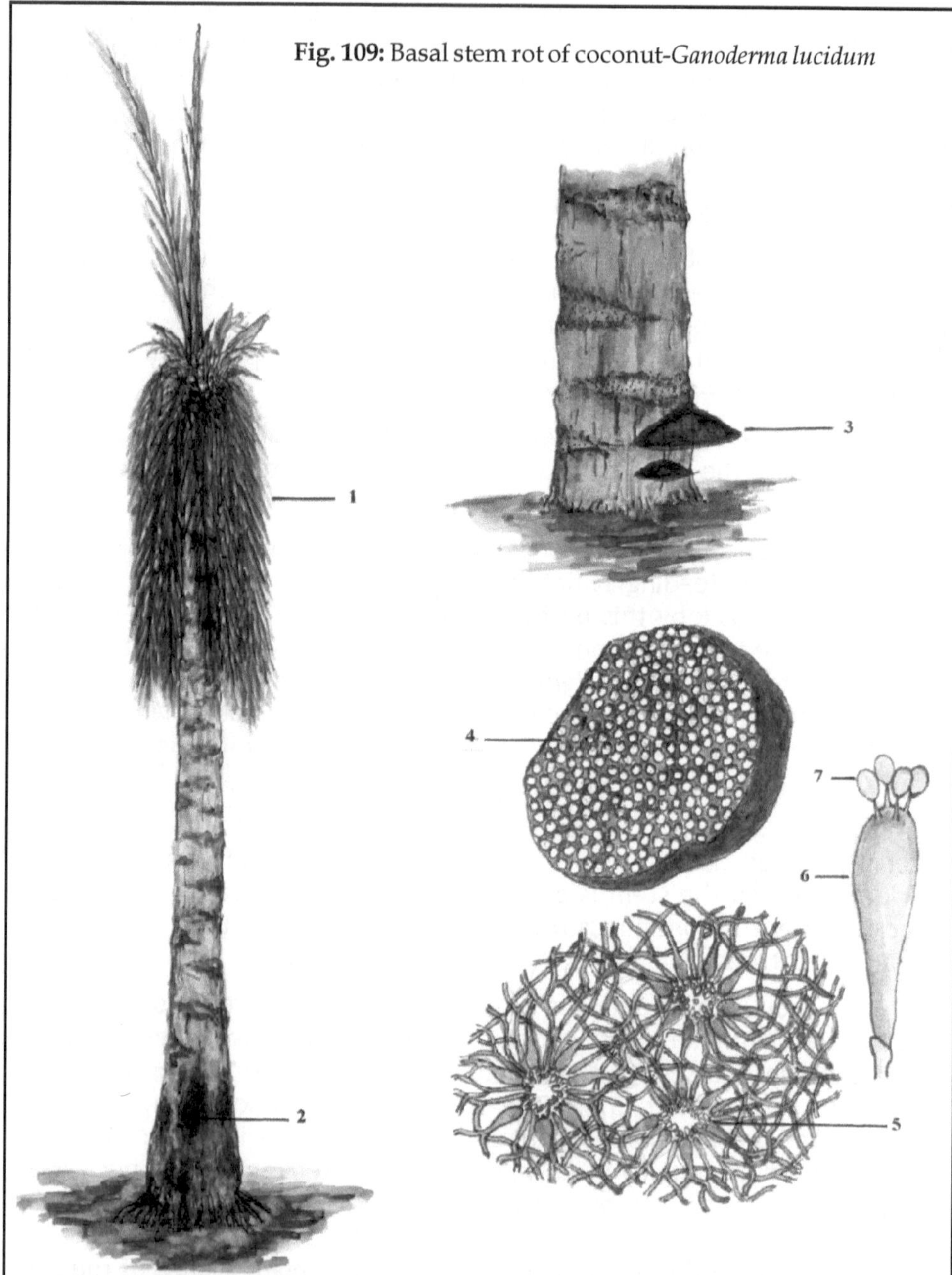

1. Disease affected tree showing drooping of outer whorls of leaves and bleeding patches on the stem 2. Bleeding patches 3. Basidiocarp on the stem of affected tree 4. Under surface of sporocarp showing the pores 5. Basidia with basidiospores lining the inside of the pores of the sporocarp 6. Basidium 7. Basidiospore.

Mode of survival, spread and epidemiology. . The sporocarps (basidiocarps) produced by the pathogen remain in the dead or dying trees, as well as in the soil and continue to produce basidiospores, which may affect healthy trees. The disease usually spreads from a particular focus of infection towards the periphery in a concentric manner. The spread is mainly through natural root grafting between diseased and healthy trees. The mycelium from germinating spores enters the host through natural openings or injuries caused during cultivation operations. The spores produced from alternate and collateral hosts may also get access to healthy coconut trees and cause new infection. The spores may also be disseminated through irrigation water.

The disease is more prevalent in coastal sandy or sandy-loam soils. Ill-drained conditions and water- logging favor disease occurrence.

Disease management

Agronomic practices (i) Dead palms, as well as palms in the advanced stages of the disease should be removed along with the entire root system and burnt (ii) Palms showing early symptoms of the disease should be isolated from the neighboring healthy palms by making trenches 1.0 m. deep and 30 cm. wide in a circle, 1.5 m. away from the stem of infected trees to prevent root contact (iii) In addition to the normal recommended dose of fertilizers at 500 : 320 : 1200 gm. of NPK per palm per year to be applied in two split doses, first in May - June and the second in August - September, neem cake at 5.0 kg. per palm per year should also be applied along with the second split of fertilizers in August - September. This will help natural multiplication of actinomycetes and *Trichoderma* sp. in the soil, which may fight and destroy the pathogen (iv) Application of farmyard manure or compost at 25 - 50 kg. per palm per year also helps in increasing the population of antagonistic organisms in the soil, which may help in fighting and destroying the pathogen (v) Soil application of manganese sulfate at 225 gm. per palm per year has also been found to reduce the incidence of basal stem rot.

Chemical control (i) In the initial stages of disease development, sulfur dust at 500 gm. per tree is spread uniformly in a circular trench, 60 cm. deep and 30 cm. wide all round the tree, 1.5 m. away from the base of the stem. The trench is afterwards covered with soil and compacted. This helps in controlling the disease to some extent, at the same time prevent spread of the disease through roots. The treatment should be repeated every 6 months (ii) Drenching the soil around the base of the affected trees with Bordeaux mixture - 1 % at 40 lit. per tree also helps in controlling the disease. The treatment should be repeated every 2 months (iii) Root feeding with Aureofungin Sol. - 2.0 gm. + Copper sulfate - 1.0 gm. in 100 ml. of water gives effective control of the disease, if the treatment is given in the early stages of disease development. The root region is exposed and an actively growing young root is selected. A slanting cut is made with a sharp knife and the cut end is immediately inserted into a polythene bag containing the antibiotic solution and tied securely to the root. Normally the root absorbs the entire solution within 24 hours. The treatment should be repeated once in four months (iv) Instead of Aureofungin Sol., tridemorph (Calixin) at 2.0 ml. in 100 ml. of water, without the addition of copper sulfate may also be used

Biological control. Application of *Trichoderma viride* and *T. harzianum* along with 50 kg. of FYM or compost has also been found to be effective in controlling the disease and preventing the spread of the disease.

2. Stem bleeding disease of coconut

Thielaviopsis paradoxa
(Ceratocystis paradoxa / Ophiostoma paradoxa)

The disease, which was first reported from Sri Lanka, is found in India, Indonesia, Malaysia, New Guinea, the Philippines and a few other coconut growing countries of the world. In Tamil Nadu, the disease is prevalent in almost all the districts, where coconut is grown. Besides coconut, the disease attacks arecanut and a few other palms.

Symptoms. The characteristic symptom of the disease is the exudation of a dark-reddish-brown fluid, which oozes out from small cracks in the outer tissues. The cracks are scattered mostly at the lower portion of the trunk up to a height of 2 - 3 meters. This exudate turns black, when it dries up on the bark. In case of severe infection, such bleeding cracks may be seen on the entire stem region almost up to the top.of the trunk. The tissues beneath the bleeding patches become yellowish-brown and decay. The parenchymatous tissues become disintegrated and small cavities are formed. The vascular tissues are not affected. The disease spreads very slowly and the affected, grown-up trees remain alive for several years. However, their growth is stunted and the trunk becomes narrower at the top. There is reduction in the number of leaves and leaf size, besides marked reduction in the yield of coconuts.

In the young trees, the spread of the disease is more rapid. General rotting of the soft inner tissues, formation of cavities inside the trunk and accumulation of a yellowish fluid are found. Such affected young trees may die within a year or so **(Fig.110)**.

Fig. 110: Stem bleeding disease of coconut-*Ceratocystis paradoxa*

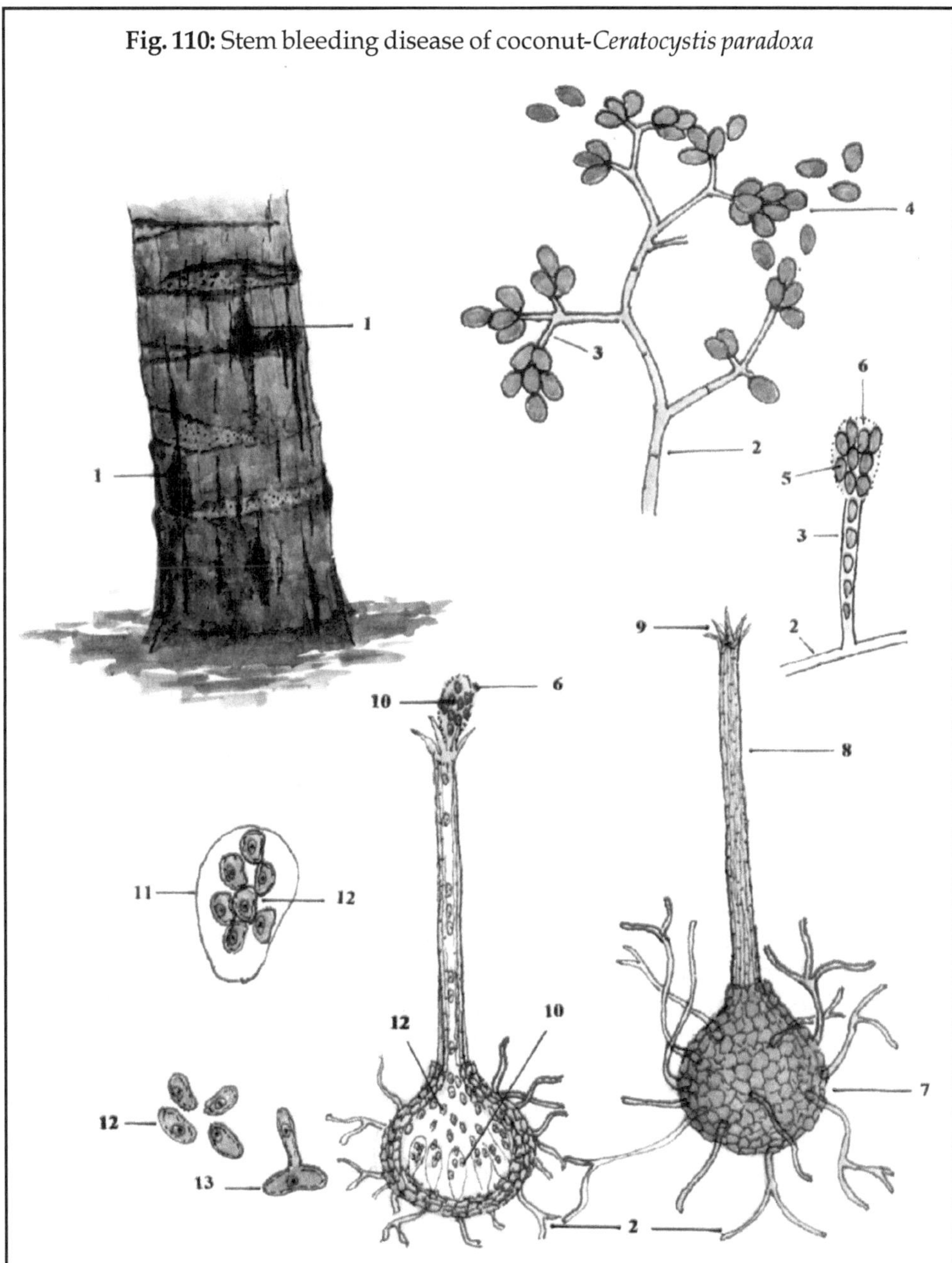

1. Bleeding from cracks in the infected stem 2. Hypha 3. Conidiophore 4. Macroconidia 5. Microconidia (Endoconidia) 6. Droplet of mucus 7. Perithecium 8. Neck of perithecium 9. Feathery tip of the neck 10. Asci inside the perithecium 11. Ascus 12. Ascospore 13. Germinating ascospore.

Causal organism. *Ceratocystis paradoxa*, the causal organism of this disease has its conidial stage in *Thielaviopsis paradoxa* in the Form class - Deuteromycetes. The mycelium of the fungus is hyaline to light greenish in color and septate. The fungus produces both micro- and macroconidia in its asexual phase.

Microconidia are produced endogenously in succession from conidiophores, which are slender, aseptate, hyaline to pale brown, about 100µ in length and usually occur in the form of synnemata. The microconidia are pushed out in long chains and are held together in a drop of mucus. They are oblong to cylindrical, thin-walled, smooth, hyaline at first, becoming black later on and measure 10.0 - 15.0 x 3.6 - 5.6µ in size. Macroconidia are produced on short lateral conidiophores, which arise almost perpendicular to the hypha and are 20 - 80µ in length. The macroconidia are oval to obovate, thick-walled, light green to black in color and measure 16 - 19 x 10 - 12µ in size. About 20 macroconidia may be produced from each conidiophore.

In its sexual phase, the fungus produces perithecia. The perithecia are partially buried in the disintegrated tissues and in-between the bark on the stem region. The perithecia are ostiolate, light-brown, have globose bases and greatly elongated necks with shredded, feathery tips. They measure 190 - 350m in diameter. The ascus walls gelatinize inside the perithecium and the ascospores are extruded through the long neck of the perithecium and are embedded in mucus that forms a droplet at the ostiolar opening. Ascospores are ellipsoid, slightly curved, hyaline, non-septate and measure 7.0 - 10.0 x 2.5 - 4.0m. **(Fig.110).**

Mode of spread. The fungus is a wound parasite and enters through natural cracks or injuries caused by insects and inadvertently by laborers. The spores are disseminated by rain, wind, insects of various kinds and tree climbers. The spread of the disease is more rapid and the severity of the disease is also more in poorly maintained gardens. Trees between the age group of 10 - 20 years are more prone to attack by this disease and the spread of the disease is more rapid in young trees rather than in older trees.

Disease management

Agronomic practices (i) The trees in the garden should be properly maintained by providing adequate irrigation, especially during the hot summer months and the trees should be manured as per recommendations, so as to keep the trees healthy and strong (ii) Causing mechanical injuries to the palms should be avoided.

Chemical control. Using a fine chisel, the affected tissues showing discoloration along with a portion of healthy tissues surrounding the affected tissues are scooped out. The cut surface is protected by applying Bordeaux paste or cold coal tar mixed with saw dust or Aureofungin Sol. Then the cavity is sealed with clay or cement to prevent water accumulation.

3. Bud rot of coconut

Phytophthora palmivora

The disease is found in most of the major coconut growing countries of the world, including Brazil, Africa, Guyana, Sri Lanka, Fiji Islands, Jamaica, Mauritius, Mexico, Nigeria, Philippines, Trinidad, West Indies and India. In India, it occurs commonly in the East and West coasts. In Tamil Nadu, the disease is found in Chingleput, Coimbatore, South Arcot and North Arcot districts. The disease, though sporadic in nature occurs in an epiphytotic form during some seasons. A severe outbreak of bud rot, which occurred in Kerala state during 1992 - 1993, resulted in the destruction of about 5,000 trees. The pathogen has a wide host range and besides coconut, attacks various palms, arecanuts, cacao, rubber, *Citrus* species, cinchona, castor, safflower etc.

Symptoms. Trees of all ages are vulnerable to attack by the disease, but young trees are more susceptible. The central shoot and the surrounding leaves are affected first. Initially the central shoot becomes pale, turn yellow and bend over. Small, discolored, water-soaked, sunken spots appear at the base of the central shoot and young leaves. The spots enlarge and spread up and down and form irregular patches. The tissues at the basal part of the shoot rot and turn brown within a few weeks. The affected shoot comes off easily, when it is pulled. Soon, the rotting extends downwards and the base of the surrounding young leaves also start rotting. The rotten tissues become a slimy mass, which emits a foul smell. Young nuts fail to develop normally and drop off prematurely. When the growing meristem is affected, the tree is killed. While the young leaves rot from the base and fall, the older leaves remain green and fall after maturity, finally leaving the tree barren **(Fig.111)**.

The causal organism. The mycelium of the fungus is coenocytic, intercellular and sends haustoria into the host cells. The much-branched mycelium grows on the surface of the infected areas also as fine, white, cottony growth. The sporangia arise from branched hyphae, which are slightly different from the normal hyphae. These sporangiophores are simple or may have one or a few branches that bear terminal sporangia singly. Sporangia are inverted pear-shaped, rarely round, hyaline, thin-walled, with an apical papilla and measure 38 - 72 x 33 - 42μ in size. They germinate in water and release large zoospores through the papilla. The zoospores after swimming in free water for a short while, come to rest and encyst. They measure 8.0 - 10.0μ in diameter. They germinate by producing a germ tube and cause fresh infection. The fungus produces oospores in its sexual phase in the affected tissues. The oospores are spherical, thick-walled and are 35 - 45μ in diameter. They germinate readily giving rise to a short stalk on which secondary zoosporangia are formed. The fungus also produces thick-walled, terminal or intercalary chlamydospores, measuring 30 - 55μ in diameter **(Fig.111)**.

Mode of survival, spread and epidemiology. The fungus perpetuates through mycelium, chlamydospores and oospores that may remain dormant in the infected parts. The fungus is known to survive in the infected crown region for more than 5 months. In the rainy season, the pathogen becomes active and produces sporangia and zoospores on the tree-top, which are disseminated by rain, wind and mechanically by some insects to other trees. Toddy tappers and rhinoceros beetles are other agents responsible for the spread of the disease.

Low temperatures of 18° - 20°C, continuous monsoon rains and wet weather favor the occurrence and spread of the disease.

Disease management

Chemical control (i) Control measures have to be initiated, as soon as initial symptoms appear on the central shoot. If the growing meristem is attacked, then the tree may not survive. The infected central shoot and a few leaves surrounding it should be cut and removed. All the discolored tissues should be scrapped, removed and cleaned. Then Bordeaux mixture - 1 % is sprayed over the entire surface of the crown. Afterwards Bordeaux paste is applied over the entire area. To prevent the fungicide from being washed off, the crown is kept covered with an inverted, wide-mouthed mud pot till the emergence of a new shoot, which may take a few months. The unaffected trees adjacent to the affected tree should also be sprayed thoroughly with Bordeaux mixture - 1 % to cover the crown region. Regular spraying with Bordeaux mixture - 1 % or copper oxychloride - 2.5 gm./ lit. of water commencing from the onset of monsoon, followed by another spraying 30 days later is an effective preventive measure (ii) Spraying with metalaxyl - 1.0 gm./ lit. of water has also been found to control the disease.

4. Gray leaf spot or leaf blight of coconut

Pestalotia (Pestalotiopsis) palmarum

The disease was first reported from British Guyana and now it is prevalent in most of the coconut growing countries of the world, including Nigeria, Sri Lanka, Malaya and India. In Tamil Nadu, it occurs quite commonly in many of the gardens.

Symptoms. The disease usually attacks the older, outer whorls of leaves. The symptoms appear as minute, scattered, isolated, yellow, roughly oval to slightly elongated spots on the leaflets, which are surrounded by a grayish band. When the disease intensity is high, hundreds of spots may be produced on a single leaflet. The spots enlarge in size and the center of the spots turns grayish-white, while the outer band becomes dark-brown. Adjacent spots may coalesce to form irregular, gray patches. Dark, minute, pinpoint-like dots, which are the fruiting bodies of the fungus (acervuli) are produced in large numbers on the spots on the upper surface of leaflets. In advanced stages, the leaves become yellow and the leaflets start drying from the tips and margins, giving the leaves a scorched appearance **(Fig.112).**

The causal organism. The mycelium of the fungus is septate, inter- and intracellular and mostly confined around the spots. The hyphae congregate beneath the epidermis of the spots to form stromata. Acervuli are formed from such stromata, which appear inverted saucer-shaped, when fully formed. Numerous, short, erect conidiophores arise from the base of the acervulus like a palisade layer. Conidia are produced from the tips of conidiophores singly in succession. Because of the pressure exerted by the conidia, which are produced in large numbers, the host epidermis is ruptured and the conidia are exposed. Conidia are 4-septate, sometimes 3-septate, eye-shaped with both the ends tapering, the median cells dark-colored, while the two end cells are hyaline and measure 17.0 - 25.0 x 4.5 - 7.5µ in size. They are pedicellate, slightly constricted at the septa and the apical cell crowned with 3, rarely 2, hyaline, divergent appendages of length varying from 5.0 - 25.0µ. **(Fig.112).**

Mode of spread and epidemiology. The disease is mostly spread by air-borne conidia. In poorly maintained and ill-drained gardens, the disease occurs very commonly. Potash deficiency predisposes the plants to infection.

Warm and humid weather conditions favor the occurrence and spread of the disease. When young trees are affected severely, the growth is badly affected.

Fig. 111: Bud rot of coconut *Phytophthora palmivora*

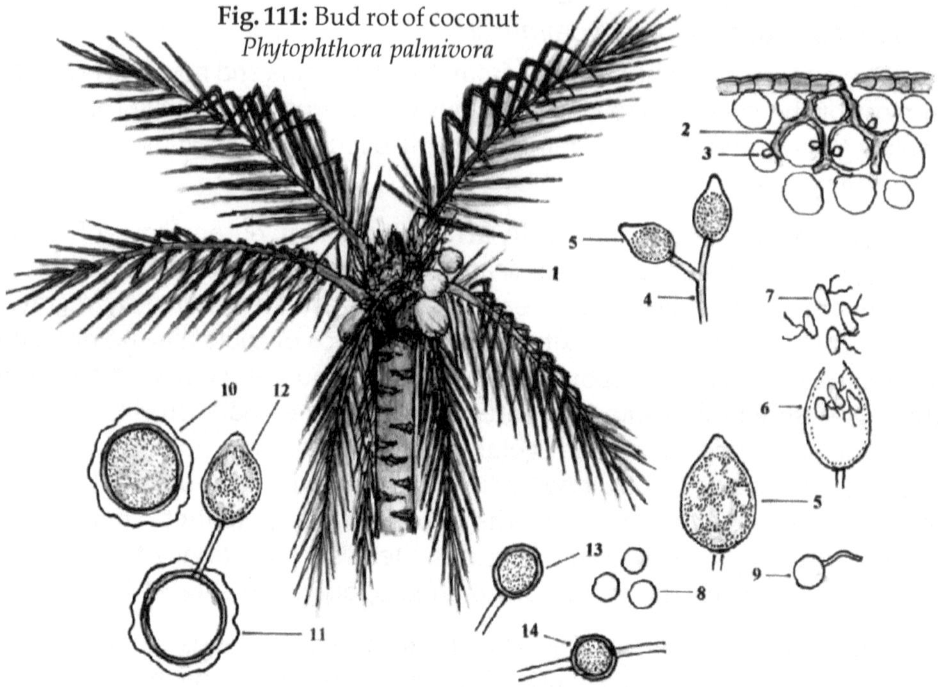

1. Infected tree with only a few outer leaves 2. Intercellular mycelium 3. Haustorium 4. Sporangiophore 5. Sporangium 6. Zoospores emerging from sporangium 7. Zoospore 8. Encysted zoospore 9. Germination of encysted zoospore 10. Oospore 11. Grerminating oospore 12. Secondary sporangium 13. Terminal and 14. Intercalary chlamydospore.

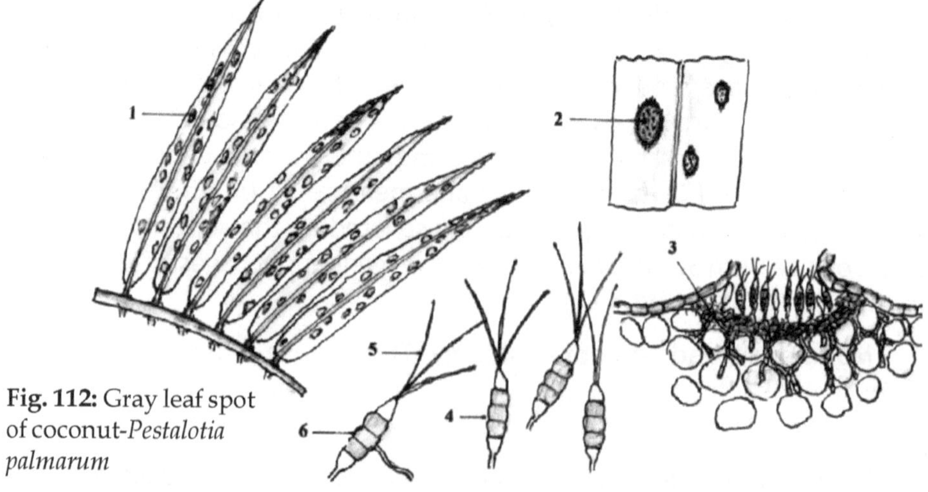

Fig. 112: Gray leaf spot of coconut-*Pestalotia palmarum*

1. Spots on the leaflets 2. Acervuli on the spots on the surface of leaflet 3. Acervulus 4. Conidia 5. Appendages 6. Germinating conidium.

Disease management

Agronomic practices (i) Severely affected, old leaves should be cut off and destroyed (ii) Coconut gardens, especially young plantations should be maintained properly by timely irrigation and fertilization.

Chemical control. Foliar spraying with Bordeaux mixture - 1 % or copper oxychloride - 2.5 gm./ lit. of water controls the disease. To prevent rolling down of the spray fluid from the surface of leaflets, wetting and spreading agents, such as Sadovit or Teepol or Triton may be added to the fungicidal fluid at the rate of 1.0 ml./ lit of the solution. Two to three sprayings should be given at 15 - 30 days interval.

Diseases of minor importance. A few other diseases are also known to attack coconut palms. *Alternaria alternata* causes '*Alternaria* leaf spot'; 'Kerala wilt' or 'Root wilt' is supposed to be caused by a *Mycoplasma-like* organism, but its etiology has not been established beyond doubt.

Tea *(Camellia sinensis)*

1. Blister blight of tea

Exobasidium vexans

The disease was first reported in India from Assam in 1868 and since then it has spread to all other tea growing states of the country. The disease is also prevalent in Sri Lanka, Burma, Formosa, Malaya, Indonesia, Japan and Taiwan. In South India, the disease occurs in most of the tea gardens every year and causes severe damage.

Symptoms. The initial symptoms appear as small, round, pinkish spots on the young leaves. In the beginning, the symptoms appear in only a few plants after the rains. Then the disease spreads rapidly and soon, almost all the plants are attacked. Within a few days after the infection, the spots enlarge up to a diameter of 0.5 - 2.0 cm.and become blister-like. The blisters may be formed on both the surfaces of the leaves, but mostly they are formed on the under surface of the leaves. The spots later turn dark red in color. Old spots become whitish or grayish in color and are covered by a whitish powdery coating. In the blisters which are formed on the under surface of leaves, the corresponding upper surface of the blisters become shallow, cavity-like depressions. The surface of the blisters is covered by whitish growth of the fungus and the spores formed give an appearance of a whitish, powdery coating. Later, the blisters turn dark-brown in color, shrink and become flattened spots. The disease attacks only the young leaves and buds. Old leaves are usually not attacked and if they are attacked they fall off. The disease may attack the petioles and tender, succulent branches also, but no blisters are formed. When a number of blisters are formed on the leaves, they become distorted and curled. As the disease attacks young and tender leaves, the yield and quality of tea is badly affected **(Fig.113).**

The causal organism. The mycelium of the fungus, which is confined to the blistered areas on the leaves and tender stem regions, are septate, inter- and intracellular. Before sporulation, the dikaryotic hyphae collect in bundles between the epidermal cells of the host. The basidia, which are produced directly from these hyphae, push through the cuticle and form a layer on the surface of the host. The basidia are hyaline, long, club-shaped and measure 30.0 - 35.0 x 5.0 - 6.0 μ in size. The basidia are interspersed with sterile hair-like appendages (cystidia). At the tip of each basidium, usually two short sterigmata are formed, each bearing a basidiospore. The basidiospores are single-celled, uninucleate, haploid, hyaline, oval or oblong in shape and measure 5.0 x 3.0 μ in size on an average. The basidiospores germinate either by budding, producing large numbers of blastospores or directly by producing a germ tube that develops into a haploid primary mycelium. Production of large numbers of basidia, basidiospores and blastospores on the surface of the blistered areas gives the appearance of a white, powdery coating **(Fig.113).**

Mode of survival, spread and epidemiology. The pathogen is capable of surviving within the host tissues throughout the year. When favorable environmental conditions return, the fungus becomes active, produces large number of basidiospores and blastospores and infects the young leaves and buds. Under favorable conditions, the fungus completes its life cycle in 11 - 28 days and in one season several generations of spores are produced. The spores are disseminated by wind and rain accompanied by wind.

Relative humidity and temperature are the main factors that decide the severity of disease outbreak. High relative humidity (above 80 %) and low temperatures (below 24°C) are favorable for disease outbreak. Cool, moist weather conditions, relatively still air and bright light are conducive for disease occurrence. The incidence of the disease is more severe in localities near about woods, wind breaks and damp low-lying areas, where there is mist. The new sprouts that start growing after picking or pruning are highly susceptible to infection.

Disease management

Agronomic practices. Disease affected leaves and shoots should be removed and destroyed.

Chemical control (i) Foliar spraying with Bordeaux mixture - 1 % or copper oxychloride - 0.25 % is widely advocated for the control of this disease (ii) Spraying with nickel chloride - 0.1 % affords good control of the disease, however, a combination of copper oxychloride - 0.25 % + nickel chloride - 0.1 % is more effective (iii) Combined spraying with copper oxychloride - 0.25 % + Agrimycin 100 - 200 ppm (0.02 gm./ liter of water) is also effective in controlling the disease (iv) Spraying with systemic fungicides, such as tridemorph - 0.1 % or triadimefon (Bayleton) - 0.1 % or bitertanol (Baycor) - 0.1 % is also found to be promising in controlling the disease (v) As the disease attacks primarily young and tender leaves, it is better to take up prophylactic spraying after each picking (vi) Repeated sprayings at 5 - 10 days interval have to be given, especially during the rainy seasons.

2. Gray blight of tea

Pestalotia theae

The disease is prevalent in Java, Sri Lanka and India. In India, it is found in most of the tea estates in North and South India. In South India, it occurs in Karnataka, Kerala and Tamil Nadu.

Symptoms. The disease is more common on old leaves that are about to fall down. However, when the plant is weakened by other causes, leaves of all ages are affected. The first symptoms appear as minute, brownish spots, which enlarge to form bigger spots. Sometimes, many small spots may coalesce to form bigger patches. The patches are light- to dark-brown, with a grayish center on the upper surface of the leaves. They are roughly circular to oval and marked with concentric zonations almost from the center to the edge. Black pustules, which are the fruiting bodies (acervuli) are produced in concentric lines on the upper surface. The diseased patch may occur at the margin or in the middle of the leaf. On young leaves, the patch is usually dark-brown to almost black, rather irregular in shape and not marked with concentric rings. The affected young leaves become distorted **(Fig.114).**

The causal organism. The mycelium of the fungus is septate, much branched and are mostly intercellular. Haustoria arising from the intercellular mycelium enter the cells of the tissue and absorb nutrition required for the pathogen. Later, the hyphae aggregate below the epidermis to form a stroma and then produce acervuli. The acervuli appear as inverted saucer-shaped structure. From the base of the acervulus numerous, minute, short, unbranched conidiophores arise. From the tip of the conidiophores, conidia are produced one by one. The conidia are eye-shaped, with both ends tapering towards the tips and are four-septate. The two terminal cells are

Fig. 113: Blister blight of tea-*Exobasidium vexans*

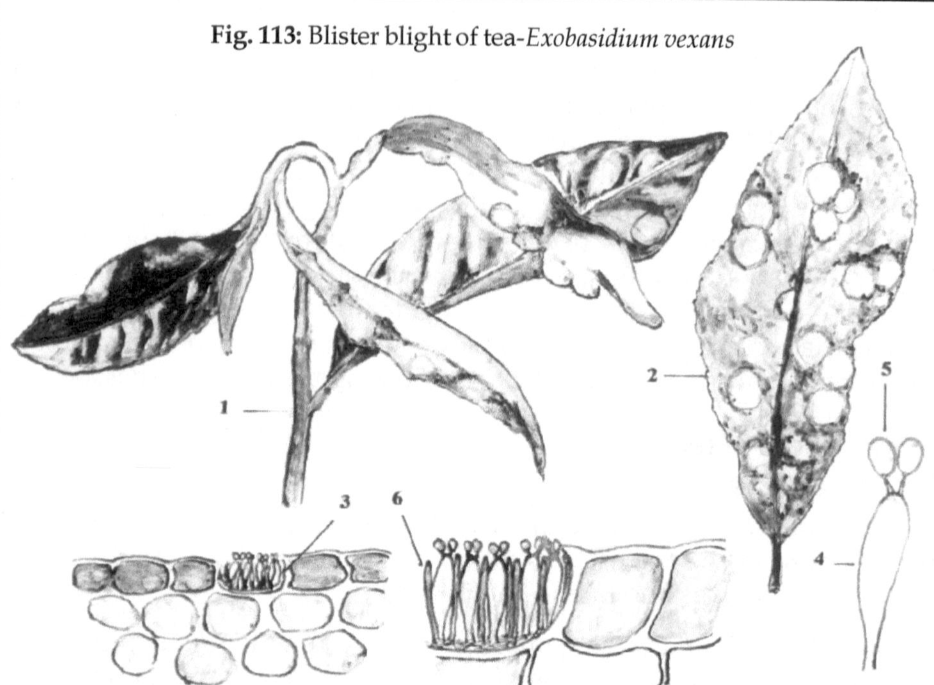

1. Stem and leaf infectin on tea shoot 2. blistered tea leaf 3. Epidermal cell of the host 4. Basidium 5. Basidiospore 6. Sterile hairs.

Fig. 114: Gray blight of tea-*Pestalotia theae*

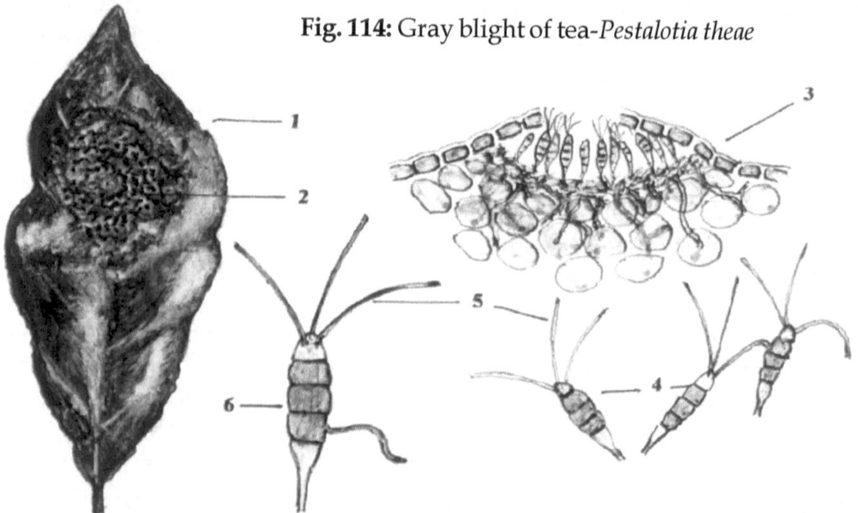

1. Leafspot on the infected leaf 2. Acervuli on the leafspot 3. Acervulus with conidia 4. Conidia 5. Appendages 6. Germinating conidia producing germ tube.

hyaline, while the three central cells are brownish. The conidia possess 2 - 3 long appendages at the apex. The spores germinate from one or more of the central cells by issuing a germ tube **(Fig.114).**

Mode of spread and epidemiology. The conidia are dispersed mostly by wind. When weather conditions are unfavorable for the pathogen, it remains dormant in the affected leaves until favorable conditions return.

High moisture and moderate temperature favor the occurrence and spread of the disease. Tea plants weakened by causes, such as mite and jassid infestation, overdose of chemical fertilizers, lack of adequate nitrogen, water-logging, sun-scorch, hail damage, lack of shade, drought and hard plucking predispose the plants to attack by the pathogen.

Disease management

Agronomic practices. Field sanitation helps to minimize the incidence of the disease. Diseased old leaves that have fallen down should be collected periodically and destroyed.

Chemical control. Spraying with Bordeaux mixture - 1 % or copper oxychloride at 2.5 gm./ lit. of water or dithiocarbamates at 2.0 gm./ lit. of water is effective in controlling the disease.

3. Black rot of tea

Corticium theae

'Black rot' is more common in the plains and is rare in the hilly tracts. The fungus attacks tea crop at all growth stages, from seedlings to grown up plants.

Symptoms. The pathogen produces large, discolored patches on the leaves, covering about half and sometimes the entire leaf area. On the upper surface of the affected area, the patches are reddish-brown in color in the early stages, similar to sun-scorch damage. Later, the color turns to a mixture of brown, yellowish-brown and gray, while the under surface is light-brown or grayish-white and usually covered with a network of cream to brown mycelium **(Fig.116).**

The causal organism. On the stem region, the fungus produces thick cords of mycelium, which are dark purplish-brown on the older portions of the stem and dull-white to light-brown on the green portions at the top. Dead leaves remain suspended on to the stem, held by the mycelial cords. Fructifications appear as powdery, white patches on the under surface of apparently healthy green leaves. The white powder consists of basidiospores **(Fig.115, 116).**

Fig. 115: Genus - *Corticium*

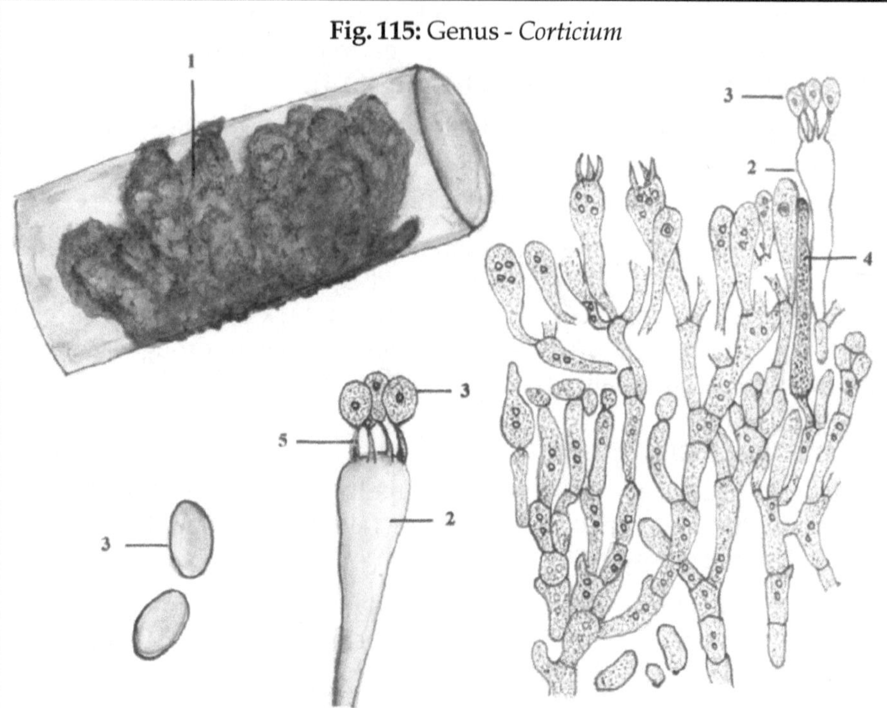

1. Crusty fructification (Basidiocarp) on wood 2. Basidium 3. Basidiospores 4. Cystidium 5. Sterigmata.

Fig. 116: Black rot of tea
Corticium theae

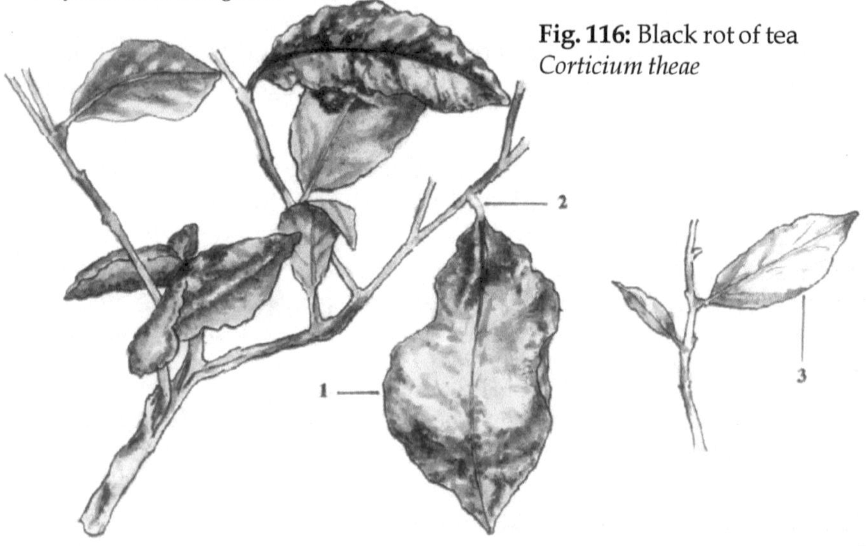

1. Diseased leaf hanging from a fungal thread 2. Fungal thread 3. White powdery fructifications on thre undersurface of the leaf.

Mode of survival, spread and epidemiology. The disease persists in the same areas for years, causing gradual deterioration of the affected plants.

The pathogen thrives best and persists for years in badly ventilated gardens, where air movements are prevented by live wind barriers, under dense shade or areas surrounded by jungle, where the air in the rainy season is hot, moist and still. Black rot is more prevalent in badly maintained, skiffed and unpruned tea gardens.

Besides air-borne spores, the fungus spreads by direct contact from bush to bush and also by the distribution of diseased materials by wind, birds, workers etc. The pathogen also thrives in several grass and weed hosts.

Disease management

Agronomic practices (i) Affected plant parts should be cut and destroyed (ii) Dense shade should be reduced and ventilation improved by cutting back the shade trees or jungle.

Chemical control. Spraying with Bordeaux mixture - 1 % or copper oxychloride at 2.5 gm./ lit. of water, twice, at an interval of two weeks during mid April to end of May prophylactically in the affected areas is recommended. Thereafter, individual bushes should be sprayed again, whenever black rot attack is observed. While spraying care should be taken to cover the under surface of the leaves and the frame thoroughly with the fungicide.

4. Red rust of tea

Cephaleuros mycoidea

The disease is prevalent in Sri Lanka, Africa, America and India. In India, the disease is widespread and is found in all the tea growing states. All the cultivated varieties of tea are susceptible to the disease. In poorly maintained tea estates the disease incidence is found to be very severe.

Symptoms. The disease attacks plants of all growth stages and all the parts of the plants are subjected to attack by the algal pathogen. Though leaves are attacked to a larger extent by the pathogen, the stems, branches and even the fruits are attacked. On the leaves, the symptoms are seen as small, reddish, rough spots on the upper surface. These spots gradually enlarge in size and become orange-red, roughly circular spots of size 1.0 -

1.5 cm. in diameter. Sometimes, the spots are sunken with raised margins. The spots appear rough and corky. When many such spots are formed on the leaves, photosynthetic activity of the leaves is adversely affected. The algal filaments penetrate the epidermal cells and grow in-between the palisade cells and absorb nourishment from the host cells by osmosis. The filaments do not grow deeper into the mesophyll tissues. The affected cells are killed and they turn brown and die. Severely affected leaves exhibit variegation in color, turn brown, dry and fall down. On the stem region, the algal growth is seen as reddish, crusty spots on the surface, which become orange-red fruiting patches. When weather conditions become favorable, fructifications appear on the fruiting patches as fine, cottony outgrowths consisting of sporangiophores and sporangia. Older patches become purplish-red and develop longitudinal cracks. The surface of the patches appears rough and corky. Severely affected shoots may die **(Fig.117).**

The causal organism. The algal filaments on the leaves are found in the epidermal cells and in-between the palisade cells. On the stem region, the pathogen first attacks the bark and then penetrates deeper into the cortex. From the orange-red fruiting patches, which are formed under favorable conditions, fine, cottony outgrowth consisting of sporangiophores and sporangia are produced. The sporangiophores are short and slender. They bear 4 - 5 sporangia at their tips. The sporangia are thin-walled and spherical. Each sporangium contains 8 - 10 endospores. On maturity of the sporangia, the outer wall breaks and motile endospores are liberated. The endospores are orange-colored, spherical or oval-shaped and have two long, fine, flagella. The endospores swim in water for sometime and cause fresh infection **(Fig.117).**

Mode of survival, spread and epidemiology. Under unfavorable conditions, the algal filaments remain dormant in the leaves and the stem region. When favorable conditions are present, the pathogen gets activated and forms fructifications. The motile endospores swim about in a film of water on the host surface for some time, penetrate the epidermis and cause infection. Rainsplash or rain accompanied by wind disseminates the spores.

Free water is necessary for germination of the spores. Low temperature, high humidity and rainfall favor the spread of the disease. The alga attacks *Tephrosia* species and *Desmodium gyroides* grown as green manure and shade plants, which serve as alternate hosts.

Fig. 117: Red rust of tea-*Cephaleuros mycoidea*

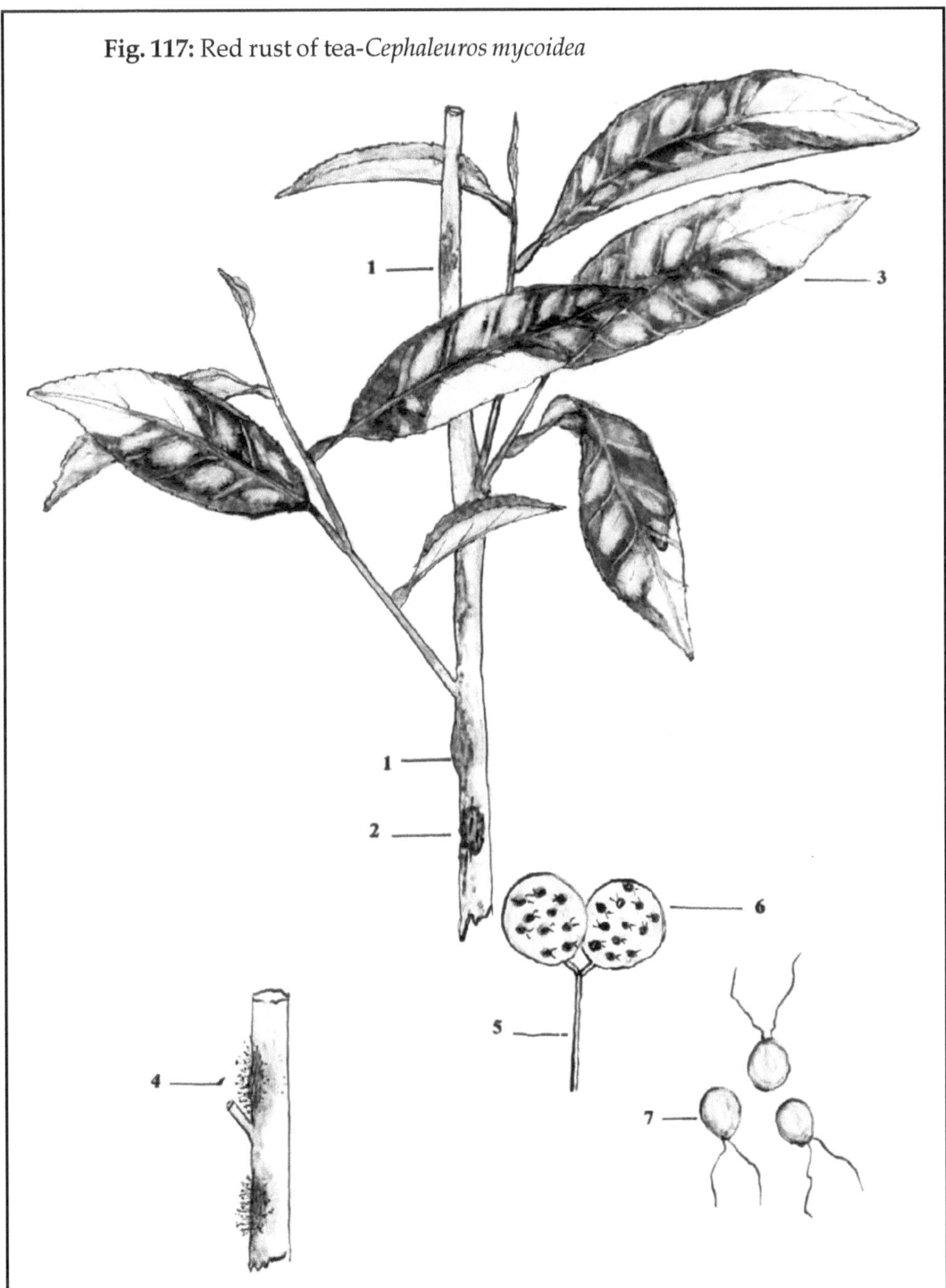

1. Orange red fruiting patches 2. Longitudinal cracks on a purplish-red old patch 3. Variegation of the leaves 4. Fructification from a fruiting patch 5. Sporangiophores 6. Sporangium containing endospores 7. Endospores.

Disease management

Agronomic practices (i) Severely affected plant parts should be pruned and destroyed. After pruning the crop should be sprayed with Bordeaux mixture - 0.75 % or copper oxychloride at 2.5 gm. per lit. of water (ii) Proper maintenance of the crop is very important, as the vitality of the plants are directly related to the incidence of the disease. So, the recommended dose of fertilizers should be applied at the appropriate time and the nutrient status of the soil should be maintained.

Chemical control. Periodical sprayings should be given, especially during the monsoon seasons with Bordeaux mixture - 0.75 % or copper oxychloride - 2.5 gm. per lit. of water.

Minor diseases. Besides the above-mentioned diseases, several other diseases attack different parts of tea plants. *Fomes lamaoensis* causes 'Brown root rot'; *Poria hypolateritia* causes 'Red root rot'; *Rosellinia arcuata* causes 'Black root rot'; *Botryodiplodia theobromae* causes '*Diplodia* root disease'; *Rhizoctonia bataticola (Macrophomina phaseoli)* causes 'Root and collar rot'; *Nectria cinnabarina* causes 'Die-back' disease; *Pellicularia salmonicolor (=Corticium salmonicolor)* causes 'Pink disease'; *Physalospora neglecta (=Macrophoma theicola)* causes 'Branch canker'; *Glomerella cingulata (=Colletotrichum camelliae)* causes 'Brown blight'.

Coffee *(Coffea arabica and C. robusta)*

1. Leaf rust of coffee

Hemileia vastatrix

This leaf disease, which is the most important and dreaded coffee disease in the world, is said to have been introduced to Ceylon probably from Africa by the Dutch settlers during the middle of the 18th century. In 1867, coffee rust became a serious disease in Sri Lanka and by 1871, the disease ravaged the coffee plantations and made coffee cultivation uneconomical. As a result, most of the coffee growers were forced to abandon coffee cultivation and turn to tea cultivation. The disease then spread to Java and Sumatra and there too did much damage to the Arabian coffee. In India, the disease first appeared in 1870 and ravaged coffee plantations of Mysore. The disease occurs in Kerala, Tamil Nadu, Karnataka and Madhya pradesh.

Symptoms. The disease, which attacks mostly the leaves, especially the younger leaves rarely attacks the berries and tender shoots. As the leaves mature, they develop some kind of resistance to infection by the pathogen.

The disease appears first as pale green or yellowish spots, which are 1.0 - 2.0 mm. in diameter and the spots gradually increase in size. Then orange-yellow pustules break out on the lower surface of the leaves. The corresponding upper surface of these spots turns brownish and eventually become necrotic. Each pustule is about one-tenth of a mm. in size and they crowd together in small spots or blotches, sometimes spreading across the whole width of the leaf. When the attack is severe, the whole plantation may appear yellowish at a distance on account of the orange rust spots on the leaves. Severe leaf shedding follows this. The berries remain small and fail to ripen. *Coffea robusta* is less susceptible to leaf rust than *Coffea arabica* **(Fig.118).**

The causal organism. The pycnial and aecial stages of *Hemileia vastatrix* have not been found so far. The mycelium of this fungus is intercellular. The globose haustoria arising from the mycelium enter the cells of the host tissue and obtain nourishment required for the pathogen. The sori produced by the fungi consist of numerous narrow, interwoven hyphae forming a stroma below the stoma. From the stroma, clavate filaments emerge through the stoma. The tips of these filaments bear numerous pedicels on which the uredospores are borne. The uredospores produced in large numbers from each of the uredosori appear as orange dust spread over the leaf spots. The uredospores are reniform and the upper surface is humped like a top of a tortoise-shell and echinulated, while the under surface, which is concave or flat is smooth. They measure 26 - 40 x 20 - 30µ. in size.

Later on, teliospores are produced in the same sori. The teliospores are single-celled, pedicellate and are round or turnip-shaped. They measure 18 - 28 x 14 - 22µ in size and have a smooth surface. They have no dormancy and as such germinate immediately by producing a basidium through a germ pore situated at the apex. The basidium is four-celled and from each cell, a basidiospore is formed. The fate of the basidiospores is not known **(Fig.118).**

Mode of survival, spread and epidemiology. Only the uredospores are responsible for the occurrence and spread of the disease. During unfavorable seasons, the fungus survives as uredospores in the affected plant parts or diseased plant parts, which have fallen down to the ground. The spores are spread mainly by means of wind, rainsplash and by some insects.

High moisture and high temperature are favorable for the disease. Intermittent rains, dew or mist, moderately high temperature, plenty of light, slight overhead shade and shelter from wind favor severe attacks. The uredospores germinate only in the presence of a film of water.

Four physiological races of the fungus have been identified

Disease management

Agronomic practices. Field sanitation helps to minimize the incidence of the disease. Fallen diseased leaves, berries and other plant parts should be collected periodically and destroyed by burning or composting.

Chemical control. Spraying with Bordeaux mixture - 0.75 % or copper oxychloride at 2.5 gm./ lit. of water or dithiocarbamates (mancozeb or Dithane M-45) at 2.0 gm./ lit. of water or triadimefon (Bayleton) at 1.0 ml./ lit. of water provides adequate control of the disease. Triadimefon has both curative and protective properties and may be very effective in controlling the disease. Depending upon the weather conditions, 4 - 5 sprayings should be given. Care should be taken to cover the undersurface of the leaves with the spray fluid.

2. Coffee berry disease

Colletotrichum coffeanum var. *virulans*

(Glomerella cingulata)

'Coffee berry disease' was first recorded in Kenya in 1922. It is of great importance in Kenya and Belgian Congo. The disease is prevalent in all coffee growing states of India.

Symptoms. The disease attacks green berries at any stage of their development, causing a dark-brown rot, which usually penetrates the inner tissues and destroys the beans. Eventually, the whole berry dries out and looks mummified like dried cherry coffee. The first lesion may occur at one side of a berry as a slightly sunken, dark-brown spot, which rapidly enlarges. Sometimes a number of small lesions appear, which quickly coalesce to form irregular, sunken areas. Often, the base of the berry and the top of the fruit stalk are affected first. On the surface of the lesions, minute, slightly projecting pustules, mostly darker than the surrounding tissue develop. These are the fruiting bodies (acervuli) and under moist conditions they may produce masses of pinkish spores **(Fig.119).**

Dry weather halts the progress of the disease and in such cases, only the skin of the berry is affected. The lesions formed become ash gray, with a dark-brown margin. The gray area is dotted with dark acervuli.

The causal organism. The acervuli produced by the fungus are erumpent and appear like inverted saucer. From the base of the acervulus, a number of minute, short conidiophores are produced. The conidiophores produce conidia one by one. The conidia are rather large, 6.0 - 33.0 x 4.0 - 8.0μ in size. They vary much in size and shape, from oval through elliptical to somewhat irregularly clavate and have granular contents. The conidium germinates by issuing a germ tube and penetrates the epidermis of the berry after putting forth an appressorium. Setae may or may not be present in the acervuli. The fungus also produces chlamydospores in short chains. They are convoluted and dark-brown in color.

Mode of spread and epidemiology. The disease occurs in high altitudes associated with cold, wet conditions. It is more prevalent in hollows, valleys and low situations at high altitudes, where cold air collects at night. The peak of an attack appears to develop in cold, cloudy weather when there is no sunshine. The conidia are dispersed mostly by rain splash.

Disease management

Chemical control. Spraying with Bordeaux mixture - 0.75 % or copper oxychloride at 2.5 gm./ lit. of water is found to be effective, provided the berries are covered by the fungicidal fluid thoroughly.

Resistant varieties. The variety 'Blue Mountain Jamaica', though not immune is found to have resistance to the disease, when it is grown at altitudes below 5,700 feet.

3. Brown eyespot and berry blotch of coffee

Cercospora coffeicola

The disease is more prevalent in Central America, South America, India and Uganda.

Symptoms. The fungus attacks the leaves and berries of all types of coffee. The leaf spots are dark-brown in color and 3.0 - 5.0 mm. in diameter, but sometimes two or three times larger. They later become gray or almost white in the center and have a reddish-brown zone round the margin. The pathogen may attack nursery plants or young plants, soon after they have been transplanted in the field. The fungus also causes dark blotches on the berries, which later turn black, shrink and drop off before they are fully ripe. The diseased tissues of the fruit skin become dry and hard making them difficult to pulp **(Fig.120).**

The causal organism. The conidiophores are short, olivaceous and emerge in clusters. Conidia arise from the conidiophores singly. They are long, sub-cylindrical, hyaline, 2 - 3 septate and measure 40.0 - 60.0 x 3.5μ in size.

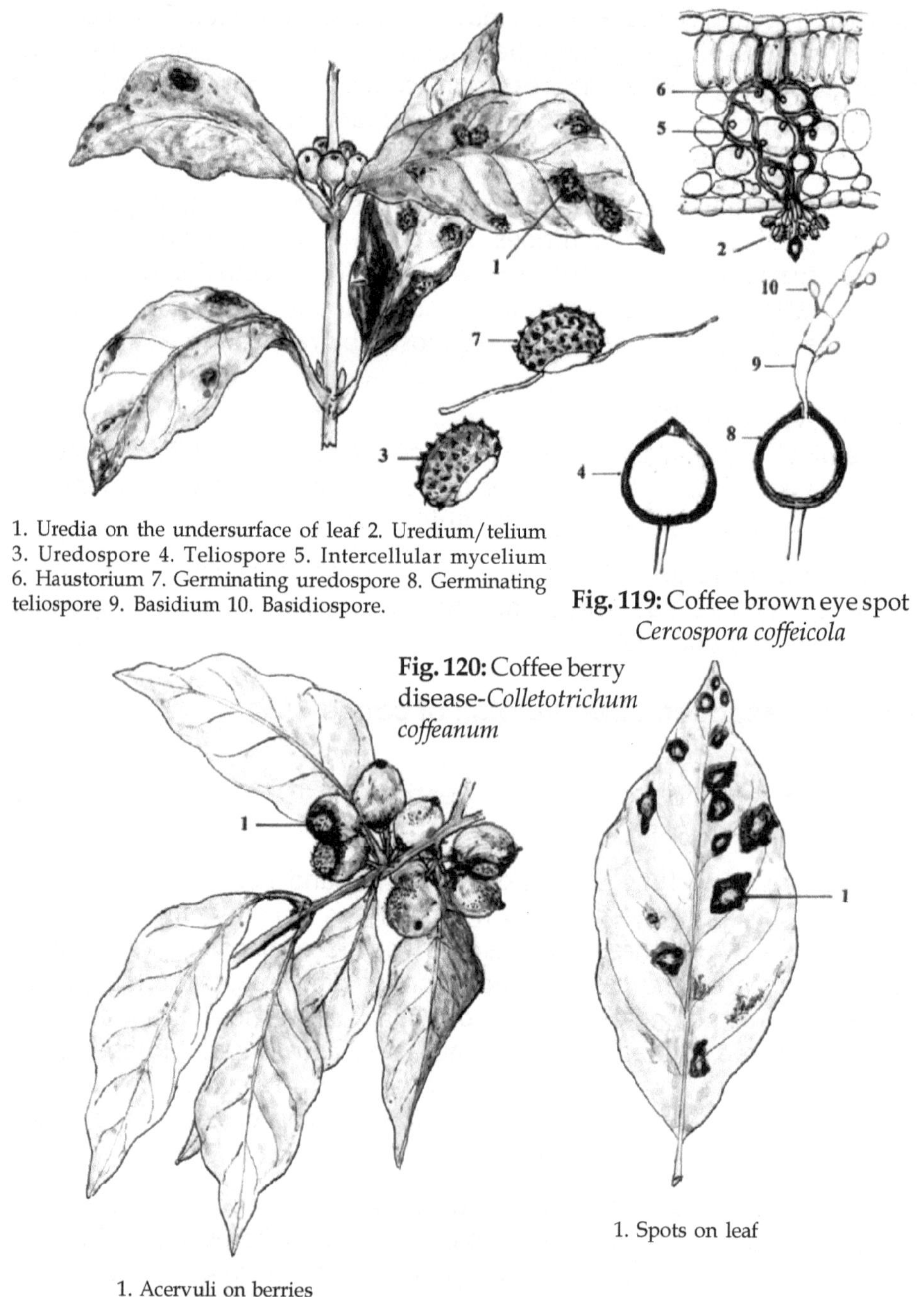

Fig. 118: Leaf rust of coffee-*Hemileia vastarix*

1. Uredia on the undersurface of leaf 2. Uredium/telium 3. Uredospore 4. Teliospore 5. Intercellular mycelium 6. Haustorium 7. Germinating uredospore 8. Germinating teliospore 9. Basidium 10. Basidiospore.

Fig. 119: Coffee brown eye spot *Cercospora coffeicola*

1. Spots on leaf

Fig. 120: Coffee berry disease-*Colletotrichum coffeanum*

1. Acervuli on berries

Mode of spread. The disease is generally found in ill-nourished, unshaded coffee at low altitudes, where moisture is deficient in the soil. The disease is mostly spread through wind-borne conidia

Disease management

Agronomic practices (i) Good overhead shade should be provided (ii) Proper crop management, such as balanced fertilization, timely irrigation, removal of diseased plant parts etc. should be followed, so as to maintain the vigor of the plants.

Chemical control. Spraying with Bordeaux mixture - 1.0 % or copper oxychloride at 2.5 gm./ lit. of water or dithiocarbamates (Mancozeb, Ferbam or Ziram) at 2.0 gm./ lit. of water gives adequate control of the disease.

Diseases of minor importance. In addition to the diseases dealt above, coffee is vulnerable to several other diseases, which may affect one or more parts of the plant or the whole plant. *Rhizoctonia solani* and *R. bataticola* cause 'Collar rot' and 'Damping off' that can kill seedlings; *Corticium salmonicolor* causes 'Pink disease'; *Fomes lignosus* causes 'Brown root' and 'Stump rot'; *Capnodium braziliense* causes 'Sooty mould'.

Arecanut *(Areca catechu)*

1. Foot rot or 'Anebe roga' of arecanut

Ganoderma lucidum

The disease was first reported in India from Karnataka state in 1807. Now, it is prevalent in the states of Karnataka, Kerala, Tamil Nadu and Assam. It is a very destructive disease of arecanut and the affected trees are killed within a short span of time. The name **'Anabe roga'** denotes a disease caused by **'mushroom'**.

Symptoms. The initial symptoms of the disease appear as slight yellowing of the leaflets of the outer whorls of leaves. Yellowing of the inner whorls of the leaves follows this and the entire crown becomes yellow. Then the leaves of the outer whorl turn brown and start drooping around the stem. As the disease advances, the remaining leaves also droop down in quick succession, while only the spindle alone stands erect without unfurling. New leaves may not be formed or there may be considerable delay in the formation of new leaves. Soon, the hanging outer leaves fall off. Eventually, the entire crown topples and falls off and the decapitated stem alone stands. When the leaves start drooping down, the fruit bunches also hang down. The nuts are very much reduced in size and the kernel quality also is very poor. Further development of inflorescence is arrested and the flowers and immature nuts are shed. The infected stem becomes very week and breaks down easily during heavy winds. Small cracks may develop at the basal portion of the stem and reddish -brown, viscous fluid oozes out through the cracks. The bleeding may extend up to a height of one meter or more. The tissues around the bleeding patches are reddish-brown in color and are rotten. The discoloration and rotting may be found up to the center of the stem and even the vascular elements show such discoloration and rotting. Most of the roots show reddish-brown discoloration and rotting. Production of new roots is also very much reduced. In advanced stages of the disease development, when the tree is dead or in a dying state, the fructifications (basidiocarps) of the pathogen may appear on the basal stem portion as brackets, just above the ground level.

Invasion by the pathogen causes disintegration of the paranchymatous and vascular tissues of the stem. As a result, the translocation of water and nutrients to the aboveground parts of the tree is affected. This leads to **'pathological drought condition'** and death of the affected tree. Further, as the vascular tissues are affected, there is continuous bleeding from the stem region, which is also responsible for causing rapid death of the tree. Trees above the age of 5 - 10 years are more prone to attack by the disease. Besides arecanut, the pathogen attacks coconut, oil palm, palmirah and other palms, mango, *Casurina*, *Cassia* and many other trees.

The causal organism, mode of survival, spread, epidemiology and disease management. Refer 'Basal stem rot of coconut' (Page 450), **(Fig.109).**

2. Fruit rot or 'Koleroga' of arecanut

Phytophthora arecae (P. palmivora)

'Fruit rot', **'Koleroga'** or **'Mahali'** is a very serious disease of arecanut and is prevalent in Sri Lanka, Sumatra Islands, Maldives and India. In India, the disease was first reported from Karnataka in 1906 and occurs in the states of Karnataka, Maharashtra, Kerala and Tamil Nadu. The disease is called **'Mahali'**, which means devastation or **'Koleroga'**, which means rotting disease ('Kole' = rotting and 'Roga' = disease) or more commonly as **'fruit rot'**. The pathogen also causes **'bud rot'**. Besides arecanut, the disease attacks coconut and various other palms, cacao, rubber, cinchona, castor, safflower, *Citrus* species and several other hosts.

Symptoms. The disease usually occurs soon after the onset of monsoon rains. The disease is characterized by rotting and excessive shedding of immature nuts. Initial symptoms appear on the nuts as water-soaked lesions towards the base of nuts and change of coloration of the nuts. The color of the nuts change from their normal green into dark green and loose their luster. The lesions gradually enlarge in size and may cover the nuts completely. The affected nuts rot and shed from the bunches. On the fallen fruits, fine, white, felt-like mycelial growth appears in patches under humid conditions, which soon covers the entire surface of the nuts. Kernels of affected nuts show central internal depressions and radial strands and they also rot finally. Fruit stalks and rachis of the inflorescence are also attacked and they rot. Sometimes the bud region is infected and rotting sets in. This is followed by rotting and withering of all the leaves and bunches, resulting in drying of the crown. Bud rot often results in the death of the affected tree **(Fig.121)**.

The causal organism, mode of survival, spread, epidemiology and disease management. Refer 'Bud rot of coconut' (Page 457), **(Fig.111)**.

Diseases of minor importance. A few other diseases are also known to attack arecanut. *Thielaviopsis (Ceratocystis) paradoxa* causes 'Stem bleeding'; *Colletotrichum gloeosporioides* causes 'Inflorescence die-back'.

Oil palm *(Elaeis guineensis)*

Many of the diseases that attack other palm trees attack oil palm also. *Phytophthora palmivora* causes 'Bud rot'; *Colletotrichum gloeosporioides* causes 'Anthracnose'; *Pestalotiopsis palmarum* causes 'Leaf spot disease'.

Cocoa *(Theobroma cacao)*

1. Black pod disease of cocoa

Phytophthora palmivora

It is the most destructive disease of cocoa and causes considerable damage to cocoa pods and economic loss. The disease was first reported from West Indies in 1897 and is now widespread in all cocoa-growing countries of the world. The disease appeared in India in 1965 after its introduction to the country and occurs in the states of Karnataka, Kerala and Tamil Nadu.

Symptoms. Cocoa pods at every stage of development may be infected. Infection may start at the stalk end tip of the pods or on the lateral sides. Initial symptoms appear as brown, discolored patches on the pods. The brown discoloration spreads rapidly in all directions. A definite demarcation is seen between the diseased and healthy areas. Soon, the entire surface of the pod becomes brown. The inner tissues also become brown and rot. The rotting may be partial or complete. Under humid conditions, a fine, white, weft of mycelial growth appears on the surface of the affected pods. The beans in the infected pods are also discolored and are of poor quality.

The pathogen also causes seedling die-back and seedling blight during the monsoon periods and the affected seedlings may die within a short time.

Under too wet conditions and when the temperature remains very low continuously, the pathogen may cause stem canker in 2 - 16 year old plants. The fungus attacks the plant at or near the soil level, where it causes water-soaked, darkening of the bark of the trunk. Such water-soaked lesions may be formed on the branches also. The infected area enlarges and may encircle the stem or the branches. The leaves and pods wilt and fall off. On grown-up trees, the infected darkened area may be on one side of the stem and becomes a depressed canker below the bark in course of time. In the early stages, the diseased bark is firm and intact, but later it becomes shrunken and cracked. The canker may spread up into the trunk and sometimes into the branches or down into the root system. As the canker spreads and enlarges, it may girdle the trunk, branches or roots, as a result the growth of the tree is very much affected. The branches may show die-back symptoms and the pod yield is reduced markedly. Severely affected plants may die eventually. The infection of the trunk and branches may spread from the black pods **(Fig.122).**

The causal organism, mode of survival, spread and epidemiology. Refer 'Bud rot of coconut' (Page 457), **(Fig.111).**

Disease management

Agronomic practices (i) Infected pods remaining on the tree or that have fallen to the ground should be removed periodically and destroyed (ii) Disease-free planting stocks raised in disease-free nurseries should be used for planting (iii) Planting should be done in light soils with adequate drainage facilities (iv) In the early stages of disease development, the affected bark may be scrapped off and the wound dressed with Bordeaux paste (v) Severely affected branches may be cut off and destroyed. The cut ends should be treated with Bordeaux paste (vi) Any type of injury to the pods, trunk, branches, roots etc., caused either mechanically or by insects predisposes the parts to infection by the fungus. So, care should be taken to avoid causing such injuries.

Chemical control (i) Spraying with Bordeaux mixture - 1 % or copper oxychloride - 2.5 gm./ lit. of water prior to the onset of monsoon rains, followed by fortnightly sprayings controls the black pod disease. While spraying, care should be taken to give a thorough covering of the pods and pod bearing branches (ii) Spraying with systemic fungicides, such as metalaxyl - 1.0 gm. or focetyl-Al - 1.0 gm. or ethazol (Terrazole) - 1.0 gm. or propamocarb (Banol or Previcur) - 1.0 gm./ lit. of water has been found to be very effective (iii) Dipping the seedlings in the above-mentioned systemic fungicides prior to planting helps to prevent seedling infection (iv) Drenching the nursery beds with the above-mentioned fungicides also prevents seedling infection.

2. Cocoa swollen shoot

Cocoa swollen shoot virus or *Theobroma virus* 1

The disease was first reported from Ghana in 1940. Now, it is prevalent in Sierra Leone, Ivory Coast, Nigeria, Trinidad and Sri Lanka. The disease threatened the very cultivation of cacao in West Africa and several thousands of infected trees had to be cut off and destroyed, so as to eliminate the disease.

Symptoms. The disease is characterized by the development of abnormal swellings in the stem and taproot, necrosis of the lateral roots and chlorosis of leaves and pods. The pods produced are small, rounded and smoother and contain fewer and smaller beans of poor quality. The affected trees decline and die soon or may linger on **(Fig.123).**

The causal organism. The virus transmitting the disease is classified under the Genus - *Badnavirus*. The virus particle lacks a membrane envelope. The bacilliform, DNA viruses measure 142 x 27 nm in size. The thermal inactivation point of the viruses is 55° - 60°C.

Mode of spread. The virus is transmitted by several species of mealy bugs belonging to *Dysmicoccus, Ferrisia, Planococcus, Pseudococcus* and *Planococcoides* in a semi-persistent manner. The vectors are spread through wind, harvested fruits, planting materials and ants. The virus also attacks species of *Cola, Ceiba* and a few other hosts.

Disease management

Agronomic practices (i) Severely affected plants, especially young plants should be removed and destroyed (ii) Disease-free planting materials should be used for planting (iii) Dead branches, twigs and mealy bugs infested branches should be cut off and destroyed (iv) Alternate hosts in the vicinity of cacao gardens should be eradicated.

Fig. 121: 'Koleroga disease' of arecanut
Phytophthora palmivora

Fig. 122: Black pod of cocoa
Phytophthora palmivora

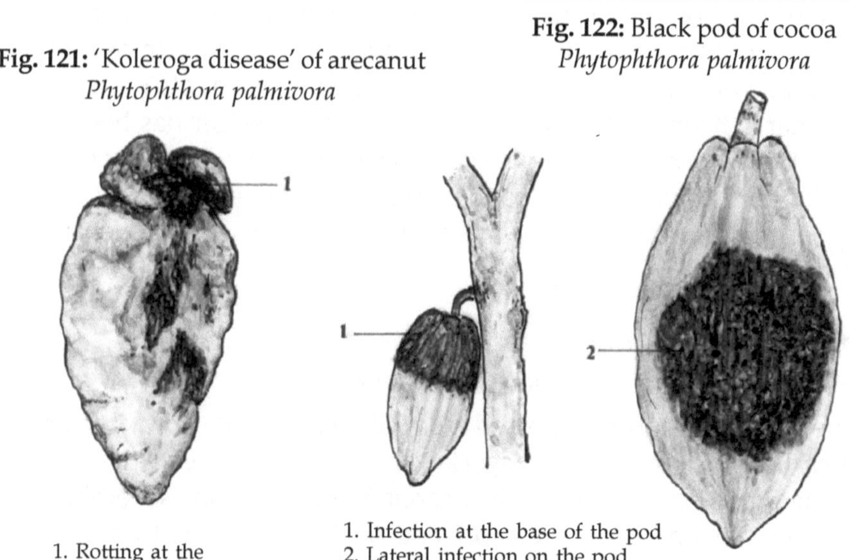

1. Rotting at the base of the nut

1. Infection at the base of the pod
2. Lateral infection on the pod

Fig. 123: Cacoa swollen shoot-*Cocoa swollen shoot virus*

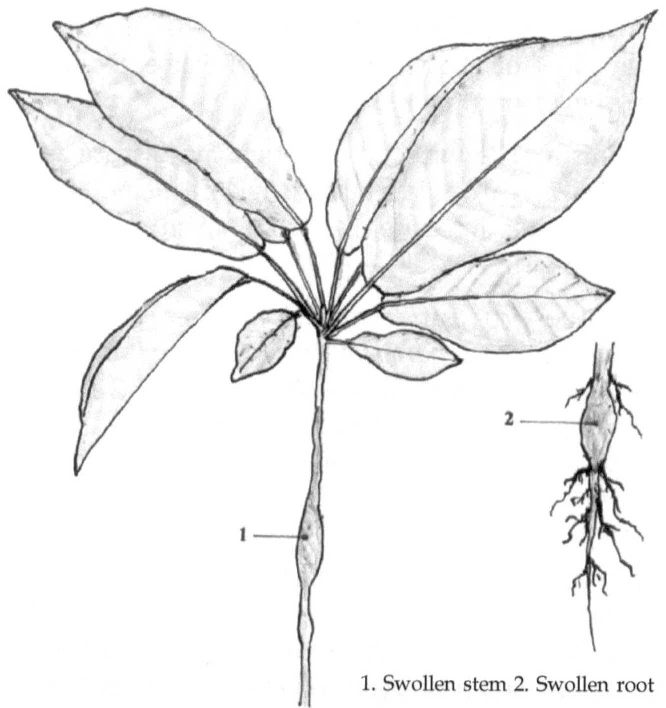

1. Swollen stem 2. Swollen root

Chemical control (i) Ant colonies should be destroyed by application of carbaryl or endosulfan dust (ii) The mealy bug vectors can be controlled by spraying with dimethoate - 2.0 ml. or monocrotophos - 1.5 ml. or methyl demeton - 2.0 ml./ lit. of water.

Minor diseases. Besides the diseases discussed above, a few other diseases also attack cacao. *Botryodiplodia theobromae (=Diplodia theobromae)* causes 'Charcoal pod rot'.

Rubber *(Hevea brasiliensis)*

1. Abnormal leaf fall, canker and pod rot of rubber

Phytophthora palmivora

The disease is found on a large scale in India, Sri Lanka, Malaysia, Burma and Java. It causes considerable damage to the trees and reduction in the yield of latex.

Symptoms. The pathogen causes different types of symptoms on different parts of the trees. The characteristic and striking symptom is the abnormal shedding of leaves prematurely, mostly during the months of June - August. The leaves are shed, either when they are still green or after becoming coppery- red. Initially, small, circular, water-soaked, grayish spots develop on the leaf lamina. Within a few days, the spots enlarge and become large, irregular spots. The spots may coalesce and form big patches covering major area of the leaves. The infection then spreads to the petioles and the affected leaves fall off. Under favorable conditions, total defoliation may occur within a fortnight, leaving behind only the bare branches. When the disease intensity is severe, young branches may also be attacked. The twigs of the affected branches turn brown, leading to die-back of the branches.

The fungus attacks the bark around the tapping cuts from where latex is collected, as well as the collar region of the trunk. In such infected areas, swellings are formed on the bark, which later develop cracks. From such infected areas, a yellowish fluid oozes out. The tissues below the bark rot and cankers are formed. When the bark is peeled off, the inner tissues appear brown or black and are rotten. Sometimes, only the bark alone is affected and becomes black and rotten. The fungus also attacks the pods. The infected pods turn brown or black, rot and are shed prematurely.

The causal organism, mode of survival, spread and epidemiology. Refer 'Bud rot of coconut' (Page 457), **(Fig.111)**.

Disease management

Agronomic practices (i) Fallen leaves, pods and other plant debris should be collected periodically and destroyed (ii) In the trunk region, the affected bark and inner rotten tissues should be scrapped out, the wound cleaned with mercuric chloride (I : 1000) solution and then treated with Bordeaux paste.

Chemical control. Spraying with Bordeaux mixture - 1 % or copper oxychloride - 2.5 gm./ lit. of water, using a high tree sprayer gives effective control. Prophylactic sprayings should be given to protect the trees from infection just prior to the onset of the South West monsoon rains, followed by another spraying when new flushes appear. In regions where rubber is cultivated on a large scale over extensive areas as in Kerala, aerial spraying should be given. For this Fycol is used at the rate of 2.0 lit./ acre mixed with soybean oil.

2. Powdery mildew of rubber

Oidium heveae

Symptoms. The initial symptoms of the disease appear mostly on the newly formed tender leaves and young branches during the months of January - March, when new flushes are formed. On the upper surface of the leaves, white, powdery coating consisting of the mycelium, conidiophores and conidia of the fungus appear in patches, while the corresponding under surface of leaves become yellow or brown. Soon, the entire area of the leaves and branches may be covered with the powdery coating. Severely affected young leaves show crinkling and curling. The edges of the leaves curl downwards and inwards. The affected leaves drop off and the petioles alone stand attached to the branches like a broom. Within a few days the petioles also fall off. On older leaves, white patches of powdery growth of the fungus develop, which soon become necrotic patches. Affected tender shoots dry from the tip downwards and cause die-back symptoms. The fungus also attacks the flowers and tender pods, which drop off. As a result of the fungus covering the leaf surface and due to necrosis of the affected areas, the photosynthetic efficiency of the leaves gets reduced markedly and consequently the vitality of the plants is also affected badly.

The causal organism. The mycelium of the fungus is much branched, septate, hyaline, thin-walled and completely ectophytic. Pyriform or globular haustoria produced from the hyphae penetrate into the epidermal cells of the host. From the superficial hyphae, erect conidiophores arise in profusion. Conidia are cut off from the conidiophores in chains of 2 - 7. They are single-celled, barrel-shaped or sub-cylindrical and measure 25 - 45 x 12 - 25μ in size. The matured conidia are at the top of the conidiophores, while the youngest are at the base.

Mode of survival, spread and epidemiology. The mycelium of the fungus remains in a dormant state underneath the bark of the trunk for a long time during unfavorable weather conditions. When new flushes come and when favorable conditions return, the fungus becomes active and produces conidiophores and conidia in large numbers.

Low humidity favors the growth of the pathogen and infection. The conidia do not require water for their germination and infection. Generally, cloudy days with warm weather and low humidity favor disease occurrence and development. The disease incidence is more after the South West and North East monsoon. Under shaded conditions and in higher elevations, the disease continues to be severe all through the year. Secondary spread of the disease is through wind-borne conidia. The conidia are very light and are carried over long distances by wind currents. The fungus also attacks *Euphorbia hirta*, a common weed.

Disease management

Chemical control. Spraying with wettable sulfur - 4.0 gm. or dinacap (Kelthane) - 1.0 gm. or tridemorph (Calixin) - 1.0 ml. or carbendazim - 1.0 gm./ lit. of water controls the disease effectively. The sprayings should be commenced as soon as new flushes come, followed by one or two more sprayings, at fortnightly intervals.

3. Brown root disease of rubber

Fomes noxius

Symptoms. The disease usually attacks old and weak trees. The leaves of infected trees turn yellow and wither. The roots are covered with white and brown mycelium, which penetrates the bark and forms a thin layer between the bark and the wood. The affected roots are encrusted with mass of soil adhering to the network of mycelium. The affected roots turn brown and rot gradually. The progress of the disease is slow and ultimately, the tree is killed. On the base of the dying or dead tree just at the soil level fructifications (sporocarps) appear.

The causal organism. The fructifications of the fungus appear as lateral brackets on the basal trunk region at the soil level. The sporocarps are perennial in nature. Sporulation occurs from the sporocarp year after year. The interior of the sporocarp becomes hard and woody and annual growth rings of newly formed hymenophores are seen. The pores on the underside of the sporocarp bear the hymenium on the inner surface from which basidia and basidiospores are produced abundantly.

Mode of survival and spread, The sporocarps (basidiocarps) can survive for several years in the old stumps and continue to produce spores and provide inoculum. The disease spreads through basidiospores carried through soil or water. The disease also spreads by direct contact of the roots of diseased and healthy trees.

Disease management

Agronomic practices (i) Old dead trees and stumps should be removed along with the root system and destroyed (ii) Application of lime to the soil at 1.0 ton/ acre suppresses the growth of the pathogen.

Chemical treatment (i) Affected bark of the tap root and stem should be scrapped off to expose the wood, the wound thoroughly cleaned with mercuric chloride - 0.1 % and then treated with Bordeaux paste (ii) Drenching the soil around the base of the trunk with Bordeaux mixture - 1 % or copper oxychloride - 2.5 gm./ lit. of water helps to suppress the growth of the pathogen.

4. Bird's eye spot of rubber

Drechslera (Helminthosporium) heveae

Symptoms. The disease attacks leaves of seedlings in the nursery, as well as plants in the field. The earliest symptoms appear as minute, circular, purple spots on the leaves. As the spots enlarge, they become whitish, surrounded by a narrow, purple border. The matured spots are usually circular and small in size, but occur in large numbers, scattered all over the leaf lamina. Severely affected leaves turn yellow, wilt and die very soon.

The causal organism. From the internal mycelium, conidiophores emerge through the stomata, either singly or in small groups. They are septate, pale brown in color, erect and sometimes geniculate and measure up to 200µ in length. Conidia are produced from the tips of the conidiophores singly. They are large, cylindrical to slightly tapering and have rounded ends, thin-walled, smooth, light-brownish in color, with 6 - 11 pseudosepta, without constriction at the septa and measure 90 - 130 x 15 - 20µ. They germinate by producing germ tubes from the end cells, but rarely from the middle cells.

Mode of spread and epidemiology. The disease spreads through wind-borne conidia.

The disease is more severe during periods of rainfall and heavy dew. Sufficient moisture to wet the conidia is necessary for the conidia to germinate and cause infection.

Disease management

Agronomic practices. Seedlings and grown-up plants should be maintained properly by application of recommended doses of manure and provision of adequate irrigation facilities.

Chemical control. Spraying with Bordeaux mixture - 1 % or copper oxychloride - 2.5 gm. or mancozeb - 2.0 gm. or edifenphos - 1.0 ml. or captafol - 1.5 gm./ lit. of water is effective in controlling the disease.

Diseases of minor importance. Rubber is also subjected to attack by a few more diseases. *Ganoderma pseudoferarum* causes 'Red root rot'; *Glomerella cingulata (=Gloeosporium albo-atrum)* causes 'Anthracnose' and 'Leaf fall'; *Corticium salmonicolor* causes 'Pink disease'.

ORNAMENTAL PLANTS

Rose *(Rosa species)*

There are several species of cultivated roses. Most of the modern roses are derived from crosses between the Chinese rose *(Rosa chinensis)* and the European roses *(Rosa gigantia, R. damasaena* and *R. moschata).*

1. Black spot of rose

Marssonina rosae

(Diplocarpon rosae)

The disease, which was first recorded in Italy in 1824, made its appearance in America in 1887. Now, the disease is prevalent in all the European countries, Australia, United Kingdom, India and in many other countries, where rose is grown on a large scale. Black spot is a most common and devastating disease of rose. All varieties of roses are found susceptible to the disease.

Symptoms. The fungus produces small, brownish-black spots on the upper, sometimes on the lower surface of the leaves. The spots are round and furnished with a radiating, fringe-like margin. Individual spots may reach a diameter of 1.0 cm. If a number of spots are produced, the spots may coalesce to form large, irregular, black lesions. The leaf tissues around the lesions turn yellow. In case of heavy infection, extensive areas of the leaves may become black, leading to large-scale leaf fall. As a result of infection, the epidermal cells dry and die and so, the spots turn black in color. Due to extensive leaf fall, the affected plants become weak and stunted and flower production is also badly affected. Tiny lesions also occur on the stem region and rarely on the flowers **(Fig.124).**

Causal organism. The fungus *Diplocarpon rosae* has its conidial stage in *Marssonina rosae* in the Form class - Deuteromycetes. The conidial stage is actually the parasitic stage, which is mainly responsible for the spread of the disease. The fungus has three types of reproductive structures viz., the summer acervuli, which produce conidia, the spermagonia on older leaves, in association with the ascigerous stage and the apothecia, on overwintered leaves. All three fructifications are developed between the cuticle and the epidermis and appear as tiny, raised specks within the blackened area, especially towards the fibrillar margin of the spots.

The conidia are produced from the sub-cutaneous acervuli and are borne on short conidiophores. The conidia are hyaline, oval to elliptical, bicellular and constricted at the septum, the cells sometimes breaking apart at the septum to function as separate spores. They measure 18.0-25.0 x 5.0-6.0µ. The tiny spermagonia contain a large number of bacilliform spermatia, 2.0-3.0µ in length. Sometimes typical bicellular conidia are also produced in the same spermagonia along with spermatia.

The sub-cuticular apothecia, which are produced on dead, over-wintered leaves, are sunken deep into the tissues. They are spherical to disciform and measure 100-250µ in diameter. The asci are oblong or sub-clavate, 70-80 x 15µ. in size and are interspersed with capitate, septate paraphyses. The ascospores closely approximate the conidia and measure 20.0-25.0 x 5.0-6.0µ. They vary considerably in shape and are hyaline and bicellular like the conidia. The apothecia open at the top in a stellate manner and the ascospores are exuded in viscid masses **(Fig.124).**

Mode of survival, spread and epidemiology. The perfect stage of this fungus is rarely found. Sometimes, the ascigerous stage is found in the overwintered leaves persisting in the plants or that had fallen to the ground and in such cases, the ascospores produced from apothecia may initiate the disease. Otherwise, the conidial stage alone is responsible for causing the disease. The conidia produced from acervuli on old leaves or on young branches late in the season initiate fresh infection.

Both conidia and ascospores are disseminated by wind or rainsplash. Insects, which are attracted by the viscid nature of the spore exudate also serve as disseminating agents. First signs of infection following inoculation with, either conidia or ascospores appear in about 10 days and acervuli containing conidia are developed 8 days later but, growth is much faster on young leaves than old.

Fig. 124: Black spot of rose-*Diplocarpon rosae (=Marssonina rosae)*

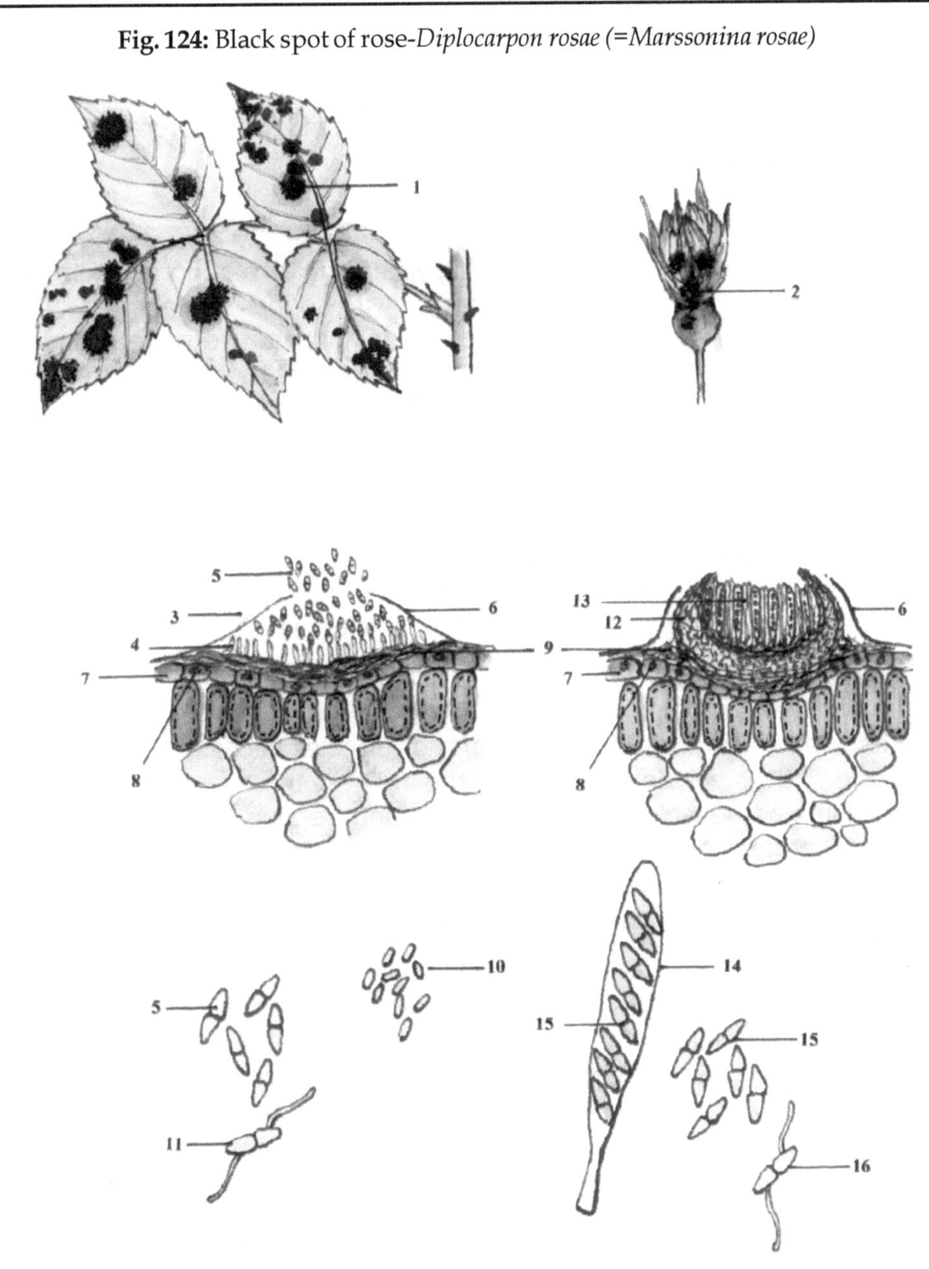

1. Black spot of rose symptoms on leaves 2. Symptoms on flower bud 3. Acervulus 4. Conidiophores 5. Conidia 6. Cuticle 7. Epidermis 8. Haustoria 9. Stroma 10. Microconidia (Spermatia) 11. Germinating conidium 12. Apothecium 13. Hymenium (Asci and paraphyses) 14. Ascus 15. Ascospores 16. Germinating ascospore.

Temperatures ranging from 20.0° - 26.5°C and a relative humidity of 92 - 97 % favor direct penetration of the cuticle by the germ tubes. The germ tubes form appressoria and after passing between the epidermal cells, the infection hyphae establish haustoria in the epidermis before proceeding to ramify between the epidermis and cuticle.

Disease management

Chemical control (i) Foliar spraying with a mixture of Bordeaux mixture - 0.75 % and liquid soap - 1.0 % or spraying with copper oxychloride - 0.25 % gives effective control of the disease (ii) Spraying with wettable sulfur - 0.4 % or carbendazim - 0.2 % is also effective (iii) Application of potash manure is found to reduce the extent of leaf infection

2. Powdery mildew of rose

Sphaerotheca pannosa

This very common disease of rose, which affects all kinds of roses, was first reported from Germany in 1819. Now, it is prevalent throughout Europe, America, Australia, India and many other countries, where roses are cultivated

Symptoms. The disease, which first appears on young leaves and shoots as gray or white spots, soon spreads until the entire leaves, long stretches of stems, buds and even the basal parts of prickles become almost covered with the white, powdery spores of the fungus. In severe attacks, plant growth may be adversely affected, the plants remain stunted, the leaves curl, wither and drop off. The affected buds fail to open. Sepals and receptacles of flower buds are often covered with the fungus. However, the affected plants rarely get killed **(Fig.125)**.

Causal organism. The fungus causing this disease is an obligate parasite. The mycelium, which is superficial, does not branch much, but grows as long hyphae and spreads over the host surface as an interwoven net. The hyphae are septate, the cells uninucleate, thick-walled and somewhat rigid. The superficial mycelium sends branched haustoria into the epidermal cells. From the mycelium, erect conidiophores are formed, which cut off 6 to 8 conidia. The conidia are elliptical or barrel-shaped and measure 22.9 - 28.7 x 13.6 - 15.8 μ. The conidia may germinate on either side of the young leaves by producing several germ tubes and cause fresh infections.

The cleistothecia are formed on the leaves, petioles, stems and around the bases of thorns, embedded in the felted patches of persistent mycelium, but are mostly found on the petioles and at the back of the midribs. They are globose to pyriform, 85 - 120 μ in diameter, with few or no appendages.

Fig. 125: Powdery mildew of rose-*Sphaerotheca pannosa*

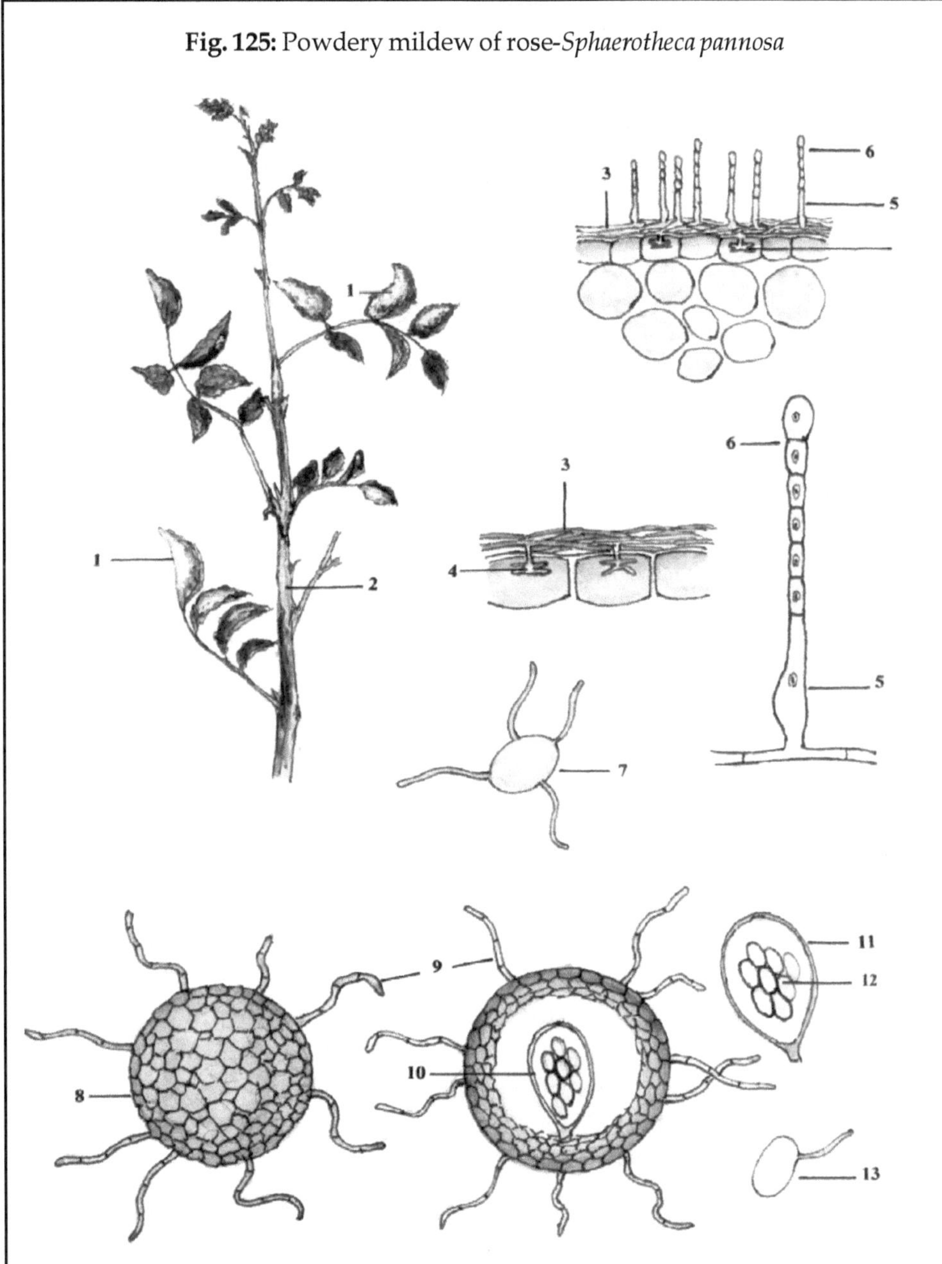

1, 2. Whitish conidial pustules on leaves and stem 3. Superficial mycelium 4. Haustorium 5. Conidiophore 6. Conidia 7. Germinating conidium 8. Cleistothecium 9. Appendages 10. Single ascus inside the cleistothecium 11. Ascus with ascospores 12. Ascospores 13. Germinating ascospore.

The appendages when present are short, tortuous, pale brown and septate. Each cleistothecium contains a single ascus. The asci are 8-spored, broadly oblong to globose and measure 88-115 x 60-75μ. The ascospores are oblong and measure 20-27 x 12-15μ in size **(Fig.125).**

Mode of survival, spread and epidemiology. Conidia produced on sheltered and protected plants, as well as the ascospores produced from overwintered cleistothecia bring about fresh infections in the spring. Sometimes, the resting mycelium may survive in the buds and cause fresh infections. After their separation from the conidiophores, the conidia are very sensitive to environmental conditions and are not adapted for long survival. Usually, they lose their germinating capacity in less than 24 hours. They are mostly wind-borne and are dispersed for short distances. They cannot withstand extreme temperatures and humidities

A temperature of 21°C and 80 - 90% relative humidity favor germination of conidia and ascospores.

Disease management

Agronomic practices. Severely affected and dried branches should be cut off and burnt.

Chemical control (i) The plants may be dusted with sulfur dust at 10 kg./ acre. But under some circumstances, especially when high temperature conditions prevail, young leaves may be adversely affected by sulfur dust and may lead to scorching (ii) Spraying the plants with wettable sulfur - 0.4% or karathane - 0.2% or carboxin - 0.1% is found to be effective in controlling the disease.

3. Rust of rose

Phragmidium mucronatum (P. disciflorum)

'Rust of rose' is common everywhere, on wild and cultivated roses. In India, it occurs in the states of Jammu and Kashmir, Himachal Pradesh, Punjab, Tamil Nadu, Uttar Pradesh and Karnataka. The disease is mostly restricted to the higher altitudes.

Symptoms. The disease appears as minute, dark-brown, pinhead-like dots on the upper surface of leaves. These are the spermagonia (pycnia) of the fungus. Orange to lemon yellow, circular pustules develop on the buds, stems, lower surface of leaves, petioles and fruits, but occur mostly on the stem region. These are the aecidia of the pathogen. Later in summer, small, yellow pustules are formed on the under surface of leaves, which represent the uredia of the pathogen. These pustules later turn dark-brown to black or fresh black pustules may develop on the under surface of leaves. These are the telia of the pathogen. The affected leaves turn yellow and drop off prematurely. Infected branches may become hypertrophied and deformed. The growth of the plant is adversely affected, as a result of premature defoliation and flower production is badly affected **(Fig.126).**

The causal organism. Spermagonia are rarely produced. They are found on the upper surface of leaves as minute, dark-brown, pinhead-like dots and produce minute pycniospores in large numbers. The 'caeomata' (aecia lacking peridium) occur singly or joined together in clusters. They break through the host epidermis exposing a sorus of bright orange aeciospores in short chains. The spores are verrucose, orange-yellow in color and 24-28 x 18-21μ in size. The uredosori are somewhat localized and surrounded by a circle of clavate hairs. The uredospores are ovate or ellipsoid, yellow, echinulated and 21-28 x 14-20μ in size, with numerous germ pores. The teliospores arise usually in the same uredosori. They are black, bulbous and have long stalks, 70-80μ in length. They are usually 6-8 septate, the terminal cell with a pointed papilla and measure 65-120 x 30-45μ in size. The teliospores require a long resting period prior to germination **(Fig.126).**

Mode of survival, spread and epidemiology. Only under certain undetermined conditions, the teliospores germinate and produce sporidia, which cause caeomatal lesions on both the leaves and young stems. But uredospores cause only leaf infection. Caeomata have been found to develop in successive years on the same branch lesions, indicating that the mycelium can survive within the host for a long time. The uredospores do not survive long during the cold winter and hot summer seasons. However, they survive to cause fresh infection in places, where the temperature is uniformly favorable, with adequate rainfall. Secondary spread is through uredia carried by wind or rainsplash.

Temperatures between 16°-21°C, with continuous moist conditions, such as heavy dew formation are necessary for infection with uredospores. Strong light depresses the rate of uredospore germination.

Disease management

Agronomic practices (i) Since the fungus is carried over mainly through perennating mycelium producing caeomata, all branches showing fungal fructifications should be cut off and destroyed (ii) All diseased leaves and branches fallen on the ground should be collected and destroyed.

Fig. 126: Rust of rose-*Phragmidium mucronatum (P. disciflorum)*

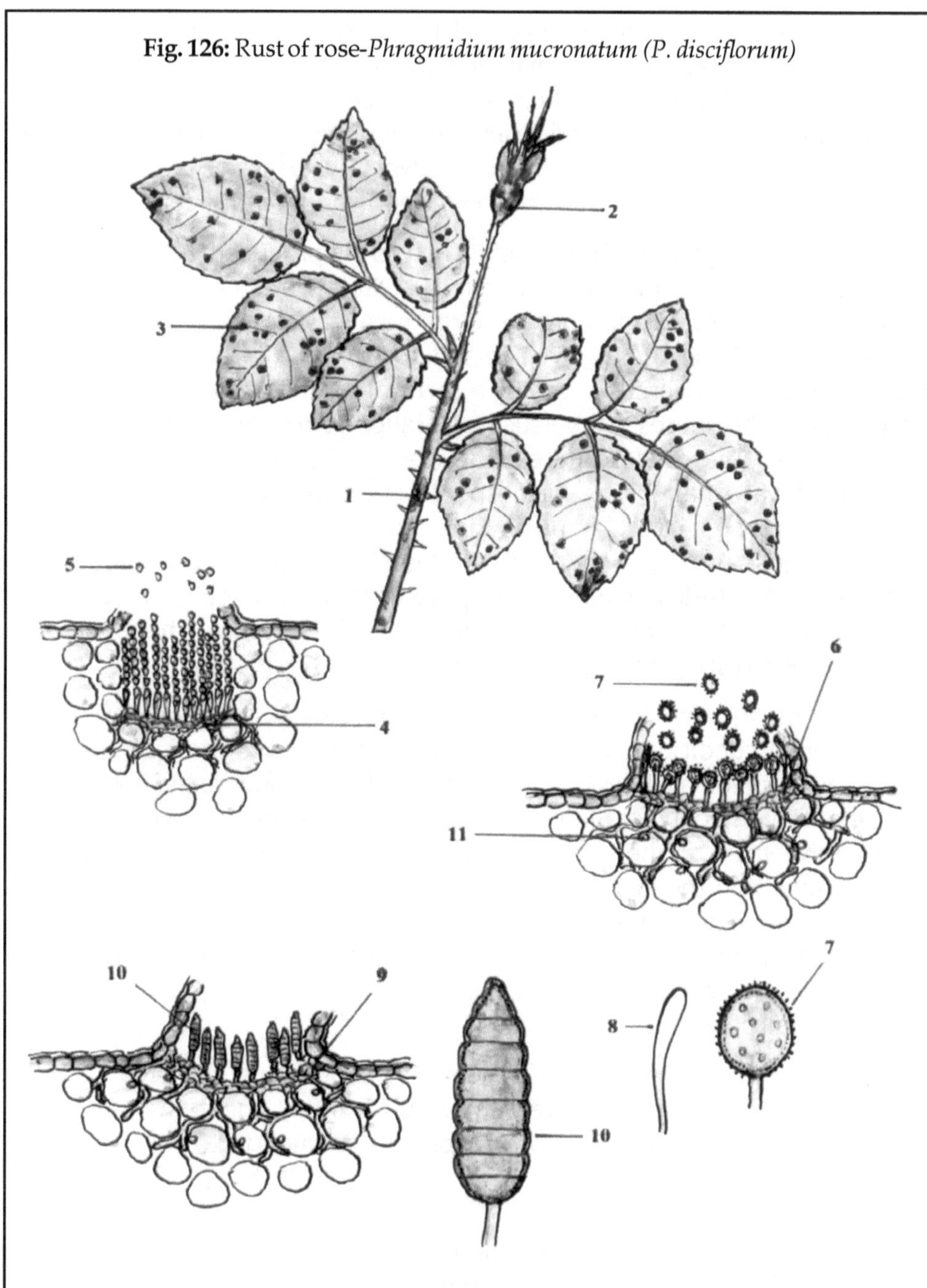

1. Aecia on stem 2. Aecia on bud 3. Uredo and teliosori on leaves 4. Aecidium 5. Aecidiospores 6. Uredium 7. Urediospore 8. Clavate hair-like structure 9. Telium 10. Teliospore 11. Haustorium.

Chemical control. The disease can be controlled by spraying with wettable sulfur at 4.0 gm. or chlorothalonil at 1.25 gm. or ferbam at 2.0 gm. or carboxin at 1.0 gm./ liter of water. Two to three sprayings may be given at fortnightly intervals, commencing from the time the symptoms are noticed.

4. Stem canker of rose

Coniothyrium fuckelii
(Leptosphaeria coniothyrium)

Symptoms. 'Stem canker' or **'graft canker'** attacks plants at all growth stages. Stem canker lesions appear as more or less confluent, reddish-brown to purplish spots on the bark in any part of the plant, such as branches, twigs, thorns or buds. Mostly, the lesions appear on one-year-old green wood. On these lesions, pycnidia are formed. The pycnidia break through and liberate sooty masses of spores. As the disease progresses, deep cracks or cankers are formed on the shoots. In the case of graft cankers, the color of the canker is darker brown than that of stem cankers and the cankers have a deeper dark margin. In the erumpent pustules, pycnidia are produced, which break through and liberate sooty spore masses. With repeated attempts by the host to heal the cankers by callus formation, the affected areas gradually become girdled with large warts and ultimately, the stem or branch dies. The cankers on the bark and wood interfere with the translocation of water and mineral nutrients to the top portions, with the result the shoots wilt and die. Insect damage, thorn pricks, leaf scars or aberrations may induce canker formation.

The causal organism. Only the pycnidial stage *(Coniothyrium fuckelii)* is more important and occurs commonly. The pycnidia are globose, ostiolate and are 180 - 200μ in diameter. The conidiophores are very minute and inconspicuous. The pycnidiospores are globose or slightly elliptical, olivaceous and are exuded in gelatinous masses. They measure 2.4 - 5.0 x 2.0 - 3.5μ in size. The perithecia appear on the cankers usually towards the autumn and the ascospores are discharged during the spring. Each ascus contains eight ascospores, which are 4-celled, brownish in color and are 10.0 - 15.0 x 3.0 - 4.0μ in size.

Mode of spread and epidemiology. Infection occurs through spores from cankers dispersed in splashing raindrops. The fungus gains entry into the host through wounds on the bark caused by insects, hailstorms, aberrations produced by friction between the shoots, but mainly through pruning wounds. The mycelium is capable of surviving on stubs left on the plants after pruning or in diseased pieces of diseased stems and branches on the ground.

The optimum temperature for disease occurrence and development is 25°-26°C. The pathogen also attacks raspberry.

Disease management

Agronomic practices (i) All diseased and dried parts should be cut off at the base and destroyed by burning. Then, the cut-ends should be treated with Bordeaux paste (ii) Pruned plant materials and plant debris should be collected and burnt (iii) Graft-work should be done carefully without leaving the stub of the scion exposed. The graft joints should be covered with wax.

Chemical control. Spraying the plants with Bordeaux mixture - 1 % or copper oxychloride - 2.5 gm. or mancozeb - 2.0 gm./ lit. of water controls the disease effectively. The plants should be sprayed after each pruning to prevent infection through the cut ends.

Minor diseases. Besides the diseases discussed above in detail, rose is attacked by many other diseases, some of which occur only sporadically and are not of much importance. *Diplodia rosarum* causes 'Die-back'; *Botrytis cinerea* causes 'Bud and twig blight'; several fungi, such as *Septoria rosae, Phyllosticta rosarum, Colletotrichum capsici, Alternaria alternata* etc. cause 'Leaf spots' and 'Blights'.

Jasmine *(Jasminum species)*

The commonly grown jasmine species are *Jasminum sambac, J. grandiflorum, J. auriculatum, J. officinale, J. angustifolium, J. paniculatum, J. pubescens, J. arborescens* and the yellow flowered *Jasminum primulinum* and *J. humule*.

1. *Alternaria* leaf spot of jasmine

Alternaria alternata (A. tenuis)

Symptoms. The disease affects both young and old leaves. On the leaves, spots appear from the margins mostly from the tips and enlarge downwards and inwards. The spots are irregular in shape, brownish in color, with a deep-colored border. In advanced stages of the disease, the central portion of the spots become necrotic, break and fall off, leaving shot holes. During humid weather the spots are covered with a dense growth of the fungus on both the surfaces of the leaves. This growth consists of conidiophores and conidia produced by the fungus. Severely affected leaves turn yellow and brown and then fall off.

The causal organism. The mycelium of the fungus is intracellular, hyaline at first, becoming darker with age, septate and branched. The conidiophores, which emerge through the stomata in the diseased spots, are erect, septate, pale brown in color, simple or branched, geniculate and are 6.0 - 45.0μ in length. Conidia are produced at the tips of conidiophores in chains of 2 - 4. They are muriform, fairly long beaked, septate, with few transverse septa, brown colored and are 23.6 - 40.2 x 11.0 - 18.6μ in size.

Mode of spread and epidemiology. The disease spreads mostly by air-borne conidia and also by raindropsplash. The fungus can survive in the diseased leaves on the plants or in the diseased plant debris in the soil.

The disease is more severe during the rainy seasons.

Disease management

Agronomic practices (i) Disease-free seedlings should be used for new plantings (ii) Diseased plant debris fallen to the ground should be collected and burnt.

Chemical control. Spraying the plants with Bordeaux mixture - 1 % or copper oxychloride - 2.5 gm. or mancozeb - 2.0 gm./ lit. of water controls the disease.

2. *Cercospora* leaf spot of jasmine

Cercospora jasminicola (C. jasmini)

The disease was first reported in India in 1946 and now, it is distributed all over the country. It attacks all species of *Jasminum*.

Symptoms. The disease attacks all aboveground parts of the plants. On the leaves, small, circular to irregular, reddish-brown spots of size 2.0 - 6.0 mm. appear on the upper surface. The spots may enlarge or they may coalesce to form large, irregular patches, which may cover almost the entire surface of the leaf lamina. The leaf margins curve inwards and the leaves become hard and brittle. Severely affected leaves fall off prematurely. Oval to elongate, brownish lesions develop on the petioles, stems, branches, sepals and petals. Vegetative buds are also affected and they become necrotic. Flower production is very much affected.. Severe incidence of the disease leads to acute defoliation and eventual death of the plants.

The causal organism. The mycelium of the fungus is septate, inter- and intracellular. Prior to sporulation, the hyphae form stromata beneath the stomata. The stromata are pale to dark-brown in color and fill the stomatal cavities. Conidiophores emerge in fascicles through the stomata.

They are pale olivaceus-brown in color, slender, sparsely septate and straight or sinuous. Conidia are produced singly from the tips of conidiophores. They are long, septate, slender, pale to olive-brown in color, obclavate to cylindrical, straight or slightly curved, thin-walled and are 20.0 - 66.0 x 2.0 - 4.0μ in size.

Mode of spread and epidemiology. The disease is air-borne. The pathogen attacks all species of *Jasminum* and many of them are perennial. So, there is a constant supply of conidia throughout the year to cause fresh infection, when conditions are favorable. The conidia produced from diseased leaves and plant parts that have fallen on to the ground also serve as inoculum.

The disease starts appearing soon after the onset of monsoon in June and continues till February. Moderate temperature and high atmospheric humidity favor the occurrence and development of the disease.

Disease management

Agronomic practices (i) Severely affected branches and twigs may be cut off and destroyed (ii) Diseased leaves and plant debris fallen on to the ground should be collected and burnt.

Chemical control. Spraying the plants with copper oxychloride - 2.5 gm. or mancozeb - 2.0 gm. or chlorothalonil - 1.5 gm. or carbendazim - 1.0 gm. or thiophanate - 1.0 gm./ lit. of water controls the disease. Spraying should commence from the time of onset of the monsoon rains and should be continued at 14 - 21 days interval till February.

Minor diseases. Jasmine is subjected to attack by several other diseases. *Alternaria jasmini* causes '*Alternaria* leaf blight'; *Uromyces hobsonii*, an autoecious rust fungus causes 'Rust'; *Fusarium solani* and *Sclerotium rolfsii* cause 'Wilt disease'; *Oidium jasmini* causes 'Powdery mildew'; *Phoma herbarum*, *Coniothyrium jasmini* and *Phyllosticta jasminina* cause 'Leaf spots'; *Glomerella cingulata (=Colletotrichum gloeosporioides)* causes 'Leaf blight'; *Xanthomonas campestris* pv. *jasmini* causes 'Bacterial leaf spot'.

Crossandra *(Crossandra species)*

Few *Crossandra* species viz., *Crossandra infundibuliformis*, *C. guineensis*, *C. mucronata* and *C. subacanlis* are commonly cultivated for their flowers. Only a few diseases are known to attack *Crossandra*. *Fusarium solani* causes 'Wilt'; *Rhizoctonia solani* causes 'Stem rot'; *Phytophthora nicotianae* causes 'Root and crown rot'; *Colletotrichum crossandrae* causes 'Leaf blight'.

Chrysanthemum *(Chrysanthemum species)*

Chrysanthemum is subjected to attack by a few diseases. *Septoria chrysanthemella* causes 'Leaf blotch'; *Fusarium oxysporum* f.sp. *chrysanthemi* and *Verticillium albo-atrum* cause 'Wilt disease'; *Erysiphe cichoracearum* and *Oidium chrysanthemi* cause 'Powdery mildew'; *Tomato spotted wilt virus* causes 'Spotted wilt'.

Dahlia *(Dahlia species)*

Many species of *Dahlia* are grown in gardens for their flowers. *Dahlia australis, D. coccinea, D. imperalis, D. pinnata, D. tenuis, D. tenuicaulis* and *D. rudis* are the commonly grown species.

1. Leaf smut of dahlia

Entyloma dahliae

The disease was first reported from South Africa in 1911. Now, it is found in many other countries, including England and India. In India, it is found in the states of Uttar Pradesh, Maharashtra, Karnataka, Rajasthan, Tamil Nadu and Himachal Pradesh.

Symptoms. The disease usually appears when the plants are well established and are in the flowering stage. The lower leaves of the plants are more prone to attack by the disease, while the upper leaves remain free from the disease altogether. Initial symptoms appear as small, yellowish-green spots, scattered all over the leaf surface. The spots are mostly circular, but the spots that are delimited by the larger veins are irregular in shape. The spots enlarge in size and may reach a diameter of 0.6 cm. or sometimes even up to 2.0 cm. They turn brown at the center and each spot is surrounded by a halo. Sometimes the brown central area may become necrotic, breaks and falls off leaving a shot hole, but still the halo persists. New spots continue to develop and they may coalesce to form large, brown patches on the leaf surface. In case of severe incidence, the leaves shrivel and die leading to acute defoliation. The petioles may also be affected and oval to elongate, brown lesions are formed.

The causal organism. The fungus is confined only to the spots on the leaves and forms sparsely branched, slender, septate mycelium in the mesophyll tissues. Later, the hyphal cells are transformed into 'smut spores' or 'chlamydospores' or 'teliospores', which occur singly or in chains in the leaf tissue. The spores occurring in chains are polygonal in shape due to mutual pressure, while those formed singly are spherical in shape. They are yellow to light-brown in color, becoming darker on maturity, thick-walled and measure

11-17μ in diameter. The spores germinate to produce a basidium (promycelium), which gives rise to a whorl of 5-6 secondary branches. These branches grow out into thin, thread-like hyphae, which eventually produce sporidia. The sporidia are narrow, needle-like, straight or slightly curved, non-septate and measure 45.0-75.0 x 2.0μ in size. The sporidia, which are produced in profusion on the surface of the leaves, present a white, powdery appearance.

Mode of spread and epidemiology. On germination of the teliospores, the promycelium comes out to the surface of the leaf and then produces sporidia. The teliospores may remain dormant in the infected leaf debris or in the soil until the next season. When favorable conditions return, they germinate and produce sporidia, which may cause fresh infection.

The optimum temperature for the germination of chlamydospores is 22°C. High humidity and low temperature favor disease development. Late planting and deficiency of lime in the soil predispose the plants to infection.

Disease management

Agronomic practices. Collection and destruction of infected plant debris from the soil, adoption of crop rotation and application of sufficient quantities of lime to the soil reduce the incidence of the disease.

Chemical control. Spraying with Bordeaux mixture - 1 % or copper oxychloride - 2.5 gm. or zineb - 2.0 gm. or carbendazim - 1.0 gm./ lit. of water, 2-3 times at fortnightly intervals controls the disease.

Minor diseases. Besides the disease discussed above in detail, a few other diseases attack dahlia. *Verticillium albo-atrum* causes '*Verticillium* wilt'; *Sclerotinia sclerotiorum* causes 'Stem rot'; *Erysiphe cichocearum* causes 'Powdery mildew'; *Cercospora dahliae* causes '*Cercospora* leaf spot'; *Choanephora infundibulifera* causes 'Blossom blight'; *Pseudomonas solanacearum* causes 'Bacterial wilt'.

Post-harvest diseases of fruits and vegetables

Besides attacking standing crops and causing direct losses to the produce or indirect losses by affecting the crop growth and productivity, many fungal and bacterial diseases attack the harvested produce and cause considerable damage. These diseases occur on the plant produce or plant products during harvesting, transportation, grading, packing and marketing and may continue till they reach the consumer. The harvested produce, such as seeds, fruits, vegetables, storage organs etc. are actually dormant structures, which may develop symptoms of diseases that had been contacted in the field and had remained latent or the infection may occur after harvest during handling. Post-harvest diseases causing deterioration and spoilage of agricultural produce may be termed as '**Market Pathology**'. These diseases may cause rotting, spotting, blotching and ghastly appearances of the produce, resulting in poor food value and low commercial value. Further, the disease causing organisms secrete certain toxic substances, such as '**mycotoxins**' and '**aflatoxins**', which may cause health hazards. Post-harvest losses to fruits and vegetables in developing countries have been estimated at 5 - 20 % or even more under some conditions.

All plant products are vulnerable to attack by post-harvest diseases. However, the more tender and succulent produce containing more water and rich nutrient contents are more prone to attack by fungi and bacteria. Post-harvest diseases are caused only by a few species of fungi and bacteria. Among the fungal organisms, some species of *Pythium, Phytophthora, Rhizopus, Mucor, Rhizoctonia, Sclerotinia, Peniciilium, Aspergillus, Fusarium, Alternaria, Botrytis, Gloeosporium, Colletotrichum, Geotrichum, Phoma, Diplodia* and *Helminthosporium* are responsible for causing such diseases, while among the bacteria, a few species of *Erwinia, Pseudomonas* and *Xanthomonas* are mostly responsible.

Fleshy fruits and vegetables are usually stored at high relative humidities to avoid shrinkage and under such conditions they are more vulnerable to attack by pathogenic microorganisms. Wounds, cuts, bruises or any kind of injuries caused during harvest and post-harvest operations favor entry of the pathogens into the produce. Entry through natural openings or by direct penetration of the cuticle and epidermis also occurs commonly when diseased produce is stored along with healthy ones. A single infected fruit or vegetable may become the source of infection to the entire lot of fruits or vegetables respectively during storage and transit.

Post-harvest diseases caused by bacteria. Succulent fleshy fruits and vegetables, such as carrots, potatoes, tomatoes, onions, celery etc. are attacked by a few species of bacteria and cause 'soft rot'. After entry into the host, the bacteria multiply enormously in the intercellular spaces. Soft rot is caused by the action of certain '**pectolytic**' and '**cellulolytic enzymes**' secreted by the bacteria on the pectic and cellulose components of the cell walls respectively. As a result of such enzymatic activity, the host tissues disintegrate, become soft and watery. The color of the tissues becomes darker and they rot. Other saprophytic bacteria, which get access to the infected produce, may aggravate the spoilage process **(e-g)** *Erwinia* *carotovora* pv. *carotovora* causes - 'Soft rot of carrots and onion'; *E. carotovora* pv. *atroseptica* - causes 'Soft rot of carrot'; *E. amylovora* - causes 'Soft rot of apple'; *Xanthomonas campestris* - causes 'Black rot of cabbage'; *X. vesicatoria* - causes 'Bacterial spot of tomato'; *Pseudomonas solanacearum* - causes 'Brown rot of potato'.

Post-harvest diseases caused by fungi. Many fungal pathogens are responsible for causing post-harvest diseases of fruits, vegetables and storage organs, such as tubers, rhizomes and corms, as well as grains. Some species of ***Rhizopus*** cause severe damage to peaches, grapes, strawberries, sweet potatoes, cucurbits, crucifers, tomatoes and eggplant fruits. *Rhizopus* species cause 'Soft rot of fleshy parts', which spreads very rapidly at high temperatures. The affected parts become soft and pulpy and often a watery fluid oozes out from such affected areas. The tissues become darker and rot. Under humid conditions, whitish, coarse, cottony and stingy mycelium of the fungus, with characteristic black sporangia covers the infected areas **(e-g)** *Rhizopus arrhizus* - causes 'Soft rot of apple, peach, pear and mango'; *R. stolonifer* - causes '*Rhizopus* rot of muskmelon, strawberry and sweet potato'.

A few species of ***Sclerotinia*** cause rotting of succulent storage organs and fruits in the field and after harvest. Under moist conditions, a soft, watery decay is caused and the affected areas are rapidly covered with a white, cottony growth of mycelium. In dry weather, the water liberated as a result of rotting of the tissues evaporates and the affected fruits or vegetables shrivel, dry and are mummified. The disease spreads rapidly by contact with diseased materials and both green and matured fruits and vegetables are susceptible. Black sclerotia develop later in the mycelial mat **(e-g)** *Sclerotinia sclerotiorum* - causes 'Watery soft rot of cabbage, carrot and beans'; *S. fructigina* and *S. fructicola* - cause 'Brown rot of stone and pome fruits'.

Various species of ***Alternaria*** cause decay on several fruits and vegetables, either before or after harvest. Brown or black, flat or sunken spots with definite margin or large diffused decayed areas may appear on the fruit or vegetable. The decay may be shallow or may extend deep into the fleshy portion. Usually a mat of mycelium that is white at first, but later turns brown to black develops on the surface of the rotten area under humid conditions **(e-g)** *Alternaria citri* - causes 'Stem end rot of citrus fruits'; *A. brassicae* - causes 'Brown rot of cauliflower and cabbage'; *A. tenuis* - causes 'Fruit rot of muskmelon'; *A. solani* - causes 'Early blight of tomato and fruit rot'.

Some species of ***Botrytis*** cause gray mould rots of fruits, vegetables and bulbs, both in the field and in storage. The decay starts at the stem end of the fruit or at any wound. The decay appears as a well-defined, water-soaked, brown area. The infection may spread deeper into the fleshy tissues. Usually, under moist conditions, a grayish or brownish-gray, granular, velvety mould growth appears on the surface of the decayed areas **(e-g)** *Botrytis cinerea* - causes 'Gray mould rot of apple, pear, strawberry, tomato, Citrus fruits and cabbage'.

Species of ***Geotrichum*** cause sour rot of fruits and vegetables. Ripe and over-ripe fruits and vegetables and those kept in moisture holding plastic bags are more prone to attack by these fungi. The produce may get infected before or after harvest through wounds. Infected areas become water-soaked, soft and pulpy. Later the skin usually breaks over the affected area and the inside is filled with a whitish, creamy fungal growth.

The decayed tissues become a sour smelling, watery mass **(e-g)** *Geotrichum candidum* - causes 'Sour rot of citrus fruits, tomatoes, carrots and many other fruits and vegetables'.

Some species of ***Fusarium*** cause 'pink' or 'yellow moulds' on vegetables and ornamentals, especially on root crops, tubers and bulbs. Fruits of cucurbits and tomatoes touching the soil or just above the soil level are usually attacked. Infection with *Fusarium* mostly takes place in the field before or during harvest, but very often occurs during storage also. Affected tissues become water-soaked and light-brown in the early stages and later turn dark-brown and somewhat dry. As the decay advances, the affected areas get sunken and the skin wrinkled. Tufts of whitish, pinkish or yellow mould growth appear on the affected areas. The rotting is much faster in the softer tissues of fruits, such as tomatoes and cucurbits and the tissues become rotten and pink in color. Pinkish mycelium may develop on the rotten areas **(e-g)** *Fusarium oxysporum* f.sp. *lycopersici* - causes 'Fruit rot of tomato'; *F. caeryleum* - causes 'rotting of potato tubers'; *F. roseum* - causes 'Crown rot of banana'; Several other *Fusarium* species cause 'Fruit rot of cucurbits, vegetables, tubers and bulbs'.

Various species of ***Penicillium*** cause 'the blue mould' and 'the green mould rots'. They are the most common and the most destructive of all post-harvest diseases. They attack almost all kinds of fruits and vegetables. They may cause even up to 90 % decay of infected fruits in storage, during transit and in the market. Entry into the tissues is mostly through wounds, but they can penetrate directly through the uninjured skin also. *Penicillium* rots initially appear as soft, watery, slightly discolored spots of varying sizes on any part of the fruit. The spots are shallow in the beginning and the infection spreads rapidly into the deeper tissues. Within a few days, almost the entire infected fruit is decayed. A whitish, mouldy growth develops on the surface on the infected area and starts producing spores. The sporulating area appears blue, bluish-green or olive-green in color, depending upon the color of the spore masses produced by the fungal species. White mycelium and a water-soaked zone surround the sporulating area. The decaying fruits emit a musty, foul odor. Later, secondary fungi and bacteria also enter the infected fruit and the fruit is reduced to a wet, soft, pulpy mass.

In addition to the direct losses caused by rotting of these agricultural produces, *Penicillin* species also produce several **'mycotoxins'**, such as **'patulin'** in the affected produce, the consumption of which causes health hazards to humans, animals and birds. The diseases caused by mycotoxins are called **'Mycotoxicoses'**. Most of the mycotoxins are produced by the common and widespread fungal species of *Penicillium, Aspergillus* and *Fusarium. Aspergillus* and *Penicillium* produce their toxins mostly in stored seeds and hay, but also on commercially processed foods and feeds including meat, cheese, spices, oilcakes etc. **(e-g)** *Penicillium expansum* - causes 'Blue mould rot of citrus fruits and muskmelon'; *P. digitatum* - causes 'Green mould rot of citrus fruits'.

Several species of ***Aspergillus*** are often responsible for the deterioration of stored grains, legumes and oilcakes. The colonies of *Aspergillus* appear black, brown, yellow or green in color, depending upon the color of the spore masses the fungi produce. The fungi produce several enzymes, which cause disintegration of the cells and rotting. Besides causing such direct damage to the produces, they produce toxins known as **'aflatoxins'** in the infected produce. In groundnuts, cottonseeds and other cereal seeds, nuts, fishmeal, oilcakes, food for humans and feed for animals and birds, aflatoxins are produced in high concentrations under warm and humid conditions. Some of these toxins, when ingested with the feed by dairy cattle are secreted in the milk in toxic form. Acute aflatoxin poisoning may be even lethal to humans, animals or birds. Some species of *Aspergillus* cause diseases in human beings, animals and birds **(e-g)** *Aspergillus niger* - causes 'Black mould rot of apple, pear and mango'; *A. flavus* - produces aflatoxin; *A. fumigatus* - causes **'Aspergillosis'** of the lung in animals and humans.

Several other fungi also cause many other post-harvest diseases. *Pythium aphanidermatum* and *P. butleri* cause 'Cottony leak of cucurbits'; *Phytophthora infestans* causes 'Fruit rot in tomato' and 'Tuber rot in potato'; *Guignardia bidwelli* causes 'Black rot of grapes'; *Gloeosporium fructigenum,*

G. musarum, G. ampelophagum, G. psidii and *G. mangiferae* cause 'Fruit rot of apple, banana, grapes, guava and mango' respectively; *Phoma destructiva* and *P. violacea* cause '*Phoma* rot of tomato and of apple and pear' respectively; *Diplodia natalensis* causes 'Stem end rot of citrus and mango'; *Sclerotium rolfsii* causes 'Stem end rot of citrus fruits'.

Mode of spread and epidemiology of post-harvest disease organisms. In most of the post-harvest diseases, the pathogens responsible for causing deterioration and spoilage, enter the fruits, vegetables or seeds in the field before harvest. The pathogens remain latent or dormant in the produce and cause diseases in storage, transit and in the market, when conditions are favorable for their growth and development. In the case of certain species of *Gloeosporium*, the spores usually germinate in moisture and infect fruits of banana, citrus, mango, guava, papaya etc. at any time during their development. The infection remains latent for months together and causes diseases after harvest when the fruits ripen. Similarly, species of *Diplodia* causing stem end rot of fruits, such as citrus and mango, enter the fruits at any stage of development and remain latent till the fruits start ripening. Tuber rot of potato and fruit rot of tomato caused by *Phytophthora infestans* usually occur in the field before harvest and the rotting continues during storage. Bacterial soft rot of potato, carrot etc. caused by *Erwinia carotovora* is initiated through lenticels, where the bacteria harbor at the time of harvest. Initial infection by *Sclerotinia sclerotiorum* occurs through snap beans at the time of harvest and subsequently the disease spreads to other beans rapidly in storage.

In many cases, the injury caused while severing the fruits or vegetables from the plant is a common point of entry of wound pathogens, which cause post-harvest diseases. Crown rot of banana, stem end rot of mango, papaya and cucurbits, black rot of cabbage etc. are mostly initiated through such wounds. Further, wounds, cuts, bruises etc. caused at the time of harvest, transit, storage, grading, packing etc. pave the way for the entry of the wound pathogens into the produce and is one of the main predisposing factors of post-harvest diseases. Similarly, injuries caused by insect pests and nematodes also predispose the produce to post-harvest diseases. *Penicillium, Rhizopus, Geotrichum* and few other pathogens are not capable of direct penetration of the intact surfaces of the host, but gain entry into fruits, vegetables, tubers, corms, rhizomes, groundnut etc. only through some kind of injuries and cause post-harvest diseases. However, many fungal pathogens are capable of entering the host by direct penetration and cause post-harvest diseases. Harvested produce may become infected by contact with already infected and diseased produce through natural openings in many cases. Physiological injuries caused by extremes of weather conditions, such as frost, heat etc. also predispose the produce to post-harvest diseases.

Management of post-harvest diseases

Control of pre-harvest infection. Many of the post-harvest diseases are carried over from the fields to the harvested produce. So, every care should be taken to protect the produce from infection in the field itself. Several tropical fruits are infected by species of *Gloeosporium* at all stages of development. The diseases caused by these pathogens are controlled by spraying with Bordeaux mixture - 1% or copper oxychloride - 2.5 gm. or mancozeb - 2.0 gm. or captafol - 1.5 gm. or chlorothalonil - 1.5 gm. or carbendazim - 1.0 gm. or thiobendazole - 1.0 gm. or bitertanol - 1.0 gm./ lit. of water at fortnightly intervals from the time the fruits start developing. Pre-harvest sprayings with systemic fungicides have been found to be very effective in eliminating latent infection in the fruits and vegetables, besides giving protection against their establishment and reduction in disease incidence after harvest. Several diseases caused by species of *Alternaria, Diplodia, Phoma, Guignardia* and *Botrytis* can be controlled by pre-harvest sprayings with systemic fungicides. Pre-harvest spraying with the antifungal antibiotic Aureofungin Sol - 1.0 gm./ 5 lit. of water has also been found to be effective in controlling a number of post-harvest diseases. Many insect pests, especially sucking pests are known to cause injuries in fruits and vegetables through which the pathogens can get access and cause diseases in fruits and vegetables in the field. Rotting of citrus fruits caused by *Penicillium digitatum* and *P. italicum* is facilitated by punctures made in the skin of the fruits by white flies. So, insect pests causing injuries and predisposing the fruits and vegetables to fungal and bacterial infection should also be controlled by application of suitable insecticides in the field.

Harvesting and handling. Utmost care should be taken at the time of harvesting, during subsequent handling, storing and transporting, so as to avoid causing any kind of injuries, wounds, cuts and bruises that may serve as ports of entry for the pathogens. Harvesting and handling of the produces should be done when the weather is dry, sunny and cool to avoid further chances of contamination and infection. Harvesting should be done preferably in the morning hours to avoid too much heat. All fruits and vegetables showing signs of disease infection should be sorted out and removed from the produce that are to be stored, so as to avoid further spread of the disease. The storage containers, ware-houses, godowns and transporting vehicles should be kept clean and disinfected with formaldehyde, copper sulfate or any other suitable disinfectant. Produce, such as sweet potatoes, onions and other such roots, bulbs and tubers can be protected from some fungal decays by curing them at 28° - 32°C for 10 - 14 days, which helps to reduce surface moisture and heal any exposed wounds by suberization or wound periderm formation. However, tuber crops, if exposed to direct sunlight for more than a few hours may be affected by severe storage rots. Sufficient care should also be taken to avoid causing injuries during transport of the produces from the field to the store-houses and from the store-houses to the markets and during loading and unloading operations.

Storage. The produces should be stored at temperatures low enough to slow down development of infection, but very low temperatures may cause chilling injuries, which may predispose the produce to fresh infection. The produce should be free of surface moisture when placed in storage. There should be adequate ventilation to prevent build-up of excessive humidity that may condense on the surface of the produce. Further, the produce should be free from insect and other pests at the time of storage and should be kept free of them while in storage to avoid creation of new wounds and development of new infections. In case of long term storage and long distance transport of fruits and vegetables, they should be kept under controlled atmosphere employing low oxygen level (5 %) or increased carbondioxide levels (5 - 20 %), so as to suppress the respiration rate of the host and the pathogen. Addition of carbon monoxide (10 %) to any of these gases improves the storage condition. These measures reduce the chances of development of post-harvest diseases. Refrigeration is one of the useful methods for long-term storage of fresh vegetables. Low storage temperature is very effective for apples and grapes, which tolerate near-freezing temperatures that check the development of pathogenic microorganisms. Pathogens, which thrive at relatively high temperatures, such as *Geotrichum* and *Erwinia*, are inhibited at a storage temperature of 0°C.

Hot air or hot water treatment is sometimes used to eradicate latent infections at the surface of some produce. Incipient infection of *Phytophthora* can be eradicated by immersing the fruits in water at 60°C for 1½ minutes. Decay of peaches and stone fruits is reduced by immersing the fruits in hot water at 51.5°C for 2 - 3 minutes or at 46°C for 5 minutes. Hot water treatment has been found to be effective in controlling *Penicillium*, *Diplodia* and *Colletotrichum* infection on mangoes and papaya. Aerated steam treatment is effective against many post-harvest diseases of mango, guava, citrus and papaya and this treatment has been found to be better than hot water treatment

Biological control, employing antagonistic microorganisms or their metabolites have been successfully used for the control of post-harvest fungal and bacterial diseases. *Trichoderma* species reduce the severity of a number of fruit rots by producing antifungal antibiotics, lytic enzymes and volatile metabolites, which inhibit sporulation or interfere with some physiological functioning of the pathogens. *Bacillus subtilis* has been found to be antagonistic to the brown rot pathogen, *Sclerotinia* of stone fruits and *Pseudomonas putida* to be antagonistic to *Erwinia* causing tuber rot of potatoes. *Trichoderma* species also give good control of ginger rhizome rot caused by *Sclerotium rolfsii*

Irradiation with some light rays has been found to reduce storage rots of some produce and are very effective in case of superficial infections. Gamma rays are capable of penetrating the tissues of fruits and vegetables and destroying pathogens established deep in the host tissues. Gamma irradiation reduces rots caused by *Aspergillus niger* by inhibiting spore germination.

Post-harvest decays can be controlled by some chemical treatments by preventing infection and suppressing the pathogens on the surface of harvested produces. Elemental sulfur is often used as dust. It undergoes sublimation and protects fruits and vegetables from attack by pathogens in storage. Spraying on the fruits and vegetables or infiltrating the fruits with calcium chloride reduces development of post-harvest infection on fruits. Other chemicals most commonly used for controlling post-harvest diseases of fruits, particularly in the case of citrus fruits include diphenyl, sodium-o-phenyl phenate, dichloran, thiobendazole, benomyl, thiophanate methyl, imazalil, chlorothalonil, cytovirin, triforine, captan, iprodione, vinclozolin, soda ash and borax. They are usually applied as fungicidal wash treatments and are more effective when applied hot between 28° and 50°C, depending upon the susceptibility of the produce to heat injury. Chlorinated water is also used to wash and treat tomatoes and certain other vegetables in commercial packing-houses. Elemental sulfur, sulfurdioxide, dichloran, captan, and benzoic acid are used for the control of storage rots of stone fruits, pome fruits, bananas, grapes, strawberries, melons and potatoes.

Some growth regulators, such as indole acetic acid (IAA) and maleic hydrazide (MH) have also been found to be effective in preventing infections by *Aspergillus* and *Rhizopus* rots in papaya.

Leaf extracts of *Eucalyptus globulus, Punica granatum* and *Datura stramonium* help to prevent fruit rot of lemon. Leaf extract of tulsi checks infection by various rot pathogens. Groundnut, mustard, castor and paraffin oils have also been found to be effective against *Rhizopus* rot of mango and *Aspergillus* rot of papaya.

In some crops, post-harvest diseases are controlled by periodic fumigations with sulfur dioxide. Sulfur dioxide fumigation eliminates most of the pathogens in grapes particularly *Botrytis cinerea,* as well as pathogens, such as *Glomerella cingulata* and *Penicillium expansum* affecting apple and pear. Vapors of biphenyl inhibit the sporulation of *Penicillium* on the surface of decayed oranges. Lemons may be fumigated with nitrogen trichloride during storage to inhibit sporulation of *Penicillium.*

Packing. The boxes, crates, baskets etc. used for packing fruits and vegetables should be disinfected thoroughly with formaldehyde, nitrogen trichloride or sulfur dioxide. Various antifungal chemicals are used for impregnating fruit wrappers and box liners. Wrapping grape clusters in tissue paper impregnated with sodium-o-phenyl phenate or sodium metabisulphite reduces post-harvest decay of grapes caused by *Aspergillus niger* and *Penicillium canescens.* Oiled paper wraps containing copper sulfate can prevent the spread of *Botrytis* rot on pears. Fruits wrapped individually in polyethylene films prevent the spread of diseases by direct contact between fruits in the same container. Keeping fruits wrapped individually in newspaper impregnated with carbendazim or thiobendazole prevents blue-green mould. Apples coated with mustard oil, paraffin wax or castor oil checks infection by a large number of pathogens. Post-harvest keeping quality of apples, pears and plums is extended by surface coating with 'Prolong', a mixture of sucrose esters of fatty acids and polysaccharide. Prolong treatment reduces the permeability of oxygen without affecting carbon dioxide permeability and thus reduces the activity of the pathogens.

❑❑❑

Chapter 3

Edible Mushrooms

'**Fungi**' including '**mushrooms**' are '**Thallophytes**', which do not possess chlorophyll and have been classified under the **Kingdom - Myceteae**. They do not have stems, roots and leaves and they lack a vascular conducting system as in the case of plants. They propagate by means of spores, either asexual or sexual. Though almost all the fungi are microscopic, the mushrooms are among the largest fungi and had attracted the attention Scientists and Naturalists of many countries dating back to centuries. The Greek word '**mykes**', which means '**cap**', was given to the mushrooms because of the '**cap-like**' appearance of the pileus and the Greek word '**Mycology**' (mykes = mushroom + logos = discourse) denotes the study of mushrooms. What we call mushrooms, are actually the sporocarps produced by several species of fungi, while the mycelium that produces them live in various types of substrata and obtain nourishment from them. Usually, the fleshy, macroscopic fruiting bodies or sporocarps produced by fungi belonging to Basidiomycetes and Ascomycetes are called '**mushrooms**'

The Romans attributed the appearance of mushrooms and truffles to lightning hurled by Jupiter to the earth. Even now in India and several other countries, people believe that mushrooms appear only after lightning and thunder. The Greeks believed mushrooms as **'Food for the Gods'**. The Romans supplemented mushrooms in the food of soldiers for strength and vitality. The much-priced mushrooms were supposed to be the food delicacy of the elite society of the Romans and Greeks in olden days. The Chinese consider mushrooms as a cure for many diseases and they believe that mushrooms increase the longevity. A number of medicines in all forms are prepared from mushrooms by the Chinese.

The mushrooms comprise a large, heterogeneous group having various shapes, size, color, appearance and edibility. Mushrooms are cosmopolitan in nature and appear in a wide variety of habitats, ranging from the Arctic regions to the tropical regions. There are parasitic, saprophytic and mycorrhizal forms of mushrooms. More than 69,000 different species of mushrooms have been recorded so far, of which about 2,000 are known to be edible. Many species of mushrooms are found wild in very diverse habitats in nature and among such wild species there are both edible and poisonous ones, which are sometimes very difficult to identify.

There are several wild species of mushrooms, which are edible, but are not amenable for cultivation under artificial conditions. In nature, they appear suddenly in some specific localities during some seasons and disappear within a short span of time. The **'morels'** belonging to the Genus - *Morchella*, viz., *Morchella deliciosa, M. esculenta, M. crassipes, M. semilibera (hybrida), M. conica* etc. are very delicious and have excellent flavor. But these cannot be cultivated under artificial conditions. These mushrooms, which occur wild in the high hilly terrains of Jammu and Kashmir, Himachal Pradesh and Uttaranchal are usually collected by the native people whenever they occur. These highly priced mushrooms are mostly exported to the European countries.

From the beginning of the 20th century, different types of mushrooms have been grown commercially in America, Europe, France, Netherlands, Canada, China, Japan and many other countries. Presently mushrooms are being artificially cultivated in about 100 countries and the annual production of edible mushrooms in the world at present is estimated to be around five million tons and is increasing at the rate of seven per cent per annum. With increasing knowledge about the food value and other good qualities of mushrooms and methodology for their cultivation, mushrooms have been accepted as a quality food by people all over the world. The increasing demand for food, especially protein rich food for the ever increasing population can be fulfilled to some extent by including mushrooms in the regular diet. A large majority of people in India are vegetarians and they use pulses as their main source of protein. In this context, the protein rich, low calorie and delicious mushroom food may serve as an excellent alternate dietary supplement for all sections of people.

According to the Chinese, as on 1993 about 14 species of mushrooms were cultivated for food and of these only six viz., the 'enoki mushroom' *(Flammulina velutipes)*, the 'Shiitake mushroom' *(Lentinus edodes)*, the 'paddy straw mushroom' *(Volvariella volvaceae)*, the 'oyster mushroom' *(Pleurotus ostreatus)*, the 'meadow mushrooms' or 'button mushrooms' *(Agaricus brunnescens* and *A. bitorquis)* were grown on industrial scale. At present about 20 species of mushrooms are being cultivated commercially and 8 - 10 species are grown on an industrial scale. In India about 300 species under 70 genera have been reported. About 80 mushroom varieties are being grown experimentally under artificial conditions. About 20 species are being cultivated commercially and 4 - 5 species on an industrial scale. The most important varieties commercially grown are furnished in **table 8**.

Table 8 : Important mushrooms grown in India

No.	Common name	Scientific name
1	Button, temperate or European mushroom	*Agaricus brunnescens (bisporus)*
2	Edulis or hot weather mushroom	*Agaricus bitorquis*
3	Oyster mushroom	*Pleurotus sajor caju, P. flabellatus, P. ostreatus, P. florida, P. citrinopileatus, P. comucopiae, P. sapidus, P. membranaceous, P. eryngii, P. fossulatus, P. eous,* and *P. platypus*
4	Paddy straw, Chinese or Tropical mushroom	*Volvariella volvaceae* and *V. diplasia*
5	White milky mushroom	*Calocybe indica*
6	Black ear mushroom	*Auricularia polytricha*
7	Shiitake mushroom	*Lentinus edodes*
8	Enoki mushroom	*Flammulina velutipes*
9	Reishi mushroom	*Ganoderma lucidum*

Varieties suitable for cultivation in Tamil Nadu viz., the pure white oyster mushroom, Co.1 *(Pleurotus citrinopileatus)*, the light pink oyster mushroom, APK.1 *(P. eous)*, the white oyster mushrooms, MDU.1 *(P.djamor)*,

MDU.2 *(P. flabellatus)*, PF *(P. florida)*, the grayish-white oyster mushroom, Ooty 1 *(P. ostreatus)*, the gray oyster mushroom, M.2 *(P. sajor caju)*, the white milky mushroom, APK.2 *(Calocybe indica)*, and the pure white button mushroom, Ooty 1 *(Agaricus bitorquis)* have been released by the Tamil Nadu Agricultural University, Coimbatore.

IMPORTANCE OF MUSHROOMS

1. **Mushrooms are rich in food value.** Mushroom is a purely vegetarian, delicious and rich food with excellent flavor and contains many of the vital nutrients required for the normal functioning of the human body. The nutritional index of mushrooms is much higher than most of the vegetables. Mushrooms contain higher percentage of protein than vegetables and fruits and are good substitutes for pulses. About 80% of protein present in mushrooms is easily digestible. Compared to other food supplements, they contain lesser quantity of carbohydrates, sugars and fat. A comparative statement of the nutritional index of mushrooms is furnished in **table 9.**

Table 9 : Composition of mushrooms as compared to some vegetables (fresh weight basis)

Food item	Calorie	Moisture	Fat	Carbohydrate	Protein
Mushroom	16	91.1	0.3	4.4	2.69
Potato	83	73.8	0.1	19.1	0.76
Beetroot	42	87.6	0.1	9.6	0.96
Eggplant fruit	24	92.7	0.2	5.5	1.51
Cabbage	24	92.4	0.2	5.3	1.84
Cauliflower	25	91.7	0.2	4.9	2.88
Green peas	98	74.3	0.4	17.7	2.61

Mushrooms contain large quantity of easily digestible protein (2.69%), which is highly essential for the build-up of the human body. Further, mushrooms are low-calorie food (16%) when compared to several vegetables and is highly beneficial for people who wish to shed excess fat from their body. The fat content in mushrooms is also very low. While cholesterol is absent, ergosterol is present and hence, it is an ideal food for heart patients. Mushrooms contain much lesser quantity of carbohydrates (4.4%) and starch is absent. So, it is an ideal food for diabetic patients also.

Table 10: Composition of different types of mushrooms (Percentage on dry weight basis)

Mushroom types	Protein	Fat	Carbohydrates	Potash	Fiber
Button mushroom	23.9	1.75	41.3	7.7	8.0
Oyster mushroom	26.6	2.00	50.7	6.5	13.3
White milky mushroom	32.3	0.70	59.8	8.9	41.0
Paddy straw mushroom	21.2	10.00	58.6	10.1	11.1

Besides containing such large quantity of protein, mushrooms contain substantial quantity of Vitamin C and Vitamin B complex, such as Thiamin, Riboflavin, Niacin, Ascorbic acid, Pantothenic acid and Folic acid, which are essential for the physiological functioning of every part of the human body. The vitamins present in different types of mushrooms are presented in **table 11.**

Table 11: Quantity of vitamins present in different types of mushrooms (mg./ 100 gm. dry weight)

Mushroom types	Thiamin	Riboflavin	Niacin	Ascorbic acid
Pleurotus species	4.8	4.7	108.7	0.0
Agaricus species	1.1	5.0	56.7	81.9
Volvariella species	1.2	3.3	91.9	20.2

Mushrooms contain appreciable quantity of minerals, which are vital for the functioning of different organs of the human body. The iron present in mushrooms is in the available form and can be readily assimilated by the human systems. The Potassium : Sodium ratio is very high in species of *Agaricus*, which is very good for patients suffering from **'hypertension'**. The minerals present in different types of mushrooms are presented in **table 12.**

Table 12: Quantity of minerals present in different types of mushrooms (mg./ 100 gm. dry weight)

Mushroom types	Calcium	Phosphorus	Iron	Sodium	Potassium
Pleurotus species	98.0	476.0	8.5	61.0	0.0
Agaricus species	23.0	1,429.0	0.2	0.0	4,762.0
Volvariella species	71.0	677.0	17.1	374.0	3,465.0

Mushrooms also contain many essential amino acids, which are required in small quantities by the human systems. Mushrooms are rather rich in some of the amino acids, such as Lucine and Tryptophan, which are deficient in cereals. The amino acids present in the different mushroom types are given in **table 13.**

Table 13: Amino acids present in mushrooms (mg. / 100 gm. dry weight)

Amino acids	Quantity
Alanine	2.40
Arginine	1.90
Aspartic acid	3.14
Cystine	0.18
Glutamic acid	7.06
Glycine	1.20
Histidine	0.34
Isolucine	0.28
Methionine	2.16
Lysine	0.39
Cerine	2.50
Threonine	1.89
Tryptophan	3.48
Tyrocine	0.78
Valine	1.63

2. **Mushrooms have good medicinal properties.** Many mushroom species have good medicinal properties and are widely used for curing a number of diseases. A few pharmacological effects of whole Reishi extracts have antitumor activity, antiviral effect, lowering blood pressure, enhancing bone marrow nucleated cell proliferation, lowering serum cholesterol levels with no effect on triglycerides, improving coronary artery hemodynamic, central depressant and peripheral anticholinergic actions on the autonomic nervous system, reducing the effects of caffeine and relaxing muscles, eliminating hydroxyl free radicals by antioxidant activities. Medicines, which are claimed to cure some of the killer diseases, such as cancer, HIV, aids, jaundice, tumors, venereal diseases etc obtained from *Ganoderma lucidum* and some species of *Lentinus, Auricularia* and *Grifola* are being sold in the world market in the form of tablets, powders, liquids and capsules. Tonics and medicines obtained from mushrooms are widely used in China, Japan, Korea, Taiwan, the Philippines and America. Some mushrooms are said

to lower the cholesterol level in humans. A drug produced in Japan from *Coryolus versicolor* is being sold worldwide for the cure of cancer. Some species of *Morchella* have got good medicinal properties and large quantities of these mushrooms are being exported to other countries from India at very high prices. In general due to the presence of different types of polysaccharides and glucans, regular consumption of mushrooms is said to stimulate the immunogenic system in the body, thereby providing resistance against most of the diseases.

3. **Mushrooms are capable of degrading agro-wastes.** Mushrooms can be grown on inexpensive substances commonly considered as agro-wastes, such as paddy and wheat straw, saw dust, composted horse and cattle manure, tobacco stems, sugarcane bagasse, cotton stubbles, vegetable wastes, pieces of hard wood etc. and these substrates are ultimately converted into enriched manure. The residual substances in the straw, vegetable wastes etc. left out after the harvest of mushrooms contain much higher quantities of N, P and K. The nitrogen content in the residual straw is enhanced to 2.0 % from 0.6 % in fresh straw. In the residual straw, antagonistic microorganisms, such as species of *Trichoderma* grow well, which can control several root rot, collar rot and wilt diseases.

4. **Mushrooms are grown indoors and do not require fertile land.** Many of the mushrooms are grown indoors, which do not require large areas of cultivable fertile land and comparatively lesser amount of water is required for the cultivation of mushrooms. From a thatched shed of about 13' length and 9' width, 2 - 3 kg. of oyster mushrooms can be produced every day, which will give a net profit of Rs.800/- per month. While milky mushrooms are grown in pits situated under shady places, which are covered by blue silpauline sheet. From two pits of size 5½' length x 3' width x 2½' depth, about 2.0 kg. of mushrooms can be produced per day, which will give a net profit of Rs.1200/- per month.

5. **Mushroom cultivation provides excellent scope for self-employment.** In the integrated system of Agriculture, mushroom cultivation provides good opportunities for self-employment and helps to increase the family income. In Tamil Nadu, plenty of raw materials, such as paddy straw, sugarcane bagasse, sorghum and pearl millet stubbles, groundnut, cotton and pulses plant materials are available all through the year and these can be profitably used for mushroom cultivation. From one acre of rice crop about 3,000 kg. of straw is obtained. If one-fourth of this

quantity of straw is used for the cultivation of oyster mushrooms, at least 1,600 - 2,000 kg. of mushrooms can be produced, which may give a net profit of Rs.8,000/- to Rs. 10,000/-.

Production of mushroom spawn. Grains of sorghum, pearl millet, maize etc produced from the farms can be used for preparing mushroom spawn, which can be sold for a much higher price.

Value added products from mushrooms. From mushrooms several other food commodities, such as mushroom pickles, mushroom soup powder, vadagam etc. can be prepared. These products can be stored for a much longer period of time and the women folks can do such works during their leisure time.

Cattle and poultry feed supplement. The residual straw after the harvest of mushrooms, besides being a good organic manure, can also be used as cattle and poultry feed. It contains the fungal mycelium and small mushroom buttons in large numbers and is a protein-rich feed for cattle. It can either be fed as such or mixed with some sugarcane molasses and mineral salts and then fed to cattle as enriched feed. This helps to enhance the quantity and quality of milk. The residual straw can be powdered, mixed with other poultry feed and fed to poultry birds. This increases the egg-laying capacity of the birds. The enzymes produced by the mushrooms during their active growth, degrade the hard lignin and cellulose present in the straw and makes the residual straw more palatable for cattle and poultry birds and makes the feed easily digestible. Up to 25 % of the regular feed of cattle may be supplemented with mushroom digested straw, without affecting the growth of cattle in any way.

Vermicompost. Further, the residual matter obtained after the harvest of mushrooms can be used in vermiculture for the production of vermicompost. Along with various plant and animal wastes, this residual matter can also be used for this purpose. The earthworms devour the straw containing the mycelium of the mushroom and hasten the process of composting. The compost thus produced contains more of nitrogen, phosphorus and potash and several other essential nutrients, such as calcium, magnesium, Vitamins, Hormones etc. This type of compost can also be used for casing while growing milky mushroom or button mushrooms.

Bio-gas. The residual straw can be used for production of natural gas also. As this straw material contains more than 2.0 % of nitrogen, they are more useful for natural gas production. It can also be mixed with cowdung and used for gas production. The residues obtained after gas production also serves as good organic manure.

Coir pith composting. Several species of *Pleurotus* are used for decomposing coconut coir waste, which is not easily degradable. Coir waste cannot be used directly as manure and so these wastes are dumped in heaps in coir factories and on roadsides. They rot in wet seasons and cause health hazards. But, they can be easily composted by using some species of *Pleurotus, Polyporus* and *Volvariella* and converted into good organic manure. However, by using coir waste alone, mushrooms cannot be grown.

To decompose 1.0 ton of coir waste, 5.0 kg. of urea and 5 mushroom spawn bottles (1.5 kg. of mushroom spawn) are required. To prepare compost from coir waste, an area of 5.0 m x 3.0 m is selected in a shady place and levelled. About 100 kg. of coir waste is spread uniformly on this area. Mushroom spawn from one bottle (300 gm.) is taken out and sprinkled evenly over the coir waste. Then a second layer 100 kg. of coir waste is spread over this and 1.0 kg. of urea is spread over this layer. Similarly 10 layers of coir waste alternated with mushroom spawn and urea are made to a height of about 1.0 m. as a rectangular bed. Water is sprinkled over the bed occasionally to keep the layers moist. In about a month's time the entire lot is composted well and the heap turns black and powdery and becomes good organic manure.

By composting coir waste, environmental pollution can be avoided and health hazards can be eliminated. This compost helps to increase the soil fertility and enhances water-retaining capacity of the soil. It provides substantial quantities of macro- and micronutrients required for the crops. It is useful for the reclamation of saline and alkaline soils and is highly suitable for dry areas. Composted coir waste can be used for growing biocontrol agents, such as species of *Trichoderma* and used for the control of soil-borne diseases. The organic matter content in the composted coir waste to nitrogen ratio is brought down from 112 : 1 to 24 : 1. The lignin and cellulose content in such compost is brought down to 4.8 % and 10.1 % from 36.0 % and 26.5 % respectively. The macro- and micronutrients present in coconut coir waste and composted coir waste is furnished in **table 14**. Among the different species of *Pleurotus, Pleurotus platypus* and *Pleurotus sajor caju* are more efficient in decomposing coir waste.

Table 14: Nutrients present in coconut coir waste

Nutrients	Coconut coir waste Raw coir Waste (%)	Composted (%)	Percentage increase of nutrients
Nitrogen	0.30	1.06	253
Phosphorus	0.01	0.96	500
Potash	0.78	1.20	54
Calcium	0.40	0.50	25
Magnesium	0.36	0.48	33
Iron	0.07	0.09	29
Manganese	1.25	20.10	1508
Zinc	7.50	15.80	111
Copper	3.10	6.20	100

6. **Mushrooms give fastest return and steady income when compared to any other crop.** The crop cycle of oyster, milky and paddy straw mushrooms is completed within a short period of 25 - 50 days, while in the case of button mushrooms it takes about 100 days for completion of the crop cycle. Within this period three harvests can be taken, thereby a steady income is obtained. The cost of production of mushrooms is also less and the profit is high. The period of crop cycle for the different types of mushrooms, the yield obtained and the profit that can be got are given in **table 15.**

Table 15: Cropping cycle and yield parameters of different types of mushrooms

Characters	Oyster mushroom	White milky mushroom	Paddy Straw mushroom	Button mushroom
Species	*Pleurotus* sp.	*Calocybe sp.*	*Volvariella sp.*	*Agaricus sp.*
Substrate	Paddy straw	Paddy straw	Paddy straw + Horse Gram powder	Compost
Growing Temperature	20° - 30°c	25° - 35°c	30° - 35°c	15° - 25°c
Crop cycle	30 - 45 days	45 - 50 days	25 - 30 days	85 - 100 days
Days for first Harvest	15 - 25 days	24 - 28 days	25 - 30 days	60 -70 days
Yield per bed	500gm./ 500gm. bed	356gm./ bed	5.0KG./30KG. bed	3.5kg./ Tray
Shelf life	1 - 3 days	3 - 5 days	1 - 3 days	3 - 5 days
Bioefficiency	80 - 100 %	143 %	20 - 30 %	30 - 35 %
Production cost (Rs./kg.)	Rs.15/-	Rs.20/-	Rs.20/-	Rs.30/-
Market price (Rs./Kg.)	Rs.50/-	Rs.70/-	Rs.60/-	Rs.80/-
Net profit (Rs./kg.)	Rs.35/-	Rs.70/-	Rs.40/-	Rs.50/-

7. **Mushrooms give maximum yield per unit area.** The area required for growing mushrooms is much less and the yield obtained per unit area is very high, when compared to other crop husbandry. The yield particulars are also given in **table 15.**

8. **Mushrooms have excellent export potential.** Several edible species of *Morchella* collected from the wild in the hilly terrains of Jammu and Kashmir, Himachal Pradesh and Uttaranchal by the local people are being exported to many European countries at premium prices and helps to earn substantial amount by way of foreign exchange. Large quantities of button mushrooms are produced in our country mostly in the cooler parts and are being exported either as raw mushrooms or in preserved conditions. Because of the availability of huge quantities of raw materials, which can be used for the cultivation of different types of mushrooms, there is ample scope for the production of much more of mushrooms for home consumption and for export.

CULTIVATION OF MUSHROOMS

Oyster Mushroom

'Oyster mushrooms' (*Pleurotus* species) are shell, fan or spatula-shaped. They occur in different shades of white, cream, pink or light brown depending upon the species. The pink types usually turn whitish at maturity. Oyster mushrooms are mostly soft and fleshy. The stipe is strongly eccentric to lateral and in some species the stipe may be absent. Many of the oyster mushrooms are edible and can be cultivated artificially under controlled conditions indoors in mushroom houses (sheds), mostly thatched sheds.

Substrates for oyster mushroom cultivation. Oyster mushrooms can be grown on many agro-wastes, such as paddy and wheat straw, maize and pearl millet cobs (after the grains are removed), fresh sorghum stems (after the leaves are stripped off), cotton waste, fresh cotton stubbles without leaves and bolls (the stubbles are beaten, crushed and soaked in water for 24 hours), groundnut shells, sugarcane bagasse and vegetable wastes. The carbon to nitrogen ratio and the cellulose to lignin ratio are important criteria for the choice of substrates. The carbon to nitrogen ratio should be 40-50 : 1 and the cellulose to lignin ratio should be about 4 : 1. In paddy and wheat straw, the carbon to nitrogen ratio is 45 : 1 and the cellulose to lignin ratio is about 4 : 1, while in coconut coir waste it is 141 : 1 and 1 : 1 respectively. So coir waste, though available in plenty, is not suitable for oyster mushroom cultivation. The bio-efficiency of paddy and wheat straw is much higher (80-150%), while it is very low in sorghum stem and cotton stubble (40-50%). Paddy straw, which is easily available in large quantities and fairly cheap, is the ideal choice for mushroom cultivation. The straw should be dried and stored properly and it should be free from any fungal contaminants. Hand threshed straw is much better than cattle or tractor threshed straw. Along with the straw, a small percentage of other agro-wastes can be added.

Mushroom shed. For growing oyster mushroom, rectangular thatched sheds with sloping roofs can be used. About 16.0 sq.m. area is required for obtaining 1.0 kg. of mushrooms daily and for getting 20.0 kg. of mushrooms per day about 300 sq.m. area may be required. The length and width of the shed may be adjusted according to the space available. However, instead of having a single long shed, it may be separated into compartments by providing suitable partitions, so that the temperature and humidity in each of the compartment can be maintained properly. Each compartment is partitioned into two portions or rooms, one portion is used as **'spawn running room'** and the other **'cropping room'**. The height of the shed should be about 4.0 m. at the center and 2.25 m. at the sides. The shed is put up along the East-West direction to avoid direct effect of the sun and to reduce the temperature inside the shed. A door is provided at the long end of the shed and a few ventilators are also provided for aeration. The windows are covered with nylon nets to avoid mushroom flies and common houseflies from entering into the shed. The entire shed is fenced all around with chicken mesh to prevent entry of rats, snakes etc.

Sand is spread uniformly to a height of 15 cm. on the floor of the mushroom shed and the inner sides of the shed are covered with jute gunny bags. Water is sprinkled on the floor and the gunny bags, two or three times a day to maintain the required temperature and humidity inside the shed. Keeping the ventilators and door open for short spells three or four times a day provides proper ventilation. Racks with 3-4 shelves are arranged all round. In the central portion of the shed, sufficient room is left for movement and the beds are placed on the shelves.

After preparation of the beds, they are first kept in the spawn running room for vegetative growth of the fungus. The temperature in the spawn running room should be maintained at 25°-30°C. No light is required. But fresh air is necessary to replace the used up air inside the room. So, the ventilators are opened frequently to let in fresh air. The gunny bags lining the shed and the sand on the floor should be kept moist by spraying water so that, the humidity is maintained at about 80%.

After the beds are covered by the white mycelial growth, they are transferred to the cropping room. The temperature and relative humidity in the cropping room is maintained at 23° - 25°C and 85 - 90 % respectively and there should be diffused light and aeration. The sudden dip in the temperature in the cropping room induces transformation of the vegetative phase to the reproductive phase and encourages the fungus to produce fruiting bodies or sporocarps, which are the mushrooms.

Spawn. Spawn is the seed material used for growing mushrooms and it is actually the mycelium of the mushroom fungus grown in specific medium. Mushroom fungus grown on a grain based medium is called as **'grain spawn'**. Spawns are of three types viz., **'primary culture'** or **'primary spawn'**, **'mother spawn'** and **'commercial'** or **'bed spawn'**.

Preparation of primary mushroom spawn. Small pieces of the inner tissue from well formed, matured mushroom are taken and aseptically transferred on to nutrient media, such as potato-dextrose-agar slants and incubated at room temperature. The fungus starts growing and within 7 - 10 days the entire surface of the medium is covered with a fine white mycelial growth. This is the **'primary culture'** or **'primary spawn'**. From this primary culture, mother spawn is produced and from the mother spawn bed spawn is produced.

The primary culture obtained from mushroom tissue is inoculated on to sorghum or maize grains to get mother spawn or grain spawn. The mother spawn thus obtained can be used 3 or 4 times in succession to prepare bed spawn. If the mother culture is used repeatedly for more than 3 or 4 times for preparing bed spawn, the quantity and quality of mushrooms produced will be adversely affected. In such cases fresh mother spawn should be prepared from tissue culture.

Preparation of mother spawn and bed spawn. Spawn can be grown on grains, such as sorghum or maize or on paddy straw. However, sorghum grains are considered to be more suitable. Sorghum grains are half cooked till one or a few grains start to split or crack open. Then the excess water is drained out and the grains are spread on gunny cloth and allowed to dry. The partially dried grains are thoroughly mixed with calcium carbonate - 2.0 % (20 gm. calcium carbonate / kg. of grains). The extent of drying of the cooked grains is very important. After mixing with calcium carbonate, the grains should be loose and should not stick to one another or form lumps. These grains are then filled to about ¾ th in glucose-saline bottles or heat-resistant polypropylene bags of size 26 cm. x 10 cm. and plugged with non-absorbent cotton wool. These grain-filled bottles or bags are sterilized by autoclaving them at 15 lb. pressure for 2 hours or 20 lb. pressure for 1½ hours. The bottles or bags are then taken out from the autoclave and allowed to cool.

Mushroom tissue culture grown on artificial nutrient medium slants or petridishes are cut into small round discs, using an inoculating needle and inoculated on to the grain-based medium aseptically and allowed to grow. From each of the primary culture grown on nutrient agar slants, 5 - 10 spawn bottles can be inoculated. These first-generation spawns obtained from tissue culture are called as **'mother spawn'**. After inoculation, the bottles are incubated in a separate room and the temperature in that room should not be more than 30°C. The mycelium of the fungus starts growing and within 10 - 15 days spreads and covers the entire bottle and appears white in color. This is called as **'spawn-running'**. Then after 3 - 5 days, when the growth is completed the spawn is ready for use. The spawns produced should be used within 30 days after inoculation.

From the grains in each of the well-grown mother spawn bottle, about 25 - 30 spawn bottles can be prepared. These are called as **'bed spawn'** or **'commercial spawn'** and are used for preparing mushroom beds.

Preparation of mushroom beds for mushroom growing. Fresh paddy straw, preferably hand threshed straw is cut into small pieces of length 3.0 - 5.0 cm. The straw bits are soaked in potable water for 6 - 8 hours. The pre-soaked straw bits are sterilized, so that the straw is completely free from any microbial contaminants, as well as insects. Sterilization can be done, either by hot water treatment or chemical treatment. In hot water sterilization, the pre-soaked straw bits are completely kept immersed in boiling water for 30 minutes. Then the straw bits are spread over clean disinfected floor and shade-dried till no water drips, when the straw bits are squeezed by the hands. At this stage the moisture present in the straw will be about 65 %, which is optimum.

In chemical sterilization, the straw bits are kept immersed in a fungicidal solution containing carbendazim and formalin for 16 hours. The fungicidal solution is prepared by mixing 7.5 gm. of carbendazim and 125 ml. of commercial formaldehyde (45 %) in 100 lit. of water. For sterilizing 10 kg. of dry straw, 100 lit. of the fungicidal solution is required. Formalin is a fumigant and the fumes can escape easily from the solution. To avoid that, the vessel containing the fungicidal solution is covered with a polythene sheet. After sterilization, the straw bits are removed, shade dried and used for bed preparation.

For preparing beds, polythene bags of size 60 cm. x 30 cm. are used. The sealed end of the bag is tied with a piece of thread and at the center of the bag two pencil size holes are put on opposite sides. Sterilized and shade dried straw bits are put inside the bag to a height of 5.0 cm. Over this layer, 25 gm. of grain-based spawn is spread uniformly. Over that another layer of straw bits is put to a height of 10 cm. followed by 25 gm. of spawn. Similarly a third and fourth layers of straw are put to a height 10 cm. each followed by 25 gm. of spawn over each layer. Finally the top layer of spawn is covered by a 5.0 cm. layer of straw. After putting each layer of straw, the bed is gently pressed and compacted. Thus, in all, the cylindrical bed will have two 5.0 cm. straw layers, one at the bottom and the other at the top and three 10 cm. layers of straw in the middle and in-between five layers of spawn and the total height of the bed will be 40 cm. The beds are then transferred to the spawn running room and arranged in the shelves or hung one below the other in layers of 3 beds from cross poles in the spawn running room. In about 15 - 20 days, the entire bed is covered with a white mycelial growth.

After the spawn-running period, the polythene bag is cut opened and the opened bed is transferred to the cropping room. Under conditions, when the temperature in the cropping room is higher than the optimum temperature and the relative humidity is lower for mushroom growth, the polythene bags are not completely removed. Instead a number of small holes are made all around the bag or a few longitudinal cuts are made in the bag with a blade to facilitate the emergence of mushrooms. The surface of the beds are kept moist by spraying water on the beds 2 - 3 times in a day, besides maintaining optimum temperature and humidity inside the cropping room. In about 3 - 4 days mushroom pinheads appear on the surface of the bed and become full grown mushrooms in another 3 - 4 days. Matured mushrooms are harvested by pulling them out along with the roots before spraying water. Then they are packed in perforated polythene bags and sent to the market. After the first harvest, the surface of the bed is scrapped off and the beds kept moist by spraying water. In about 7 - 10 days a second flush of mushrooms can be harvested. By repeating this practice, a third harvest can also be obtained. The entire cropping is over in 35 - 40 days.

Enhancing oyster mushroom production. Addition of neem seed cake at 5.0 per cent to the sterilized straw before preparing the beds has been found to increase mushroom production by 20 - 30 per cent. Besides increase in mushroom production, in such beds pest incidence is also low.

Table 16: General characters of oyster mushroom varieties *(pleurotus* species*)*

Characters	Co.1 (white Oyster)	M.2 (Grey Oyster)	APK.1 (Pink Oyster)	MDU.1 (White Oyster)	Ooty 1 (Grayish-White Oyster)
Species	*P. Itrinopileatus*	*P.sajor-caju*	*P.eous*	*P.djamor*	*P.ostreatus*
Year of Release	1986	1975	1995	1993	1998
Color	White	Gray	Deep rose buds and white at maturity	Bright white	White
Texture	Fleshy	Fleshy	Fleshy and tough	Fleshy	Fleshy
Spawn run	15 days	20 days	7 - 12 days	12 -16 days	33 days
Crop cycle	35 - 40 days	40 - 45 days	35 - 40 days	35 days	55 -60 days
Yield / bed of 500 gm substrate	395 gm.	328 gm.	620 Gm.	538 gm.	531 gm.
Bio-efficiency	79.0 %	65.6 %	124.0 %	108.0 %	107.0 %
Shelf life	48 hours	48 hours	60 hours	24 hours	36 hours

Milky mushroom

'Milky mushroom' *(Calocybe indica)*, also known as **'white milky mushroom'** or **'white summer mushroom'**, because of its milky white color, is umbrella-shaped like the button mushroom. Its stipe is fairly thick and both the stipe and pileus are quite fleshy. A variety APK.2 was released from the Regional Research Station, Aruppukottai under the Tamil Nadu Agricultural University, Coimbatore. This mushroom is amenable for artificial cultivation under controlled conditions in simple mushroom sheds. It can be grown under a wide range of temperatures, has a longer shelf life (up to 5 days), gives higher yield (356 gm. / 250 gm. bed), has higher protein and fiber content, has good market potential and fetches a higher price than oyster mushroom. Because of these advantages cultivation of this mushroom is gaining much importance.

Substrates for milky mushroom cultivation. Milky mushroom can be grown on several agro-wastes, such as paddy straw, wheat straw, maize and sorghum stems, sugarcane bagasse, groundnut plants after harvest and other vegetable wastes. However, for commercial cultivation of this mushroom, paddy straw has been found to be quite ideal. Along with straw, a small quantity of other agro-wastes can be added.

Mushroom shed. Milky mushroom can be grown in sheds just as oyster mushroom with separate compartments for spawn-running and cropping. The temperature in the spawn running room is maintained at 30°C and the humidity at 80 % or more. Spawn running is over in 10 - 12 days and the entire bed turns white as a result of mycelial growth. Cultivation of milky mushroom involves an additional process known as **'casing'** after spawn running. After casing the beds are transferred to the cropping room where the temperature is maintained at 25° - 30°C and humidity at 85 %. Provision is made to allow aeration and diffused light inside the room.

Preparation of spawn and mushroom bed. Preparation of primary spawn, mother spawn and bed spawn, as well as mushroom beds are quite similar to that of oyster mushroom.

Casing. Casing is done primarily for initiation of fruiting bodies of the fungus.

Advantages of casing

(i) When the fungus is allowed to grow in the nutrient-rich substrate, it continues its vegetative growth instead of reverting to the reproductive phase as long as there is sufficient nutrients for its growth. However, when the bed is covered with a casing material with less nutrient content, the fungus is prompted to revert to the reproductive phase and starts producing fruiting bodies, which grow to become mushrooms.

(ii) As the rapidly growing hyphae begin its reproductive phase, the casing material serves as a moisture-retaining reservoir and provides required water to the hyphae to produce fruiting bodies.

(iii) The casing material prevents drying of the mushroom bed and helps to maintain high level of humidity immediately around the bed and in the air inside the cropping room.

(iv) As the hyphae start producing mushroom buds, certain chemical and physiological changes take place in the mushroom bed, as a result certain vital chemical substances are released in a gaseous state. The casing material prevents their escape and makes them available to the growing hyphae and encourages the formation of fruiting bodies.

(v) The casing material harbors certain species of *Pseudomonas* and a few other beneficial bacteria, which helps the development of mushrooms.

(vi) The casing material helps to anchor the developing mushrooms and provides physical and mechanical support as the mushrooms grow.

Materials used for preparing casing compounds. Several materials are used for preparing casing compounds. The casing material or compound should have good water holding capacity (not less than 80 %) and the rate of water loss should also be less.

It should be slightly alkaline in reaction and the pH should be around 7.0 - 8.0

It should have less organic matter content and its nutrient status should be low.

It should be porous in nature, so as to allow aeration and should not become hard when dry.

It should be free from contaminants, such as fungi, insects and nematodes.

It should be free from heavy metal ions of mercury, cadmium etc., which are toxic to mushrooms.

It should have moderate cation exchange capacity and low electrical conductivity.

The following materials or compounds are used for casing

(i) Garden soil or clay loam.
(ii) Red soil mixed with sand and calcium carbonate (2 %).
(iii) Well decomposed farm yard manure + sand (2 : 1).
(iv) Well decomposed farm yard manure + ash (1 : 1)
(v) Humus
(vi) Well decomposed coir waste.

Sterilization of casing material or compound. Before casing, the casing material has to be sterilized to eliminate contaminants, if any present in it. It can be done, either by heat treatment or chemical treatment.

Heat sterilization. The casing material can be heat sterilized by using steam. The material is put in perforated wooden boxes or packed in cloth bags and steamed for one hour. Then it is allowed to cool and used for casing after about 24 hours.

Chemical sterilization. The casing material is spread uniformly over clean cement floor and moistened with 5 % formalin solution (5.0 lit. of commercial formaldehyde in 95 lit. of water) and covered with polythene sheet for 10 - 14 days to prevent escape of formalin vapor. A quantity of 25

- 30 lit. of formalin solution is required to moisten 1.0 cu.m. of casing material. After this period, the casing material is stirred thoroughly till it is completely free from any residual formalin vapor. Then this material is used for casing.

Method of casing. After the completion of spawn running, the bed is cut horizontally into two equal halves without removing the polythene bag. The casing material is mixed with water to make a paste and then plastered over the two cut ends to a height of 1.0 - 2.0 cm. Afterwards the beds are placed in the racks in the cropping room with the cased surface facing upwards.

Maintenance of beds. In the cropping room, water is sprinkled over the casing material regularly, so as to maintain the moisture level at 50 - 60 %. In about 8 - 10 days, small mushroom buds start appearing as pin-heads over the casing and in another 6 - 8 days they become full grown and ready for harvest.

Harvesting. Matured mushrooms are harvested before they start shedding white-colored spores. They are picked by twisting and the mushrooms along with the stubs are removed completely. It is better to pick all the mushrooms in the bed at a time. Each well-developed mushroom weighs 55 - 60 gm. on an average. The first harvest is obtained 24 - 28 days after bed preparation. After the first harvest is over, the surface of the casing is gently scratched, compacted and water is sprayed. In about 40 - 50 days after bed preparation 2 - 3 harvests are obtained. From each bed of 250 gm. dry straw, about 360 gm. of mushrooms are obtained.

Under natural conditions milky mushrooms can be stored for about 5 days without any deterioration in quality and hence they can be transported over long distances for marketing.

Enhancing milky mushroom production. For commercial cultivation and to enhance the production of milky mushroom, a special type of cropping shed is recommended. This rectangular, arched shed may be 18' x 12' or 15' x 10' (3 : 2 ratio) in size according to the area available and is situated in a shady place. The area where the shed is to be constructed is dug to a depth of 2.5'. To prevent soil sliding into the pit, a wall is constructed inside the pit on all the four sides with bricks or hollow cement blocks. The wall may extend to a height of 1' - 2' above ground level. If no wall is put up, the sides are plastered with clay. Over the pit an arched shed is constructed. The sides of the arched shed are 8' and the center of the arch 9' above ground level. The shed is completely covered with fairly thick, blue-colored silpauline sheet. Inside the shed racks with shelves are put up to accommodate the beds. After casing the beds are transferred to this cropping room and arranged in the shelves. To allow good aeration ventilators are provided and to evacuate heated air from inside exhaust fans are provided. Arrangements are also made to allow some light inside the shed during the daytime.

Paddy straw mushroom

'Paddy straw mushroom' *(Volvariella* spp.*)* is otherwise called as **'Chinese mushroom'** or **'tropical mushroom'.** Because it is grown on a very large scale in China, it is known as 'Chinese mushroom' and because of its tolerance to higher temperatures, it is also called as 'tropical mushroom'. It can be grown successfully in the tropical and sub-tropical regions and is cultivated on a large scale in countries, such as China, Japan, the Philippines, Taiwan and Indonesia. The sporocarps (fruiting bodies) of this mushroom first appear as grayish-white buttons, resembling the eggs of birds. The generic name *Volvariella* for the fungi denotes the presence of a wrapper-like membrane (universal veil), which completely envelops the sporocarp during the young stage of development. At maturity, the button enlarges, the pileus expands and becomes an umbrella-like fruiting body after tearing off the universal veil. The torn universal veil remains as a cup-shaped body called **'volva'** (pl. volvae) around the base of the stalk (stipe). Though these mushrooms are of excellent taste and flavor, because of their very short shelf-life and low bio-efficiency, these mushrooms are not cultivated widely.

Among the different species of *Volvariella,* the white-colored *Volvariella bombicina,* the gray-colored *Volvariella volvaceae* and *V. esculenta* and the brownish-colored *Volvariella diplasia* are grown to a larger extent. *Volvariella bombicina* grows well on hard-wood shavings or scrapings or pieces, while the other species are capable of growing on many other substrates.

Substrates for growing paddy straw mushroom. *Volvariella* sp. can be grown on various substrates, such as well-dried paddy straw and several other cereal straw, cotton waste, vegetable waste, banana trash, sugarcane bagasse, groundnut shells etc. It can also be grown on coconut waste to some extent. However, as the name indicates, it grows best in paddy straw. The other agro-waste materials can be supplemented with paddy straw to a small extent.

Mushroom shed. No special type of mushroom shed is required for growing paddy straw mushroom. It can be grown in any type of ordinary thatched shed or even under shades of densely foliaged trees.

Preparation of spawn. For preparation of spawn, well-dried, clean straw is used. Spawn can be prepared in 500 ml. glucose-saline bottles or in heat resistant polypropylene bags of size 26 x 10 cm. as in the case of oyster mushroom. For preparing one bottle or one bag of spawn, 100 gm. of straw and 15 gm. of coarsely ground horsegram or redgram powder are required. The straw is cut into small bits of length 2.5 - 3.0 cm. and put inside the bottles to fill them up to ¾ of the volume. Then water is poured inside the bottle and the straw bits are allowed to soak in the water for 12 - 16 hours and then the excess water is completely drained off. Then a hole is made at the center of the substrate by means of a rod and 15 gm. of horsegram or redgram powder is put inside the hole. The mouth of the bottle or polypropylene bag is plugged with non-absorbent cotton wool, covered with paper and tied with a string. Then the bottles or bags are autoclaved at 20 lb. pressure for one hour to sterilize the whole thing. After sterilizing, the bottles or bags are taken out, allowed to cool and inoculated with the mushroom culture grown on agar medium in petridish aseptically in the culture room. After inoculation, the spawn bottles or bags are incubated at room temperature. In about 15 - 20 days, the fungus grows and covers the substrate completely, which can be seen as a fine, white, cottony growth with large number of reddish-brown chlamydospores. This fully grown spawn is used for inoculating the mushroom beds. Instead of paddy straw, mushroom spawn can also be grown on sorghum or maize grains as in the case of oyster mushroom. But here also 15 gm. of coarsely ground redgram or horsegram powder is added to the grains before autoclaving.

Preparation of mushroom beds. Paddy straw mushrooms can be grown by adopting different cultivation methods viz.,

1. Conventional bed method
2. Twisted straw bed method
3. Hollow cylindrical bed method or
4. Improved cage cultivation method.

1. Conventional bed method or raised bed method

A raised, square, level platform of size 1.0 to 1.5 m^2 is made about 25 cm. above the ground level by using bamboo or wooden planks or bricks, with small gaps in-between, so that water may not stagnate on the platform. Mushroom beds are formed on the platform.

Preparation of conventional mushroom bed. The following materials are required for preparing each mushroom bed :

Paddy straw - 10 kg.

Spawn bottle or bag - 2 Nos.

Coarsely ground redgram or horsegram powder - 50 gm.

Well-dried, preferably fairly old (about 6 months old), clean, hand-threshed paddy straw is quite suitable for preparing mushroom bed. The straw is first of all tied into small bundles of about 250 - 300 gm. each, with the basal stem portion on one side. The top portion of the bundle is trimmed to a length of 1.0 - 1.5 m. according to the size of the platform. To form one bed, 32 such bundles are required.

The straw bundles are soaked in clean water for 6 - 8 hours. Then, they are taken out and placed on a clean surface, so that the excess water is drained off. Four such pre-soaked straw bundles are placed on the platform side by side in a parallel row, with the basal portion of all the bundles facing one side. While the two bundles placed on both the ends of the platform are kept as such, the two bundles in the central portion are untied and the straw spread uniformly in-between the two bundles at the ends. Over this, another four bundles are placed in a parallel manner, with the basal portion of the bundles facing the opposite side in a parallel manner. All these eight bundles together forms the first layer of the mushroom bed.

Spawn prepared with paddy straw is in the form of a lump and so, cannot be taken out easily from the spawn bottle or from the polythene bag. The spawn bottle has to be broken to take out the spawn. Two such spawn bottles or bags are required to spawn one bed. The lump of spawn is divided into small pieces of size 2.0 - 4.0 cm. Then the pieces are placed all round on the surface of the first layer at 10 cm. apart and 3.0 - 5.0 cm. inside from the edge of the bed. Small quantities of sterilized red gram or horse gram powder is spread over the pieces of spawn. After that, the second layer of the bed is formed over the first layer. Four bundles of the pre-soaked straw are placed over the first layer in a parallel manner, with the base of all the bundles facing one side. But, the bundles are placed at right angle to the bundles in the first layer. Then another four bundles are placed over these bundles in a parallel manner, with the basal portion facing the opposite side. These eight bundles together forms the second layer of the bed. On this layer spawn bits are placed as in the first layer and dhal powder is spread over the spawn bits. Similarly, a third layer is formed over the second layer and here the straw bundles are placed at right angle to the second layer and parallel to the first layer, and then spawned. Finally the fourth layer is formed over the third layer. Four straw bundles are placed at right angle to the third layer and another four bundles parallel to these bundles, with the basal portion facing the opposite side. Here, the two bundles at both the ends are placed as such, while the central two bundles are untied and the straw spread uniformly in-between the two bundles at the ends. The fourth layer forms a cover and no spawning is done over this layer. Any straw bits protruding out of the bed are trimmed. Then the bed is pressed with a wooden plank and compacted to encourage rapid growth of the fungal mycelium. Finally, the entire bed is covered with a polythene sheet to prevent too much evaporation of water from the bed.

For the first 5 or 6 days after preparing the bed, it will be sufficiently moist and so, there is no necessity to sprinkle water on the bed. After that period, the surface of the bed starts drying. Then the polythene sheet is removed and water is sprayed on the bed so as to moisten the straw. The bed is kept open for one or two hours to provide aeration. Then the bed is again covered with the polythene sheet. Afterwards, once in every two or three days the bed is sprinkled with water and aerated. Sprinkling excessive quantities of water on the bed or allowing the bed to become dry is avoided. It is better to use a sprayer and only sufficient quantity of water to moisten the straw is sprayed uniformly on the surface of the straw.

In about 10 - 12 days after preparation of the bed, mushroom buds start appearing on the sides and on the top of the bed. In another 3 - 4 days, the buds become big and appear as eggs. This is the proper stage for the harvest of the mushrooms. If they are allowed to remain in the bed after this period, they enlarge rapidly and become umbrella-shaped. At this stage they become more fibrous and less tasty. After the first harvest, spraying of water is continued once in two or three days to keep the bed sufficiently moist and 2 - 3 successive crops of mushrooms develop at an interval of 7 - 10 days. A total quantity of 1.0 - 1.5 kg. of mushrooms is obtained from each bed.

Instead of spawn prepared with straw bits, grain spawn can also be used. In this case, one bottle of grain spawn or one bag of grain spawn is sufficient to spawn one bed. Grain spawn is quite loose and can be taken out without breaking the bottle. The spawn is divided into three equal portions. One portion is spread uniformly all round on the first layer of the bed, 3.0 - 5.0 cm. inside the outer edge of the bed. Then dhal powder is sprinkled over the spawn. Similarly, the second and third layers are formed followed by spawning and finally the third layer is covered by the fourth layer.

2. Hollow cylindrical bed method.

Materials required for preparing hollow cylindrical bed

Paddy straw - 5.0 kg.

Spawn bottle or bag - 2 nos.

Red gram or horse gram powder - 50 gm.

Tin can or cement pipe, about 30 cm. In diameter - 1 no.

In this method also, the beds are formed preferably on a raised platform. The straw rope is soaked in clean water for 6 - 8 hours, followed by immersion in boiling water for 30 minutes and then the excess water is drained off by keeping it on a clean surface. After the straw is sufficiently dried, it is wound around the tin can or cement pipe uniformly and 4 - 6 layers are formed. Then the tin can or cement pipe is removed and thus, a hollow cylindrical bed is obtained. Spawning is done both on the inside and outside of the hollow bed, over each layer and dhal powder is sprinkled over the spawn. The bed is compacted by pressing with a wooden plank and covered with a polythene sheet. All other procedures to be followed afterwards are quite similar to the conventional raised bed method. After the period of spawn-run, which may take about 10 - 15 days, mushrooms will appear on the outside, as well as on the inside of the hollow cylinder and also on the top portion of the bed. A quantity of 0.5 - 0.75. Kg. of mushroom can be obtained from each bed.

3. Improved cage cultivation method.

A wooden cage of size 1.0 m. x 50 cm. x 25 cm. is made. Fresh, clean straw is taken and cut into lengths of 25 cm. With the cut pieces, small bundles, about 30 cm. thick are made. Sixty such bundles are required to prepare one bed and for this 5.0 kg. of straw is needed. The straw bundles are soaked in clean water for 6 - 8 hours and then sterilized by immersing them in boiling water for 30 minutes. Excess water is drained off by spreading them on a clean surface. After the straw bundles are sufficiently dry, bed is prepared inside the cage. Ten bundles are laid uniformly on the bottom of the cage and grain spawn is spread uniformly on this layer and over the spawn, dhal powder is sprinkled. Over this layer another 10 bundles are arranged uniformly and then spawned as before. Similarly the third, fourth and fifth layers are arranged one above the other and spawned. The sixth layer forms the cover. The bed is compressed lightly and then covered completely with a polythene sheet. In about 10 - 15 days after spawning, pinheads appear and enlarge. Once the mushrooms become egg-shaped and before they open up, they are harvested. Spraying of water is continued once in 2 or 3 days, so as to keep the beds sufficiently moist and within a week's time a second crop of mushrooms will appear.

Precautions to be taken during cultivation of paddy straw mushroom

Clean, fairly old straw (preferably 6 months to one-year-old straw) should be selected for paddy straw mushroom cultivation. Too old, rotten and straw with very thick stem should not be used.

Strict hygienic measures should be followed at every stage during the cultivation. At the time of drying the pre-soaked straw or after the preparation of the bed, if small, pinhead-like sclerotial bodies of some fungi are seen, then production of mushrooms is very much reduced. To eliminate such contaminants, the pre-soaked straw should be immersed in boiling water for 30 minutes before bed preparation. Afterwards, the straw is immersed in a fungicide cum insecticide solution containing carbendazim at 1.0 gm. or mancozeb at 2.0 gm./ lit. of water + malathion at 1.0 ml./ lit. of water. The pre-soaked, sterilized straw should be spread on clean, dry gunny cloth, sterilized by dipping in disinfectant solution of lysol or formalin for draining the excess water.

About 2 hours before preparing the beds, the mushroom shed should be thoroughly sprayed with carbendazim at 1.0 gm./ lit. of water or formalin at 20 ml./ lit. of water + dichlorvos at 1.0 ml./ lit. of water. At the time of bed preparation, hands should be washed thoroughly in dettol or soap solution. After the first harvest of mushrooms, if growth of any other fungi is noticed, the beds are sprayed with mancozeb at 2.0 gm./ lit. of water. If any insects, such as mushroom flies are noticed, the beds are sprayed with malathion at 1.0 gm./ lit. of water.

The mushroom beds should not be allowed to dry and they should not be too moist or wet. The beds should be maintained at a moisture level of 60 - 70 % throughout the cropping period. The optimum temperature for the growth of paddy straw mushroom is 30° - 35°c.

BUTTON MUSHROOM

'Button mushrooms' otherwise called **'European mushrooms'** are preferred to all other types of mushrooms, because of their excellent taste, flavor, texture, shape and keeping quality. Button mushroom was first grown in France in 1670 ad mainly in quarries and caves with low temperatures and low light intensity on horse dung. In the 19th century, button mushrooms were grown in Sweden, England, Canada, the United States of America and other temperate countries, and various techniques were developed for increasing the production of button mushrooms.

Button mushrooms form about 80 % of the world's total mushroom production. On an average, 19 lakh tons of button mushrooms are produced every year all over the world. In India, cultivation of button mushrooms was started only in 1961 in the Himachal Pradesh. Now, button mushrooms are grown in the temperate regions of India and about 35,000 tons of button mushrooms are produced every year. In Tamil Nadu, they are grown in the cooler, hilly regions of Nilgiris, Kodaikanal, Yercaud and a few other places to a lesser extent.

Among the button mushroom species, *Agaricus bisporus* and *A. bitorquis* are grown commercially on a large scale. *Agaricus bisporus* requires a cool climate, with a temperature of 15° - 20°c, while *A. bitorquis* requires a slightly higher temperature of 20° - 25°c.

The prerequisites for successful cultivation of button mushrooms are selection of varieties suitable for growing in the locality, use of good quality seed-spawn, well-prepared compost, proper bed preparation and proper maintenance of mushroom sheds.

For the cultivation of button mushrooms on a commercial scale, high initial investment is required. Special roofed shed for composting the straw, pasteurization chamber for sterilizing the substrate at the appropriate temperature with aerated steam, separate rooms for sterilizing the casing materials, well equipped laboratory for preparing mushroom spawn, separate rooms for growing mushrooms, equipments and machinery for adjusting and maintaining the temperature, humidity, aeration and proportion of carbon dioxide in the mushroom growing room at the optimum level are quite essential.

However, for cultivation of button mushrooms on a small scale, certain simplified techniques can be adopted for the preparation of compost and mushroom production.

Substrate for button mushroom cultivation.

Compost is the suitable medium of desirable characteristics and contains specific nutrients, which promotes the growth of the particular mushroom with the practical exclusion of other organisms. The process of preparing the compost required for the growth, and development of these mushrooms is called composting.

Special type of compost is required for the cultivation of button mushrooms. Two types of compost viz., 1. **Natural compost** and 2. **Synthetic compost** can be used.

Preparation of natural compost. The main constituent in this compost is horse dung. Horse dung is collected from horse sheds along with the wheat straw litter and urine. The horse manure is composed of 3 parts of horse dung and one part of wheat litter along with plenty of horse urine. One ton of this horse manure is mixed thoroughly with 100 kg. of chicken manure and 3.0 kg. of urea. This manure mixture is heaped in a trapeze-shaped heap, 1.0 m. in width and 1.0 - 1.5 m. in height. The length of the heap may be varied depending upon the availability of substrate materials and the quantity of compost required. After 3 - 4 days, ammonia gas starts emanating from the manure heap. The heap is then dismantled and turned over thoroughly. The urine, which comes out from the heap is collected and sprinkled over the composting materials to prevent loss of nutrients and the whole thing is again heaped as before. Similarly, the heap is dismantled 3 - 4 times, at 3 days interval and remade. By the time the composting heap is turned for the 2nd or 3rd time, most of the ammonia gas liberated may escape and the heap will be almost free from any smell of ammonia gas. At the time of the 2nd and 3rd turning of the composting material, 10 kg. of gypsum is added to the mix each time to condition the compost. The compost will be ready in 12 - 15 days. The prepared compost may contain about 65 % moisture, 18 - 20 % total nitrogen and the pH may be around 7.5. The nitrogen in the compost is in the form of protein, amino acids, hexamine, a small quantity of ammonia (0.1 - 0.2 %) and lignin. The mycelium of *Agaricus* sp. contains the enzyme, phenol oxidase, which breaks down the lignin present in the straw. A well-prepared compost contains actinomycetes, hemicola and other thermophiles, but contains less of bacterial population. This compost is almost a selective substrate for the growth of *Agaricus* sp. Composting is done in a roofed shed at room temperature.

Preparation of synthetic compost. Natural compost was subsequently replaced with synthetic compost consisting of wheat straw supplemented with various organic and inorganic nitrogen sources. The compost is prepared either by long or short-term methods of composting.

Long-term method of composting (LMC) It takes 26 - 28 days and involves 7 - 8 turning at varying intervals. It does not require pasteurization. Small growers lacking boiler facilities adopt this method. The compost prepared by this method usually gives comparatively low yields of mushrooms. It is also more liable to attacks from pests and diseases during cropping.

Short-term method of composting (SMC). This method takes about 18 days. It is done in two phases. Phase I is done out doors and takes 3 - 12 days for completion. Phase II involves pasteurization in a peak heat room and takes 3 - 7 days.

Materials required for preparation of synthetic compost (SMC)

Wheat straw - 1000 kg

Chicken manure - 400 kg.

Brewers grain - 72 kg

Urea - 14.5 kg

Gypsum - 30 kg

Phase I : Preliminary stacking is done on a concrete floor four days before composting by putting straw and chicken manure in layers. Sufficient water is added to wet the straw. The stack is usually made of the size 1.75m x 1.50m x any length. On the second day, turn the stack, sprinkle water, if necessary, and press hard.

Day 1 : (Aerobic stack period); Open the stack, add urea and water on dry patches and make new stack with wooden boards.

Day 2 : First turning is given

Day 4 : Second turning is given; and gypsum is added

Day 6 : Third turning is given

Day 8 : Fill in trays for pasteurization, add BHC or lindane dust (5%).

In a heap, only the centre portion is fermented under ideal conditions of heat, moisture and aeration. The top and sides are too cold and the bottom lacks air. While giving a turn, one foot from top, sides and bottom are removed and kept separately. The centre portion of the original stack is placed at the bottom of the new stack; the top and sides make the centre and the bottom is placed at top and sides. The temperature in the compost heap goes even up to 80°C. Proper recording of temperature is necessary at different points. Turning should be given as indicated above to avoid over or under composting. Over-composted substrate will be deficient in certain nutrients and vitamins while under-composted one will retain ammonia and encourage *Coprinus* sp. infection during spawn run and cropping.

Phase II: Peak heat or pasteurization: It is done in a properly insulated room, since the temperature of the compost is required to be raised to 60°C. The compost is loosely filled in trays to a depth of 15 cm. After light pressing, the trays are arranged in tiers leaving a vertical space of about 10 cm between trays, 30 cm on each side of the room, 75 cm in the centre and 5.0 cm between stacks for circulation of hot air or steam. Avoid cooling of compost during filling by heating the room.

Pasteurization is then carried out as follows

Day 1 : Door and ventilators are closed after stacking the trays, the room temperature is brought to 48° - 49°C by dry heat. Fresh air is inducted with a reverse mounted exhaust fan for 2 hours.

Day 2 : Raise the temperature of the compost to 54.4°C by injecting steam (air temperature 43° - 48°C), maintain proper aeration, temperature is recorded every hour.

Day 3 : It is now time to go in for the 'kill'. Put in full fresh air for 15 minutes and the bed temperature is gradually increased to 58° - 60°C by injecting steam (air temperature 60°C) after closing the vents. Keep the temperture at 60°C for one hour and introduce fresh air to bring the bed temperature below 57.6°C and air temperature to 46° - 49°C.

Day 4 : Heat produced by fermentation of compost will be enough to keep the temperature at 50° - 53°C. Inject fresh air as required.

Day 5 : Compost temperature will drop gradually with the introduction of fresh air. When the temperature of the compost at a depth of 3 inches from the surface is around 25° - 30°C, the bed is ready for spawning.

Preparation of bed and spawning The size of tray that is convenient to handle is 90 x 50 x 15 cm the empty weight of which will be of the order of 8 - 9 kg. A quantity of 32 kg of compost can be filled into this tray making the total weight to roughly 40 kg. Spawning is done at the rate of 0.45 % of weight of compost. After spawning, the compost is pressed hard to make it compact. The trays are then arranged in the cropping room in tiers and are covered with newspaper sheets sprayed with mancozeb - 0.2 % solution to moisten the bed and to prevent the growth of any other fungal contaminants.. The newspapers should be sprinkled with water periodically. Compost is impregnated with mycelial threads completely within 15 - 20 days showing whitish strands. The dark brown color of the compost changes to light brown. Little ventilation is required during spawn run.

Casing. Casing is done 12 - 15 days after spawn running. For casing, red soil mixed with sand and calcium carbonate (2.0 %) is commonly used. Other casing compounds containing humus, well-decomposed farmyard manure, decomposed coconut coir waste in combination with other materials may also be used. The casing compound is sterilized, either by heat sterilization or by chemical sterilization. The casing compound should be slightly alkaline, with a pH range of 7.0 - 8.0. The sterilized casing material is mixed with water to make a thick paste and then applied on the surface of the bed to a thickness of about 2.0 cm. After casing, mancozeb - 0.2 % solution is sprayed over the bed for 2 days and subsequently clean water is sprayed whenever necessary, so as to maintain the moisture level of the bed at 50 - 60 %.

In about 2 - 3 weeks, pinhead-like mushroom buds begin to appear on the bed. At this stage, the mushroom shed should be well-aerated, so as to prevent build-up of carbon dioxide inside the shed. Further, the temperature in the mushroom shed should be lowered to 15° - 20°c. And the humidity should be maintained at 85 %. When the mushroom buds reach the size of pea seeds, water is sprayed on the buds without allowing any water to stagnate on them. In another 3 - 5 days, the mushrooms will be ready for harvest. The first harvest may be obtained about 45 days after spawning.

Harvesting. Harvest of the mushrooms should be taken up before the pileus starts opening. The entire mushroom along with the stipe is harvested. In about 7 - 10 days intervals, 3 - 4 harvests can be made. After the second harvest, the casing material is scratched and stirred, and then levelled. Water is then sprayed on the surface of the casing. By doing this, good yield is obtained during the 3rd and 4th harvests. From 10 kg. of compost, about 3.5 kg. of button mushrooms are obtained in about 70 days. Areawise, from one sq. m. of bed, 10 - 15 kg. of mushrooms are obtained. At the time of harvest, the compost, soil etc. adhering to the base of the stipe are gently removed and the mushrooms packed in perforated polythene bags for marketing.

Constraints in the cultivation of mushrooms. Besides climatic factors, which are the limiting factors in the cultivation of edible mushrooms, several other problems are being faced from the time of preparation of spawn till harvesting and marketing of mushrooms. Mushrooms are naturally weak fungal organisms and cannot compete with several other commonly found saprophytic and facultative parasitic fungi. Several species of *Trichoderma, Aspergillus, Penicillium, Rhizopus, Fusarium* etc. may contaminate the spawn and they may occur on the mushroom beds also and restrict the growth of the mushroom mycelium. Some insects are also known to affect the mycelium, as well as the mushrooms.

Fungal, bacterial and virus diseases affecting mushroom production. Several fungal organisms are found to grow on the mushroom beds, competing with the growth of the mushroom mycelium and restrict or arrest the growth of the mycelium, thereby adversely affecting mushroom production. Species of *Aspergillus, Penicillium* and *Trichoderma* grow as green or black moulds, while species of *Chaetomium* grow as whitish-green moulds. These mouldy growths may sometimes cover the entire bed, as a result the growth of the mushroom mycelium is completely suppressed. Species of rhizopus, the bread mould, grow profusely on the bed surface and appear black after sporulation. Species of *Sclerotium* produce dark-brown sclerotial bodies on the surface, as well as inside the mushroom bed. Species of *verticillium, Stemonitis, Hypomyces, Neuospora* and few other fungal organisms also grow on the mushroom beds and casing as contaminants.

Control of fungal contaminants. Strict hygienic measures should be taken at each and every stages of mushroom cultivation, from selection of grains for spawn production to harvest of the mushrooms. The fungal contaminants on the bed and the casing can be controlled by spraying with mancozeb - 0.2 % or carbendazim - 0.1 %. Spraying should be done to cover the entire surface of the bed or casing. If contamination occurs regularly in the spawn-running room or in the cropping room, then preparation of beds should be stopped for a few days. The rooms are fumigated with potassium permanganate and formalin solution in the ratio 1 : 2 and the rooms kept closed for one or two days. Afterwards, the rooms are kept opened and aerated, so that the formalin vapor is completely removed from the rooms. The sides of the rooms, roof and the racks are sprayed thoroughly with mancozeb - 0.2 % or carbendazim - 0.1 % and only after that beds should be kept in the rooms.

In case the moisture content in the straw or compost is more and if there is liberation of ammonia gas, then a few species of *Coprinus* may start growing and in case of severe contamination with these fungi, the yield loss of mushrooms may up to 90 %. These black, tall and lanky, flimsy mushrooms are called **'inky mushrooms'** because of exudation of a black, ink-like liquid. These mushrooms should be uprooted and burnt as soon as they are noticed and before the pileus opens.

After the final harvest is over, the mushroom beds should be removed from the cropping room, dumped into manure pit and covered with soil.

A few bacteria belonging to species of *Pseudomonas* also affect mushrooms and cause spoilage. Use of very old and sticky spawn for preparing mushroom beds, too much of moisture in the straw or compost, presence of maggots of mushroom flies, spraying excessive quantities of water on the mushroom beds, on mushroom buds or grown-up mushrooms etc. predispose the beds and mushrooms to bacterial attack. The bacterial attack is more in old beds after the 2nd and 3rd harvest, which may contain excessive moisture. The bacterial diseases can be controlled to a large extent by following strict hygienic measures and proper water management.

A few viruses are also known to attack mushrooms and cause considerable damage. As a result of virus infection, the mushrooms produced are much smaller in size, malformed, rotten and emit a bad odor. Virus diseases can be controlled by adopting proper hygienic measures, destruction of affected beds by burning, use of freshly prepared mother spawn for preparing bed spawn and by controlling insects attacking mushrooms.

Insects attacking mushrooms and their control. Some insect and non-insect pests are found to attack mushroom beds or mushrooms and cause extensive damage. The mushroom flies belonging to the Order - Diptera are important pests of mushrooms. The Sciarid flies *(Bradysia paupera, B. tritici* and *Lycoretla avripila)* and the Phorids *(Megaselia sandhui)* are commonly found to attack mushrooms beds and cause serious damage. The flies, which are found naturally in refuse and rotten organic matter throughout the year, get access to mushroom beds and lay eggs in the mushroom beds through the holes. The maggots hatching out from the eggs feed on the substrate material and mycelium of the mushroom fungus and cause considerable damage. The beds attacked by the maggots are prone to infection by several fungal and bacterial organisms, leading to rotting of the mushroom beds.

Spring tails *(Seira tricolor* and *Lepidocyrtus cyancus)* belonging to Order - Collembola are also found to attack mushroom beds and the mycelium and cause considerable damage. A small, black mushroom beetle belonging to the Order - coleoptera also cause damage to the mushroom beds and mushrooms.

Mushroom mites, which are non-insect pests also cause damage to the mushroom beds and mushrooms.

Control of insect and non-insect pests. Judicial water management is most important in preventing the occurrence of most of the pests attacking mushroom beds and mushrooms. Excessive quantities of water should not be sprayed on the mushroom beds and on the mushroom buds or on the grown-up mushrooms. Provisions should be made for the circulation of sufficient quantity of air inside the cropping room. Suitable wire mesh should be fitted on to the ventilators to prevent entry of mushroom flies into the cropping room. Immediately after the final harvest of mushrooms is completed, the beds should be removed, dumped into manure pits and covered with soil. Spraying with dichlorvos - 0.1 %, which has got contact and fumigant action is effective in controlling all the pests without affecting the mushroom fungi. Mites can be controlled by spraying with wettable sulfur - 0.4 % or dicofol - 0.1 % on the walls, roof and on empty racks. But, these fungicides should not be sprayed on the mushroom bed or on the mushrooms.

Preservation of mushrooms. Mushrooms, like fruits and vegetables are highly perishable. They grow in flushes and every 8 to 10 day they are harvested in abundance. In between the flushes the production is very low. The demand never coincides with the supply and the day when there is good production the demand may be low and vice versa. To prevent such a glut in the fresh market it is necessary to preserve them.

Dehydration of mushrooms. Mushrooms are harvested at a mature stage. The stalk is trimmed. Mushrooms can be dried in the sun or in a mechanical dehydrator at 50-60°C till the mushroom attains a constant weight. The weight of the dried mushrooms are reduced to nearly one tenth of the fresh weight. Dried mushrooms are hermetically sealed so that they retain their flavor for six to seven months during storage. The major disadvantage of dried mushroom is its browning. They are reconstituted again by immersing in hot water.

Canning of mushrooms. Mushrooms can be canned either whole, sliced or in smaller pieces. Mushrooms to be canned should be harvested early and without any blemishes. Cleaned mushrooms are blanched (immersing mushrooms in a boiling solution of 0.1 to 0.2 per cent citric acid for about 5 minutes and cooling immediately in cold water), and immersed in a brine solution (consisting of 2 per cent common salt, 2 per cent sugar and 0.3 per cent citric acid) filled in cans upto the brim. The cans are sealed and sterilized at 15 lb/ sq inch. for 25 to 30 minutes.

Mushrooms are also preserved in the form of pickles and ketchup.

Table 17: List of Fungicides and dosage for spraying.

Chemical name (Technical name)	Trade name	Dosage (Based on the formulation)
Wettable sulfur	Thiovit, Cosan, Sulfex, Elosan etc.	0.4%
Thiram	Thiride, Arasan, Hexathir, Panoram,Tersan etc.	0.2%
Ferbam	Fermate, Fermocide, Coromet, Hexaferb etc.	0.2%
Ziram	Ziride, Zerlate, Hexazir, Milbam, Cuman L etc.	0.2%
Zineb	Dithane Z 78, Lonocol, Hexathane, Blicin, Unizeb, Polyram, Parzate etc.	0.2%

contd...

Maneb	Dithane M 22, Manzate etc.	0.2%
Mancozeb	Dithane M 45, Manzate, Pencozeb etc.	0.2%
Nabam	Dithane D 14, Parzate etc.	0.2%
Bordeaux mixture	Bordeaux mixture	0.75-1.0%
Copper oxychloride	Fytolon, Cupramar, Blitox, Perenox, Cuprocide, Micop, Parrycop, Blue copper etc.	0.25%
Chloranil	Spergon	0.125-0.15%
Dichlone	Phygon	0.125-0.15%
Chloroneb	Demosan	0.125-0.15%
Chlorothalonil	Daconil, Bravo, Termil, Kavach, Speldrum etc.	0.125-0.15%
Dinocap	Karathane, Crotothane, Arathane, Capryl, Mildex etc.	0.125-0.15%
Dichloran	Botran	0.125-0.15%
Captan	Captan 50, Esso fungicide 406, Orthocide, merpan, Hexacap, Vancide etc.	0.125-0.15%
Captafol	Difolatan, Foltaf, Sanspor, Difosan etc.	0.125-0.15%
Iprodione	Rovral, Chipco, Glycophene etc.	0.125-0.15%
Flutolanil	Moncut, Prostar etc.	0.125-0.15%
Vinclozolin	Ornalin, Ronilan, Vorlan etc.	0.125-0.15%
Folpet	Phaltan	0.125-0.15%
Du-ter	Super tin, Farmatin, Tubotin etc.	0.125-0.15%
Brestan	Brestan	0.125-0.15%
Brestanol	Tinmate	0.125-0.15%
Benomyl	Benlate, Tersan 1991 etc.	0.1%
Carbendazim	Bavistin, Derosal, Bengard, Zoom, MBC, Tagstin, Agrozim etc.	0.1%
Thiabendazole	Mertect, Tecto, Storite etc.	0.1%
Thiophanate	Topsin, Cercobin, Fungo, Enovit etc.	0.1%
Thiophanate methyl	Topsin M, Cercobin M, Enovit M etc.	0.1%
Carboxin	Vitavax	0.1%
Oxycarboxin	Plantvax	0.1%

Tridemorph	Calixin, Beacon, Bardew etc.	0.1%
Edifenphos	Hinosan	0.1%
Fosetyl-Al	Aliette	0.1%
Iprobenphos	Kitazin, EBP etc.	0.1%
Pyrazophos	Afugan, Curamil etc.	0.1%
Metalaxyl	Ridomil, Acylon, Apron etc.	0.1%
Diamethirimol	Milcurb	0.1%
Ethirimol	Milstem	0.1%
Bupirimate	Nimrod	0.1%
Fenarimol	Rubigan	0.1%
Nuarimol	Trimidal	0.1%
Triadimefon	Bayleton	0.1%
Triadimenol	Bayton	0.1%
Bitetanol	Baycor	0.1%
Butrizol	Indar	0.1%
Propiconazole	Tilt	0.1%
Etaconazole	Vangard	0.1%
Myclobutanil	Rally	0.1%
Difenoconazole	Score	0.1%
Imazalil	Fungaflor	0.1%
Propamocarb	Banol, Previcur etc.	0.1%
Triforine	Cela, Funginex, Saprol etc.	0.1%
Dodine	Cyprex, Guanidol, Melprex, Syllit etc.	0.1%
Tricyclazole	Beam, Bim etc.	0.1%
Pyroquilon	Fongorene	0.1%

0.1% = 1.0 gm or 1.0 ml / liter of water

0.125% = 1.25 gm or 1.25 ml / liter of water

0.15% = 1.5 gm or 1.5 ml / liter of water

0.2% = 2.0 gm or 2.0 ml / liter of water

0.3% = 4.0 gm or 4.0 ml / liter of water

Bordeaux mixture 1.0% = 1.0 kg of powdered copper sulfate crystals + 1.0 kg of slaked lime in 100 liters of water.

❐❐❐

References

Anaja, K.R. 2003. Experiments in Microbiology, Plant Pathology and Biotechnology. *New Age International (P) Limited, Publishers, New Delhi*. pp. 607.

Anon. 1997. Hand book of Agriculture. *Directorate of Publications and Information on Agriculture, Krishi Anusandhan Bhavan, Pusa, New Delhi*-110012. pp. 1304.

Ashok Kumar Sinha. 2007. Fundamentals of Plant Pathology. *Kalyani Publishers, Ludhiana*. pp. 462.

Alexopoulos, C.J. and Mims, C.W. 1979. Introductory Mycology. *Wiley Eastern Limited, New Delhi*-110002. pp. 631.

Alexopoulos, C.J., Mims, C.W. and Blackwell, M. 1996. Introductory Mycology. *John Wiley and Sons, Inc., New York*. pp. 869.

Chaube, H.S. and Pundir, V.S. 2005. Crop diseases and their management. *Prentice Hall of India Pvt. Ltd., New Delhi*. pp. 703.

Chopra, G.L. and Verma, V. 1985. A Textbook of Fungi, *Pradeep Publications, Jalandhar*. pp. 586.

Darwin Christdhas Henry, L. 2008 'Certain studies in the cultural, physiological and post-harvest aspects of *Volvariella volvaceae*'. *Thesis submitted in fulfillment for the award of Doctorate Degree in Agriculture (Plant Pathology) of the Annamalai University. Chidambaram, Tamil Nadu.*

Dubey, R.C. and Maheshwari, D.K. 2002. A Text Book of Microbiology. *S. Chand and Company Limited*

Edwin J. Butler and Jones, S.G. 1955. Plant Pathology. *Macmillan and Company Limited, New York*. pp. 979.

Ganapathy, T. Rabindran, R. and Sabitha Doraiswamy. 2002. An Illustrated Glossary of Plant Pathology. *AE Publications, Coimbatore*-41. pp. 251.

Gaumann, E.A. and Wynd, F.L. 1952. The Fungi. *Hafner Publishing Company, New York*. pp. 420.

George N. Agrios. 1997. Plant Pathology. *Academic Press, California, USA*, pp. 635.

Haarer, A.E. 1956. Modern coffee production. *Leonard Hill (Books) Ltd., 9-Eden Street, N.W.I., London*. pp. 467.

Lewin Devasahayam, H. and Darwin Christdhas Henry, L. 2009. Illustrated Plant Pathology - Basic concepts. *New India Publishing Agency, Pitampura, New Delhi*-110088. pp. 470.

Mani, A., Selvaraj, A.M., Narayanan, L.M. and Arumugam, N. 1998. Microbiology (General and Applied). *Saras Publications, Nagercoil, Tamil Nadu*. pp. 166.

Mehrotra, B.S. 1967. The Fungi. *International Publishing House, Allahabad*. pp. 331.

Mehrotra, R.S. and Ashok Aggarwal, 2003. Plant Pathology. *Tata McGraw-Hill Publishing Company Limited, New Delhi*-110008. pp. 846.

Nita Bhal. 1988. Hand book on Mushrooms. *Oxford and IBH Publishing Co., Pvt., Ltd., New Delhi*. pp. 129.

Padoley, S.K. and Mistry, P.B. 1982. A manual of Plant pathology. *S. Chand* and *Co., Ltd. New Delhi*. pp. 131.

Pandey, B.P. 2001. Plant Pathology. *S. Chand* and *Company Limited, New Delhi*-110055. pp. 492.

Powar, C.B. and Daginawala, H.F. 2001. General Microbiology (Vol. II). *Himalaya Publishing House, Mumbai*-400004. pp. 680.

Rangaswami, G., 1988. Diseases of Crop Plants in India. *Prentice Hall of India Pvt., Ltd., New Delhi*. pp. 498.

Rangaswami, G. and Mahadevan, A. 1999. Diseases of Crop Plants in India. *Prentice Hall of India Pvt., Ltd., New Delhi*. pp. 536.

Salle, A.J. 1961. Fundamental Principles of Bacteriology. *McGraw-Hill Boo Company, Inc., New York*. pp. 812.

Sarmah, K.C. 1960. Diseases of Tea and Associated Crops in North-East Indsia.

Sendilvel, V. Kavitha, K. Nakkeeran, S. Raguchander, T. and Marimuthu, T. 2004. Glimpses of Plant Pathology. *AE Publications, Coimbatore*-41. pp. 276.

Seshagiri Rao, D. 1972. A handbook of Plant Protection. *S.V. Rangaswamy & Co., Pvt., Ltd., Bangalore*-20 pp. 841.

Singh, R.S. 1983. Plant Diseases. *Oxford and IBH Publishing Co., New Delhi*-110028. pp. 608.

Singh, R.S. 1984. Introduction to Principles of Plant Pathology. *Oxford and IBH Publication Co., New Delhi*. pp. 532.

Singh, R.S. 1989. Plant Pathogens - The Prokaryotes. *Oxford and IBH Publishing Co., Ltd., New Delhi*. pp. 215.

Singh, S.S. 1998. Crop Management. *Kalyani Publishers, Ludhiana*. pp. 507.

Verma, H.N. 2003. Basics of Plant Pathology. *Oxford and IBH Publishing Co. Ltd., New Delhi*. pp. 218.

Vidhyasekaran, P. 2006. Principles of Plant Pathology. *CBS Publishers and Distributors. New Delhi*-110302. pp. 166.

Walker, J.C. 1969. Plant Pathology. *Tata McGraw Hill Publishing Co., Ltd., New Delhi*. pp. 819.

□□□

Salle, A.J. 1961. Fundamental Principles of Bacteriology. McGraw Hill Book Company, Inc., New York. pp. 8[illegible]2.

[illegible], K.C. 1960. Diseases of [illegible] and Associated Crops in North-East India.

[illegible]

[illegible] pp. [illegible]

[illegible] Bacteriology [illegible], New Delhi [illegible] pp. 508.

[illegible] Introduction to Principles of Plant Pathology. Oxford and IBH Publishing [illegible]

[illegible] Plant Pathogens – The Prokaryotes. Oxford and IBH [illegible]

[illegible] Crop Management. [illegible] Publishers, Ludhiana. [illegible]

[illegible] Plant Pathology [illegible] Oxford and IBH Publishing Co. [illegible] pp. 218.

[illegible] Plant Pathology. CBS Publishers and [illegible], New Delhi [illegible]

[illegible]

Glossary of Scientific Terms

Abscission. The shedding of leaves or other plant parts as a result of physical weakness in a specialized layer of cells (the abscission layer) that develops at the base of the structure.

Acervulus (pi. Acervuli). A subepidermal, inverted saucer-shaped, asexual fruiting body, consisting of a cushion-like mass of hyphae and palisade-like conidiophores that cut off conidia from their tips; characteristic of Melanconiales of Fungi Imperfecti.

Acquired resistance. Resistance response developed by a normally susceptible host following a predisposing treatment such as inoculation with a virus, fungus, bacterium or treatment with certain chemicals (induced resistance or acquired immunity); this resistance is not inherited.

Acquisition feeding time. The feeding time (acquisition access time) required for a vector to feed on an infected plant so as to acquire a virus for subsequent transmission of the virus to healthy plants (to become viruliferous)

Acropetal. Upward from the base to the apex of a shoot of a plant; in fungi, production of spores in succession in the direction of the apex, so that the apical spore is the youngest.

Acuminate. Pointed; tapering at the end.

Aecium (pi. Aecia). A cup-shaped fruiting body of the rust fungi consisting of binucleate hyphal cells, with or without a peridium that produces spore chains consisting of aeciospores, alternating with disjunctor cells, following successive conjugate division of the nuclei (Stage I of the rust iungus).

Aflatoxin. Mycotoxins produced by the fungi belonging to Genus - *Aspergillus*, which is carcinogenic (substance or agent that causes cancer) and sometimes may even be lethal. They are the derivatives of coumarine.

Agar. A jelly-like polysaccharide derived from certain seaweeds (algae) that is used at a concentration of 1.5 - 2.0 % for solidifying culture media, which are used for culturing microorganisms; the term is also applied to the medium itself (agar medium).

Amphitrichous. Single flagellum at both the polar ends.

Antheridium. The male gametangium of heterogamous fungi.

Antibiotic. A chemical compound produced by a microorganism, which inhibits or kills some other specific microorganisms.

Apothecium (pi. Apothecia). An open cup- or saucer-shaped ascocarp.

Appressorium (pi. Appressoria). A swollen, flattened, pressing organ arising from the tip of a hypha or germ tube that facilitates attachment to the host. A minute infection peg usually grows from it and enters the epidermal cell of the host.

Ascocarp or Ascoma. Sexual fruiting body of an ascomycetous fungus that produces asci and ascospores.

Ascogonium (pi. Ascogonia). The female gametangium of the Ascomycetes.

Ascostroma (pi. Ascostromata). A stromatic ascocarp bearing asci directly in locules within the stroma.

Ascus (pi. Asci). A sac-like cell containing definite number of ascospores, usually eight, formed by free cell formation, mostly after karyogamy and meiosis; characteristic of Class - Ascomycetes.

Ascus mother cell. The binucleate crook cell in the Ascomycetes, in which karyogamy occurs and then develops into the ascus.

Aseptate. Non-septate, coenocytic; lacking cross walls or septa.

Atrophy. Reduction in size of an organ or the entire organism.

Autoecious (syn. Monoecious). A parasitic fungus, which can complete its entire life cycle on a single host species as in the case of some rusts.

Avirulent Lacking virulence; unable to cause disease.

Azoosporangium. A sporangium that contains non-motile spores.

Azygospore. A zygospore that develops parthenogenetically

Basidiocarp. A sexual fruiting body of the basidiomycetous fungi, in or on which basidia and basidiospores are produced.

Basidium (pi. Basidia). A club-shaped, zygote cell bearing a definite number of basidiospores, usually four, at the end of minute sterigmata formed at the apex of the basidium, following karyogamy and meiosis.

Basipetal. Downward from the apex toward the base of a shoot; development in the direction of the base, so that the apical part is oldest; formation of spores in succession in which the apical spore is the oldest ant the youngest spore is at the base.

Biological control. Manipulation of natural enemies, such as parasitoids, predators or other microorganisms in an attempt at reducing the pest numbers and keep them at much reduced levels.

Biotic agent A living organism that is the cause of damage.

Blight A disease characterized by general and rapid killing of tissues of leaves, flowers and stems.

Blotch. A disease characterized by large and irregularly shaped, necrotic spots or blots on leaves, shoots and stems.

Broad-spectrum pesticides. Pesticides effective against a variety of pests and diseases.

Budding. A form of asexual reproduction typical of yeast, in which a new cell is formed as a small outgrowth (bud) from the parent cell.

Bud wood. Wood consisting of strong, young shoots bearing buds suitable for use in budding.

Callus. Superficial, unspecialized tissue produced by plants in response to wounding; parenchymatous tissue of cambial origin that are formed in response to wounding; a mass of thin-walled, undifferentiated cells usually developed as a result of wounding or infection; culture on nutrient media.

Canker. A necrotic, often sunken or cracked lesion surrounded by callus on a stem, branch or twig of a plant.

Catenulate. Conidia produced in chains.

Cellulolytic. Enzymes, such as cellulase capable of decomposing cellulose.

Chemotherapy. Control of plant diseases with chemotherapeutants, such as systemic fungicides and antibiotics that are absorbed and are translocated internally.

Chlamydospores. Thick-walled or double-walled, asexual resting spores formed from hyphal cells, either terminal or intercalary or by the transformation of one or more conidial cells, that can serve as an overwintering reproductive structure.

Chlorosis. Yellowing of normally green tissue due to chlorophyll destruction or failure of chlorophyll formation.

Chronic. Slow developing, persistent or recurring symptoms.

Clavate. Club-like, narrowing in the direction of the base.

Cleistothecium (pi. Cleistothecia). A completely closed ascocarp containing asci and ascospores.

Coalesce. Grow or join together into one body or spot.

Coelomycetes. A group of fungi in the Deuteromycetes that produce their spores within covered or closed structures, such as pycnidia or acervuli.

Coenocytic. Non-septate or aseptate; the nuclei are distributed in the cytoplasm without being separated by cross walls or septa

Columella (pi. Columellae). A sterile structure within a sporangium or other fructification; often an extension of the stalk bearing the sporangium or fructification.

Conidiophores. Simple or branched specialized hyphae arising from somatic hyphae, bearing at the tip conidiogenous cell that produces conidia singly or in chains.

Conidium (pi. Conidia). Asexual, non-motile spore formed by abstriction and detachment of part of a hyphal cell at the end of a conidiophore that germinates by a germ tube. Sporangia germinating by issuing germ tubes, instead of producing zoospores may also be called as conidia.

Conjugate nuclear division. The simultaneous division of the two nuclei in a dikaryon, giving rise to four daughter nuclei; these generally become separated by a septum into two cells, with the sister nuclei migrating into different daughter cells.

Coremium (pi. Coremia; syn. Synnema). Compact or fused, generally upright conidiophores with branches and terminal spores forming a head-like cluster.

Cotyledon. Seed leaf, one in monocots and two in dicots; primary embryonic leaf within the seed, in which nutrients for the emerging seedling are stored.

Cross protection. The process whereby a normally susceptible host is infected with a less virulent pathogen (usually a virus) and thereby the host becomes resistant to the infection by a more virulent strain of the same virus.

Cystidium (pi. Cystidia). A sterile element or structure occurring in the hymenium of certain Basidiomycetes; cystidia are larger than the other hymenial elements and protrude beyond them.

Deciduous. Detach and falling down easily.

Demicyclic. A rust fungus that lacks the uredinial (repeating) stage, but typically has stages 0, I, III and IV.

Determinate. The stage at which vegetative growth ceases and the reproductive structures are formed.

Diagnosis. Identification of the nature and cause of the illness, ailment or disease on the basis of its signs, symptoms, etiology, pathogenesis etc.

Dichotomous branching. Pair-wise forking; often repeatedly.

Dictyospore. A spore with both vertical and horizontal septa as in *Alternaria.*

Dikaryon. Having two sexually compatible haploid nuclei per cell, either in the spores or in the hyphal cells, which is common in the Basidiomycetes.

Dikaryophase. The division of the dikaryon simultaneously, leading to the presence of such pair of sexually compatible nuclei in each of the cell.

Dioecious. Species, which produce separate male and female thalli. The male thalli produce only male sex organs, while the female thalli produce only female sex organs. Single dioecious fungi cannot reproduce sexually by itself.

Diploid. Having two complete sets of chromosomes ('2n' chromosomes), as a result of fusion of two compatible male and female, haploid nuclei, each having 'n' chromosomes.

Disinfectant. A physical or chemical agent that frees a seed, plant, organ or tissue from infection.

Disinfestant An agent that kills or inactivates pathogens in the environment or on the surface of a seed, plant or plant organ before infection takes place.

Dissemination (syn. Dispersal). The transportation of inoculum from diseased plants to healthy plants and from one location to another location.

Echinulate. Having small, pointed processes or spines projecting from the cell walls.

Ecology. The study that deals with the effect of environmental factors, such as soil, climate, culture etc. on the occurrence, severity and distribution of plant diseases or plant pathogens.

Economic threshold. The pathogen density or the disease or pest intensity at or above which the value of crop losses would exceed the cost of management practices.

Ellipsoid. Spores, which are rounded-oblong; having long sides parallel and ends almost hemispherically.

Embryo. The rudimentary plant or germ as formed in a seed.

Enation. Tissue malformation or an abnormal outgrowth, usually on the leaf and sometimes on the stem also, often induced by virus infection.

Encyst. To form a cyst or protective covering.

Endemic disease. A disease permanently established in moderate or severe form in a defined area, commonly a country or part of a country.

Endobiotic. Living entirely within the host tissues.

Endoconidium. A conidium produced inside a hypha or conidiophore.

Endosperm. The tissue containing food storage within the embryo sac but outside the embryo of seeds of plants.

Enzyme. Complex organic substances formed by living cells, such as cellulase that act as biological catalysts and accelerate specific transformation or decomposition of materials such as cellulose, digestion of food etc.

Ephemeral. Short-lived.

Epibiotic. Producing only the reproductive organs on the outside of the host while the thallus remains inside the host tissues.

Epicotyl. The portion of the axis of a plant embryo or seedling between the cotyledons and the first leaf formed after the cotyledons.

Epidemic (Epiphytotic). A sudden, rapid spread of a disease; a widespread and severe temporary increase in the incidence of an infectious disease, particularly within a season. The term **'epiphytotic'** is also equally applicable

Epidemiology. The study of factors influencing the initiation, development and spread of infectious diseases.

Epidermis. Outer layer of cells in plants below the cuticle and forming the external integument in plants.

Epiphyllous. Found on the upper surface.

Epiphytic. Living on the surface of plants, but not as a parasite.

Eradicants. Systemic fungicides that can enter into the host tissues to a little extent and eradicate the dormant structures of the pathogen, as well as the active pathogen from the host.

Erumpent Bursting or erupting through the substrate surface.

Etiology. The determination and study of the cause of a disease.

Eucarpic. The reproductive organs arise from only a portion of the thallus, while the remainder continues its normal somatic life.

Extramatrical. Outside the host tissues.

Exudate. Liquid excreted or discharged from diseased tissues, from roots and leaves, or by fungi.

Facultative parasites. Organisms, which are normally saprophytes, but when environmental conditions are favorable to them and when suitable hosts are present they become parasites. They can be grown on artificial culture media

Facultative saprophytes (Facultative saprobes). Organisms, though normally parasitic on living hosts, start leading a saprophytic life in the absence of suitable hosts. They can be grown on artificial culture media.

Falcate. Curved; sickle-shaped.

Fascicle. A small group or bundle.

Flaccid. Wilted; lacking turgor or rigidity.

Flagellum (pi. Flagella). A whip-, hair- or tinsel-like structure that serves to propel a motile cell.

Forma specialis (f.sp.). A group of races and biotypes of a pathogen species, differing in some physiologic properties, that can infect only plants within a certain host Genus or species.

Free cell formation. The process by which the eight nuclei, each with a portion of adjacent cytoplasm are cut off by walls in the immature ascus to become ascospores.

Free water. Unbound water; a film of water on the plant surface.

Fructification. Production of any complex fungal fruiting body or spore bearing organ that contains or bears spores.

Fungicidal. Fungicides capable of killing fungal spores or mycelium. Applicable also to physical agents, such as heat, ultraviolet light, X-rays, gamma radiation etc., as well as to chemicals that are lethal at low concentrations.

Fungistatic. Arresting fungus growth but not killing the fungus concerned.

Fusiform. Spindle-shaped, tapering at each end as in the case of macroconidia *of Fusarium.*

Gall (syn. Tumor). Abnormal swelling or localized outgrowth, often roughly spherical, but unlike any organ of the normal plant, produced by a plant as a result of attack by a fungus, bacterium, nematode, insect or other organisms.

Gametangia. Hyphal cells of Zygomycetes containing gametes or nuclei that behave as gametes, which fuse to form zygospores in sexual recombination.

Gamete. Sex cell; a reproductive cell usually haploid that unites with another gamete of the opposite sex to form a zygote.

Geniculate. Bent like a knee.

Germ tube. The initial hyphal growth from a germinating fungal spore, which develops into a mycelium.

Gill. Thin, radial membrane or lamella in the cap of a gill fungus belonging to Family - Agaricaceae of Basidiomycetes that produces basidiospores (e-g. mushrooms)

Girdle. To encircle and cut through a stem or the bark and outer few rings of wood, disrupting the phloem and xylem.

Glabrous. Smooth; without hairs

Gloeospore. Slimy spore.

Guard cells. Paired, specialized epidermal cells that contain chloroplasts and surround a stoma.

Gummosis. Pathologic condition characterized by excessive formation of gums as a result of cell degeneration.

Guttation. Exudation of watery, sticky liquid from hydathodes, especially along the leaf margins.

Halo. A spot surrounded by a chlorotic zone, which is characteristic of spots produced by some microorganisms on the leaves and sometimes on the other parts.

Haploid. A cell or organism whose nuclei have a single complete set of chromosomes.

Haplophase. The part of the life cycle of the organism in which the cells are haploid.

Haustoria. (sing. Haustorium). Specialized absorbing organs produced by the mycelium, which penetrate the host cells and absorb nourishment from the host; Mostly produced by obligate parasites, but also produced by some facultative parasites.

Helical. Shaped like a corkscrew.

Hermophroditic. Species, which produce distinguishable male and female sex organs in the same thallus i.e., each individual produces both recognizable male and female sex organs. This is also called as 'monoecious'.

Heteroecious. Organisms, which require two botanically different hosts for completion of its life cycle, as in the case of some rust fungi.

Heterogametangia and heterogametes. The gametangia and gametes produced by the sex organs, which are morphologically quite different.

HeterokonL A biflagellate zoospore with two flagella of unequal size or unequal length.

Heterothallic fungi. Fungi producing compatible male and female sex organs or gametes on morphologically similar but physiologically distinct mycelia.

Hilum. A scar indicating the point of attachment.

Holocarpic. In the case of either asexual or sexual reproduction, the entire thallus is converted into reproductive structures and segmented into spores.

Homothallic fungi. Fungi producing compatible male and female sex organs or gametes on the same mycelium.

Honeydew. A sugary exudate secreted by the ergot pathogen and sucking insects, such as aphids, white flies and scale insects.

Host plant. Living plant attacked by or harboring a parasite or pathogen and from which the invader obtains part or entire requirement of its nourishment.

Hyaline. Colorless, transparent.

Hyalosporae. Organisms belonging to Fungi Imperfecti producing hyaline spores.

Hybridization (Cross breeding). The crossing of two individuals of different genotypes to produce 'F_1' progeny with genes for the characteristics of both the parents.

Hydathodes. Structures with one or more openings at the leaf edge that discharge water from the interior of the leaf to the surface.

Hymenium. A spore-producing layer of a fruiting body of Ascomycetes or Basidiomycetes.

Hyperplasia. Abnormal multiplication of cells; abnormal increase in the number of cells, often resulting in the formation of galls or tumors.

Hypertrophy. Abnormal enlargement of cells; abnormal increase in the size of cells in a tissue or organ, often resulting in the formation of galls or tumors.

Hypha (pi. Hyphae). One of the simplest branched filaments of the mycelium of a fungus that is composed of one or more cylindrical cells and that increases in length by growth at its tip. New hyphae arise as lateral branches.

Hypocotyl. Portion of the stem below the cotyledons and above the root.

Hypophyllous. Found on the lower surface.

Hypotrophy. Under development of a tissue or plant due to abnormally reduced cell enlargement.

Immune. Cannot be infected by a specific pathogen; having absolute resistance to a particular disease.

Immunity. The state of being immune; total exclusion of the potential pathogen; may be natural due to innate genetic characters or may be acquired by some predisposing treatment that modifies the chemistry of the plant.

Immunization. The process of inducing complete resistance to a particular disease.

Imperfect stage (Anamorph stage). The part of the life cycle of a fungus in which no sexual spores are produced.

Incubation period. The period of time between penetration of a host by a pathogen and the first appearance of symptoms, signs of disease or both in the host.

Indefinite. Not sharply delimited.

Indeterminate. Continuing to grow vegetatively while producing reproductive structures or flowers.

Indigenous pathogen. Native to an area; not introduced.

Infection. The entry of a pathogenic organism or virus into a host and establishment of a permanent or temporary parasitic relationship.

Infection peg. A very fine hypha with a deposition of substances, such as lignin, callose, cellulose, suberin etc. around it, that is thrust through the cuticle or epidermis of a host cell to cause infection.

Inhibit. To hold in check; to arrest the growth.

Initial inoculum (syn. Primary inoculum). Inoculum, usually from an overwintering source, that initiates disease in the field, as opposed to inoculum that spreads disease during the season.

Inner veil. The hyphal membrane covering the gills of young mushrooms.

Inoculate. To introduce the inoculum into the host for the purpose of producing infection for testing susceptibility to infection.

Inoculation feeding time. The length of time a vector feeds on the test host in transmission experiments.

Inoculum potential The number of infective agents or particles (inoculum) present in the environment of the uninfected host.

Inoperculate. Not opening by a lid or operculum.

Integrated control. An approach that attempts to use all available methods of control of a disease or of all the diseases and pests of a crop plant, such as physical, cultural, biological and chemical methods.

Intercalary. Formed along and within the mycelium and not at the hyphal tips.

Intercellular. Hyphae growing in-between the host cells.

Internally seed-borne. The pathogen, either the fungus thallus or bacterium or virus is located or deep-seated inside the functional part of the seed and cause primary infection.

Interstitial cells. Small, flat, dikaryotic cells formed in-between pairs of true aeciospores that aid in the release of aeciospores.

Intracellular. Hyphae penetrating into the host cells.

Invasion. The penetration, establishment and colonization of a host by a pathogenic organism.

Isthmus. Pads of gelatinous material formed in-between every pair of sporangia, which function as disjunctor cells that aid in the discharge of the sporangia.

Karyogamy. The fusion of two nuclei brought together by plasmogamy into one diploid or zygote nucleus.

Khaira. A disease caused due to deficiency of zinc in rice.

Kresek. The vascular wilt phase of bacterial leaf blight of rice caused by *Xanthomonas oryzae* pv. *oryzae*.

Lamella (pi. Lamellae). A gill; the vertical radial plates or gills on the lower surface of the 'mushroom' of Agaricaceae, which bear basidia and basidiospores on both the surfaces.

Lamina (Leaf blade). The expanded, flat part of a leaf.

Latent infection. The state in which a host is infected by a pathogen, but does not show any visible symptoms.

Leaf crinkle. Crinkling and puckering of the leaves (e-g) *urdbean leaf crinkle virus*.

Leaf curl. Distortion resulting from unequal growth or expansion of leaf tissues along the two sides of the midrib or of the palisade and mesophyll layers.

Leaf roll. Curling of the leaf lamina, generally towards and parallel with the midrib.

Leaf scorch. Leaf necrosis, usually marginal, due to phytotoxicity or nutrient deficiency.

Leaf spot. A plant disease lesion typically restricted in development in the leaf after reaching a characteristic size.

Lenticel. A natural opening in the stem of woody plants, bark, tuber, some fruits or root for the exchange of gases between the plant and the atmosphere.

Lesion. Localized, usually well-defined, abnormal change in the structure of an organ or diseased tissue due to disease or injury; localized diseased area or wound.

Liberation. Spore discharge; the process of aerial dispersal, in which the fungal spores are freed from its sporophore, pustules or lesions; also called 'take off'.

Life cycle. The stage or successive stages in the growth and development of an organism that occur between the appearance and reappearance of the same stage, such as spore of the organism.

Ligule. Membranous appendages at the base of leaves of Graminaceous plants.

Local invasion. Infection involving only localized part of a plant.

Local lesions. Small, restricted lesion, often the characteristic reaction of different cultivars to specific pathogens, especially in response to mechanical inoculation with a virus.

Locule. A cavity within a stroma.

Macerate. To cause disintegration of tissues by separation of cells; to soften by soaking.

Macroclimate. The climate operating over wider regions and larger heights.

Macroconidium (pi. Macroconidia). The larger of the two kinds of conidium formed by certain fungi.

Macrocyclic. A rust fungus that typically exhibits all five stages of the rust life cycle, viz., stages 0,1, II, III and IV.

Macroscopic. Visible to the naked eye without the aid of a magnifying lens or a microscope.

Mechanical transmission. Transmission of a virus from an infected host to a healthy one without the intervention of a vector; either by physical contact and friction of one plant against the other or experimentally by rubbing sap from an infected plant on the leaves of a healthy one or by other mechanical means of inoculation.

Meiosis. Process of nuclear division, in which the number of chromosomes per nucleus is halved, i.e., converting the diploid state to the haploid state; a series of two nuclear divisions, usually in quick succession in which the chromosome number is reduced by one-half.

Metabasidium. The portion of the basidium in which meiosis takes places; also called the promycelium in certain Basidiomycetes.

Metabolism. The process by which cells or organisms utilize nutritive substances to build up living matter and structural components or to break down complex cellular material into simple substances to perform special functions.

Microclimate. The climate of a usually small site or habitat, usually the weather conditions prevailing between and immediately above the plants of an individual crop.

Microconidium (pi. Microconidia). A type of small conidium as compared to a macroconidium; some type of microconidia act as spermatia.

Microcyclic. A rust in which the teliospore is the only binucleate spore in the life cycle.

Micrometer (Micron) (im, i). A unit of length 1/1,000 of a mm or 1/10,000 of a cm.

Microsclerotia. A type of small sclerotia produced by some fungi, such as *Verticillium albo atrum*, which can withstand adverse environmental conditions.

Middle lamella. The cementing layer between adjacent cell walls, generally consisting of pectinaceous materials, except in wood tissues, where pectin is replaced by lignin.

Midrib. Central, thickened vein of a leaf.

Mildew. A fungus disease characterized by the appearance of a powdery coating of white or grayish mycelial growth and spores on the surface of infected plant parts; mildew is usually synonymous with powdery mildew caused by members of the Order - Erysiphales of Ascomycetes. Downy mildew is characterized by the appearance of white or grayish, fine, cottony downy growth caused by members of the Family - Peronosporaceae of Phycomycetes.

Mitosis. Nuclear division, in which the chromosome numbers remain the same in the daughter nuclei.

Mode of action. The way or method by which a pesticide may alter or adversely affect the physiological or biochemical events in an organism resulting in a toxic effect, usually ending in death of the organism.

Monocyclic. Having one cycle per season or year.

Monoecious (syn. Autoecious). Rust fungi that have all stages of their life cycle on a single species of plant; also refers to organisms that produce male and female reproductive organs on a single individual.

Monotrichous. Bacteria having single polar flagellum.

Morphology. Study of form, structure, architecture and development of an organism.

Mosaic. Symptoms of many virus diseases, appearing as an abnormal pattern of dark green, light green and chlorotic or yellow areas on leaves.

Motile. Ability of self-propulsion by means of flagella, cilia or amoeboid movement.

Mottling. A leaf symptom showing small but numerous spots or blotches of more distinct discoloration comprising of light and dark green areas in an irregular pattern usually caused by a virus.

Mould. Any profuse or wooly, conspicuous, superficial fungus growth, consisting of mycelium and / or spore masses on various substrates, especially on damp or decaying matter or on surfaces of plant tissue.

Mummy. A dried, shrivelled fruit.

Mutant An individual acquiring a new, heritable characteristic as a result of mutation caused accidentally or due to exposure to certain mutagenic substances or irradiation, which causes a change in the gene of chromosome that leads to one or more discrete heritable differences from the parent.

Mycelium (pi. Mycelia). Mass of hyphae constituting the body (thallus) of a fungus.

Mycorrhiza (pi. Mycorrhizae). A symbiotic association between the hyphae of certain fungi and the absorptive organ, typically the roots of some plants.

Mycotoxins. Toxic substances produced by several fungi in infected seeds, feeds or foods, capable of causing illness of varying severity and even death to animals and humans that consume such substances.

Nanometer (nm). A unit of length 1 / 1,000 of a micrometer (micron) or l/ 10,00,000 of a mm.

Natural host (syn. typical host). A host in which the pathogenic microorganism or parasite is commonly found and in which the pathogen can complete its development.

Natural immunity. Immunity due to innate genetical character.

Natural openings. Openings, such as stomata, lenticels, hydathodes and nectarthodes.

Necrosis. Disintegration and death of cells or tissue, usually a clearly delimited part of a plant, accompanied by black or brown darkening due to local toxic or microbiological action.

Nectarthode. An opening at the base of a flower from which nectar exudes.

Node. Enlarged portion of a shoot at which leaves or buds arise.

Non-persistent transmission. A type of insect transmission in which the virus is acquired by the vector after very short acquisition feeding times and which is transmitted during very short inoculation feeding periods. The vector remains viruliferous for only a short period unless it feeds again on an infected plant.

Non-persistent virus. A virus that persists in its vector only for a short period of time, usually less than 4 hours after acquisition.

Obligate parasite. An organism that can live and obtain food only from living protoplasm of the host; it cannot be grown in artificial culture media.

Obligate saprophyte. An organism that can obtain its food only from dead organic matter; it is incapable of infecting another living organism.

Oblong. With rounded or blunt tips.

Oidiophore. Specialized hypha that bears oidia.

Oidium (pi. Oidia). Spores produced by simultaneous segmentation of oidiophore.

Oogamous. A type of fertilization, in which two heterogametangia viz., an antheridium and an oogonium come in contact and the contents of one flow into the other through a pore or tube.

Oogonium. The female gametangium of heterogamous fungi belonging to Oomycetes, containing one or more gametes.

Oosphere. An undifferentiated egg; a female gamete.

Oospore. A thick-walled, resting spore produced by sexual reproduction in the Oomycetes.

Operculum. Well-defined circular hinged cap-like structure at the tip of the discharge papilla, through which the spores escape outside.

Ostiole. A pore-like opening in the papilla or neck of a perithecium, pseudothecium, pycnium or pycnidium, through which spores are released.

Ovary. Enclosed structure or egg contained within the ovary that becomes a seed after fertilization.

Pandemic. A widespread and destructive outbreak of a disease simultaneously in several countries.

Panicle. An indeterminate inflorescence, the main axis of which is branched, with pedicillate flowers on the secondary branches.

Papilla. A nipple-like elevation.

Paraphysis (pi. Paraphrases). A sterile, upward growing, basally attached hyphal element present in the hymenium of the fruiting bodies, such as perithecium, apothecium etc. produced by some fungi.

Parasite. Organism that lives in intimate association with another organism on which it depends for its nutrition.

Parenchyma. Soft tissue of living plant cells with undifferentiated, thin, cellulose walls.

Parthenogenesis. Reproduction by the development of an unfertilized female gamete alone.

Pasteurization. The process of heating a liquid or solid food to a controlled temperature to enhance the keeping quality and destroy harmful microorganisms.

Pathogen. A disease-producing organism, agent or factor.

Pathogenic. Capable of causing disease in a host or range of hosts.

Pathogenicity. The ability of the pathogen to cause disease.

Pathovar (Pathotype). A subdivision of a plant pathogenic bacterial species characterized by its pattern of virulence or avirulence to a series of differential host varieties; pathovar (pv) for bacteria is equivalent to *forma specialis* (f.sp.) for fungi.

Pedicel. Small, slender stalk bearing an individual flower, inflorescence or spore.

Peduncle. Stalk or main stem of an inflorescence, part of an inflorescence or a fructification.

Penetration peg. The specialized, narrow, hyphal strand produced from the underside of appressorium that penetrates the host cell.

Perennial mycelium. Mycelium, which persists in a host plant from one season to the next.

Perfect stage. The sexual stage in the life cycle of a fungus in which spores such as oospores, zygospores, ascospores or basidiospores are formed after nuclear fusion or by parthenogenesis.

Peridium (pi. Peridia). The outside covering or wall of a fructification.

Periphysis (pi. Periphyses). Short, hair-like growth, in the form of a fringe lining the inside of an ostiole or of a pore in the stroma.

Periphysoids. Lateral periphyses.

Periplasm. A layer of protoplasm surrounding the oosphere of certain Oomycetes.

Perithecium (pi. Perithecia). A flask-shaped or subglobose, thin-walled, fungus fruiting body (ascocarp), containing asci and ascospores, which are expelled or released through a pore (ostiole) at the apex.

Peritrichous. Having a number of flagella distributed over the whole surface.

Persistent transmission. A type of insect transmission in which the viruses are acquired by the vector after a long acquisition feeding period, and in which there may be a latent period following acquisition before the vector can transmit the viruses. The vector remains viruliferous for a long period, often throughout its life span. The viruses sometimes multiply within the vector.

Petiole. The slender stalk portion that supports the blade of a foliage leaf.

Phanerogamic parasites. Parasites belonging to the higher plants.

Phaeosporae. Organisms belonging to Fungi Imperfecti producing colored spores.

Phaeospore. A dark colored or brown spore; categories of Fungi Imperfecti, having dark colored spores.

Phialide. A type of bottle-shaped conidiogenous cell, that produces blastic conidia in a basipetal fashion without detectably increasing in length.

Phloem. Food-conducting tissues, consisting of sieve tubes, companion cells, phloem parenchyma and fibers.

Phragmobasidium (pi. Phragmobasidia). A basidium typically divided into four cells by transverse septa.

Phyllody. The replacement of floral parts by leaf-like structures.

Physiologic race. Biotype of a species of a pathogen that are alike in morphology, but differ in certain cultural, physiological, biochemical, pathological or other characteristics and hence differ in their ability to infect particular varieties of the susceptible host species; they are also referred to as 'physiological form' or 'biological form'.

Phytopathogen. An organism or virus able to induce disease in plants.

Phytotoxic. Harmful or poisonous to plants or plant growth (usually used to describe toxicity developed due to application of herbicides, insecticides, fungicides etc.)

Pileus. The fleshy, umbrella-like or cap-like upper portion of the basidiocarp of Agaricales.

Pistil. The ovule bearing organ of the plant consisting of the ovary and its appendages, such as style, stigma etc.

Pith. Parenchymatous tissue occupying the center of the stem.

Planogamete. A motile gamete.

Plasmodesma (pi. Plasmodesmata). A fine strand of cytoplasm passing through a pore in the cell wall, thus usually connecting the protoplasm of adjacent cells.

Plasmogamy. Union of protoplasts of two cells that brings the nuclei close together within a single cell.

Pleomorphic. Having more than one independent form or spore stage in the life cycle.

Polycyclic. Completing many life cycles or disease cycles in one year.

Polyploid. Having three or more complete sets of chromosomes.

ppm. Parts per million (parts in 10^6 parts).

Predisposing factors. Factors, such as genetic-, cultural- or environmental factors, which by their actions render an organism susceptible to a particular disease; conferring a tendency to disease occurrence.

Prevalence of a disease. The total number of cases of a particular disease at a given moment of time, in a given population.

Primary infection. The first infection following a rest or dormant period of a pathogen.

Primary inoculum. Propagules of the pathogen, usually from an overwintering source, that cause primary infection or that initiate the disease in the field, as opposed to inoculum that spreads the disease during the season.

Primordium (pi. Primordia). The rudiment of beginning stage of any structure.

Probasidium. The primary basidial cell; teliospores and smut spores.

Prokaryotes. Organisms that lack a clearly defined nuclear membrane.

Proliferation. A rapid multiplication of microorganisms; formation of new cells, tissues or organs, specifically a hyperplastic development of plant diseases in which organs continue to develop after they have reached the point beyond which they normally do not grow.

Promycelium (pi. Promycelia). A germ tube or a short hyphal filament or basidium produced meiotically by the germinating rust spores (teliospores) or smut spores (chlamydospores), from which basidiospores are formed.

Propagative virus. A virus, which multiplies in its vector.

Propagule. Any unit or part of an organism, such as spore, sclerotium, chlamydospore, mycelial fragment etc., capable of independent growth to produce a new individual.

Pseudomorph. An indefinite or irregular structure.

Pseudomycelium. A series of cells adhering end to end to form a chain; produced by some yeast.

Pseudoparaphyses. Sterile threads attached both to the roof and to the base of an ascocarp.

Pseudoseptum (pi. Pseudoseptae). Plug-like partition of cellulin or other substances in the hyphae or spores of some fungi, such as *Drechslera.*

Pseudothecium (pi. Pseudothecia). Contraction of pseudoperithecium; an ascocarp similar to a perithecium, but unlike typical perithecium, produces asci in unwalled locules or cavities.

Pulverulent. Break through the host surface.

Pustule. A blister-like fungal spore mass, breaking through a plant epidermis; a pimple-like eruptive fruiting structure, such as uredinium of a rust fungus.

Pycnidiospore. An asexual spore produced in a pycnidium.

Pycnidium (pi. Pycnidia). An asexual, thick-walled, spherical or flask-shaped fruiting body lined inside with conidiophores that produce pycnidiospores (conidia) from their tips; characteristic of Sphaeropsidales of Fungi Imperfecti.

Pycnium or spermagonium. Globose or flask-shaped, haploid fruiting body of rust fungi bearing pycniospores or spermatia and receptive hyphae, representing stage '0' of the rust fungi.

Quarantine. Legislative control of the transport of plants or plant parts from one particular place to another, from one state to another or from one country to another to prevent the spread of pests or pathogens.

Quiescent. Dormant or inactive.

Race. A subgroup or biotype within a species or variety, morphologically identical, but distinguished from other races by virulence, symptom expression or host range.

Rachis. Elongated main axis of an inflorescence.

Radicle. Part of the plant embryo that develops into the primary root.

Receptive hyphae. A simple or branched hypha-like structure of a rust fungus produced by a pycnium (spermagonium) that protrudes out through the ostiole and receives the nucleus of a pycniospore (spermatium) of the opposite mating type so as to initiate the dikaryotic phase of the fungus.

Relative humidity (RH). The ratio of amount of water (water vapor) contained in a sample of air to that which could be contained in the same volume, if it were saturated.

Resistance. The ability of an organism to exclude or overcome completely or to some extent the harmful effect of the pathogen or other damaging factor.

Respiration. A metabolic process occurring in all living organisms, in which a series of chemical reactions take place and by the oxidation of carbohydrates and fat, energy is made available to the organisms.

Resting spore. A sexual or asexual, thick-walled spore of a pathogen that is resistant to extremes of temperature and moisture, and which often germinates only after a resting period of dormancy from its formation.

Reticulate. Covered with net-like ridges; reticulate spore.

Rhizoid. A short, thin hypha growing in a root-like fashion toward the substratum and obtain nourishment for the fungus.

Rhizome. A mostly horizontal, jointed, fleshy, often elongated, usually underground stem with nodes, buds and scales.

Rhizomorph. An aggregation of hyphae resembling a root and having a well-defined apical meristem capable of transporting nutrients and spreading infection over long distances; often differentiated into a rind of small, dark cells surrounding a core of elongated, colorless cells.

Rhizosphere. The soil closely surrounding living plant roots, in which the microbial population is enhanced by root exudates.

Rhynchosporous. Having beaked spores.

Ring spot A disease symptom characterized by yellowish or necrotic, more or less concentric rings, enclosing green tissue, as in some virus diseases.

Root exudates. The various compounds that are exuded from growing or expanding sections of roots.

Rosette. Short, bunchy habit of plant growth with small, circular, cluster of leaves.

Rouging. Removal of undesirable plants, usually off-types or infected plants in a crop.

Rugose. Wrinkled and roughened appearance of leaves as in some virus diseases.

Russetting. Formation of brownish, roughened areas on the skin of fruits or tubers because of hyperplastic reaction caused by some fungi.

Rust. A disease caused by one of the Uredinales that often produces spores of a rusty color.

Saccate. Sac-like; bag-like.

Sanitation. The removal and burning of infected plant parts, decontamination of tools, equipments, hands etc.

Saprobe (syn. Saprophyte). Organism that obtains nourishment from non-living organic matter.

Scab. A hyperplastic symptom characterized by rough, crusty lesions formed by excessive cork production.

Scald. A lesion that appears as an apparent consequence of scalding with hot water, usually bleached and sometimes translucent.

Scar. A clear mark in the conidiophore at the place from where a conidium had dislodged.

Sclerenchyma. A protective or supporting plant tissue made up of cells with thickened, lignified walls

Sclerotium (pi. Sclerotia). A hard resting body, resistant to adverse environmental conditions and may remain dormant for long period of time and germinates when favorable conditions return; it consists of closely packed, more or less isodiametric or oval cells, known as pseudoparenchyma tissue.

Scolecospore. An elongated, slender, needle- or worm-like or filiform spore.

Scorch. Any symptom that suggests the action of fire or flame on the affected part, often seen at the leaf margins.

Secondary inoculum. Inoculum produced by infection that took place during the same growing season or inoculum produced as a result of infection by the primary inoculum, which causes secondary infection.

Secondary mycelium. The dikaryotic mycelium of Basidiomycetes, resulting from plasmogamy.

Seed. Ripened ovule consisting of an embryo and stored food enclosed by a seed coat.

Seed-borne. Carried on or in the seed.

Seed disinfection. Treatment of seeds to destroy harmful organisms or to protect the seeds against infection.

Seed dressing or seed treatment. Application of a biological agent, chemical substance or physical treatment to seed, seed material or plant, so as to protect the seed, seed material or plant from pathogens or to stimulate germination or plant growth; the process of coating or impregnating seeds with a protectant pesticide.

Seed protectant. A chemical applied to seed before planting to protect seeds from diseases or pests.

Seed rot. Seed decay caused by pathogenic fungus.

Seed transmission. Perpetuation of a pathogen from one generation to another through seed.

Semi persistent transmission. Virus transmission by an insect vector that is intermediate between non-persistent and persistent transmission.

Senescence. Process or state of growing old.

Septate. Hyphae divided into compartments or cells by cross walls or septae.

Sessile. A leaf, leaflet, flower, floret, fruit, ascocarp, basidiocarp etc. without a stalk, petiole, pedicel, stipe or stem.

Seta (pi. Setae). Stiff, bristle-like or hair-like structure occurring in the fruiting body (acervuli) of some fungi belonging to Melanconiales.

Sett. A piece of stem with nodal buds or a piece of tuber used as planting material.

Sexual reproduction. Reproduction involving fusion of two haploid nuclei of opposite mating types (karyogamy) to form a diploid nucleus followed by meiosis back to the haploid nuclei state at the same point in the life cycle, resulting in genetic recombination.

Sexual spore. A spore produced as a result of sexual reproduction.

Shoestring. A symptom of extreme narrowing of the leaf often caused by a virus.

Shot holes. A symptom of certain leaf spotting pathogens, in which necrotic areas of limited size fall out from the lesions on the leaf lamina, leaving small, almost circular holes, which appear as holes made by bullet shots.

Sieve tubes. Plant cells specialized for the transport of sugars, connected to one another by perforated ‘sieve plates’ to form a ‘cellular tube’ or ‘sieve element’ in the phloem vessels of leaves, stems and roots.

Slime mould. Saprophytic organisms that form vegetative amoeboidal plasmodia and spores as found in the Myxomycetes.

Smut. A group of fungi belonging to Order - Ustilaginales in the Basidiomycetes that typically releases masses of black, thick-walled, dusty, resting spores or teliospores or smut spores at maturity.

Soft rot. A rot of fleshy fruit, vegetable or ornamental, in which the tissue becomes macerated by the enzymes of the pathogen.

Soil drench. Application of a solution or suspension of a chemical to the soil, especially pesticides to control soil-borne pathogens.

Soil-inhabitant. Pathogenic microorganism that maintains its population in soil over a long period of time.

Soma. Filamentous body or thallus of a fungus.

Somatic cells. Cells of the thallus that are not involved in sexual reproduction.

Sooty mould. A black, sooty coating on the foliage and fruits formed by the dark hyphae of the fungi that live in the honeydew secreted by phloem-feeding insects, such as aphids, plant hoppers, mealy bugs, scales etc.

Sorus. Compact fruiting structure, especially spore masses produced in or on the host plant, such as the rust and smut fungi.

Source of inoculum. The place, which harbors inoculum and from which it may be disseminated and cause infection.

Spawn. Fungal culture consisting of mycelium that is used for mushroom cultivation.

Species (pi. Species). The unit of classification of a group of closely related individuals resembling one another in certain inherited characteristics of structure and behavior, and relatively stable in nature; the individuals belonging to a species are subordinate to a Genus, but above a race; it is designated by a binomial consisting of the generic name and the specific epithet.

Spermagonium (pi. Spermagonia, syn. pycnium). A globose or flask-shaped, haploid, fruiting body composed of receptive hyphae and spermatia (pycniospores) produced by the rust fungi (stage '0' of the rust fungi).

Spermatium (pi. Spermatia). A non-motile, haploid, male sex gamete produced in the spermagonium.

Spermatization. Plasmogamy by the union of a spermatium with a receptive hypha of the opposite mating type to form a dikaryotic mycelium.

Spike. An elongated inflorescence with sessile flowers on the main axis.

Spikelet. Spike-like appendage comprising of one or more reduced flowers and associated bracts, a unit of inflorescence as in grasses.

Spindle-shaped. Fusoid, narrowing towards the tip.

Sporadic. A disease that breaks out occasionally without being constantly destructive.

Sporangiolum (pi. Sporangiola). A small sporangium containing few spores.

Sporangiophore. A specialized branch of fungal hyphae bearing sporangia.

Sporangiospore. Asexual spores borne inside a sporangium.

Sporangium (pi. Sporangia). A sac-like fungal structure, in which the entire contents are converted into an indefinite number of asexual spores.

Spores. Reproductive structures produced by fungi and some other organisms constituting of one or more cells, which are synonym to seeds of higher plants; bacterial cells modified to survive any adverse environmental conditions.

Sporidium (pi. Sporidia). The basidiospores produced on a promycelium, as in the rusts and smuts.

Sporocarp. Spore bearing fruiting body.

Sporodochium (pi. Sporodochia). A specialized, cushion-shaped stroma consisting of masses of closely woven hyphae without any lateral union and produce conidia from the hyphal tips.

Sporophore. A spore producing or spore bearing structure, such as a conidiophore, ascocarp or basidiocarp.

Sporulation. The process of producing spores.

Spot. A disease symptom characterized by a limited necrotic or chlorotic area on leaves, flowers and stems of plants.

Spreader. A chemical substance added to a spray fluid to assist even and uniform distribution of the active material on the sprayed surface.

Stamen. Male reproductive structure of a flower, composed of a pollen-bearing anther and a stalk.

Starch. A polysaccharide consisting of glucose units, the principal food substance stored in plants.

Stem pitting. A symptom of some virus diseases characterized by longitudinal depressions or pits on the stem of plants.

Sterigma (pi. Sterigmata). A small, slender, pointed projection that supports a spore.

Sterile fungi (Fungi Imperfecti). A group of fungi that are not known to produce any kind of spores.

Sterilization. Total destruction of all living organisms by various means, including heat, chemicals or irradiation.

Sticker. A chemical substance added to a plant protectant spray fluid to increase its tenacity.

Stigma. The portion of a flower that receives pollen and on which the pollen germinates and causes infection.

Stipe. Stalk.

Stipule. Small, leaf-like appendage at the base of a leaf petiole, usually occurring in pairs.

Stolon (syn. Runner). Fungal hyphae that grow horizontally along the surface of the substratum as in the case of *Rhizopus* species.

Stoma (pi. Stomata). Structure composed of two guard cells and the opening in-between them in the epidermis of a leaf or stem.

Stone fruit. Fruit with a stony endocarp (e-g. cherry, peach, plum etc.).

Strain. A distinct form of an organism or virus within a species, differing from other forms of the species biologically, physically or chemically, but identical morphologically.

Streak. A disease characterized by elongated lesions or areas of discoloration, usually of limited length, on leaves with parallel venation or on stems.

Striate. Marked with delicate lines, grooves or ridges.

Stripe. A disease characterized by elongated areas of discoloration of indefinite length on stems or on leaves with parallel venation.

Stroma (pi. Stromata). A compact somatic structure, much like a mattress or cushion made up of loosely woven hyphae, more or less parallel to one another known as prosenchyma tissue; usually fructifications are formed on or in the stroma.

Stunting. Reduction in height of a vertical stem axis, resulting from a progressive reduction in the length of successive intemodes or a decrease in their number.

Style. Slender part of many pistils located between the stigma and the ovary, through which the pollen tube grows.

Stylet. Stiff, slender, hollow feeding organ of plant parasitic nematodes or sap-sucking insect pests, such as aphids, leafhoppers etc.

Stylet-borne transmission (syn. non-persistent transmission). A type of virus transmission, in which the virus is acquired and transmitted by the vector after a short feeding period and is transmitted by the vector for only a short period of time.

Subglobose. Almost global or spherical.

Subhyaline. Somewhat or imperfectly clear or nearly colorless.

Substrate. The material or substance on which a microorganism feeds and develops.

Sunscald. Injury of plant tissues burnt or scorched by direct sun.

Suppressive soil. Soil in which certain diseases are suppressed because of the presence of microorganisms antagonistic to the pathogen in that soil.

Susceptible. Capable of being easily infected or infested; lacking the inherent ability to resist disease or attack by a pathogen; not immune.

Swarm spores. Zoospores or swarm cells or swarmer.

Symbiosis. Mutually beneficial association of two different kinds of organisms.

Sympodial. Proliferation of axes, in which each successive spore or branch develops behind and to one side of the previous apex where growth has ceased.

Symptom. A visible abnormal change in a host and its behavior as a result of infection by a pathogen.

Symptomatology. The study of symptoms of disease and signs of pathogen throughout the plants.

Symptomless carrier. A plant although infected with a pathogen (usually a virus) produces no obvious symptoms.

Synnema (pi. Synnemata). A group of conidiophores unite more or less closely and cemented together into columns and produce conidia from the tips of the conidiophores. It is also called as **'Coremium'** (pi. Coremia).

Systemic. Term applied to pesticide when applied to soil or foliage are absorbed by the plant parts and translocated to other parts through the vascular system, and imparting protection from disease or pest attack; in case of plant disease, a single infection at one site resulting in the spread of the pathogen throughout the plant

Systemic acquired resistance. A broad, physiological immunity in plants that results by the application of certain natural and systemic chemical compounds, which trigger some kind of responses in the host and induce resistance.

Target spot. A lesion consisting of a dark brown circular area containing a series of brown, concentric rings appearing like a target board; typical of infection caused by *Alternaria* sp.

Teliomorph. The sexual form in the life cycle of a fungus.

Teliospore or teleutospore. A thick-walled resting spore of the rusts and smuts in which karyogamy occurs.

Telium (pi. Telia). The fruiting body (sorus) of a rust fungus that produces teliospores (stage III of the rust fungus).

Thallus. Vegetative body of fungi devoid of stem, roots and leaves or soma of a fungi

Therapeutants. Systemic fungicides that can enter into the host, translocated inside the host tissues, affect the deep-seated infection and can cure diseases even after their establishment in the host.

Tiger-stripe. A leaf symptom in which clear yellow areas separate marginal and interveinal areas of dark necrotic tissue from the normal green areas along and adjacent to the main veins, giving a yellow and black appearance as in a tiger skin.

Tinsel. Ciliated

Tissue. Group of cells, usually of similar structure, that performs the same or related functions.

Tolerance. Ability of a plant to endure an infectious or non-infectious disease, adverse conditions or chemical injury without serious damage or yield loss.

Tomentose. Densely covered with short, fine hairs.

Toxin. An organic poisonous substance produced by a microorganism, which even at low concentrations deleteriously and irreversibly affects the normal processes of living organisms.

Tracheid. Elongated conducting cell of a xylem, with tapering or oblique end walls and pitted walls.

Translocation. Movement of water, nutrients, chemicals or food materials within a plant.

Translucent. Semi-transparent; allowing some light to pass through, but not transparent.

Transmission. The transfer of an infectious agent (usually a virus) from one plant to another; the dissemination of pathogens and the inoculation of suscepts.

Transovarial transmission. Transmission of the virus through the eggs of the insect vector to its progeny; transmission of a virus from an adult organism to its progeny through the ovaries and eggs.

Transpiration. The evaporation of moisture from a living plant, mainly through the stomata of the leaf. This moisture represents a surplus from that taken in by the roots and which is not required for photosynthesis.

Transstadial. Viruses retained in the vector even after moulting.

Trichogyne. The receptive neck of the ascogonium, which is often long and hair-like.

Truncate. Ending abruptly as though with the end cut off horizontally.

Tuberculate. Having small swellings or warts on the surface.

Tumor. An uncontrolled overgrowth of tissue or tissues.

Turgidity. State of being rigid or swollen as a result of internal water pressure.

Umblicate. With a central depression.

Undulate. With wavy surface at the margin.

Unilocular. Single, simple, undivided cavity.

Uniseriate. Arrangement of ascospores in an ascus in one series.

Unitunicate. An ascus in which both the inner and outer walls are more or less rigid and do not separate during spore ejection.

Universal veil. A thin, veil-like membrane that covers certain types of young mushrooms; upon expansion of the mushroom, the universal veil tears and its remnants may be seen in the form of scales on the pileus and in the form of a volva at the base of the stipe.

Uredinium or uredium. The sorus of rust fungi producing urediniospores or uredospores (stage II of the rust fungus)

Uredospore. An asexual, binucleate, repeating spore (stage II) of the rust fungi.

Ustilospore. Smut spore of Ustilaginales, from which the basidiospores are formed.

Utriform. Bag-like.

Vacuole. Mostly spherical organelle within a cell bounded by a membrane and containing dissolved materials, such as metabolic precursors, storage materials or waste products,

Variegation. Pattern of two or more colors in a plant part, as in a green and white leaf.

Vascular bundle. A strand of conductive tissue, usually composed of xylem and phloem vessels (in leaves, small bundles are called veins).

Vascular tissue. A general term referring to either or both xylem and phloem tissue.

Vascular wilt disease. A disease in which the pathogen is almost entirely confined to the vascular system of the host during pathogenesis, and in which wilting is a characteristic symptom.

Vector. A living organism (e-g. insects, mites, birds, higher animals, nematodes, parasitic plants, human etc.) capable of carrying and transmitting a pathogen and disseminating diseases, especially virus diseases.

Vegetative. Asexual, somatic.

Vegetative phase. Refers to a non-reproductive phase in fungi and plants.

Vein. A vascular bundle forming the framework of fibrous tissue in a leaf; a blood vessel conducting blood towards the heart in many animals.

Vein banding. Symptom of a virus disease, in which the areas adjacent to the veins of the leaf remains green, in contrast to the remaining areas, which may be chlorotic.

Vein clearing. The cells adjacent to the veins become translucent or chlorotic, while the interveinal areas remain green.

Vermiform. Worm-shaped.

Verrucose. Slightly roughened surface.

Vesicle. In fungi, a small, thin-walled, bladder-like structure produced by a zoosporangium and in which the zoospores are released or are differentiated.

Vessel. A xylem element or series of such elements whose function is to conduct water and mineral nutrients to the aboveground parts of the plant through pit openings in end walls.

Viable. Capable of living; able to germinate.

Vigor (of seeds). Physiological potential for rapid and uniform germination and fast growth of seedlings under normal field conditions.

Viricide. A substance that completely and permanently inactivates a virus.

Viroid. A pathogenic agent that has virus-like properties composed of a circular, single-stranded molecule, containing a small amount of RNA and lacking a protective protein coat.

Virulence. Degree or qualitative measure of pathogenicity; relative capacity of a pathogen to cause disease.

Virulent. Capacity of causing a severe disease; strongly pathogenic.

Viruliferous. An insect vector containing virus and is capable of transmitting it into a susceptible host.

Virus. An ultramicroscopic, filterable, intracellular, infective agent of obligate nature, consisting of a core of infectious nucleic acid, either RNA or DNA, usually surrounded by a protein coat and capable of causing various diseases in living organisms.

Volatile. A chemical compound, which evaporates or vaporizes from a liquid to a gaseous state at ordinary temperatures on exposure to the air.

Volunteer plant. A plant from a previous season's crop that regenerates in a subsequent crop or plant developing from a self-sown or lost seed.

Vulnerability. Inability of a plant or organism to resist attack by a phytopathogen or parasite and to counteract the effect of the attack.

Volva (pi. Volvae). A cup-like structure at the base of the stipe of certain mushrooms, which is a remnant of the universal veil.

Warty. Having a rough surface.

Water-soaked. A disease symptom, in which an area of plant cells becomes darker in color owing to the filling of intercellular air spaces with cell sap, which comes out to the surface as exudate.

Wet rot. Any kind of rot, in which the tissue is rapidly and completely disintegrated, with release of water from the lysed cells.

Wetting agent. A substance added to the plant protection chemical formulation or to the pesticidal spray fluid to reduce the surface tension of the applied droplets and to facilitate uniform spreading of the spray fluid over the sprayed surface.

Whiplash. Long, whip-like flagellum

White rot. Rotting of wood in trees invaded by lignin-destroying fungi that leaves a white cellulose residue.

Wilt. Loss of turgidity and drooping of plant parts generally caused by insufficient water or excessive transpiration in the plant; a disease symptom caused by the wilt fungi.

Witche's broom. Broom-like growth or massed proliferation caused by the dense clustering of branches of woody plants.

Wound parasite. A parasitic organism, which can invade a host only if it can first become established in damaged tissues.

Xylem. A plant tissue consisting of tracheids, vessels, parenchyma cells and fibers; wood.

Xylem-limited bacteria (XLB). Endophytic bacterial parasites that live in plants exclusively in the xylem cells or tracheary elements (e-g. *Clavibacter xyli* sub.sp. *xyli)*

Yellows. A plant disease characterized by pronounced chlorosis, yellowing and stunting of the host plant.

Zoogametes. Motile gametes; planogametes.

Zoospores. Non-sexual, motile sporangiospores or swarm spores produced from either sporangia or zoosporangia, having one or two flagella.

Zygosporangium. A sporangium that arises from a zygospore at the end of a stalk, which contains non-motile spores.

Zygospore. The zygote formed as a result of fusion of two morphologically identical gametangia.

Zygote. A diploid cell formed by the union of two, compatible gametes, as well as the individual produced from such a cell.

□□□

Subject Index

a

B

C

D

E

F

G

I

K

M

N

O

P

S

T

U

V

W

X

Y

Z

□□□